Studies in Neuroscience, Psychology and Behavioral Economics

Series editors

Martin Reuter, Bonn, Germany
Christian Montag, Ulm, Germany

More information about this series at http://www.springer.com/series/11218

Martin Reuter · Christian Montag
Editors

Neuroeconomics

Springer

Editors
Martin Reuter
Differentielle and Biologische Psychologie
Rheinische Friedrich-Wilhelms-Universität Bonn
Bonn
Germany

Christian Montag
Department of Molecular Psychology, Institute for Psychology and Education
Ulm University
Ulm
Germany

and

Key Laboratory for NeuroInformation, School of Life Science and Technology, Center for Information in Medicine
University of Electronic Science and Technology of China
Chengdu
People's Republic of China

ISSN 2196-6605 ISSN 2196-6613 (electronic)
Studies in Neuroscience, Psychology and Behavioral Economics
ISBN 978-3-642-35922-4 ISBN 978-3-642-35923-1 (eBook)
DOI 10.1007/978-3-642-35923-1

Library of Congress Control Number: 2016948772

Printed on acid-free paper

This Springer imprint is published by Springer Nature
The registered company is Springer-Verlag GmbH Germany
The registered company address is: Heidelberger Platz 3, 14197 Berlin, Germany

Preface

Dear scientists, students, and readers,

The present book, *Neuroeconomics*, was originally intended as the cornerstone of the Springer book series *Studies in Neuroscience, Psychology and Behavioral Economics*. It was not our idea to write (or edit) a book before or close to retirement. So it took a while before Springer, who had contacted MR to write a book on Neuroeconomics, persuaded us to do it. One prerequisite for MR was that CM agreed to do this jointly with MR, and CM did agree!! We have worked together side by side for many years and become very close friends. Springer ultimately convinced us, not only to publish this book, but to edit a whole book series. Meanwhile a first book, *Internet Addiction—Neuroscientific Approaches and Therapeutical Interventions*, which we edited, was published early in 2015, forming the first publication of this series and more books will appear in the future.

Science has become more and more interdisciplinary and so new scientific disciplines emerge—like neuroeconomics, which is a joint venture between neuroscientists, psychologists, and behavioral economists. The focus of interest in neuroeconomics lies on human decision-making under an economic perspective. "Economics" refers not only to monetary transactions, but also to all kinds of costs and benefits associated with decisions. Before a decision is reached and an actual action is exerted, cognitive and affective processes are active and these processes originate in the brain. Therefore, if one is interested in the question why people behave impulsively or rather rationally and in a manner guided by self-interest, the role of the brain has to be taken into account. Differences in the hard-wiring of the brain or functional differences in brain activity help to explain variation in human decision-making. Research topics like this are at the core of the young discipline of neuroeconomics.

Neuroeconomics has adopted and expanded games and paradigms from behavioral economics and psychology, and uses concepts from diverse disciplines like addiction research (e.g., reward or temporal discounting) and applies nearly all kinds of neuroscience techniques to the study of human decision-making.

In 2009, the Center for Economics & Neuroscience (CENs) was founded at the University of Bonn. There are three arms to the CENs; comprising a behavior economics lab (Armin Falk), an imaging center (Bernd Weber), and MR's genetic laboratories. The collaboration with the colleagues of the CENs has alerted us to the problems related to interdisciplinary work in the field of neuroeconomics. Even the statistical methods preferred to analyze a given data set differ between psychologists and economists. Finding a common language is sometimes cumbersome, but at the same time offers researchers the chance to learn from our colleagues. This is further outlined by an example: behavioral economists are particularly fascinated by the opportunities offered by the neurosciences. However, they typically have not come across these techniques during their undergraduate and postgraduate training. We have often been asked for a scholarly introduction to molecular genetics, the field in which we are specialized. A comparable demand exists for other neuroscientific techniques. Thus, we decided not only to publish a textbook on neuroeconomics, but to enrich the book by a broad methods section in which the most common neuroscientific techniques, ranging from molecular techniques including genetics and hormone analysis, to brain imaging are introduced in a scholarly manner by experts in the field. This methods section is so far unique for a neuroeconomics book. We are convinced that many scientists and students will find they have an interest in this methods section, even if they are not primarily interested in neuroeconomics. We hope that this potential readership becomes aware of this special feature of the book.

The book comprises eight sections, starting with an introduction into neuroeconomics (1) followed by an overview on frequently applied experimental paradigms (games) in neuroeconomic research (2). In the next section, the molecular basis of human decision-making is addressed (3). Here, the focus is on the role of hormones, neurotransmitters, and (their underlying) genes that have been reported to be of relevance for the field. Section four focusses on environmental and situational factors (4) and section five on social contexts influencing human decision-making (5). From the synopses of sections 3, 4, and 5, it becomes apparent that the successful prediction of human behavior must include nature and nurture, as well as situational factors related to the decision (e.g., framing effects). Section six presents translational and developmental approaches to neuroeconomics including, among other valuable contributions, chapters on decision-making in children and among patients suffering from mental illness (6). An article on neuromarketing demonstrates how knowledge from neuroeconomics research can be applied in real life. For this reason, this chapter has been labeled *Applied Neuroeconomics* (7). Hopefully this section can be extended in the future; we are very confident that the applicability of basic neuroeconomic research will increasingly be acknowledged. The culmination of the present book constitutes the above-mentioned methods section, in which eight different neuroscience techniques are introduced (8).

The completion of this book took longer than planned, but now that it is finished, we are very satisfied with the product. We are happy to have received contributions from so many highly regarded experts in this field. Thank you all for your strong contributions and for your patience. Big thanks also go to Éilish Duke, for her critical redaction of our chapters.

We thank all the readers interested in this work. We hope that we meet your expectations and are thankful for your criticisms and comments.

We also want say thank you to our beloved wives [Anette (MR) and Susanne (CM)] and to our families, friends, and colleagues for their never ending support and love over all the years. Their support came long before this book project began.

Bonn, Germany
Ulm, Germany
June 2016

Martin Reuter
Christian Montag

Contents

Chapter 1
Neuroeconomics—An Introduction

Martin Reuter and Christian Montag

Abstract The present chapter provides an introduction into the young discipline of neuroeconomics and into the present *Neuroeconomics* book. Historical aspects, core concepts and future research avenues are presented.

1.1 Historical Aspects

Neuroeconomics is a very young scientific discipline that constitutes an interdisciplinary symbiosis of economics, psychology and the neurosciences. The general aim of neuroeconomics is to study human decision-making with a focus on the neural mechanisms thereof. The official establishment of the discipline was marked by the foundation of The Society for Neuroeconomics in 2005.

Research in this field is prolific and of high quality, however, scepticism remains, especially among those scientists who retain a purist vision of their respective disciplines. History has taught us that great achievements are made possible only by the combined expertise of scientists from different fields. For example, only through such successful interdisciplinary research could man have

M. Reuter (✉)
Department of Psychology, University of Bonn, Bonn, Germany
e-mail: martin.reuter@uni-bonn-diff.de

M. Reuter
Center for Economics and Neuroscience (CENs), University of Bonn, Bonn, Germany

C. Montag
Department of Molecular Psychology, Institute of Psychology and Education, Ulm University, Ulm, Germany
e-mail: christian.montag@uni-ulm.de

C. Montag
Key Laboratory for NeuroInformation, Center for Information in Medicine, School of Life Science and Technology, University of Electronic Science and Technology of China, Chengdu, People's Republic of China

M. Reuter and C. Montag (eds.), *Neuroeconomics*, Studies in Neuroscience, Psychology and Behavioral Economics, DOI 10.1007/978-3-642-35923-1_1

flown to the moon; astronomers working in isolation could never have achieved this dream, but with input from other disciplines (e.g. informatics, mathematics, physics, etc.) mankind's dream of walking on the moon became a reality.

Cognitive neuroscience, which emerged during the 1970s, is the youngest member of the neuroeconomic trio, while the disciplines of psychology and economics have been around for over one hundred years. For decades the two disciplines seemed to live in an uneasy parallel, arguably ignoring each other. This is surprising, given that the understanding of human behaviour is intrinsic to both disciplines. Of note, the scientific worldviews and the methodological approaches utilized by each discipline differ dramatically. Whereas economists try to establish a formal theory explaining human behaviour in an axiomatic way, psychologists build and refine theories through an empirical approach. Roughly speaking, economists have traditionally favoured a theoretical—and psychologists an empirical—approach. Since the launch of the journal of *Experimental Economics* in 1998 (and in view of the chairs for behavioural economics newly created at Universities throughout the world), it is clear that this strict differentiation between the theoretical economics and empirical psychology no longer holds. Nevertheless, such historical traditions are of importance; even today the two disciplines show marked differences that are far-reaching, which manifest in different theoretical foundations and methodological and statistical approaches, all of which serve to undermine successful interdisciplinary research efforts.

Whereas economics had not made direct acquaintance with the neurosciences prior to the establishment of the new discipline of neuroeconomics, the idea of investigating the role of the brain in human behaviour is an old one in psychology. For decades psychologists have used electroencephalography (EEG; see the method Chap. 19 by Debener et al. in this book) to investigate cognitive and emotional processes. Therefore, the invention and scientific application of magnetic resonance tomography (MRI; for an introduction see the method chapters on MRI by Markett (fMRI; Chap. 20), Gaser (MRI; Chap. 21) and Rüber (DTI; Chap. 22) in this book) in the 1990s proved a logical step for psychologists interested in subcortical processes that are not explicitly measurable through EEG. The subdiscipline of Biological Psychology makes use of all kinds of techniques that characterize the neurosciences, incorporating, in addition to EEG and MRI, genetics, psychophysiology, endocrinology, etc. In order to help bridge the gap between the "subdisciplines" of neuroeconomics, the present book deliberately features a broad methods section, which gives a scholarly introduction into neuroscience techniques relevant to this field (see PART VII of this book).

1.1.1 Economic Models and Their Parallels in Psychology

As mentioned above, economic models of human behaviour are axiomatic and try to establish algorithms valid for all participants across different situations. This idea is mirrored in classical experimental psychology, with the difference that experimental psychology uses experimental conditions to analyse behaviour. In Personality Psychology, however, the central tenet recognizes that large variability exists across participants, even in strictly controlled experimental settings or natural environments; a phenomenon referred to as individual differences.

A prominent economic model in neoclassical economics is *utility maximization*. According to utility maximization, people make their best choices according to their desires, knowledge and resources. The term *utility* does not refer to a good's quantity or monetary value per se in determining the decisions of an agent, but to the utility they obtain from the item. According to Marshall (1920, p. 78) "utility is correlative to desire and want", but desire and want can only be inferred indirectly by "the price which a person is willing to pay for the fulfilment or satisfaction of his desire". Although utility maximization makes correct predictions in a wide range of settings and situations including politics, markets and social life, its validity has been questioned, e.g. by *Prospect Theory* (Kahneman and Tversky 1979). Whereas the concept of *expected utility, which* originates from *Utility Maximization Theory,* postulates that alternative choices are valued by weighting the hedonic utility of possible outcomes against the chances of those outcomes actually occurring (e.g. in gambles), *Prospect Theory* claims that people do not always show a numerical evaluation of probabilities, but that outcomes are valued according to two aspects: a reference point (reference-dependent value) and an absolute utility. The reference-dependent value is thought to represent the valuation of past experiences and future aspirations and is therefore related to learning (past) and motivation (future). Most prominently, *Prospect Theory* explains why people grant more weight to losses than to gains, a phenomenon called *Loss Aversion*. There is empirical evidence across different cultures and ethnicities that, on average, losses are valued about twice as large as equal-sized gains. Of note, *Prospect Theory* has gained empirical support from the neurosciences. Using an fMRI study, Tom and colleagues have demonstrated that different brain activity patterns are correlated to the amount of gains and losses (Tom et al. 2007). Interestingly, they did not identify different brain circuits coding for gains and losses, but instead identified a unique system—the ventral striatum—that has become famous in the neurosciences as the brain's reward centre. Gains were expressed by an increase—and losses by a decrease—of the BOLD (blood oxygen level dependent) response in the ventral striatum.

Utility Maximization Theory focuses on economic decisions taken by a single agent in isolation. In contrast, *Game Theory* has extended the idea of utility maximization to social situations, e.g. it makes predictions of how the choices of other peoples influence the choice of an individual. Behavioural economics (partly influenced by psychology) has developed a battery of different games (e.g. Trust Game, Public Goods Game, Prisoner's Dilemma, etc.; *for an introduction to economic games* see Chap. 2 *by Civai and Hawes in this book*), which test the assumptions made by *Game Theory*. However, the empirical data do not always yield support for the theory. Naturally, people take into account the choices—or putative choices—of others when making their decisions, but their behaviour is often incongruent with the traditional economic view of man as a *homo economicus*. It is stated that the homo economicus makes decisions guided by self-interest (i.e. the maximization of personal benefit), that his decisions are completely rational and that all information necessary for making a choice is available. Results from the dictator game where player 1 (the dictator) has to split an endowment with an anonymous person (player 2) show that people do not behave in a manner congruent with that expected according to the view of man as a homo economicus (i.e. to take all the money and to award no money at all to player 2) (Camerer 2003). Instead, cross-cultural studies have shown that the "dictator" is far more cooperative, with mean allocations to the receiver (player 2) of about 28 % (Engel 2011). Based on the fact that this game, in its original version, is played as a 'one shot' game, i.e. the dictator has no reason to fear punishment from player 2 in a subsequent interaction; the dictator game is thought to be a measure of pure altruism.

In addition to the influence of others on people's choices (*Game Theory*), there are other crucial factors that influence economic behaviour. One of the most prominent factors studied in neuroeconomics is the relationship of the time lag between the decision and its consequences, referred to as *temporal discounting*. Interestingly, psychologists have investigated this topic for decades as *delay of gratification* (Mischel et al. 1989). In his seminal 'Marshmallow Study' at Stanford University in 1972, Mischel devised an experiment in which children were afforded the opportunity to 'earn' marshmallows. If the children could resist eating the first marshmallow they were offered, they were promised a second one, i.e. they would receive two marshmallows instead of one. The duration each child resisted the temptation to eat the initial marshmallow was analysed, and it was subsequently investigated whether or not delaying gratification correlated with future success. While the majority of the approximately 600 child participants attempted to resist the urge to eat the first marshmallow, only one-third delayed gratification long enough to get the second marshmallow. Analyses suggested that the age of the children was a crucial factor in influencing the child's success on this task. With increasing age, the ability to defer gratification increases. These findings have since been extended to adult samples, using various kinds of reinforcement. Intelligence (positive association) and gender (females were superior in resisting an immediate small reward in favour of a delayed bigger reward; evolutionary factors are discussed to account for this gender effect) turned out to be further prominent predictors of the ability to defer gratification. Under the label *temporal discounting* this phenomenon was investigated by means of functional magnetic resonance imaging (fMRI). McClure et al. (2004a) could identify distinct neural systems responding to immediate and delayed rewards. Whereas the limbic system is activated by immediate rewards ($t = 0$) the prefrontal cortex responds to both immediate and delayed rewards ($t > 0$), but more so when the delayed option is chosen. These findings hold true for monetary reinforcement as well as for primary rewards, e.g. sex (McClure et al. 2004b). The dissociation between cortical and subcortical brain regions with respect to immediate rewards supports the role of the limbic system (comprising the ventral striatum that is also named "the reward centre") for drives and instincts and the role of the prefrontal cortex for impulse control and cognitive processes. The latter are essential for evaluating offers and for deferring rewards until a future time point. There is plenty of evidence that the more a person discounts a delayed reward, the more likely that person is to exhibit a range of behavioural problems, including clinical disorders (e.g. drug addiction, impulse control disorders). The ventromedial prefrontal cortex (vmPFC) has shown to be involved in impulse control and in individual propensity to engage in risky behaviours (Bechara et al. 2000, 2002).

1.2 What We Have Learned from Animal Models

The crucial question when referring to findings from animal research is whether results can be extrapolated to humans. Preclinical trials—typically conducted in rodents—for the development of new drugs targeted at the treatment of human diseases, clearly answer this question with "yes". Excellent animal models for a range of psychological phenomena, e.g. anxiety, are available and do allow for predictions of the anxiolytic effects of a certain substance in humans. Even for those more complex behaviours relevant to the field of neuroeconomics, animal model exist. For example, Chen et al. (2006) were able to demonstrate that Capuchin monkeys are able to use tokens to purchase food from experimenters and that they prefer to trade with those experimenters who offer the best deals for their "money". In other words, even New World monkeys understand the principles of the market. Nonetheless, it is evident that the transfer from animal model to human is not always successful or feasible. Ethical concerns are a crucial consideration in this respect.

The invention of imaging techniques [e.g. MRI, positron emission tomography (PET)] has made it possible to study the human brain during task performance. Although PET imaging requires the administration of a radioactive ligand into the central blood system, it is a safe technique that can be used for research purposes with humans. More invasive techniques, such as microdialysis (a sampling technique for the continuous measurement of free, unbound concentrations of neurotransmitters or hormones in the extracellular fluid of brain tissue) or single-cell recordings (for assessment of the firing rate of neurons) in the living brain are, of course, not possible in healthy humans for ethical reasons. However, the neurosciences have provided many groundbreaking animal studies with broad relevance to neuroeconomics. As mentioned above, reinforcement and reward are crucial for decision-making, although other context variables also have a tremendous influence on our choices. The biological system most prominently related to reward is the dopamine (DA) system (Schultz et al. 1997). Its relevance was first identified in the context of studies on drug addiction. It was suggested that the dopaminergic system is the final common pathway of reward since almost all substances with the potential of causing addiction act via the DA system, either directly or indirectly (Spanagel and Weiss 1999). These findings could be extended to naturally occurring rewards (primary reinforcers like food or sex). The crucial question of how the

DA system could encode signals of reward is best studied in animal studies (for a review see Schultz 2013).

In a seminal study by Tobler et al. (2005), the activity of midbrain dopamine neurons in Macaque monkeys was recorded while cues signalled the probability of receiving a primary reinforcer (juice) of varying magnitude. This experiment tried to explain how the brain disentangles the probability and magnitude of reward. Keeping the probability of reward constant, the firing rate of the DA neurons increased monotonically with the expected liquid volume. The DA neurons were also able to encode the expected reward value, i.e. the combination of magnitude and probability. In a further step the authors conducted an experiment in which the reward outcomes were explicit rather than probabilistic. They used conditioned stimuli that explicitly predicted various amounts of liquid ($p = 1$). For example, a conditioned stimulus normally indicates the deliverance of 0.15 ml juice. They subsequently followed the conditioned stimulus with an unpredicted stimulus; either a smaller (0.05 ml) or larger (0.50 ml) volume of liquid; in response to which the firing rate of the dopaminergic neurons decreased or increased respectively. In a final experiment, Tobler et al. used one stimulus that predicted the delivery of either a small or a medium volume of juice with equal probability and a second stimulus that predicted a medium or a large volume with equal probability. Results indicated that for both conditioned stimuli, the deliverance of the, respectively, larger stimulus resulted in an increase—and that the deliverance of the, respectively, smaller stimulus resulted in a decrease—of the neuronal firing rate. Surprisingly, the identical medium volume delivery had opposite effects on neuronal activity, depending on the prediction. The prediction is in turn influenced by a *framing effect*. A medium amount of juice is attractive when compared to a small volume of juice, but unattractive in comparison to a large volume. The authors argue that, given the infinite number of reward values that are possible, this is an adaptive process. Thus, the firing rates of the dopaminergic neurons adapt to the coding of reward value in order to have a greater capacity for coding the likelihood of reward outcomes. Results showed that dopaminergic neurons encode a combination of magnitude and probability; the so-called expected reward values and that the response of the dopaminergic neurons depends on framing effects (for a concise review on the behavioural dynamics and neural basis of the framing effect please see Chap. 9 by X.T. Wang et al. in this book).

The effects of expected reward have a discrete neural signature in human decision-making, as demonstrated in a seminal study by Preuschoff et al. (2006). Using a simple gambling task in an fMRI setting, the authors varied expected reward and risk in an uncorrelated manner. Risk is a consideration because many decisions in daily life have to be made under conditions of uncertainty. Expected reward and risk were both represented in dopaminergic innervated brain regions, however, there was a temporal dissociation in their processing. The brain first processes information related to reward expectancy and later risk information. Besides the aforementioned study by Preuschoff et al., there are numerous examples in the literature of findings from animal studies being mirrored in neuroeconomic studies on humans. For example, Roiser et al. (2009) investigated the influence of framing effects on human decision-making and its neural activation patterns. They found that amygdala activation was stronger in those trials where participants made choices in congruence with—compared with those made counter to—the frame, but that this effect was only apparent in subjects carrying the short allele (s-allele) of the serotonin transporter polymorphism (5-HTTLPR; for more information on genetics see Chaps. 4 and 23 by Reuter and Montagin this book), a genetic variant related to neuroticism, depression and anxiety (Roiser et al. 2009).

1.2.1 Validation of Theoretical Models on Human Decision Making in Animals

As described above economists have developed theories (e.g. utility maximization; game theory, etc.) to predict human decision-making. Researchers from cognitive psychology and mathematics have established such theoretical models, albeit with a different focus. The best studied of these models try to explain choices via the simplest form of decision an individual can make—the choice between two alternatives. The focus here lies on the interdependency of choice probability and response time (RT). The most familiar expression of this relationship is the *speed–accuracy trade-off*, which characterizes the decision-maker's dilemma of being forced to negotiate between the competing demands of response speed and response accuracy (Bogacz et al. 2010). Many decisions are based on information that accumulates over time. Although the probability of making a correct or favourable decision increases with the amount of information we have gathered, sometimes we are forced to make quick and ill-informed decisions (e.g. to prevent harm). The development of *Sequential Sampling Models* has increased the theoretical understanding of such decision processes, however, it was the empirical validation in animal models (i.e. single-cell recordings in monkeys) that initially helped to test and refine these models. David Sewell and Philip Smith (see Chap. 14) provide a thrilling and comprehensive introduction to a research area in which theoretical mathematical frameworks and computational neuroscience meets empirical neurophysiological animal research. Through recent advances in imaging techniques, these models have now also been successfully tested in humans (Forstmann et al. 2010).

1.3 Ecological Validity

One of the most severe criticisms of neuroeconomic research is the frequent lack of ecological validity in studies. What can we learn from human decisions that are registered in fMRI scanners; a loud environment where movement is extremely restricted and where social interaction partners are presented—if at all—via video glasses (Mäki 2010). Imaging techniques like MRI, and PET are fantastic tools for allowing us to register brain activity, even in subcortical brain regions, while stimuli are processed. However, these techniques are not made for field studies, in which participants are observed in their natural environment. However, history has demonstrated that experimental approaches applied in the laboratory can indeed provide valuable insights into human behaviour and have thereby helped to legitimize the discipline of experimental economics. The same success is demonstrable for neuroeconomic studies using imaging techniques. Neuroeconomics permanently strives to establish ecological validity in any way possible. Implementing monetary reward in the behavioural games is one of these provisions. Decisions must be related to real consequences for the decision-maker, in order to be ecologically valid. It can be assumed that engagement in an economic game, which is played for monetary stakes, allows even the (fMRI) scanner environment to fade into the background.

Imaging studies are still common in neuroeconomics and have greatly boosted the success of the discipline. However, alternative neuroscientific techniques that are not limited to scanner facilities or laboratories are becoming increasingly prevalent. Molecular genetics is a key example in this instance. Behaviour can be studied in participants' natural environment and the participant subsequently provides a cell sample (e.g. by means of a noninvasive buccal swap) for genetic analyses. This approach ensures that participants are not influenced by the experimenter while exhibiting their natural behaviour. Most economists embarking upon neuroeconomic study are initially unaware that molecular genetics can provide information on the brain. Genes code for neurotransmitters, hormones, receptors and enzymes relevant for brain metabolism. Static genetic variants, called polymorphisms, exert a permanent influence on these gene products, by modulating the expression or the structure of gene products. In recent years a new field has grown from molecular genetics: epigenetics. Epigenetics dispels the ancient myth that genes are like an unstoppable computer programme, started at the moment when the semen and egg of our parents have fused. Prior to the introduction of epigenetics, genetic research often occasioned strong resentment among the general population, as it was considered synonymous with fatalism—a thing you cannot change. Epigenetic research has served to change this view of genes as destiny. Epigeneticists have demonstrated that the environment can and does influence our genes; not the static genetic polymorphisms, but rather the *expression* of our genes, by changing the methylation patterns of the genes. Thus, the relationship between genes and behaviour/environment is *bidirectional* (for a more detailed introduction, please see the genetics Chap. 23 in the methods section of this book and Chap. 4 "Genes and Human Decision Making").

Genetic approaches are not limited to field studies, but are also suitable for laboratory experiments. Neuroeconomics studies have used this method successfully and it will certainly become more and more important in the field. In a seminal study, Israel et al. (2009) have reported an association of a single nucleotide polymorphism (SNP; rs1042778) on the oxytocin receptor gene (OXTR) and prosocial fund allocations in the dictator game. This finding was replicated in an independent sample and serves to corroborate animal and human studies in demonstrating the pivotal role of the hormone oxytocin for prosocial behaviour (for a review see Ebstein et al. 2009).

1.4 Future Perspectives in Neuroeconomics

No matter how strongly neuroeconomists strive to improve prediction models on human decision-making through use of neuroscientific methods, criticism will always be present. It is impossible to convince every sceptic that biological variables can help us to better understand human behaviour and that neuroscientific approaches are helpful in verifying and refining theoretical economic models. On the whole, however, most criticism pertains to serious concerns, which must be taken seriously. The exciting possibilities offered by neuroscientific methods carry with it the risk of overselling the findings (Rubinstein 2008). The mass media contributes to this by exaggerating its reports of solid scientific work. We take this opportunity to discuss two such examples. We recently published a neuroeconomics study entitled "Investigating the genetic basis of altruism: The role of the COMT Val158Met polymorphism" (see a detailed description of this study in Chap. 4 in this book). The newspapers wrote articles on this study with headlines like this: "Altruism gene makes people generous". It is obvious that altruism is not a monogenetic phenotype, but is subject to influence both from many genes, and from environmental effects. Therefore, there cannot be "an (a single)" altruism gene. The second example demonstrates that researchers sometimes tend to oversell their scientific findings. Kuhnen and Chiao (2009) published an article based on a sample of 65 participants entitled "Genetic determinants of financial risk-taking". The Scientific American reported this study with the headline "My genes made me invest: DNA implicated in financial risk-taking". One can debate the connotations of the word "determinants", but it is obviously related to "determinism", implying that there are no other sources of variance relevant for risk-taking, besides the 5-HTTLPR polymorphism investigated in this study. For the sake of Scientific American, it must be noted that the word "implicated" reflects the scientific value of this study very well, much better, in our opinion, than the phrase "genetic determinants". Thus, a modest interpretation of scientific results in the field of neuroeconomics is essential to increase the respectability of the discipline.

It is obvious that the methodological spectrum of neuroscientific techniques has dramatically increased over the last years. Neuroeconomics is no longer limited to fMRI studies. We see EEG-, genetic-, endocrinological-, and TMS—studies, to name but a few methods, and the use of such methods will dramatically increase in future research. The paradigms and games used in neuroeconomic research will also become more and more elaborate in the endeavour to disentangle the subcomponents involved in economic decision making. Finally, the introduction of field studies will further enrich the spectrum by allowing researchers to test laboratory hypotheses in "real life".

References

Bechara A, Tranel D, Damasio H (2000) Characterization of the decision-making deficit of patients with ventromedial prefrontal cortex lesions. Brain 123(Pt 11):2189–2202 (Nov 2000). Erratum in: Brain 132(Pt 7):1993 (Jul 2009)

Bechara A, Dolan S, Hindes A (2002) Decision-making and addiction (part II): myopia for the future or hypersensitivity to reward? Neuropsychologia 40(10):1690–1705

Bogacz R, Wagenmakers EJ, Forstmann BU, Nieuwenhuis S (2010) The neural basis of the speed-accuracy tradeoff. Trends Neurosci 33(1):10–16

Camerer CF (2003) Behavioral game theory: experiments in strategic interaction. Princeton University Press, Princeton, NJ

Chen MK, Lakshminaryanan V, Santos LR (2006) The evolution of our preferences: evidence from capuchin monkey trading behavior. J Polit Econ 114(3):517–537

Ebstein RP, Israel S, Lerer E, Uzefovsky F, Shalev I, Gritsenko I, Riebold M, Salomon S, Yirmiya N (2009) Arginine vasopressin and oxytocin modulate human social behavior. Ann NY Acad Sci 1167:87–102

Engel C (2011) Dictator games: a meta-study. Exp Econ 14:583–610

Forstmann BU, Anwander A, Schäfer A, Neumann J, Brown S, Wagenmakers EJ, Bogacz R, Turner R (2010) Cortico-striatal connections predict control over speed and accuracy in perceptual decision making. Proc Natl Acad Sci USA 107(36):15916–15920

Kahneman D, Tversky A (1979) Prospect theory: an analysis of decision under risk. Econometrica: J Econom Soc 47(2):263–291

Kuhnen CM, Chiao JY (2009) Genetic determinants of financial risk taking. PLoS One 4(2):e4362. doi:10.1371/journal.pone.0004362

Israel S, Lerer E, Shalev I, Uzefovsky F, Riebold M, Laiba E, Bachner-Melman R, Maril A, Bornstein G, Knafo A, Ebstein RP (2009) The oxytocin receptor (OXTR) contributes to prosocial fund allocations in the dictator game and the social value orientations task. PLoS ONE 4(5):e5535

Mäki U (2010) When economics meets neuroscience: hype and hope. J Econ Methodol 17:107–117

Marshall M (1920) Principles of economics: an introductory volume, 8th edn. Macmillan, London

McClure S, Laibson D, Lowenstein G, Cohen J (2004a) Separate neural systems value immediate and delayed rewards. Science 306:503–507

McClure SM, Ericson K, Laibson DI, Loewenstein G, Cohen JD (2004b) Time discounting for primary rewards. J Neurosci 27(21):5796–5904

Mischel W, Shoda Y, Rodriguez ML (1989) Delay of gratification in children. Science 244:933–938

Preuschoff K, Bossaerts P, Quartz SR (2006) Neural differentiation of expected reward and risk in human subcortical structures. Neuron 51(3):381–390

Roiser JP, de Martino B, Tan GC, Kumaran D, Seymour B, Wood NW, Dolan RJ (2009) A genetically mediated bias in decision making driven by failure of amygdala control. J Neurosci 29(18):5985–5991

Rubinstein A (2008) Comments on neuroeconomics. Econ Philos 24:485–494

Schultz W (2013) Updating dopamine reward signals. Curr Opin Neurobiol 23(2):229–238

Schultz W, Dayan P, Montague PR (1997) A neural substrate of prediction and reward. Science 275(5306):1593–1599

Spanagel R, Weiss F (1999) The dopamine hypothesis of reward: past and current status. Trends Neurosci 22(11):521–527

Tobler PN, Fiorillo CD, Schultz W (2005) Adaptive coding of reward value by dopamine neurons. Science 307:1642–1645

Tom SM, Fox CR, Trepel C, Poldrack RA (2007) The neural basis of loss aversion in decision-making under risk. Sci 315(5811):515–518

Part I
Games in Experimental Economics

Chapter 2
Game Theory in Neuroeconomics

Claudia Civai and Daniel R. Hawes

> *(Spade): If you kill me, how are you going to get the bird? If I know you can't afford to kill me till you have it, how are you going to scare me into giving it to you?*
> *(Gutman): "Well, sir, there are other means of persuasion besides killing and threatening to kill."*
> *(Spade): "Sure, but they're not much good unless the threat of death is behind them to hold the victim down. See what I mean? If you try something I don't like I won't stand for it. I'll make it a matter of your having to call it off or kill me, knowing you can't afford to kill me."*
> *(Gutman): "I see what you mean. That is an attitude, sir, that calls for the most delicate judgment on both sides, because, as you know, sir, men are likely to forget in the heat of action where their best interests lie and let their emotions carry them away."*
> —The Maltese Falcon

Abstract Game theory and contemporary decision theory provide the mathematical foundation of economics. Neuroeconomics, which principally concerns itself with the integrative study of brain, mind and behavior, builds on this mathematical foundation while also drawing heavily from the repository of experimental paradigms that have grown out of economic game theory and behavioral economics. Game theory is central to neuroeconomics primarily because it constitutes a formal mathematical framework with which to bridge insights occurring at different levels of neuroeconomic analysis. In particular, game theoretic principles can be used to express neuroscientific ideas about the brain, psychological concepts regarding the human mind, and economic predictions of human behavior, thereby making these different ideas more rigorously relatable to each other. In this chapter we provide a nontechnical introduction to game theory and its relation to neuroeconomics. It has been written as an overview of the basic concepts most likely to be encountered in neuroeconomic research. The first part of the chapter introduces the reader to the

C. Civai (✉)
Donders Institute for Brain, Cognition and Behaviour, Radboud University, Nijmegen, Netherlands
e-mail: c.civai@donders.ru.nl

D.R. Hawes
Institute of Food and Resource Economics, University of Bonn, Bonn, Deutschland
e-mail: daniel.hawes@uni-bonn.de

M. Reuter and C. Montag (eds.), *Neuroeconomics*, Studies in Neuroscience, Psychology and Behavioral Economics, DOI 10.1007/978-3-642-35923-1_2

basic concepts and philosophical underpinnings of game theory in relation to neuroeconomics. The second part is an introduction and discussion of common games, including the games featured in the other chapters of this book.

2.1 Introduction

In the well-known movie "The Maltese Falcon" (Wallis and Huston 1941) Humphrey Bogart's character (Spade) attempts to 'call bluff' on his adversaries' threats. He does so by what amounts to game theoretic reasoning—namely by putting into relation each man's objectives, their individual beliefs, and their shared knowledge of the situation—and deduces that the adversary, Gutman, cannot possibly afford to kill him for risk of never learning the Maletese Falcon's whereabouts. Gutman of course concludes with the cautioning appeal to keep in mind the potential for internal conflict between reason and emotion, and how rational objectives may become lost in its wake.

Neuroeconomics relates to this scene as an extension of economic analysis, traditionally preoccupied with the prediction of behavior, into the realm of the brain and the mind (the internal processes that give rise to behavior). Neuroeconomics is inherently interdisciplinary, but draws most heavily from economic decision theory, psychology, and neuroscience. Game theory, as the mathematical foundation of economic decision theory, is central to neuroeconomics, because of the relatively new disciplines' intellectual history (i.e., because neuroeconomics lends from economics, and economics uses game theory), but also because it has the potential to function as the mathematical foundation upon which neuroeconomic theory might be developed further into a discipline of its own right (i.e., the properties of game theory make it an attractive candidate for trying to develop an integrative theory of brain, mind, and behavior).

The purpose of this chapter is to introduce researchers coming to neuroeconomics from fields other than economics to the basic principles and ideas that feature in game theory. While outlining the basic tenets of game theory, we also attempt to draw philosophical and historical connections that link game theory to the goals and of neuroeconomics.

Game theory, generally speaking, is the study of mathematical objects whose states and properties interact. At the level of neuroscience, game theory could be used to describe the interactive or competitive firing of neuron populations. At the level of psychology, game theory could be used to describe the emergence of mental states in relation to interacting cognitive processes, or the emergence of behavior from competing mental states. Traditionally, in economic analysis, game theory has been used to predict the behavior of multiple, interacting, intelligent decision-makers. Hence, game theory can be viewed as a general tool for the logical consideration of strategic and nonstrategic relationships at multiple levels of analysis, as long as the concepts being used have representation as numbers and the relationships are closed or bounded correspondences. Economic examples, which represent the mathematical objects of game theory as the behaviors of intelligent decision-makers, are generally intuitive, instructive, and easy to develop, wherefore this consideration forms the common approach in most textbooks, and also in this chapter. However, we hope to draw the reader's attention to the fact that game theory itself, being a mathematical language and logic system, has far wider application and that fundamental issues regarding measurement of utility through games, and how this relates to theory development in physics, remains outside of the scope of this chapter. The classic book Theory of Games and Economic Behavior (in particular chapter 1) remains invaluable on this account (von Neumann and morgenstern 1944).

As neuroeconomics becomes more consolidated as a stand-alone discipline, we expect the importance of game theory for neuroeconomics to increase, and have reasoned hope for the potential of game theory to function as a bridge across different levels of brain–mind behavior analysis. This latter part, concerning the future of game theory in neuroeconomics, remains personal conjecture and is only hinted at or mentioned in passing throughout this chapter; however, it informs much of our thinking in choosing which elements of game theory to focus on for this introduction, and how to present the basic principles. It makes sense, therefore, to provide a quick note on our definition of neuroeconomics.

Neuroeconomics as an interdisciplinary scientific approach aimed at discovering/creating a unified theory of human behavior and cognition, via integrative study of mind, brain, and behavior relationships.

Practically, the neuroeconomic enterprise combines concepts, methods, and technological tools from neuroscience (see Chap. 8) with formal analysis of decision-making, typically drawing heavily from economic decision theory as well as psychology.

As a procedure, neuroeconomics aims to experimentally link the neurophysiological and behavioral constituents of the decision process to each other, and then conceptually relate these links to psychological concepts of mental activity via formal models, typically in the tradition of economics and decision theory.

With this definition of neuroeconomics in place, the formal models developed within game theory enable mathematically precise descriptions of the decision process, which in turn allow prediction, specification, and comparison of neural activity presumed to underlie decision-making. Additionally, experimental paradigms developed within experimental game theory (i.e., games) are prominently featured in neuroeconomics research, where the goal is often to differentiate between competing economic models of the decision process (Glimcher and Rustichini 2004; Rustichini 2005; Camerer et al. 2005), or to investigate competing descriptions of mental processes thought to underlie a particular kind of decision or behavioral phenomenon.

Because it is a common source of confusion, the final note of this introduction points to game theories' conceptual birth in expected utility theory and connection to the economic idea of rational decision-making (Mongin 1997): As will be discussed later, strict economic rationality is not necessary for game theoretic inquiry, and many game theoretic applications in neuroeconomics are explicitly geared toward understanding decision processes for which individuals systematically violate the predictions of rational choice theory and expected utility maximization in one way or another. Furthermore, the interested reader is encouraged to consider differences in the economic meaning of rationality compared to the noneconomic colloquial meaning of rationality, for example, by consulting Anthony Downs' book "An economic theory of democracy." (1957).

The remainder of this chapter elaborates on the basic elements described in the above introduction. It concludes with a compendium of games most commonly encountered in the neuroeconomics literature, including a digest of each of the games featured in this book.

2.1.1 Basic Terms and Definitions

This chapter is a nontechnical introduction to game theory[1]; however, the main strengths of game theory derive precisely from its mathematical exactness and rigorous definitions, wherefore some minimal recourse to formal terminology appears unavoidable. We therefore begin by introducing some basic terms and definitions that will appear throughout this chapter, first among which are those that describe an economic *Game*.

A formal game consists of three basic objects:

1. **Players** are independent decision-makers, mathematically represented in terms of their utility functions; i.e., by a function that assigns ordinal preference rank to all possible outcomes of the game.
2. **Actions** are full descriptions of the actions each player may take during the game. These action descriptions may differentiate between different points of time, situational circumstances or—more generally—*stages* of the game.
3. **Payoffs** are full descriptions of the outcome (and consequently utility) experienced by each player for each possible combination of actions that may occur during the game.

Additional objects used to describe games may include

(a) **Information Sets** are full descriptions of the information available to each player in each stage of the game. Information pertains to the actions available to the players, their utility functions, as well as the current history and possible trajectory of stages in the game.
(b) **Environments** are nonstrategic mechanisms capable of influencing any of the above elements of the game, typically in a probabilistic manner. For example, the environment may impose random restrictions on what type of actions are

[1]Comprehensive, technical treatments can be found for example in Myerson (2013) and Osborne and Rubinstein (1994).

available at a given stage of the game, or may influence the payoff distribution for game outcomes.

(c) **Strategies** are probability distributions over all actions for each stage of the game. Strategies are full descriptions of what action should be chosen for each possible stage of the game. A strategy can describe a specific action for each stage, in which case it is referred to as a pure strategy. Alternatively, a strategy may assign probabilities to multiple actions for any given stage, in which case it is referred to as a mixed strategy. Importantly, a strategy specifies an action or mixture of actions for every possible contingency of the game.

In nondegenerate[2] games, the actions taken by all of the players collectively interact to determine the payoffs and consequent utilities experienced by each player individually. This gives rise to the strategic considerations, which lie at the heart of game theory.

The major objectives of game theoretic inquiry may therefore be described as the aim to:

1. Formulate applicable game descriptions for real-world decision problems;
2. Develop and apply solution concepts to positively describe or normatively prescribe the strategic reasoning processes that are used by players in game situations;
3. Develop, refine, and apply equilibrium concepts that describe stable patterns of strategic reasoning and behavior between the players.
4. Identify the existence and properties of such equilibriums.

The above elements of a game are described and presented in either a so-called *Extensive Form*, or as a so-called *Normal Form* representation.

Extensive Form representations are depictions of games by the way of mathematical graphs. In these graphs the nodes represent the different stages of the game, and the edges (the lines connecting the nodes) represent the actions that lead to each stage. Extensive Form games specify exactly the possible sequence of actions and their resulting outcomes for a game, and are therefore particularly useful when the order in which players choose during the game is relevant to its outcome. An annotated example is given in Fig. 2.1a.

Normal Form representations of games are more commonly used for games in which the order of moves is irrelevant, as is the case in so-called one-shot, simultaneous move games, i.e., games in which the resulting outcome is determined by considering all chosen actions simultaneously.[3] Normal Form games are presented as matrices, with each cell of the game matrix containing a vector of final

[2]That is, games that can not be reduced to sets of mere one-player decision problems in which the actions of other players are irrelevant to each players utility.

[3]Think for example of a secret ballot election, in which each player enters a vote into a computer, and the computer then decides the election outcome by considering all such submitted votes.

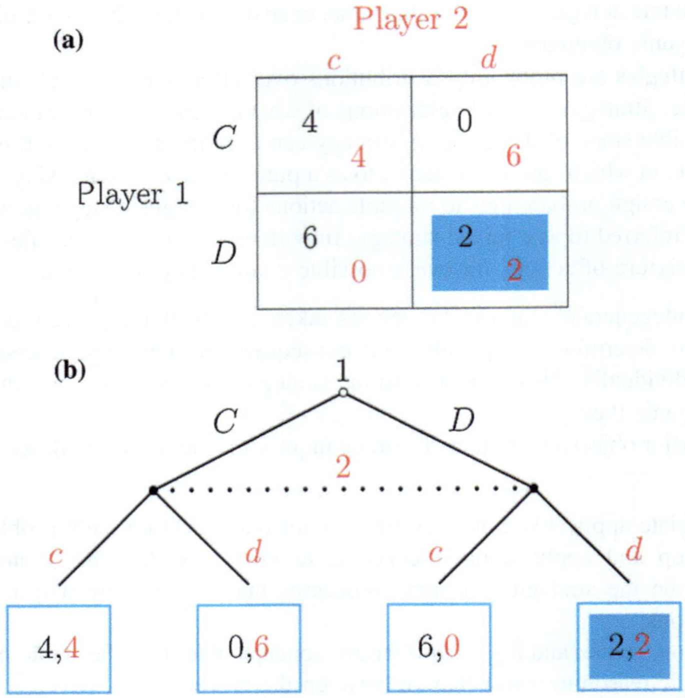

Fig. 2.1 **a** Generic Prisoner's Dilemma (normal form). *Rows* indicate actions available to Player 1 ({C, D}) and columns indicate actions of Player 2 ({c, d}). *Cells* list the utility of each combination of actions. Utility to player 1 is listed first in *black font*. Utility to player 2 is listed second and using *red font*. We have *highlighted* the Nash equilibrium outcome in the *bottom right cell*, corresponding to choice of D, and d by the two players. **b** Generic Prisoner's Dilemma (extensive form). For illustration we also show a strategic form version of the PD. Nodes indicate players' "turns" in the game. Player 1 moves first at the initial node of the game (open circle). Edges indicate actions available at each node ({C, D} for player 1, {c, d} for player 2). *Dotted lines* connecting nodes indicate information sets. Two nodes belonging to the same information set cannot be distinguished by the player. Hence, player 2—at the time of her move—does not know whether player one has chosen C or D. Final nodes contain the utility to each player. We have again highlighted the Nash equilibrium for this game. Although the extensive form and strategic form of the prisoner's dilemma are equivalent, not all extensive form games can be transformed into an equivalent normal form game

payoffs for each possible combination of actions (for all players) that may occur in the game.[4] An annotated example is given in Fig. 2.1b.

The well-known Prisoner's Dilemma (PD) provides a useful example for demonstrating the concepts introduced thus far, and both Fig. 2.1a, b are depictions of this game. The PD is a one-shot, simultaneous move game between two players. Each player in the game has two actions available to him or her. We call these

[4]Note that all extensive form games can be transformed into normal form, but not all normal form games may have an equivalent extensive form; hence there also exists a qualitative difference.

actions {C} and {D} for player 1, and {c, d} for player 2. The intuition behind the PD was originally suggested by Flood and Dreshner in the 1950s, later formalized by Albert Tucker (Poundstone 1992), and goes as follows: two criminals are being questioned by the police; both are facing the choice of whether to defect ({D} or {d}) on their partner in crime by providing incriminating evidence to the police, or to remain cooperative ({C}, or {c}) by not revealing any information during interrogation. The possible actions to each player are therefore to cooperate or to defect ({C, D}) for player one, as well as for player two ({c, d}). In the case that both prisoners remain cooperative ({C, c}) the police lacks evidence for a full conviction and both criminals go to jail for only a short period of time; in the case that both prisoners defect on each other ({D, d}), they both land in jail. The interesting scenario that creates the dilemma results from what happens when one player defects while the other remains cooperative ({C, d} or {D, c}): in this case the defector goes free entirely, while the defected upon goes to jail for an extended period of time. The possible outcomes to the game are therefore ({C, c},{C, d},{D, c}, and {D, d}), and the payoffs are given by the utility experienced for the differing amounts of jail time (in the example we let these utilities be 0, 2, 4, and 6). Assuming that both players prefer less time in jail to more time in jail (i.e., they prefer higher utility values among the outcomes in Fig. 2.1b), each player in this game fares best when he or she is the only one to defect, and worst if he or she is the only one to cooperate. Additionally, each player also prefers the outcome of mutual cooperation to the outcome of mutual defection.

Standard game theoretic reasoning predicts an equilibrium outcome for the Prisoner's Dilemma in which both players defect. This is because the outcome from meeting cooperation by the other player with defection (i.e., the best outcome) is preferred to the outcome from mutual cooperation; at the same time meeting defection by the other player with defection also is preferred to being the only cooperator in the game (i.e., the worst outcome). Hence defecting is a best response to whatever the other player does, and thus mutual defection becomes an equilibrium response in the game.

The terms *best response* and *equilibrium* deserve special note in this context, as they relate to a very particular solution concept in game theory, known as *Nash Equilibrium*, Nash Solution Concept, or Best Response Equilibrium. This concept defines equilibriums for finite noncooperative[5] games as instances in which each player is playing a *best response* given the strategies being played by all other players in the game. A strategy, in turn, is defined as a best response if and only if its expected utility is at least as large as that obtained from any other possible strategy the player may play.[6] Hence, in Nash equilibrium no player has any

[5]Game theory divides games into cooperative and noncooperative types. All of the games in this chapter are noncooperative games, and we therefore do not spend a lot of time discussing the difference between the two classes of games, which can be found elsewhere (e.g., Myerson 2013).

[6]The concept is named after John Nash, among other contributions, for his work on such equilibrium points in *n*-person games (Nash 1950).

incentive to unilaterally change his or her chosen response in the game, since it is a best response to what all other players are choosing at the time.

Returning to the PD, it is also worth noting that defection is a so-called dominant strategy for both players. This means that defecting is optimal regardless of what other players choose. Not all games have dominant strategies, however, and they are not necessary for Nash equilibrium. We will see examples of such games later in this chapter.

2.1.2 Rationality and Expected Utility Maximization

Standard game theoretic solutions to decision situations such as the Prisoner's Dilemma prescribe the optimal behavior of a rational decision-maker with unlimited analytical resources under the goal of utility maximization. However, human analytical resources are necessarily limited, introspection may be noisy, preferences may be uncertain, and any host of context-dependent affective responses may contribute to the decision process. Furthermore, decision-makers may rely on heuristics and intuitions (i.e., emotions or gut feelings) in order to reduce cognitive costs (Simon 1972, 1990; Tversky and Kahneman 1974; Damasio et al. 1991; Gigerenzer 2004; Glöckner and Betsch 2008; Glöckner and Hochman 2011). Such psychological features and behavioral strategies may be behaviorally indistinguishable from rational choice behavior in some contexts, but lead to systematic biases and inconsistencies in others. Investigation of such behavior has a long history in economics (e.g., Allais 1979; Ellsberg 1961; Loomes and Sugden 1982), and has given rise to extensions (e.g. Friedman and Savage 1948; Aumann 1997), and modifications such as bounded rationality (Selten 1990) and Prospect Theory (Kahneman and Tversky 1979). Consequently, economic theory has produced a wealth of models and theory refinements that individually address instances in which standard economic model possess poor predictive validity.[7] To the extent that none of these competing models and refinements approaches the economic ideal of a unified theory of human behavior, neuroeconomics is seen as a disciplined approach to answering unresolved questions in economics: for example by juxtaposing competing models [e.g., by investigating whether a unified system, or separate systems compute the value for present versus future rewards (McClure et al. 2004)]; by investigating the rules that govern for whom, when and under which circumstances one model, or model specification, is more applicable than another [e.g., by generating insights into the neurobiology underlying different types of decision-makers (Coricelli and Nagel 2009; Bhatt et al. 2010), and learned strategic behaviors (Hawes et al. 2012)]; or by helping to establish the precise role

[7]Note that Economic Theory is primarily concerned with generalizable predictive validity, and that content validity regarding the process via which a decision is actually made, is at best secondary, but probably irrelevant, to as-if modeling in economics.

of affect and social sensitivity during decision-making (Van't Wout et al. 2006; Civai et al. 2010).

Despite being strongly linked to expected utility theory as formulated by Von Neuman and Morgenstern, game theory features prominently in the neuroeconomic endeavors described above, as game theory itself allows for any number of postulatory models of players behaviors, and the details of the behavioral models used to describe the decision-maker may differ across studies. For example, players may be thought to be entirely rational[8] and self-interested, while updating their beliefs in complete accordance to Bayes rule, eventually converging on Nash equilibrium behavior (Kalai and Lehrer 1993). Alternatively, players may be modeled as possessing *bounded rationality*, may be compassionate to others, or may update beliefs with respect to subjective probability representations (Rabin 1993; Camerer et al. 2011). Similarly, different definitions for what comprises equilibrium behavior (i.e., different types of equilibriums) exist in game theory, therefore extending and generalizing the concept of Nash equilibrium (e.g., Perfect Nash Equilibrium in dynamic games, Correlated Equilibrium, Trembling Hand Equilibrium, etc.).

Despite this flexibility of game theory, rational choice and expected utility theory remain the principal departing points for most neuroeconomic investigations of behavior, where, at the very least, expected utility theory functions as a benchmark for comparison for any neuroeconomic investigation of alternative models. Because of this, and because the games discussed in this book are conceptually founded in the expected utility framework, our concluding compendium considers each game with respect to expected utility maximization and Nash equilibrium. Before explaining precisely what these terms mean, we want to add brief mention of some alternative frameworks that feature prominently in neuroeconomics, but are only peripherally considered in the compendium of games that concludes this chapter.

2.1.3 Alternatives to Expected Utility Theory

Insights developed under the frameworks of Bounded Rationality, Prospect Theory, and under the research agenda often labeled Behavioral Economics, are particularly relevant to the research domains we have mentioned above, and have been consequentially influential in neuroeconomics, and neuroeconomic research on games: *Bounded Rationality* as introduced originally by Simon (1955) and Gigerenzer and Selten (2002) takes into account cognitive limitations of the decision-maker. The theory posits that human beings are limited in their capacity for solving complex problems and for processing large quantities of information, wherefore useful game

[8]Rationality in the economic sense means that the decision-maker has preferences that are complete and transitive, or acyclical. This means that the decision-maker can rank all alternatives according to an ordinal metric, and that within this ranking whenever A is preferred to B, and B is preferred to C, then A will be preferred to C also.

theoretic descriptions must include explicit consideration of how players decide upon which information to consider, which information to ignore, and when to rely on simpler heuristics for decision-making. Such questions of which types of information are attended to during the decision process and how integration occurs for information derived from multiple sources, or pertaining to multiple dimensions of the decision problem, lay at the core of neuroeconomic research. Additionally, researchers have recently begun to investigate stable individual differences in how individuals use information during decision-making (e.g., Burks et al. 2009; Engelmann and Tamir 2009).

Prospect Theory, developed by psychologists Kahneman and Tversky (1979) in the context of risky decision-making is similarly central to neuroeconomics. In a nutshell, Prospect Theory argues that utilities of losses are computed differently from utilities derived from gains, and that objective probabilities undergo a subjective weighting during the decision process. The precise nature in which value and probability are represented in the brain continues to be a core theme of neuroeconomic research. Various psychological phenomena have been explained within Prospect Theory, such as the *framing effect*, which considers differences in choice behavior depending on whether the same problem is described in terms of gains or losses, and the *endowment effect*, also known as the fact that people ascribe more value to things that they own. These cognitive biases have also been subject of neuroeconomics investigation (De Martino et al. 2006; Knutson et al. 2008).

Finally, a large body of recent work spearheaded by researchers such as Colin Camerer, George Loewenstein, Matthew Rabin, Ernst Fehr, and Simon Gächter has tried to go even further than Prospect Theory in its attempt to import insights from Psychology into economic thinking. This research has drawn strongly from insights in Personality Psychology and Social Psychology to investigate the way in which humans respond to incentives and formulate beliefs during complex decision-making, especially social interactions. Influential work in this field has tried to explain the role of nonreciprocal altruism, perceptions of fairness, inequity aversion, and competitive motives during the choice process, and has thus provided novel insights to neuroeconomics regarding the investigation of a person's utility (e.g., Rabin 1993; Fehr and Schmidt 1999; Camerer 2003). Work in this field also investigates social phenomena such as the occurrence of economic bubbles in bidding competitions, or how beliefs about other player's mental abilities influence the level of reasoning and information processing that is applied to game situations. As evidenced by other chapters of this book, economic games, such as multi-person auctions or the Ultimatum Game (see Sect. 2.2.6 of this chapter), feature prominently in this type of research, and are also used as tools in neuroeconomic investigations of the role of social context and cognitive skills during decision-making.

2.2 Compendium of Common Games

The main body of this book chapter has introduced game theory as a powerful and flexible, philosophically rich framework for neuroeconomic research on decision-making. We have purposefully emphasized the distinction between game theory and the economic (*as-if*) notion of rationality, since we have found that conflation of these two concepts can confuse non-Game Theorists about the applicability of game theory to the *as-is* investigation of human behavior. Hoping to have thus preempted such confusion, we will however continue our discussion of games with reference to the common framework within which players are thought to be *rational expected utility maximizers* in the Von Neumann and Morgenstern sense. This means, first, that players consider a probability distribution of possible outcomes and their attendant payoffs. Second, the expected utility of an action is then determined by integrating the utility of the payoffs with respect to these probabilities. In the discrete case, this means that the utility from each possible outcome following an action is weighted by the probability assigned to it by the action. Finally, the sum of these weighted utilities then forms the players expected utility. Players are **expected utility maximizers** if they compute expected utilities in this way and always choose the action that results in the highest expected utility. Players are considered **rational** in the economic sense if they have complete, and transitive, preferences.

Furthermore, unless stated explicitly, we will consider only *Nash equilibrium* solutions to the games we discuss. **Nash equilibriums** are defined for finite non-cooperative games, and constitute instances in which each player is playing a *best response* given the strategies being played by all other players. A strategy, in turn, is a best response if, and only if, its expected utility is at least as large as that obtained from any other possible strategy.

2.2.1 Matching Pennies

Intuition: The Matching Pennies game features two players with identical choice between two actions. One player receives a preferred outcome whenever both players match on their actions, and incurs a less preferred outcome whenever they mismatch on their actions. The other player incurs the preferred outcome as a result of the mismatch, but incurs the less preferred outcome from a match. In the notation of the example (Fig. 2.2), matches occur when player 1 chooses U and player 2 chooses L, or when player 1 chooses D and player 2 chooses R. All other combinations produce mismatches.

In the matching pennies game, the actions that lead to a preferred outcome for one player, lead to the less preferred outcome for the other. This property is typically referred to as zero-sum (since the properties of the game remain under payoffs

		Player 2: *l*	Player 2: *r*
Player 1	*U*	−1, 1	1, −1
	D	1, −1	−1, 1

Fig. 2.2 Generic Matching Pennies. Player 1 chooses between U(p) and D(own), while player 2 chooses between l(eft) and r(ight). Player 1 wins when actions match ({U, l} and {D, r}), player 2 wins under the alternative outcomes

that add up to zero for each outcome), and the Matching Pennies game is often cited as a quintessential example of such games.

Equilibrium Solutions: The Matching Pennies game is a two-action version of popular games such as "Rock, Paper, Scissors", "Morra", or "oddsandevens". Like with these common games a player's most promising strategy in Matching Pennies is to remain unpredictable to the opposing player while also exploiting any predictability they may observe with their opponent. More concrete, the Matching Pennies game does not have best response equilibrium in pure strategies. Using the notation of the normal form example in Fig. 2.2, whenever player 1 plays a strategy in which U is played with a probability $p > 0.5$ player 2 may improve his payoffs by always choosing R. And whenever player 1 chooses U with probability $p < 0.5$, player 2 can exploit this by always choosing L. The same is true in likeness for player 2 choosing either L or R with probability different from $p = (1 - q) = 0.5$. Crucially, whenever player 1 randomizes between U and D with equal probability ($p = q = 0.5$) player 2 is entirely indifferent between any probability over R and L. Likewise player 1 has equal expected utility from U and D whenever player 2 chooses L and R with equal probability $p = q = 0.5$. Consequently, the only Nash equilibrium exists in mixed strategies for both players randomizing between their available actions with probability $p = q = 0.5$.

Insights: According to Nash equilibrium a player playing multiple games of Matching Pennies against a single opponent should try to randomize his or her actions, while also being sensitive to whether the opponent's choices follow any sort of predictable pattern. Two types of predictability are of interest here, the first being predictability based on a perceived pattern (e.g., every fifth choice of player 1 is U), the second being predictability based on random processes[9] with probabilities that deviate from $p = q = 0.5$. Exactly how subjects identify patterns or keep track

[9] Independent and identically distributed.

of shifting probabilities during repeated decisions remains an active question for neuroeconomic inquiry. Interestingly, recent research in monkeys has shown that simple reinforcement learning may be capable of explaining optimal matching pennies behaviors in equilibrium (Lee et al. 2004). More generally, reinforcement learning seems to underlie preference learning and decision in various social and nonsocial contexts, suggesting particular relevance of these models to neuroeconomics (Frank and Claus 2006; Seo and Lee 2012).

2.2.2 Prisoner's Dilemma

Extensive and normal form descriptions, as well as the intuition and equilibrium solution to the prisoner's dilemma have been provided in the main text and in Fig. 2.1a, b.

Insights: Few games have been as extensively studied as the Prisoner's Dilemma, yet substantial disagreement remains regarding whether or not humans meaningfully deviate from rational behavior in experimental and real world PD scenarios. Certain seems that in experiments and in real life, human decision-makers cooperate (and avoid the Prisoner's Dilemma outcome) more often than can be explained by mere noisy decision making[10] (Cooper et al. 1996).

Unclear is what kind of cognitive architecture promotes such cooperative behavior. Given the relative simplicity of the Prisoner's Dilemma, explanations that impose cognitive constraints on the decision-makers' ability to identify the expected utility maximizing solution would appear spurious. Instead, human decision-makers appear to systematically compute larger expected utility from cooperative behavior given certain parameterizations of the game.

One interpretation that has been offered for this is that human subjects view even one-shot prisoner's dilemma games as potentially repeated interactions. Once the probability of repeated interaction is increased beyond zero, the space of Nash equilibrium solutions is capable of sustaining any given rate of cooperative behavior, and cooperation becomes explainable within the common framework. This explanation possesses considerable face validity, in the light of experiments involving actual repeated games and the possibility of building reputations, which support the basic idea that humans are responsive to the prospect of repeated interaction and reputation building. The same explanation can be made differently by recourse to decision-makers placing some positive probability of being observed by at least one person with whom they may interact in the future. This makes the idea conceptually very difficult to repudiate (but see Cooper et al. 1996). However, the important question of whether this kind of behavior is supported by a cognitive architecture that supports truly altruistic behavior remains of interest to neuroeconomics. Special interest has been given to one particular way in which such an

[10]Economic models exist, which relate such behavior to rational choice under uncertainty (e.g., Neyman 1985; Kreps et al. 1982).

architecture may be expressed, namely in the form of other regarding preferences, or social utility. The importance of other regarding preferences lies in the observation that an ostensible prisoner's dilemma may not always possess the conceptual incentive structure of a prisoner's dilemma to the actual decision-maker. Since one of the avenues for understanding decision-making rests obviously in understanding how incentives and feasibility constraints are represented in the mind of the decision-maker, cooperative behavior (in prisoner's dilemma games and beyond) remains an active area of research in this respect (Fehr and Camerer 2007; Sanfey 2007; Houser and McCabe 2009).

2.2.3 Public Goods Game (with and Without Costly Punishment)

Intuition: The Public Goods Game is an n-person Prisoner's Dilemma (with $n > 2$) in which each person faces the choice between contributing to a shared community pool, or free riding on the contributions of others. Contributions to the pool are multiplied and shared equally among all group members, however, the return to the individual is always smaller than the contribution, so that contributing adds to the overall welfare of the group, but subtracts from the welfare of the individual (Fig. 2.3). Like the

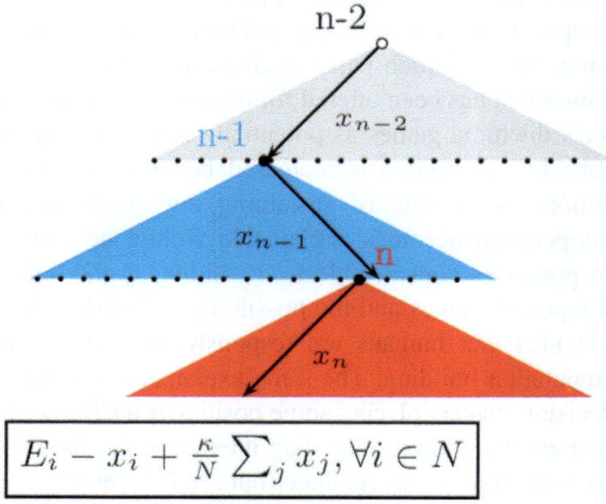

Fig. 2.3 Generic Public Goods Game. *Solid triangles* indicate a player's choice of a continuous value x within some interval. Each player moves without knowledge of what value has been chosen by the other players in game (illustrated by the *dotted lines* along the intervals). For an arbitrary number N players, payoffs are determined as a function of some initial endowment E and a multiplier κ in $[1, N)$. Each player receives exactly their initial endowment less their donation to the public good, x, as well as their share of the κ-multiplied sum of total contributions by the group as a whole

Prisoner's Dilemma, the Public Goods Game is a paragon for games that investigate the emergence of collaboration and altruistic behavior. The game is often played with an added option of costly punishment. In the Public Goods Game with costly punishment, players play a second round during which they may incur a small cost in order to subtract earnings from the final payoffs of other group members.

Equilibrium Solution: In the Public Goods Game, the individual's payoffs are increasing in the contributions made by others, and decreasing in the contributions he or she makes herself. The dominant strategy for the individual is therefore to not contribute. In equilibrium, nobody contributes, and group payoffs are minimized.

In the one-shot game with costly punishment, the same equilibrium solution obtains through backward induction. In the final round, the action of punishing reduces the individual's payoffs (i.e., it is costly), hence no player should ever punish. Consequently, the threat of being punished for free riding should not enter the player's decision in the first round.

Insights: As in the two-person Prisoner's Dilemma actual behavior in experiments and real life often violates the equilibrium prediction for the Public Goods Game. In particular subjects consistently contribute amounts larger than zero to the public good (Chaudhuri 2011). Additionally, subjects frequently punish free riders, even in the one-shot game approximations, and (supposedly anticipating this) humans contribute more in games in which the threat of punishment exists (Fehr and Gächter 2000; Dreber et al. 2008). Interestingly, punishment has also been found to be used against contributors in some experiments (so-called antisocial punishment) (Rand et al. 2011).

As discussed for the Prisoner's Dilemma, a neural architecture promoting cooperative behavior is likely to be sensitive to social outcomes, such as payoff equity, distributive fairness, and reciprocity in behaviors. The Public Goods Game with punishment is especially interesting in this regard, as it indicates player's willingness to pay for a more even, or fair distribution of outcomes. Chapter 5 discusses such punishment behaviors.

Interesting insight comes also from a set of carefully designed behavioral experiments by David Rand and colleagues involving several versions of the Prisoner's Dilemma, including the *n*-person public goods version (Rand et al. 2012). Across these experiments the authors show cooperation to occur spontaneously under conditions of low reflection and intuitive decision-making. Inducing subjects to reflect consciously on their decisions, on the other hand, reduced cooperative behavior. These data suggest that cooperative behavior may not be entirely explainable by social preferences that promote socially beneficial outcomes, but that such behaviors depend on the mode of cognitive engagement with the decision. To the extent that different networks are involved in these different types of cognitive engagement, these results present important questions for neuroeconomics. Chapter 5 discusses in particular the role of the prefrontal cortex for behaviors involving reciprocal fairness.

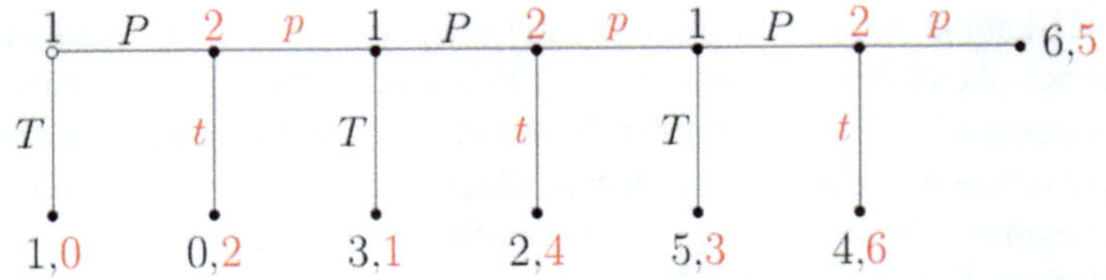

Fig. 2.4 Generic Centipede Game: at each node, the moving player may T(ake) the reward, or P(ass) the move along. Overall rewards increase with each passing decision. Conceptually, each choice of passing costs the passing player 1 utility point, but gains the other player 2 utility points in the depiction shown here

2.2.4 *Centipede Game*

Intuition: The Centipede Game is a classic example of a group of games that can be solved using backward induction. At its core it describes a finitely often repeated, sequential prisoner's dilemma. In one popular version of the game one or more players take turns deciding whether to take a given monetary reward, or to pass the move on to the next player. Whenever a player decides to take the money, the game ends that round, the taking player receives a large reward and all other players end with smaller or even zero rewards. Whenever the move is passed on, the overall size of the monetary pie grows, and the next player gets to make the decision of taking the money or passing. Critically, if the moving player in round n of the centipede game decides to take the money, then the payoff in round n to the player who passed in round $n - 1$ is smaller than his or her payoff would have been had he or she taken the money in round $n - 1$ instead of passing on the move (Fig. 2.4).

Equilibrium Solution: The dilemma of the centipede game, and games like it, arises because the game is played finitely many rounds, i.e., there will be a final stage in which one player has a strictly dominant choice in taking the money, and the players know this about the game. Using this information, it can be deduced that the last player to move will take the money. This implies that the second to last player may improve his expected utility by not passing on the second to last turn. By extension the third to last mover should not pass on the third to last turn, and eventually this iteration implies that the very first mover should take the money and end the game on the very first turn. This solution is so-called sub-game perfect, meaning that it considers each of the sub-game[11] components of the centipede game and identifies a strategy for each player that is an equilibrium strategy for each of the sub-games.

[11]A sub-game is strictly defined as any subset of a game which has a uniquely identified initial node (i.e., the initial node does not share an information set with any other node), and contains all nodes that follow the initial node in the complete game as well as all successor nodes to these nodes. An additional condition is that all the nodes included in an information set of the sub-game must also be included in the sub-game.

Insights: The Centipede Game is significant to economic theory, because its Nash solution implies a notion of rationality that seems counterintuitive, or even alienating to most human decision-makers (e.g. McKelvey and Palfrey 1992). In fact, the term "Paradox of Backward Induction" has been used to describe the intuition that most decision-makers would prefer passing on earlier moves, and would expect their counterparts to do likewise at earlier positions in the game. In fact, this intuition has fueled rich intellectual debates on the nature of economic rationality, and backward induction in games (e.g., Aumann 1992, 1995; Binmore 1996). For example, it may be deemed rational for a player to have beliefs about his opponents that violate strict economic rationality once this player has been observed to make a move that would be inconsistent with it. In other words, once the first player passes on the first move of the Centipede Game it is not irrational for the second mover to believe that the first mover will pass again on his next move. Consequently, the second mover may prefer passing on her move, given that she may now rationally believe that the first mover will also pass again. Knowing this, it may be rational and strategic for the first mover to initially pass; thus essentially sending a signal to the second mover. Experimental games show that human subjects rarely take the money on the very first round of the Centipede Game (McKelvey and Palfrey 1992).

Actual behavior in the Centipede Game holds numerous insights for neuroeconomists. For one, the backwards induction solution disciplines the way in which a player must iteratively think about the potential outcomes of his actions. Hence, behavior in games that require backward induction may be indicative of the extent of iterated reasoning that a player applies to a decision. Additionally, the notion of rationality we have described indicates the important role of second order beliefs for behavior in the game. That is, it is important for the behavior of player 1, what he thinks that player 2 thinks about the beliefs of player 1. The process of self-referential belief formation and its neural correlates have been investigated by Bhatt and Camerer (2005), and experiments investigating a subject's depth of reasoning (and the depth of reasoning he or she attributes to others) appear to hold important insights for understanding multi-person strategic interactions (Coricelli and Nagel 2010; De Martino et al. 2013).

2.2.5 Stag Hunt Game

Intuition: The Stag Hunt Game is the prototype of the social contract, and a trust dilemma: In *A Discourse on Inequality* written in 1754, Jean-Jacques Rousseau describes a situation in which two individuals go on a hunt, and each of them has to choose one of two possibilities: hunt a hare, which leads to a small personal gain, regardless of the behavior of the other player, or hunt a stag, which leads to a large gain, but will only be successful if both players join forces in doing so. Catching the stag and obtaining the large payoff for both players therefore requires spontaneous cooperation and mutual trust (Fig. 2.5).

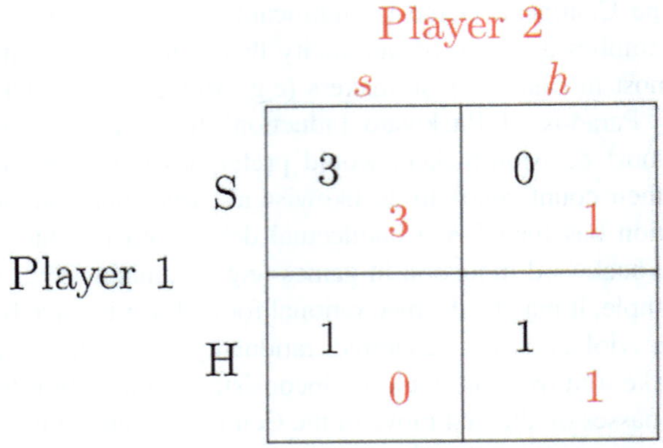

Fig. 2.5 Generic Stag Hunt. Each player chooses between a hunting a S(tag), or a H(are). Coordinated stag hunts ({S, s}) generate the highest payoffs for both players. A single stag hunt ({S, h} or {s, H}) does not produce positive utility for the player choosing S(tag). Choosing H(are) always produces an intermediate utility to the player, irrespective of the other player's action

Equilibrium Solution: The game may look similar to the Prisoner's Dilemma, in that there are two players that have to choose simultaneously whether to cooperate or not. However, this game does not have a strictly dominant strategy, and possesses two Nash equilibriums in pure strategies (as well as mixed strategy equilibrium): Strategies in which both players always hunt the stag, or always hunt the hare, are both equilibrium solutions to the game. Depending on the actual payoffs used in the game, a mixture can be found, for which the other player will become indifferent to which action they choose, and when both players mix their actions in this way, a mixed strategy equilibrium is reached.[12] The pure strategy equilibriums obtain because hunting the stag is a best response when the other player hunts the stag, and hunting the hare is a best response when the other player hunts the hare. Hunting the stag, which involves social risk taking, however, also constitutes the social optimum, as it gives the higher payoff to both players.

Insights: The Stag Hunt Game allows the investigation of individuals' attitude toward social risk taking involved in coordinating on mutual trust in the game. This feature of the game is comparable to many social and physical environments in which joint coordination of trusting behavior is required for the attainment of goals. In repeated games, people additionally gain knowledge of other player's intentions to cooperate through repeated interactions, and applying Theory of Mind (ToM) (Premack and Woodruff 1978) can infer other people's cognitive states, make predictions on their moves and act consequently. As for the Trust Game (Sect. 2.2.8), having some knowledge about the other player helps to shape one's strategy: if we know that the other agent is risk-taking, we might predict that he/she

[12]For the values used in the schematic Fig. 2.5, the mixed strategy equilibrium has both players hunting the stag with probability 1/3 and the hare with probability 2/3.

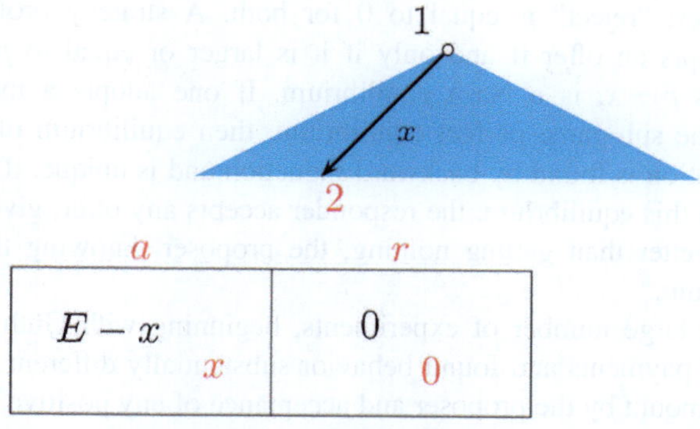

Fig. 2.6 Generic Ultimatum Game (continuous donation). Player 1 is the proposer and chooses some value x along a continuous interval. After proposing x, player two may accept or reject. Acceptance leads to player 1 receiving the remainder of his initial endowment ($E - x$), and player 2 receiving x. If player 2 rejects, both players receive 0

will probably choose to hunt the stag, and we can act consequently. From a neuroscientific perspective, a study from Yoshida et al. (2010) shows that, on one hand, knowledge acquired by updating the beliefs on the other agent through repeated interactions requires the activation of medial prefrontal cortex (the higher the activation the higher the uncertainty of belief inference); on the other hand, choosing a strategy according to the level of sophistication of thinking, meaning how many orders of belief we apply for choosing our strategy, involves the dorsolateral prefrontal cortex, whose activity positively correlates with the level of sophistication.

2.2.6 *Ultimatum Game*

Intuition: The Ultimatum Game is the classic "take-it-or-leave-it" scenario in game theory. It features two players, the proposer and the responder. The proposer is endowed with an amount of money, e.g., 10 dollars. The proposer moves first, and makes an offer to the responder on how to divide the money; the responder has to accept or reject the offer knowing that, accepting, the money will be divided as the proposer has decided, whereas rejecting neither player will receive anything. For example, if the proposer offers 2 dollars and the responder accepts, the proposer gets 8 dollars and the responder gets 2 dollars; if the responder rejects, they both gets 0. The game is one-shot, meaning the interaction between the two players occurs only once, eliminating the opportunity for reciprocity, and stripping rejections from their potential power of being negotiating tools (Fig. 2.6).

Equilibrium Solution: If the total amount available is x, then the proposer must choose an amount p [0, x] to keep for himself. The responder chooses between two solutions for determining the outcomes: "accept" is equal to $x - p$ for him and to p for the proposer; "reject" is equal to 0 for both. A strategy profile where the responder accepts an offer if and only if it is larger or equal to $p - x$, and the proposer offers $p - x$, is a Nash equilibrium. If one adopts a more restrictive concept, i.e., the sub-game perfect equilibrium, then equilibrium of this game of perfect information is found by backward induction and is unique, if a zero offer is not allowed. In this equilibrium, the responder accepts any offer, given that getting something is better than getting nothing; the proposer, knowing that, offers the minimum amount.

Insights: A large number of experiments, beginning with Güth et al. (1982), used monetary payments and found behavior substantially different from the offer of a minimal amount by the proposer and acceptance of any positive amount by the responder. It has been repeatedly demonstrated that proposers tend to make fair offers and responders tend to reject offers that are considered to be unfair, such as 1 or 2 dollars out of 10; theories of social preferences have accounted psychological phenomena such as inequity aversion (e.g., Fehr and Schmidt 1999) or negative reciprocity (e.g., Rabin 1993) for the rejections.

The deviation of actual behavior from Rational Choice has opened the discussion on the ultimate psychological and neural mechanism that leads to a rejection. Negative emotions in response to unfairness have been claimed as responsible for the behavior (Pillutla and Murnighan 1996), and evidence in support to this theory has been collected extensively (e.g., Sanfey et al. 2003; Van't Wout et al. 2006); however, recent findings challenge this account, in that emotional arousal seems to be not necessary in order to trigger rejections (Civai et al. 2010). Moreover, a crucial role is played by expectations, which interact with the unfairness level to influence significantly the rejection rate, as shown by Chang and Sanfey (2013). Overall, the UG, together with Dictator Game (DG), Trust Game (TG) and their manipulations, remains one of the most effective tasks to investigate the cognitive and neural basis of social norm's compliance.

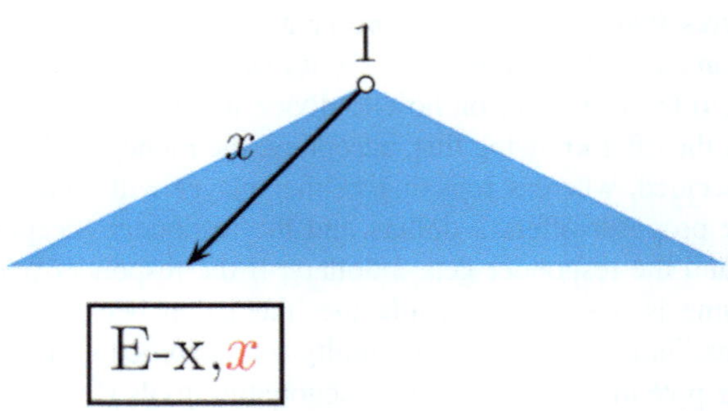

Fig. 2.7 Generic Dictator Game (continuous donation). A degenerate game in which player 1 distributes an initial endowment E between herself and player 2. Player 1 receives $E - x$, while player 2 receives the donated amount x

2.2.7 *Dictator Game*

Intuition: Forsythe et al. (1994) introduced the Dictator Game (DG), a variant of the UG, which features two players, the dictator and the recipient. The dictator is endowed with an amount of money, e.g., 10 dollars. The dictator is asked to divide his/her endowment with the recipient; the recipient's role is entirely passive, as he/she must accept the offer made by the dictator. For example, if the dictator allocates 2 dollars to the recipient, the dictator gets 8 dollars and the recipient gets 2 dollars. Unlike the UG, in the DG the recipient has no strategic input, and, in this sense, the DG is a degenerate game. This game is useful to investigate the behavior of the UG's proposer when ruling out the strategic thinking, as in the DG the dictator does not have to worry about potential rejections (Fig. 2.7).

Equilibrium Solution: The dictator maximizes his payoffs by keeping the largest amount possible.

Insights: Rational Choice theory predicts that the dictator will offer 0; however, as for the UG, this is not the case in experimental settings. Although the amounts and frequencies of nonzero offers in the DG are less and lower than in the UG, demonstrating that strategic motivations have an effect in the UG, giving behavior in the DG is consistently and significantly larger than zero, and players are often observed to give up to the 50 % of their endowment to the recipient (Forsythe et al. 1994). Interestingly, anonymity and social distance play a very important role in shaping the amount of giving (Hoffman et al. 1994, 1996; Charness and Gneezy 2008): the larger is the degree of anonymity and the social distance, the lower is the average amount of money given by the dictator.

As for the UG, the DG is a useful tool to investigate fairness concerns. In particular, given its peculiar nature of degenerate game, where the recipient has no active role and cannot decide on anything, this game is useful to disentangle actual altruistic motivations from the strategic thinking that may characterize the proposer's offers in the UG. Interestingly, the fact that people still give some money to the recipient opens the discussion about different types of altruism (for a specific discussion, see Chaps. 11 and 12 of this volume).

2.2.8 *Trust Game*

Intuition: The simple Trust Game (TG) (Berg et al. 1995) features two players, the investor and the trustee. The investor starts with a certain amount *X*, e.g. 10 dollars, and has to decide how much to keep for him/herself, and how much to give to the trustee. The investor knows that the amount received by the trustee will be triplicated, so that if T is the investment, the trustee receives 3T. At this point, the trustee must decide how much to return to the investor, from 0 to 3T. The TG is considered to be an extended DG, as in this case the investment earns a return, and it is widely employed to investigate the issue of trust/social capital (Fig. 2.8).

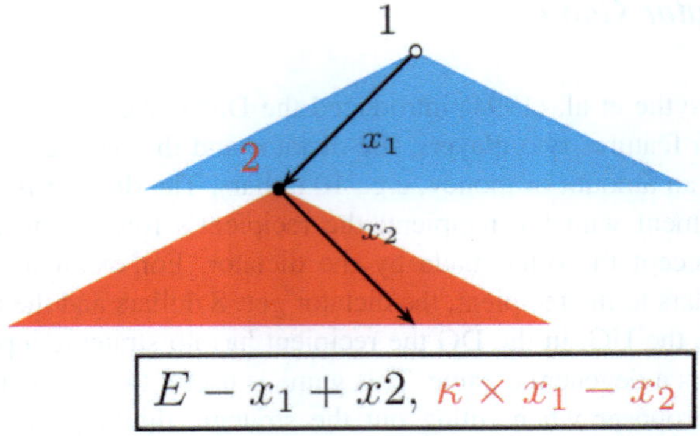

Fig. 2.8 Generic Trust Game (continuous donation): Player 1 entrusts some amount x_1 between 0 and E to player 2. This amount is multiplied by κ, after which player 2 may return some amount x_2 between 0 and κ times x_1 to player 1. Player 1 receives the part of the endowment he kept for himself plus the amount retuned by player 2. Player 2 receives the κ—multiplied amount of player 1s initial trust less the amount returned

Equilibrium Solution: Both players are better off with full investment, but the investor has to take the risk in trusting the trustee. In the one-shot version of the game, the sub-game perfect equilibrium would be for the trustee to keep all, wherefore the best response of the investor would be to invest nothing.

Insights: Although Rational Choice predicts the opposite, investors show a surprisingly high amount of blind trust: the average amount invested is half of *X*. The average amount of return is 1/3 of 3T, so that the investor ends up, on average, with T. Cochard et al. (2004) found that investors invest more and trustees return a higher percentage in the finitely repeated trust game with fixed matches (7 rounds) as compared with the one-shot version; in particular, they found two different types of trustee, i.e., those returning 2/3 of T and those returning 0.

The TG has been widely employed to investigate the psychological and neural underpinnings of trust among strangers; interestingly, it has been possible to show how trust is modulated both by the characteristics of the trustee and by those of the investors. For example, studies manipulating the moral character of the trustee showed that investors were much more likely to make risky investments with "good" partners (e.g., Delgado et al. 2005); on the other hand, it has been shown that administering the neuropeptide oxytocin increases trust in the investor, who is willing to invest more money in the trustee. In particular, Kosfeld et al. (2005) manipulated the nature of the trustee, having a human trustee in one condition and a computer in another, and showed that this increase was not due to a general increase in the willingness to take risks (computer condition), but was specific to the social risks involved in interpersonal interaction (human condition).

References

Allais M (1979) The foundations of a positive theory of choice involving risk and a criticism of the postulates and axioms of the American school (1952). In: Allais M, Hagen O (eds) Expected utility hypotheses and the Allais Paradox 27–145. Springer, Dordrecht
Aumann RJ (1992) Irrationality in game theory. Economic analysis of markets and games. Essays in honor of Frank Hahn 214-27
Aumann RJ (1995) Backward induction and common knowledge of rationality. Game Econ Behav 8(1):6–19
Aumann RJ (1997) Rationality and bounded rationality. Springer, Berlin, pp 219–231
Berg J, Dickhaut J, McCabe K (1995) Trust, reciprocity, and social history. Game Econ Behav 10 (1):122–142
Bhatt M, Camerer CF (2005) Self-referential thinking and equilibrium as states of mind in games: fMRI evidence. Game Econ Behav 52(2):424–459
Bhatt MA, Lohrenz T, Camerer CF et al (2010) Neural signatures of strategic types in a two-person bargaining game. Proc Natl Acad Sci USA 107(46):19720–19725
Binmore K (1996) A note on backward induction. Game Econ Behav 17(1):135–137
Burks SV, Carpenter JP, Goette L et al (2009) Cognitive skills affect economic preferences, strategic behavior, and job attachment. Proc Natl Acad Sci USA 106(19):7745–7750
Camerer C (2003) Behavioral game theory: experiments in strategic interaction. Princeton University Press, Princeton, NJ
Camerer C, Loewenstein G, Prelec D (2005) Neuroeconomics: how neuroscience can inform economics. J Econ Lit 9–64
Camerer CF, Loewenstein G, Rabin M (eds) (2011) Advances in behavioral economics. Princeton University Press, Princeton, NJ
Chang LJ, Sanfey AG (2013) Great expectations: neural computations underlying the use of social norms in decision-making. Soc Cogn Affect Neurosci 8(3):277–284
Charness G, Gneezy U (2008) What's in a name? Anonymity and social distance in dictator and ultimatum games. J Econ Behav Organ 68(1):29–35
Chaudhuri A (2011) Sustaining cooperation in laboratory public goods experiments: a selective survey of the literature. Exp Econ 14(1):47–83
Civai C, Corradi-Dell'Acqua C, Gamer M et al (2010) Are irrational reactions to unfairness truly emotionally-driven? Dissociated behavioural and emotional responses in the Ultimatum Game task. Cognition 114(1):89–95
Cochard F, Van Nguyen P, Willinger M (2004) Trusting behavior in a repeated investment game. J Econ Behav Organ 55(1):31–44
Cooper R, DeJong DV, Forsythe R et al (1996) Cooperation without reputation: experimental evidence from prisoner's dilemma games. Game Econ Behav 12(2):187–218
Coricelli G, Nagel R (2009) Neural correlates of depth of strategic reasoning in medial prefrontal cortex. Proc Natl Acad Sci USA 106(23):9163–9168
Coricelli G, Nagel R (2010) The Neuroeconomics of depth of strategic reasoning. Hist Econ Ideas 18(1):123–131
Damasio AR, Tranel D, Damasio H (1991) Somatic markers and the guidance of behavior: theory and preliminary testing. In: Levin HS, Eisenberg HM, Benton AL (eds) Frontal lobe function and dysfunction. Oxford University Press, Oxford, pp 217–229
De Martino B, Kumaran D, Seymour B et al (2006) Frames, biases, and rational decision-making in the human brain. Science 313(5787):684–687
De Martino B, O'Doherty JP, Ray D et al (2013) In the mind of the market: theory of mind biases value computation during financial bubbles. Neuron 79(6):1222–1231
Delgado MR, Frank RH, Phelps EA (2005) Perceptions of moral character modulate the neural systems of reward during the trust game. Nat Neurosci 8(11):1611–1618
Downs A (1957) An economic theory of democracy. Harper, New York
Dreber A, Rand DG, Fudenberg D et al (2008) Winners don't punish. Nature 452(7185):348–351

Ellsberg D (1961) Risk, ambiguity, and the Savage axioms. Q J Econ 75(4):643–669

Engelmann JB, Tamir D (2009) Individual differences in risk preference predict neural responses during financial decision-making. Brain Res 1290:28–51

Fehr E, Camerer CF (2007) Social neuroeconomics: the neural circuitry of social preferences. Trends Cogn Sci 11(10):419–427

Fehr E, Gächter S (2000) Cooperation and punishment in public goods experiments. Am Econ Rev 90(4):980–994

Fehr E, Schmidt KM (1999) A theory of fairness, competition, and cooperation. Q J Econ 114 (3):817–868

Forsythe R, Horowitz JL, Savin NE et al (1994) Fairness in simple bargaining experiments. Game Econ Behav 6(3):347–369

Frank MJ, Claus ED (2006) Anatomy of a decision: striato-orbitofrontal interactions in reinforcement learning, decision making, and reversal. Psychol Rev 113(2):300

Friedman M, Savage LJ (1948) The utility analysis of choices involving risk. J Polit Econ 56 (4):279–304

Gigerenzer G (2004) Fast and frugal heuristics: the tools of bounded rationality. In: Koehler DJ, Harvey N (Eds) Blackwell handbook of judgment and decision making. Wiley-Blackwell, New York, pp 62–88

Gigerenzer G, Selten R (eds) (2002) Bounded rationality: the adaptative toolbox. Mit Press, Cambridge, MA

Glimcher PW, Rustichini A (2004) Neuroeconomics: the consilience of brain and decision. Science 306(5695):447–452

Glöckner A, Betsch T (2008) Multiple-reason decision making based on automatic processing. J Exp Psychol Learn Mem Cogn 34(5):1055

Glöckner A, Hochman G (2011) The interplay of experience-based affective and probabilistic cues in decision-making. Exp Psychol 58(2):132–141

Güth W, Schmittberger R, Schwarze B (1982) An experimental analysis of ultimatum bargaining. J Econ Behav Organ 3(4):367–388

Hawes DR, Vostroknutov A, Rustichini A (2012) Experience and abstract reasoning in learning backward induction. Front Neurosci 6

Hoffman E, McCabe K, Shachat K et al (1994) Preferences, property rights, and anonymity in bargaining games. Game Econ Behav 7(3):346–380

Hoffman E, McCabe K, Smith VL (1996) Social distance and other-regarding behavior in dictator games. Am Econ Rev 86(3):653–660

Houser D, McCabe K (2009) Experimental neuroeconomics and non-cooperative games. In: Glimcher PW, Fehr E, Camerer C et al (eds) Neuroeconomics: decision making and the brain. Academic Press, London, pp 47–62

Kahneman D, Tversky A (1979) Prospect theory: an analysis of decision under risk. Econometrica 47(2):263–291

Kalai E, Lehrer E (1993) Rational learning leads to Nash equilibrium. Econometrica 61(5):1019–1045

Knutson B, Wimmer GE, Rick S et al (2008) Neural antecedents of the endowment effect. Neuron 58(5):814–822

Kosfeld M, Heinrichs M, Zak PJ et al (2005) Oxytocin increases trust in humans. Nature 435 (7042):673–676

Kreps DM, Milgrom P, Roberts J et al (1982) Rational cooperation in the finitely repeated prisoners' dilemma. J Econ Theory 27(2):245–252

Lee D, Conroy ML, McGreevy BP et al (2004) Reinforcement learning and decision making in monkeys during a competitive game. Cogn Brain Res 22(1):45–58

Loomes G, Sugden R (1982) Regret theory: an alternative theory of rational choice under uncertainty. Econ J 92(368):805–824

McClure SM, Laibson DI, Loewenstein G et al (2004) Separate neural systems value immediate and delayed monetary rewards. Science 306(5695):503–507

McKelvey RD, Palfrey TR (1992) An experimental study of the centipede game. Econometrica 60 (4):803–836

Mongin P (1997) Expected utility theory. In: Davis J, Hands W, Maki U (eds) Handbook of economic methodology. Edward Elgar, London, pp 342–350

Myerson RB (2013) Game theory: analysis of conflict. Harvard University Press, Cambridge

Nash JF (1950) Equilibrium points in n-person games. Proc Natl Acad Sci USA 36(1):48–49

Neyman A (1985) Bounded complexity justifies cooperation in the finitely repeated prisoners' dilemma. Econ Lett 19(3):227–229

Osborne MJ, Rubinstein A (1994) A course in game theory. MIT Press, Cambridge, MA

Pillutla MM, Murnighan JK (1996) Unfairness, anger, and spite: Emotional rejections of ultimatum offers. Organ Behav Hum Dec 68(3):208–224

Poundstone W (1992) Prisoner's Dilemma. Doubleday, New York, NY

Premack D, Woodruff G (1978) Does the chimpanzee have a theory of mind? Behav Brain Sci 1 (04):515–526

Rabin M (1993) Incorporating fairness into game theory and economics. Am Econ Rev 1281–1302

Rand DG, Arbesman S, Christakis NA (2011) Dynamic social networks promote cooperation in experiments with humans. Proc Natl Acad Sci USA 108(48):19193–19198

Rand DG, Greene JD, Nowak MA (2012) Spontaneous giving and calculated greed. Nature 489 (7416):427–430

Rustichini A (2005) Neuroeconomics: present and future. Game Econ Behav 52(2):201–212

Sanfey AG (2007) Social decision-making: insights from game theory and neuroscience. Science 318(5850):598–602

Sanfey AG, Rilling JK, Aronson JA et al (2003) The neural basis of economic decision-making in the ultimatum game. Science 300(5626):1755–1758

Selten R (1990) Bounded rationality. J Inst Theor Econ 146(4):649–658

Seo H, Lee D (2012) Neural basis of learning and preference during social decision-making. Curr Opin Neurobiol 22(6):990–995

Simon HA (1955) A behavioral model of rational choice. Q J Econ 69(1):99–118

Simon HA (1972) Theories of bounded rationality. In: McGuire CB, Radner R (eds) Decision and organization. North-Holland Publishing Company, Amsterdam, pp 161–176

Simon HA (1990) Bounded rationality. In: Eatwell J, Milgate M, Newmann P (eds) The new Palgrave: utility and probability. W. W. Norton & Company, New York, NY, pp 15–18

Tversky A, Kahneman D (1974) Judgment under uncertainty: heuristics and biases. Science 185 (4157):1124–1131

Van't Wout M, Kahn RS, Sanfey AG et al (2006) Affective state and decision-making in the ultimatum game. Exp Brain Res 169(4):564–568

Von Neumann J, Morgenstern O (1944) The theory of games and economic behavior. Princeton University Press, Princeton, NJ

Wallis HB, Huston J (1941) The Maltese Falcon [Motion Picture]. Warner Bros, USA

Yoshida W, Seymour B, Friston KJ et al (2010) Neural mechanisms of belief inference during cooperative games. J Neurosci 30(32):10744–10751

Part II
Molecular Basis of Human Decision Making

Chapter 3
Hormones and Economic Decisions

Amos Nadler and Paul J. Zak

Abstract Hormones are chemical messengers released into the body that change the probability of behavior. Because hormones are both measurable and manipulable they lend themselves to experimental methodology that can establish causal relationships. Neuroeconomics studies have shown hormones' influence on decision-making using quantifiable treatment and outcome variables in economic and social contexts. This chapter provides background and methodology for hormonal research in neuroeconomics and reviews significant studies on how oxytocin, testosterone, arginine vasopressin, dopamine, serotonin, and stress hormones impact decisions, and how research can be used to improve decisions and the business of life.

3.1 Introduction

The traditional approach to economic research employs a set of simplifying assumptions on human behavior from which to describe and predict choices. These assumptions provide a rigid framework for analyzing decision-making and facilitate models that generate predictions that can be empirically tested. For the second half of the twentieth century, theoretical modeling of people as 'rational agents' was economics' state-of-the-art methodology. However, as much as simplifying assumptions make questions analytically tractable, they also obviate the richness in behavior from the very element under analysis: human beings (Vercoe and Zak 2010).

A. Nadler
Ivey Business School, Western University, 1255 Western Road, London, ON N6G 0N1, Canada
e-mail: anadler@ivey.ca

P.J. Zak (✉)
Claremont Graduate University, 150 E 10th St, Claremont, CA 91711, USA
e-mail: paul@neuroeconomicstudies.org; paul.zak@cgu.edu

M. Reuter and C. Montag (eds.), *Neuroeconomics*, Studies in Neuroscience, Psychology and Behavioral Economics, DOI 10.1007/978-3-642-35923-1_3

Neoclassical economic models of decision-making involve rational and predictable economic agents who have specific and well-defined utility functions based on ordered, transitive preferences. Yet observation of actual people shows the opposite—rapid, heuristic-driven decisions, and decades of psychological and neuroscientific studies show that memories, cues, primes, and emotions affect decisions (Camerer et al. 2005). Fortunately, economics has continued to evolve by producing richer and more accurate models of human behavior, with more realistic assumptions, greater conformability with empirical data, and deviations from rationality. Models have advanced and new tools have been (cautiously) adopted, such as functional magnetic resonance imaging (fMRI), neurophysiology, genetics, and direct manipulation of hormones. While ignored until recently, hormones have been shown to initiate and mediate changes in the central and peripheral nervous systems and affect economic decisions (Zak 2013; Kandasamy et al. 2014; Camerer et al. 2005).

Economists have traditionally been constrained by the type of data at their disposal and relied on archival (i.e., secondary) data to generate findings. In contrast, the natural sciences produce models largely based on primary data from which causal relationships are identified—data informs the model, not the other way around. Instead of the formation of a multitude of models bearing similar explanatory power, this inductive method offers what Francis Bacon called a "selective process of elimination among a number of alternative possibilities" (1895, III, p. 340). The experimental approach to studying economic behavior produces primary data from which to draw conclusions. In particular, hormonal and physiologic research offers a robust inductive method to study biological influence on economic decision-making and produce neurally informed models of human behavior (Zak 2010). Most importantly, direct manipulation of hormones and physiologic states has begun to identify biological mechanisms motivating specific behaviors.

Over the course of evolution mammals developed two integrated communication systems—one faster, and one slower—to respond to changing environments and regulate homeostasis (i.e., physiologic stability). The nervous system communicates rapidly through neurotransmitters and neuromodulators while the hormonal system uses molecular messengers which cause both temporary and permanent changes in the body . In essence, hormones are chemical messengers that change the probability of a behavior or biological function and are the focus of this chapter.

Hormones have long been known to influence physiologic states, physical development, and genetic transcription in humans and animals yet recent developments show hormones impact cognition, mood, and, most recently, economic decision-making. Neuropeptides, such as oxytocin and arginine vasopressin, and steroid hormones, such as testosterone and estradiol, play a central role in humans in a variety of behavioral and social domains (McCall and Singer 2012). Economics' founding father, Adam Smith, connected emotions and morality to prosperity; Thorstein Veblen stated that economics should be thought of as a field of biology and that only social science shaped by biology could be considered 'scientific' (Hodgeson 1998). Alfred Marshall, one of the exponents of mathematical rigor within economics, wrote that, "the Mecca of the economist lies in economic biology rather than in economic dynamics" (1890). By putting human beings back at the center of economic analysis, neuroeconomics returns economics to its roots.

3.2 Hormones Defined

Hormones are chemical messengers that circulate in biofluids (e.g., blood) to regulate physiologic activity and maintain homeostasis (the ability to maintain and regulate internal physical equilibrium regardless of external changes) by acting on targets organs. Hormone production is regulated by the brain and mostly produced by organs in the periphery of the body, such as the kidneys, pancreas, and gonads (although some hormones and their precursors are made in the brain itself). Most hormones have receptors in the brain that allow them directly affect neural activity.

For example, in men, testosterone (T) is produced primarily by the Leydig cells in the testes in response to hormonal signaling from the brain (Midzaka et al. 2009). T is synthesized from cholesterol and released into the bloodstream. Further, it crosses the blood-brain barrier in small quantities due to high lipid solubility, modulating neural activity in the brain that changes, for instance, the threshold for aggression (Schwartz and Pohl 1992). T is also made by women, though at 5–10 % the levels in men, and has similar effects (Carré et al. 2010). For those interested in the assessment of hormones and related methodological issues we recommend Chap. 24, "Hormones" by Robert Miller and Clemens Kirschbaum in this book.

3.3 Mechanisms of Action

Once produced and attached to cell receptors, hormones initiate changes in the body at two levels: genomic[1] and non-genomic (also called classical and nonclassical, respectively). Genomic action occurs when a hormone attaches to a target cell receptor and initiates genetic transcription; this occurs on a time scale of hours to years (Falkenstein et al. 2000). For example, testosterone influences transcription of genes by interacting with receptors on the outside of the cell, initiating muscle growth and secondary sexual characteristics in males (Beato 1996). Non-genomic action, on the other hand, occurs by changing characteristics of cells themselves,

[1]Hormones differ in their endogenous release patterns and active half-lives (Santen and Bardin 1973). Genetic factors mediate receptor availability and molecular metabolism, meaning that no two people are precisely the same in the way they respond to the same hormone; this aspect can be measured in some studies (Crabbe et al. 2007).

and acts much more rapidly, on the scale of seconds to minutes (Lösel et al. 2003; McEwen 1991; Falkenstein et al. 2000). Evidence of this has been demonstrated experimentally with T, whereby infusion reduces anxiety in many animals, including humans (van Honk et al. 2005; Frye and Seliga 2001).

Studying hormonal influences on decision-making requires a comprehensive, multi-pronged approach. The 'basal model' proposed by Mazur and Booth (1998), uses endogenous ('within the body'), or basal levels as predictors of behavior, whereby participants' unaltered hormone levels are measured and used as explanatory variables. This model assumes that measurements over time represent short-term fluctuations near characteristic levels. For example, Sapienza et al. (2009) study basal levels among MBA students to predict risk aversion and career choice.

An approach better suited to identifying causation, is the exogenous ('outside the body') manipulation method, whereby participants are given a specific amount of a drug to increase their levels of a hormone (or block the action of a hormone or neurotransmitter on receptors). This methodology tests 'activational' properties of a hormone, provides a clear treatment that can be compared to placebo, and lends itself to rigorous manipulation verification through biofluid assay. For example, Kandasamy et al. (2013) test the influence of cortisol (a stress hormone) on risk preferences by increasing participants' stress hormone levels through hydrocortisone dosing and measuring changes in risk preferences over time.

A related method is through precursor manipulation, which can enhance or impair availability of the specified molecule. For example, a study tested the effects of dopamine (abbreviated 'DA') inhibition by administering participants naltrexone (a DA receptor blocker) in an asset trading experiment to test how impaired DA will affect behavior and market bubbles (Efremidze and Zak, in press). Also, modulation of chemical precursors are also used in experiments, such as tryptophan depletion or enhancement (Crockett and Fehr 2013).

Another approach evaluates changes in levels as predictors of behavior. This can be applied to both endogenous and exogenous contexts, where instead of using basal levels, percent change or absolute change from baseline is used as an explanatory variable. Apicella et al. (2014) test the influence of changes in T on willingness to compete and find a positive relationship.

In addition to explaining behavior, changes in hormones are used to measure physiologic response *to* an event, such as competition, winning, and losing (Booth et al. 1989). Understanding the process fully—the hormonal response to an event and the subsequent change in behavior motivated by a change in hormones—clarifies the hormonal role in dynamic human decision-making.

Findings are generally not as strong for basal levels relative to exogenous methods (Schipper 2015; Apicella et al. 2008; Cueva et al. 2015). Also, matching endogenous range with exogenous amounts can be challenging due to difficulties in measurement of endogenous amounts and clear understanding of receptor sensitivity and regulation under various conditions.

Experimentally manipulating a single hormone can produce changes in behavior, yet behavior is jointly driven by downstream interactions between hormones, neurotransmitters (chemical messengers in the brain), neural activity, and physiologic tone. Thus, conclusions drawn from pharmacologic manipulations are conditional on basal physiologic states and interactions with other biological factors (Zak 2004, 2005, 2011; Breedlove et al. 2007, p. 123). As a result, a thorough assay of basal physiology is necessary, which can also be assessed using electroencephalograms (EEGs), electrocardiograms (ECGs), and galvanic skin response (GSR). Unfortunately, neurotransmitters are difficult to measure directly without invasive and risky approaches (such as lumbar punctures to harvest spinal fluid) and are therefore typically indirectly assessed in human studies through, for example, urine collection to measure breakdown products.

A convergent approach is necessary to fully understand how hormones affect human decision-making. The first step is to assess basal physiologic state. Step two measures endogenous hormonal response to stimuli. The final step establishes causation by exogenously administering or inhibiting a hormone and measuring changes in behavior. This comprehensive approach is necessary because all physiologic systems are noisy and this approach helps avoid false-positive results. In this way, endogenous effects are demonstrated as well as causal relationships established. This method is summarized in Fig. 3.1.

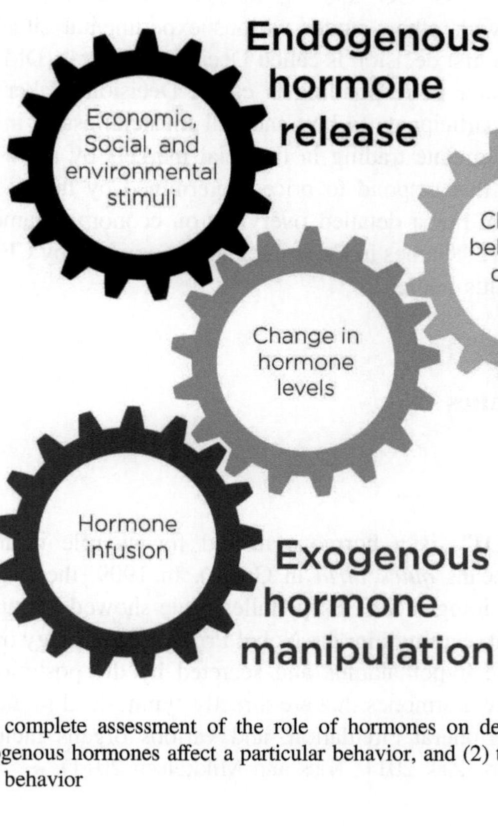

Fig. 3.1 A complete assessment of the role of hormones on decisions requires showing that (1) the endogenous hormones affect a particular behavior, and (2) that manipulating the hormone changes the behavior

3.4 Laboratory Experiments

Despite the abundance and history of hormone research in medicine and biology, the literature on hormones and economic decision-making is in early stages (Zak 2004). Hormonal neuroeconomics experiments were pioneered after animal research suggested related hormonal roles in human behavior. Findings have been both mixed and consistent in terms of convergence and replicability—some studies show contradicting results and non-replicability for the same hormones, while other hormones show corroborating patterns. In addition to testing for behavioral effect, hormone research requires understanding the precise pathways, half-lives, inhibiting and promoting qualities, and effect on other hormones, all of which are an ongoing scientific endeavor. Among the multitude of hormones produced in the human body, a subset is studied most closely and the majority of behavioral research has focused on oxytocin, testosterone, estrogen, glucocorticoids (stress hormones), arginine vasopressin, serotonin, and dopamine.

Tasks commonly used in neuroeconomics research include the trust game (TG) that measures trust and reciprocity, the ultimatum game (UG) that measures generosity or selfishness and theory of mind, the dictator game (DG) that measures unilateral altruism, and double auctions of assets (DAA) where participants trade an asset of known value. The TG, UG, and DG measure by the amount of money people choose to share with others under various experimental situations. The participant who makes the first decision is called Decision Maker 1 (DM1), and the person receiving the transfer or responding is called Decision Maker 2 (DM2). DAA experiments allow participants to buy and sell financial assets in simplified dynamic markets which simulate trading in financial markets by allowing one to make real-time decisions that respond to prices determined by the traders themselves (Smith et al. 1988). For a detailed overview on economic games we recommend reading of Chap. 2 “Games in Experimental Economics” by Claudia Civai and Daniel R. Hawes in this book.

3.5 Specific Hormones

3.5.1 Oxytocin

Oxytocin (abbreviated ‘OT’) is a hormone named for its role in mammalian reproduction (oxytocin means *quick birth* in Greek). In 1909, the English pharmacologist and neurophysiologist Sir Henry Hallett Dale showed it causes uterine contractions and isolated it, earning him the Nobel Prize in Physiology or Medicine in 1936. Produced by the hypothalamus and secreted by the posterior pituitary gland, OT is one of the few hormones that are directly synthesized in the brain and released in the brain, peripheral circulation, and various organs including gastrointestinal tract and heart (Zak 2011; Kiss and Mikkelsen 2011).

Starting in the late 1970s animal research showed OT release was associated with positive social behaviors and a groundbreaking study showed OT injection initiated maternal behavior in rats toward biologically unrelated offspring (Pederson and Prange 1979; Gimpl and Fahrenholz 2001). Convergent evidence pointed toward likely analogous influences on human behavior, leading to incorporation in economic decision tasks that provide active, measurable, and meaningful measures of greed, prosociality, trust, and profit maximization. By measuring and manipulating hormones in these tasks, an understanding the role of OT in human social behaviors has begun to emerge.

The first neuroeconomics study measuring and endogenously manipulating OT in relation to economic decisions is by Zak et al. (2004) who show that the receipt of trust signal is associated with higher peripheral OT. This study indicates intentional trust in the TG was associated with higher OT levels as measured in blood compared to individuals who received the same amount of money determined by random draw. Further, the level of OT predicted the amount of money that was reciprocated to the person who had shown trust. An analysis that extended the sample size using the same protocol corroborated the findings (Zak et al. 2005a).

A subsequent study was designed to causally connect OT to trusting behaviors by manipulating intranasal administration. In a double-blind protocol, 24 International Units (IU) of synthetic OT or an equal quantity of placebo were administered to participants who made decisions in four rounds of the TG with random rematching with other participants each round (Kosfeld et al. 2005). Those who received OT exhibited more than double the trust (as measured by monetary transfers) compared to those who received the placebo. Decisions in control tasks, such as choices among lotteries, as well as assessments of cognitive function were unchanged between conditions. It is important to point out that OT infusion did not increase allocations of money when the return on investment was determined by chance, showing that administration affected only social decision-making.

In similar studies, Baumgartner et al. (2008) found that the OT group did not send more money to DM2, although they did send more money after being given feedback about others' monetary transfers. Klackl et al. (2012) and Ebert et al. (2013) find that OT infusion did not increase trust in the TG and the latter propose OT acts as a modulator of social interaction rather than a prosocial neuropeptide. Yao et al. (2014) show no main effect of OT trust restoration, yet the design and objectives differ markedly, making the study incomparable to Kosfeld et al. (2005). In a study of basal OT (i.e., without exogenous manipulation) Christensen et al. (2014) test whether endogenous OT affects trust in an iterated DG and found no association between repeated sampling and trusting decisions.

Subsequent research found that 40 IU of intranasal OT increased generosity in the UG, complementing findings regarding its effects on the TG (Zak et al. 2007). In this study, OT did not affect decisions in unilateral DG. The authors included the DG as a control task because it does not require an understanding of another's intentions to make a decision. This finding shows the importance of context and structure to hormonal neuromodulation. A study by Mikolajczak (2010) showed that OT does not affect people indiscriminately—participants in the OT condition made larger monetary transfers but only to others who had trustworthy characteristics. This suggests that multiple brain systems are involved in decisions and OT interacts with neural inputs that influence choice, and extensive work shows OT plays an influential role in proximal and distal prosocial behaviors.

Endogenous OT release was stimulated using a video featuring a child with terminal brain cancer and a change in OT correlated with the subjective experience of empathy for the characters in the video (after controlling for distress) (Barraza and Zak 2009). High levels of empathic concern predicted larger donations to the charity that produced the video. This study identified a psychological mechanism behind the effect of OT on behavior, and connected its findings to large literatures in psychology and moral philosophy regarding the role of empathy in prosocial behaviors (see Zak 2011). The causal effect of OT on distal prosocial behaviors was demonstrated by Barraza et al. (2011) who show a causal yet conditional effect of OT of increased donations to known charities among those who donated. Other studies show exogenous OT increases gaze to the eyes (Guastella et al. 2010), enhances social memory (Guastella et al. 2010), as well as improving emotional recognition among youth with autism (Guastella et al. 2010).

The mechanism by which hormones are released throughout the body and, most importantly, pass the blood-brain-barrier and reach the brain, is central to this area of research (McEwan 2004). Carson et al. (2015) show that plasma OT concentrations significantly and positively predict cerebrospinal fluid (CSF) OT concentrations while Kagerbauer et al. (2013) show no correlation. In a small study, Striepens et al. (2013) tested the effects of exogenous OT on CSF and plasma and show differential time-courses following administration: plasma concentrations peaked 15 min after administration and CSF levels were significantly higher 75 min later with no significant correlation between the two biofluids. Relatedly, OT half-life is between 3 and 4.5 min (Rydén and Sjöholm 1969). Given the broad distribution of OT receptors throughout the body it is likely that release between brain and other organs is related but not necessarily coupled. The mechanism by which exogenous administration increases physiologic levels is still under investigation with several feasible explanations.

Further studies have brought closer attention to measurement methodology in hormone research. The extraction method long used in biomedical research was eschewed in the early 2000s, which complicates clear interpretation of data from those studies. Extraction is the process of removing distinct physical products prior to assay extraction by separation of OT from its biological matrix (e.g., saliva). This facilitates avoiding measuring compounds resembling the substance of interest as unextracted samples contain molecules similar to OT and can drastically distort its measurement—extracted and unextracted OT measurements differ by orders of magnitude. McCullough et al. (2014) discuss the evolution of measurement standards and the incommensurate results stemming from lack of standardization and distortions in measurement caused by measuring unextracted samples. Christensen et al. (2014) test differential sensitivity between ELISA (enzyme-linked immunosorbent assays) and RIA (radioimmunoassay) to extraction as part of their experimental paper and, although both methods show significant sensitivity, the former has higher sensitivity to extraction.

The path of oxytocin research is exciting and exemplifies the importance of assessing if, why, and how endocrinological mechanisms impact decisions. In addition to understanding the basic biology underlying OT release, half-life, and physiologic distribution, researchers need reliable behavior paradigms for testing the effects of OT on decision-making. As with any type of biological research, OT studies require adaptation and inclusion of new technology and a movement toward standardization to advance research to fully understand its function.

3.5.2 *Testosterone*

Testosterone (abbreviated 'T') is a gonadal hormone present in both sexes, with a receptor for it in every cell in the body. T varies seasonally as well as daily, producing a diurnal cycle that peaks in the early morning and declines throughout the day (Brambilla et al. 2009). Gonadal hormones can be both slow and fast acting, causing both long- and short-term effects. Long-term effects are caused by passive diffusion into cells where they bind to steroid-receptor complex, then DNA, and change gene transcription; fast action is caused by direct action on neurons with corresponding receptor cells. T affects not only the path of neuroanatomic and physiologic development (organizational effects) in mammals but also behavior throughout the lifespan (Thilers et al. 2006). Men's T levels peak around age 20 and decrease with age.

T has been extensively studied in medicine in relation to physical development, puberty, fertility, and pathology. It has been shown to affect mood, aggression, sexuality, and more recently, financial behavior (Nadler et al. 2016; Cueva et al. 2015). In addition to affecting—and being affected by—aggressive behavior, gonadal hormones are related to competition, spatial tasks, memory, certain sensation seeking scales, and risk preferences (Cherek et al. 1996; Roberti 2004; Apicella et al. 2008; Sapienza et al. 2009; Goudriaan et al. 2010).

3.5.3 *Basic Biology of Testosterone*

Steroid hormones are synthesized from precursors in the smooth endoplasmic reticulum, processed further in the mitochondria, and returned to the smooth endoplasmic reticulum for completion. Both steroid and steroid-like hormones are not stored in vesicles and simply diffuse out of cells after synthesis at a rate governed by production. Leydig cells produce T in response to hormonal signaling from the pituitary gonadotropin luteinizing hormone (LH) (Midzaka et al. 2009; Haider 2004; Mendis-Handagam 1997). Adult Leydig cell production depends on the pulsatile secretion of LH into peripheral circulation by the pituitary gland (Ellis et al. 1983). T (which is produced in the ovaries in women) is a precursor to estrogen, and estrogen is a metabolite of T (the testes produce some estrogen). Despite the stereotype, T is not an exclusively male hormone, as estrogen is not exclusively female, and their proportions vary by gender and (wildly) by species.

3.5.4 Types of Testosterone

There are three types of T frequently analyzed in medical and behavioral literature: total T, Free T, and DHT (androstenedione is discussed primarily in medical literature[2]). T circulates in the body primarily bound (98 %) to serum proteins, mostly sex hormone-binding globulin (SHBG) and albumin; only 1–2 % of serum T is not protein-bound (Dunn et al. 1981). Due to the fact that SHBG binds T with high affinity and the off time of T bound to SHBG is remarkably slow, SHBG-bound T is considered unavailable for dissociation to act onto target tissues via classical androgen receptor mechanisms (Pardridge et al. 1979). Albumin-bound T is low-affinity and dissociation is rapid (Manni et al. 1985). Consequently, both albumin-bound T and free T are considered available for androgen action, and are called 'bioavailable' or 'non-SHBG-bound' T (Matsumoto and Bremner 1984).

3.5.5 Total T

Serum T, also known as Total T (TT), plays an important role in the clinical evaluation of numerous common endocrine disorders, such as hypogonadism, and delayed or precocious puberty in males, as well as a variety of conditions in females. Routine assays began approximately 40 years ago and required chromatographic separation. Today, assays are more precise, specific, require less blood, and nonradioactive methods (Stanczyk and Clarke 2010; Matsumoto and Bremner 1984).

3.5.6 Free T

Forty-four percent of circulating T is bound to sex hormone-binding globulin (SHBG), 50 % to albumin, and 3–5 % to cortisol binding globulin, leaving about

[2] Androstenedione (A), also known as 'Andro', is a steroid hormone produced in the gonads and adrenal glands in men and women. Androstenedione an intermediate step in the biochemical pathway that produces T and estrone and estradiol. This hormone was at the center of controversy of baseball players and androgen use in the 1990s. Leder et al. find that sufficiently high doses (300 mg) oral A increase serum T and estradiol in some healthy men, supporting the rationale of the ban issued by the World Anti-Doping Agency (2000).

2–3 % T free (FT). Bioavailable T is the free circulating in addition to the albumin-bound portion. Salivary T represents the portion of plasma T that diffuses passively across salivary glands (Arregger et al. 2007). It is from this small percent of total T that FT can be aromatized via 5-alpha reductase to dihydrotestosterone.

3.5.7 Dihydrotestosterone

Dihydrotestosterone (DHT) is converted from T through the action of 5-alpha reductase in peripheral tissue. Both T and A are precursors of DHT (Ito and Horton 1971). As mentioned above, of total T, there is but a small amount of free T available for conversion. As discussed earlier, androgens act at transcriptional levels of gene expression via classical androgenic processes by passively diffusing through cell membranes and 'locking' into their respective receptors. Yet evidence has accumulated that some steroids may also alter neuronal excitability through interactions with specific neurotransmitter receptors at the scale of milliseconds to seconds (Rupprecht and Holsboer 1999). DHT binds faster (Hemat 2004) and remains in the cell longer (Grino et al. 1990) than TT due to higher receptor affinity, thereby likely to have more significant behavioral effects.

3.5.8 T in Behavioral Experiments

Behavioral studies involving T include basal levels, changes in endogenous levels, and exogenous administration. Burnham (2007) found that DM2 males with higher T rejected low offers more than their lower T counterparts in the UG. Endogenous variations in T have been shown to increase patience for monetary rewards for non-impulsive participants, while reducing discount rates for impulsive participants, showing an inverted-U-shape (Takahashi et al. 2006). Higher T males were also more likely to make utilitarian decisions in 'trolley car' problems (lives sacrificed to save other lives) compared to lower T males (Carney and Mason 2010).

Apicella et al. (2014) find that participants who showed an increase in T were more willing to compete. These findings may suggest 'hormonal typology' among individuals, making behavior partly explainable by baseline levels and reactivity to particular hormones. Zak et al. (2005b) showed that men had a rise in DHT when distrusted by receiving small amounts of money as DM2 in the TG. High DHT levels were associated with little or no reciprocation, and partially explain the gender gap in reciprocity in the TG in which women reciprocate more money on average than men.

The challenge with relying on endogenous levels or changes is that hormone release is noisy and subject to interindividual heterogeneity, thereby yielding unreliable control and identification. Using extensively studied exogenous treatments helps solve these problems as their pharmacokinetics and associated risks have been extensively studied in medical research (Swerdloff et al. 2000).[3] For example, typical starting dosage is 50 mg of T (and often increased to 100 mg), providing researchers with a basis for administering quantities with a predictable effect on circulating levels. Providing insight about immediate changes of exogenous T, Eisenegger et al. (2013) test the effects of T (and estradiol) administration in young men and demonstrate a rapid rise with a peak in serum T reached at 3 hours post-administration. Tuiten et al. (2000) show sublingual T caused sharp rise in serum T within 15 min (with a return to baseline after 90 min) and lagged increases in genital arousal in women. In another pharmacokinetics study of females, van Rooij et al. (2011) show sublingual T showed dose-dependence and peak levels reached in 15 min and return to baseline within 2.5 h. Yet despite vast medical literature on androgens, behavioral studies with exogenous T are limited.

Zak et al. (2009) manipulated T pharmacologically in male participants playing the UG. In this within-subject s study, participants received 10 g of Androgel® (1 % testosterone gel) on one visit and a placebo gel on another. Blood samples were obtained before and after substance administration to quantify the rise in T as well as to assess parametric effects of T on behavior. The authors showed that T decreased DM1 offers as well as increased the minimum acceptable offers by DM2 s. Both these effects scaled positively with three measures of T (T, free T, DHT). This finding is consistent with a known physiologic mechanism in which high levels of T inhibit the release of OT (Insel et al. 1993). A similar paper with females only by Eisenegger et al. (2010) purported to show the opposite effect. In the study, 0.5 mg of sublingual T or placebo was administered to women in a between-subjects study, absent an assessment of the rise in T (baseline T was measured). The authors found no main effect of T on UG offers or rejections when compared to placebo, yet found that T increased UG offers if one controls for the substance participants believed they received (Eisenegger et al. 2010, online supplementary material).

Additional manipulation studies include Bos et al. (2010), who tested the relationship between T and distrust among women in a placebo-controlled within-subject design and found that T reduced ratings of trustworthiness when viewing pictures of men's faces. Boksem et al. (2013) study the effects of exogenous on trust and reciprocity and show it inhibits trust and promotes reciprocity. Hermans et al. and Goetz et al.'s fMRI studies show that T increases neural reactivity to threat. Wibral et al. (2012) study lying and find that T reduces it relative to placebo. The role of T in asset trading behavior was assessed using a DAA paradigm (Nadler et al. under review) who found the sessions with high T traders resulted in larger asset bubbles compared to placebo sessions due to higher T participants bidding higher prices. Cueva et al. (2015) show that participants who

---

[3]Due to the rise of easily obtainable drugs and associated advertising to remedy "low testosterone syndrome" or "andropause", a large and growing proportion of men is currently using Androgel® (and similar generics), and many inject even higher doses (Baillargeon et al. 2013; Handelsman 2013). In fact, the proliferation of these drugs among financial professionals allows our experiment to mimic the "testosterone shock" in real-world asset markets such as the NYSE.

received exogenous testosterone had a higher willingness to invest in high-variance stocks. Nave et al. (under review) show exogenous T reduced men's ability to inhibit instinctive and incorrect responses in the Cognitive Reflection Task while having no impact on mathematical skills, task engagement, or motivation in a large (n = 243) sample of young men. This result, supported by previous evidence of T increasing impulsivity, suggests T 'nudges' decision-making toward rapid and intuitive processing.

A related mechanism for these findings is that T correlates with willingness to engage in competition and decrease risk aversion via allosteric modulation of $GABA_A$ receptors (Reddy and Jian 2010; Carré and McCormick 2008). In addition, dopamine, which is associated with risk-taking, positively co-varies with T and may contribute to the sensation seeking aspect of financial trading though further work is needed to identify T's effect on risk per se (Szczypka et al. 1998).

3.5.9 2D-4D Ratio

The ratio of second to fourth finger (2D:4D) has been (inconsistently) shown to negatively correlate with prenatal T exposure and that men have a lower 2D:4D ratio than women (Manning et al. 2004, 1998). Apicella et al. (2008) found that 2D:4D does not significantly correlate with economic risk-taking and Apicella et al. (2015) show males displaying *higher* ratios than women among Hadza tribe members. Contrarily, Sapienza et al. (2009) reported that 2D:4D ratio and salivary T negatively correlated with risk aversion and that high T individuals chose higher risk professions (finance, broadly defined). Coates (2012) found that 2D:4D ratio predicted high-frequency traders' long-term profitability as well as duration of employment in the profession. Brañas-Garza and Rustichini (2011) find that lower 2D:4D ratios are associated with greater risk-taking and higher abstract reasoning scores among females (188 participants, 72 female).

3.5.10 Arginine Vasopressin

Arginine vasopressin (abbreviated AVP') is a hormone synthesized in the hypothalamus and stored in vesicles in the posterior pituitary. One of its primary functions is water regulation in the body and has been shown to have behavioral influences. Despite being molecularly similar to OT and lends itself to the same endogenous and exogenous approaches to being studied, AVP has different behavioral influences. Whereas OT facilitates bonding and trust, AVP is associated with reactive aggression, stress-responses, and mate- and nest- guarding (Bester-Meredith et al. 2005; Young and Wang 2004; Young et al. 1999). Coccaro et al. (1998) show that AVP is positively associated with aggressive behavior for men with personality disorders. AVP administration increases physical arousal, biasing individuals to respond aggressively to neutral stimuli (Shalev et al. 2011; Ebstein et al. 2009; Thompson et al. 2004). In regard to social perception, Uzefovsky et al. (2012) show AvP administration leads to significant decrease in men's recognition of others' emotional states, while Kenyon et al. (2013) showed no significant effect of AVP in the same task.

Rilling et al. (2011) found that AVP increased reciprocation after cooperation with human partners as well as functional connectivity between the amygdala and the anterior insula. Israel et al. (2012) found no influence of AVP on cooperative behavior in a public goods game. Similarly, blood levels of AVP were unrelated to distrust in the TG (Zak unpublished data). In addition, an AVP infusion study produced no differences when compared to a placebo for a variety of economic decisions (Zak 2011). Determining how AVP affects economic decisions will require additional studies.

3.5.11 Dopamine

Dopamine (abbreviated 'DA') is a neurotransmitter (a chemical released by nerve cells to communicate with other nerve cells) with a central role in human functioning considered the 'gas pedal' to pursuing reward. Cell bodies of DA neurons are mostly in the midbrain and release DA with nerve impulses (Moore and Bloom 1978). The DA system broadly encodes abstract information about reward. The majority of midbrain dopamine neurons respond in unison to unpredicted rewards, with remaining neurons unresponsive to stimuli (Tobler et al. 2005).

The DA system encodes value and provides a signal of 'pure reward' vis-à-vis its expectation. Put differently, it responds not to absolute rewards, but to their reward relative to its expectation—the difference between them is known as reward prediction error (RPE). Delivery of unexpected rewards increases phasic midbrain activity while its absence decreases it (Schultz 2004). Together, the DA system forms a reinforcement learning system that guides behavior and attention toward optimal reward guided by experience and predictions of unknowns (For a comprehensive review see Schultz 1998). Based on an this extensive literature of animal studies, neuroeconomics research has successfully shown DA's role in predicting and responding to monetary reward (Preuschoff et al. 2007).

Pharmacological studies include increasing DA as well as blocking its receptors and precursors. Menon et al. (2007) tested differential effects on BOLD responses and found amphetamine treatment—which increases DA—caused larger BOLD reward prediction errors in the midbrain. Efremidze and Zak (in press) test effects on learning by blocking DA receptors with naltrexone in a DAA experiment and find interrupted reinforcement learning and larger and longer lasting price bubbles. This result is consistent with Pessiglione et al. (2006) who show that participants given L-DOPA (DA-promoting drug) performed better than those who received haloperidol (drug that binds to DA receptors as an agonist but induces the opposite response). Sevy et al. (2006) showed that tyrosine (a DA precursor) depletion impaired performance of the Iowa Gambling Task by increasing the weight to temporally proximal outcomes. Thematically similar, Scarná et al. (2005) demonstrate that participants underweighted magnitude of bad outcomes when given the same branch chain amino acid (BCAA) mixture.

Despite the multitudes of studies on this neurotransmitter, Rogers (2011) summarizes the complicating factors associated with interpreting DA studies, which include; (i) uncertainty about the pre versus postsynaptic actions; (ii) lack of specificity of medications between receptor subtypes (pharmacological studies are hamstrung by limited specifity of agents administrable to humans); and (iii) uncertainty about interaction of effects with participants' 'baseline' abilities. Also, ensuring experimental double-blind treatment is difficult given the sometimes nauseating effects of tryptophan depleting BCAA liquid given to participants (Crockett and Fehr 2013).

As mentioned earlier, biological systems are interconnected, and specific molecules can promote as well as inhibit other molecules, as illustrated by DA and serotonin; Daw et al. (2002) provide a summary of their opponency, and Cools et al. (2011) discuss their complementarity.

3.5.12 Serotonin

In 1948 Maurice Rapport, Arda Green, and Irvine Page at the Cleveland Clinic discovered a vasoconstrictor substance in blood serum affecting vascular tone and named it serotonin (Abbreviated '5-HT' due to its chemical formula). About 90 % of it is in the gastrointestinal tract where it regulates intestinal movements and the rest is synthesized in serotonogenic neurons where it regulates mood, appetite, and sleep and plays important roles in cognition and learning (Berger et al. 2009).

Experiments testing 5-HT on decisions suggests it affects risk-taking with probabilistic outcomes, time discounting, impulsivity, and cooperation. In a study of the effects of increasing 5-HT, Murphy et al. (2009) find a significant three-way interaction between treatment, size of possible gains, and size of possible losses, and suggest 5-HT modulates non-normative decision-making under uncertainty. Doya (2002) proposes that 5-HT controls the time scale of reward prediction in a theoretical model of integrated neuromodulatory learning, which is experimentally supported by Crockett et al. (2010) who deplete 5-HT's amino acid precursor (tryptophan) and find increased impulsive choice in a discounting task. They also find it jointly increased impulsive choice and altruistic punishment (i.e., rejecting an unfair albeit nonzero offer), suggesting that 5-HT modulates self control and impulsivity. In the same experiment, Crockett et al. (2008) show the 5-HT depletion protocol increased rejection rates among unfair offers. Wood et al. (2006) find that 5-HT depletion caused significant reduction in cooperation in the PD on day 1 of a cross-over, within-subject study.

3.5.13 Stress Hormones

The body responds to physical and psychological demands by releasing hormones, and virtually all people can attest to their focusing and motivating effects in the urgency of a deadline or exigency of a crisis. Stress hormones prepare the body to engage in challenging tasks by focusing attention, increasing cardiovascular tone, energy availability, and suppressing the immune system (Born et al. 1990; Axelrod and Reisine 1984). Stress hormones have long been known to affect physiology and new research shows their impact on economic and social decision-making.

Mammalian physiology has complex and specific responses to the multitude of threats and scenarios organisms are likely to face. Specific stress hormones are released as follows: Adrenocorticotropic hormone (ACTH) from the anterior pituitary, glucocorticoids (GCs) from the adrenal cortex, epinephrine from the adrenal medulla, and norepinephrine from sympathetic nerves. Instead of a one-size-fits-all stress response, the body has an evolved 'set' of hormonal responses ranging from fast release and brief influence to slower release with effects of longer duration. The neural path, known as the sympathetic adrenomudullary system, acts immediately upon exposure to stress, and releases adrenaline (epinephrine) and noradrenaline (norepinepherine) from the adrenal medulla (Elmadjian et al. 1957). The sympathetic nervous system reacts by temporarily increasing heart rate, blood pressure, and perspiration, then returning them to baseline within approximately 10 min (Het et al. 2009). The slower system involves reactions along the hypothalamic-pituitary axis (HPA), starting with the release of corticotrophin-releasing hormone, which stimulates release of hormones from the adrenal cortex and precipitating changes in physiologic state that last 10–60 min and sometimes longer (Sapolsky et al. 2010).

Stress hormones have also recently been shown to affect decision-making and risk preferences. A study using the Iowa Gambling Task (IGT) showed that men with elevated levels of cortisol, a long-acting stress hormone, performed more poorly while women show an inverse relationship, performing best with slightly elevated levels (van den Bos 2009). Putman et al. (2009) showed that elevating stress hormones pharmacologically increased risky decision-making involving potentially large rewards as well as risk-seeking choices when probability of loss was high. A related study using the Balloon Analogue Risk Task (BART) showed that under high levels of stress, men tended to increase risk-taking while women reduced it (Lighthall 2012). Kandasamy et al. (2013) also show that chronically high stress hormones increase risk aversion, though acute elevation of stress hormones does not.

Stress hormones have been shown to play a role in discounting future gains, with men showing a negative relationship between discounting and stress hormones, and women showing a positive relationship (Takahashi 2010). Moral reasoning, too, has been shown to be sensitive to stress, with higher stress correlating with less utilitarian choices among hypothetical personal moral dilemmas such as making a life-or-death decisions (Youssef et al. 2012).

ACTH in particular has been shown to be a neurochemical signal that sustains and increases visual attention (Born et al. 1990). Testing this in a consumer neuroscience paradigm, Lin et al. (2013) show that attending to a public service announcement (PSA) causes significantly higher ACTH release in both men and women. The authors posit that ACTH sustains attention to the PSA and is requisite for preparing action in market decisions.

Together, these results suggest stress hormones are instrumental to species perpetuation by focusing attention on salient and relevant information and new evidence shows tilting decisions toward outcomes more likely to ensure survival. Further, these finding are in line with the proposition that stress hormones have an inverted-U effect on cognition, affect and behavior—deficient and excessive amounts of stress hinder cognition while intermediate levels improve it (McEwan and Sapolsky 1995). Future work will likely explore the interaction between stress and other hormones to better understand their mutual influences on economic decisions.

3.5.14 Female-Specific Hormonal Influences

Sex differences exist in the brain, both in morphology and function (Cahill 2006). For example, receptor affinity for glucocorticoids is half as great in females than males, which has implications for the direction and magnitude of their influence between genders (Madeira and Lieberman 1995). Several studies show sex differences in the serotonin system as well as the analgesic effect of opioid peptides (Nishizawa 1997; reviewed in Craft 2003). The prefrontal cortex, responsible for executive function, has sex hormone receptors including the highest concentration of estrogen receptors in the brain (Bixo et al. 1995). For a review of differences in gender due to sex hormones see Collaer and Hines (1995).

In addition to differences in neuroanatomy and function, hormonal variations drive behavioral differences between men and women in decision-making. Buser (2012) finds that women show less trust in the TG than men in menstrual and premenstrual phases, but have similar trust to men in the middle stages of their cycles. As mentioned above, Zak et al. (2004) showed that women in the luteal phase of their menstrual cycle were less trustworthy in the TG than either men or women in the follicular phase. Women's satisfaction with life varies over the menstrual cycle through the interaction of estradiol, progesterone, and OT (Grosberg et al. in review).

Senior et al. (2007) found that women allocate more resources to dominant-looking men during the follicular phase of their cycles, and allocate less to non-dominant men during the luteal phase, suggesting a hormonal role for resource allocation as a sexual signal. Miller et al. (2007) found that women working in gentleman's clubs earned more money during fertile phases of their cycle. Zethraeus et al.'s (2009) randomized study of postmenopausal women found no effect of T or estrogen in a modified DG, UG, TG, and risk aversion. Further work is needed in ascertaining the hormonal role in gender differences in development as well as activational properties of specific hormones.

3.6 Field Studies

Lab work builds the basic science underlying neurobehavioral influences on decision-making and is the right place to start. However, we cannot faithfully assume that people's behavior in the lab will perfectly represent behavior outside the lab; thus the field is next frontier that offers radical improvement in ecological validity. Fieldwork requires running studies outside tightly controlled environments as well as learning how to interpret the surviving influences of biological vectors in complex scenarios. Translating data from lab experiments is an integral aspect of science and thus further work is needed to bring theory and reality into greater congruence and be able to improve practice (See Harrison and List (2004) for a thorough exposition of economic field experiments). However, the significant challenge of adapting field experiments to lab studies is ethical feasibility of pharmacologic manipulation. For example, one could not use the same double-blind T protocol used by Nadler et al. on actual traders at a trading firm without introducing substantial risk to the individual traders, employer, and the market.

In one of the few field studies involving hormones and economic behavior, Coates and Herbert (2007) studied the relationship between T, stress hormones, and trading performance in professional stock traders. They found that higher morning T was associated with higher average returns (relative to recent trading performance), and that cortisol increased with market volatility. The T finding is consistent with greater risk-taking producing larger returns (following the security market line), while the cortisol finding matches the neurophysiologic measurements of foreign exchange traders in Lo and Repin (2002). Lo and Repin's experiment tested reactivity of securities traders by their quantifiable physiologic responses driven by cognitive-emotional interactions during live trading and found heterogeneity in responses based on trading experience.

Zak (2012) reports that a variety of rituals, such as soldiers marching, rugby teams' pre-match warm up, and a war dance by indigenous peoples in Papua New Guinea are associated with increases in both OT and T. In these studies, OT was associated with a sense of group affiliation, while changes in T appeared to rise due to potential competition, consistent with the challenge hypothesis of T (Wingfield et al. 1990). This indicates that such rituals facilitate in-group bonding in response to out-group aggression, demonstrating the interactive effect of hormones on behavior.

More fieldwork is needed to assess the influence of hormones on decision-making as it occurs in complex environments, and especially to translate laboratory findings into problem-solving applications and intelligently inform policy.

3.7 Summary, Conclusions, and Future Directions

Hormones do much more than regulate homeostasis and initiate physiologic and developmental changes. From moderating trust in strangers, to affecting how much people pay for financial assets, to shifting risk tolerance, hormones play an important role in our economic and social lives.

However, modeling hormonal influences on behavior is complicated. Many hormones assert influence in nonlinear ways, often displaying inverted-U relationships between the quantity of a hormone and behavior (Zak 2010). Further, hormones are released in pulses, can respond rapidly to environmental stimuli, and vary dramatically over time meaning that hormonally based models will be complicated (Santern and Bardin 1973). Finally, hormones vary in their timeframe of effect and are influenced by and interact with agonist and antagonist hormones (Breedlove et al. 2007, p. 123; Jackson et al. 1997) further complicating model building. For example, despite clear results in the asset trading paradigm, T has been shown to act as an allosteric modulator of $GABA_A$ receptors as well as positively correlating with DA, which complicates simplistic modeling of its influence through a simple, single channel. Another important question is whether specific hormones affect high-level cognition, or whether they modulate lower level processes and manifest as experimental shift variables (e.g., T increases risk-taking; OT increasing trust). For these reasons we caution that adopting a mechanistic perspective linking hormones to decisions that does not consider holistic neural activity is likely to produce inaccurate predictions.

Neuroeconomics studies of hormonal influences on decisions follow the approach in the biological sciences where convergent evidence corroborates hypotheses using a variety of methods. When studying hormones and decision-making, we believe it is critical to use the three-stage process outlined above: establish baselines, measure endogenous hormone changes, and assess the impact of pharmacologic manipulation on specific, measurable behavior. Further, hormones can be radio-tagged to show how they affect neural activity using PET, and genetic assays can reveal relationships between different alleles and hormone function, methods that are covered elsewhere in this book.

An application of hormonal effects on behavior is in the design of institutions. Institutions can mitigate risk, off-load cognition for difficult decisions, and provide external resources. Well-functioning institutions can reduce stress, increase OT, and optimize performance and satisfaction. However, one of the significant risks of this research is simplistic interpretation and direct application to complex situations without careful translation. For example, one of the most common questions we get regarding the T and asset trading experiment is, "If testosterone causes men to trade irrationally, and if women have less testosterone than men, why don't you just have women trade along with the men?" There is scientific (and anecdotal) evidence why this 'solution' is unlikely to effectively reduce risk-taking among men, and actually more likely to increase it (see Ronay and von Hippel 2010).

Nobel Prize winner Ronald Coase stated that, "the degree to which economics is isolated from the ordinary business of life is extraordinary and unfortunate" (2012). Neuroeconomics provides positive and normative improvements to the business of life by identifying the mechanisms underlying economic decisions. By judiciously applying its methods to extant problems, neuroeconomics can improve institutions, individual happiness and performance, and society.

References

Arregger LNC, Alejandro L, Tumilasci OR, Aquilano DR, Cardoso EML (2007) Salivary testosterone: a reliable approach to the diagnosis of male hypogonadism. Clinic Endocrinol 67:656–662

Apicella C, Dreber A, Campbell B, Hoffman M, Little A (2008) Testosterone and financial risk preferences. Evol Hum Behav 29:384–390

Apicella C, Dreber A, Mollerstrom J (2014) Salivary testosterone change following monetary wins and losses predicts future financial risk-taking. Psychoneuroendocrinology 39:58–64

Apicella CL, Tobolsky VA, Marlowe FW, Miller KW (2015) Hadza hunter-gatherer men do not have more masculine digit ratios (2D:4D). Am J Phys Anthropol 159: 223–32

Axelrod J, Reisine T (1984) Stress hormones: their interaction and regulation. Science 224:452–459

Bacon F, Selby FG (1895). The advancement of learning (Macmillan)

Baillargeon J, Urban R, Ottenbacher KK, Goodwin J (2013) Trends in androgen prescribing in the United States, 2001 to 2011. JAMA Intern Med 173(15):1465–1466

Barraza J, Zak P (2009) Empathy toward strangers triggers oxytocin release and subsequent generosity. Ann NY Acad Sci 1167:182–189

Barraza J, McCullough M, Zak PJ (2011) Oxytocin infusion increases charitable donations regardless of monetary resources. Horm Behav 60:148–151

Baumgartner T, Heinrichs M, Vonlanthen A, Fischbacher U, Fehr E (2008) Oxytocin shapes the neural circuitry of trust and trust adaptation in humans. Neuron 58(4):639–650

Beato M, Chavez S, Truss M (1996) Transcriptional regulation by steroid hormones. Steroids 61 (4):240–251

Berg J, Dickhaut J, McCabe K (1994) Trust, reciprocity, and social history. Games Econ Behav 10:122–142

Berger M, Gray J, Roth B (2009) The expanded biology of serotonin. Annu Rev Med 60:355–366

Bester-Meredith JM (2005) Manipulations of vasopressin alter aggression differently across testing conditions in monogamous and non-monogamous peromyscus mice. Aggressive Behav 31:189–199

Bixo M, Backstrom T, Winblad B, Andersson A (1995) Estradiol and testosterone in specific regions of the human female brain in different endocrine states. J Steroid Biochem Mol Biol 55:297–303

Boksem MA et al. (2013) Testosterone inhibits trust but promotes reciprocity. Psychol Sci:0956797613495063

Booth A, Shelley G, Mazur A, Tharp G, Kittok R (1989) Testosterone, and winning and losing in human competition. Horm Behav 23:556–571

Born J, Bathelt B, Pietrowsky R, Pauschinger P, Fehm H (1990) The influence of display characteristics on visual exploratory search behavior. J Consum Res 101:34–38

Bos P, Terburg D, van Honk J (2010) Testosterone decreases trust in socially naïve humans. PNAS 107(22):9991–9995

Brambilla D, Matsumoto A, Araujo A, McCinlay J (2009) The effect of diurnal variation on clinical measurement of serum testosterone and other sex hormone levels in men. J Clin Endocrinol Metab 94(2):907–913

Brañas-Garza P, Rustichini A (2011) Organizing effects of testosterone and economic behavior: not just risk taking. PlosOne 6(12):e29842

Breedlove M, Rosenweig M, Watson N (2007) Biological psychology. Sinauer, Sunderland, MA, USA

Burham T (2007) High-testosterone men reject low ultimatum game offers. Proc R Soc B 274 (1623):2327–2330

Buser T (2012) Digit ratios, the menstrual cycle and social preferences. Games Econ Behav 76:457–470

Cahill L (2006) Why sex matters for neuroscience. Nat Rev Neurosci 7(6):1–8

Camerer C, Loewenstein G, Prelec D (2005) Neuroeconomics: how neuroscience can inform economics. J Econ Lit 43(1):9–64

Carney DR, Mason MF (2010) Decision making and testosterone: when the ends justify the means. J Exper Social Psychol 46(4):668–671

Carré J, McCormick C (2008) Aggressive behavior and change in salivary testosterone concentrations predict willingness to engage in a competitive task. Horm Behav 54:403–409

Carré J, Gilchrist J, Morrissey M, McCormick C (2010) Motivational and situational factors and the relationship between testosterone dynamics and human aggression during competition. Biol Psychol 84:346–353

Carson D et al. (2015) Cerebrospinal fluid and plasma oxytocin concentrations are positively correlated and negatively predict anxiety in children. Mol Psychiat 20(9):1085–1090

Cherek DR, Schnapp W, Moeller FG, Dougherty DM (1996) Laboratory measures of aggressive responding in male parolees with violent and nonviolent histories. Aggressive Behav 22(1):27–36

Christensen J, Shiyanov P, Estrepp J, Schlager J (2014) Lack of association between human plasma oxytocin and interpersonal trust in a prisoner's dilemma paradigm. PLoS ONE 9(12): e116172

Coase R, Wang N (2012) Saving economics from the economists. Harvard Bus Rev 90(12):12–36

Coates J (2012) The hour between dog and wolf: risk-taking, gut feelings and the biology of boom and bust. HarperCollins, London, UK

Coates J, Herbert J (2007) Endogenous steroids and financial risk taking on a london trading floor. PNAS 105(16):6167–6172

Coccaro E, Kavoussi RH, Cooper T, Ferris C (1998) Erebrospinal fluid vasopressin levels: correlates with aggression and serotonin function in personality-disordered subjects. Arch Gen Psychiatry 55:708–714

Collaer M, Hines M (1995) Human behavioral sex differences: a role for gonadal hormones during early development? Psychol Bull 118(1):55–107

Cools R, Nakamura K, Daw N (2011) Serotonin and dopamine: unifying affective, activational, and decision functions. Neuropsychopharmacology 36:98–113

Crabbe P, Bogaert V, De Bacquer D, Zmierczak H, Kaufman J (2007) Part of the interindividual variation in serum testosterone levels in healthy men reflects differences in androgen sensitivity and feedback set point: contribution of the androgen receptor polyglutamine tract polymorphism. J Clin Endocrinol Metab 92(9):3604–3610

Craft R (2003) Sex differences in opioid analgesia: from mouse to man. Clin J Pain 19:175–186

Crockett M, Fehr E (2013) Social brains on drugs: tools for neuromodulation in social neuroscience. SCAN 1(5):nst113

Crockett M, Clark L, Tabibnia G, Lieberman M, Robbins T (2009) Serotonin modulates behavioral reactions to unfairness. Science 320(27):1739

Crockett M, Clark L, Tabibnia G, Lieberman M, Robbins T (2010) Impulsive choice and altruistic punishment are correlated and increase in tandem with serotonin depletion. Emotion 10 (6):855–862

Cueva C, Roberts RE, Spencer T, Rani N, Tempest M, Tobler PN, Herbert J, Rustichini A (2015) Cortisol and testosterone increase financial risk taking and may destabilize markets. Sci Rep 5:11206

Daw N, Kakade S, Dayan P (2002) Influential exposition of opponency between dopamine and serotonin in terms of coding average reward and punishment rates, and prediction errors. Neural Netw 15:603–616

Doya K (2002) Metalearning and neuromodulation. Neural Networks 15(4):495–506

Dunn J, Nisula B, Rodbard D (1981) Transport of steroid hormones: binding of 21 endogenous steroids to both testosterone-binding globulin and corticosteroid-binding globulin in human plasma. J Clin Endocrinol Metab 53(1):58–68

Ebert A et al. (2013) Modulation of interpersonal trust in borderline personality disorder by intranasal oxytocin and childhood trauma. Soc Neurosci 8(4):305–313

Ebstein R, Israel S, Lerer E, Uzefovsky F, Shalev S, Gritsenko I et al (2009) Arginine vasopressin and oxytocin modulate human social behavior. Ann NY Acad Sci 1167:87–102

Efremidze L, Sarraf G, Miotto, K, Zak, PJ (in press) The Neural Inhibition of Learning Increases Asset Market Bubbles: Experimental Evidence. Journal of Behavioral Finance.

Eisenegger C, Naef M, Snozzi R, Heinrichs M, Fehr E (2010) Prejudice and truth about the effect of testosterone on human bargaining behaviour. Nature 463(21):356–361

Eisenegger C, von Eckardstein A, Fehr E, von Eckardstein S (2013) Pharmacokinetics of testosterone and estradiol gel preparations in healthy young men. Psychoneuroendocrinology 38:171–178

Ellis G, Desjardins C, Fraser H (1983) Control of pulsatile LH release in male rats. Neuroendocrinol 37(3):177–183

Elmadjian F, Hope JM, Lamson ET (1957) Excretion of epinephrine and norepinephrine in various emotional states. J Clin Endocrinol Metab 17(5):608–620

Falkenstein E, Tillmann H-C, Michael C, Martin F, Martin W (2000) Multiple actions of steroid hormones—a focus on rapid, nongenomic effects. Pharmacol Rev 52:513–555

Frye C, Seliga A (2001) Testosterone increases analgesia, anxiolysis, and cognitive performance of male rats. Cogn Affect Behav Neurosci 1(4):371–381

Gimpl G, Fahrenholz F (2001) The oxytocin receptor system: structure, function, and regulation. Physiol Rev 81(2):629–683

Grino B, Griffin J, Wilson J (1990) Testosterone at High Concentrations Interacts with the Human Androgen Receptor Similarly to Dihydrotestosterone*. Endocrinology 126:1165–1172

Grosberg D, Merlin R, Zak P (n.d.) What makes women happy: oxytocin release correlates with life satisfaction (in review)

Goudriaan AE et al. (2010) The influence of high-normal testosterone levels on risk-taking in healthy males in a 1-week letrozole administration study. Psychoneuroendocrinol 35(9):1416–1421

Guastella AJ et al. (2010) Intranasal Oxytocin Improves Emotion Recognition for Youth with Autism Spectrum Disorders. Biol Psychiat 67(7):692–694

Haider SG (2004) Cell biology of Leydig cells in the testis. Int Rev Cytol 233:181–241

Harrison G, List J (2004) Field experiments. J Econ Lit 1009–1055

Hemat RAS (2004) Principles Of Orthomolecularism

Het S, Rohleder N, Schoofs D, Kirschbaum C, Wolf O (2009) Neuroendocrine and psychometric evaluation of a placebo version of the trier social stress test. Psychoneuroendocrinology 34:1075–1086

Hodgeson G (1998) On the evolution of Thornstein Veblen's evolutionary economics. Camb J Econ 22:415–431

Holdcraft RW, Braun RE (2004) Hormonal regulation of spermatogenesis. Inter J Androl 27 (6):335–342

Insel T, Young L, Witt D, Crews D (1993) Gonadal-steroids have paradoxical effects on brain oxytocin receptors. J Neuroendocrinol 5:619–628

Israel S, Weisel O, Ebstein R, Bornstein G (2012) Oxytocin, but not vasopressin, increases both parochial and universal altruism. Psychoneuroendocrinology 37:1341–1344

Ito T, Horton R (1971) The source of plasma dihydrotestosterone in man. J Clinic Invest 50 (8):1621
Jackson T, Richer J, Bain D, Takimoto G, Tung L, Horwitz K (1997) The partial agonist activity of antagonist-occupied steroid receptors is controlled by a novel hinge domain-binding coactivator L7/SPA and the corepressors N-CoR or SMRT. Mol Endocrinol 11(6):693–705
Kagerbauer SM et al. (2013) Plasma oxytocin and vasopressin do not predict neuropeptide concentrations in human cerebrospinal fluid. J Neuroendocrinol 25(7):668–673
Kandasamy N, Hardy B, Coates J (2013) Cortisol shifts financial risk preferences. PNAS
Kandasamy N et al. (2014) Cortisol shifts financial risk preferences. Proc Natl Acad Sci U S A 111 (9):3608–3613
Kenyon A, Alvares G, Hickie I, Guastella A (2013) The effects of acute arginine vasopressin administration on social cognition in healthy males. J Horm
Kiss A, Mikkelsen J (2011) Oxytocin-anatomy and functional assignments: a minireview. Endocr Regul 39(3):97–105
Klackl J, Pfundmair M, Agroskin D, Jonas E (2013) Who is to blame? Oxytocin promotes nonpersonalistic attributions in response to a trust betrayal. Biol Psychol 92(2):387–394
Kosfeld M, Heinrichs M, Zak P, Fischbacher U, Fehr E (2005) Oxytocin increases trust in humans. Nature 435(2):673–676
Lighthall N, Sakaki M, Vasunilashorn S, Nga L, Somayajula S, Chen E et al (2012) Gender differences in reward-related decision processing under stress. SCAN 476–484
Lin P-Y, Grewal NS, Morin C, Johnson WD, Zak PJ (2013) Oxytocin increases the influence of public service advertisements. PloS one 8(2):e56934
Lo W, Repin D (2002) The psychophysiology of real-time financial risk processing. J Cogn Neurosci 14(3):323–339
Losel R, Wehling M (2003) Nongenomic actions of hormones. Nature 4:46–56
Madeira M, Lieberman A (1995) Sexual dimorphism in the mammalian limbic system. Prog Neurobiol 45:275–333
Manni A, Pardridge W, Cefalu W, Nisula B, Bardin C, Santner S et al (1985) Bioavailability of albumin-bound testosterone. J Clin Endocrinol Metab 61(4):705
Manning J, Scutt D, Wilson J, Lewis-Jones D (1998) The ratio of 2nd to 4th digit length: a predictor of sperm numbers and concentrations of testosterone lutenizing hormone and oestrogen. Hum Reprod 13(11):3000–3004
Manning J, Stewart A, Bundred P, Trivers R (2004) Sex and ethnic differences in 2nd to 4th digit ratio of children. Early Hum Dev 80(2):161–168
Marshall A (1890) Principles of economics. Prometheus
Matsumoto AM, Bremner WJ (1984) Modulation of pulsatile gonadotropin secretion by testosterone in man*. J Clinic Endocrinol Metabol 58(4):609–614
Mazur A, Booth A (1998) Testosterone and dominance in men. Behav Brain Sci 21:353–397
McCall C, Singer T (2012) The animal and human neuroendocrinology of social cognition, motivation and behavior. Nat Neurosci 15(5):681–688
McCullough M, Churchland P, Mendez A (2013) Problems with measuring peripheral oxytocin: Can the data on oxytocin and human behavior be trusted? 37:1485–1492
McEwen B (1991) Genomic and non-genomic effects of steroïds on neuronal activity. Trends Pharmacol Sci 12:141–147
McEwan B (2004) Brain-fluid barriers: relevance for theoretical controversies regarding vasopressin and oxytocin memory research. Adv Pharmacol 50:531–592
McEwan B, Sapolsky R (1995) Stress and cognitive function. Curr Opin Neurobiol 5(2):205–216
Mendis-Handagama S (1997) Luteinizing hormone on Leydig cell structure and function. Histol Histopathol 12(3):869–882
Menon M, Jensen J, Vitcu I, Graff-Guerrero A, Crawley A, Smith M (2007) Temporal difference modeling of the blood-oxygen level dependent response during aversive conditioning in humans: effects of dopaminergic modulation. Biol Psychiatry 62:765–772
Midzaka A, Chena H, Papadopoulos V, Zirkina B (2009) Leydig cell aging and the mechanisms of reduced testosterone synthesis. Mol Cell Endocrinol 23:23–31

Mikolajczak M, Gross J, Lane A, Corneille O, de Timary P, Luminet O (2010) Oxytocin makes people trusting. Not Gullible. Psychol Sci 21(8):1072–1074
Miller G, Tybur J, Jordan B (2007) Ovulatory cycle effects on tip earnings by lap dancers: economic evidence for human estrus? Evol Hum Behav 28:375–381
Moore R, Bloom F (1976) Central catecholamine neuron system: anatomy and physiology of the dopamine system. Ann Rev Neurosci 1:129–169
Murphy SE et al. (2009) The role of serotonin in nonnormative risky choice: the effects of tryptophan supplements on the "reflection effect" in healthy adult volunteers. J Cognitive Neurosci 21(9):1709–1719
Nadler A, Jiao P, Johnson C, Alexander V, Zak P (2016) The bull of Wall street: experimental analysis of testosterone and asset trading, under review
Nishizawa S et al (1997) Differences between males and females in rates of serotonin synthesis in human brain. Proc Natl Acad Sci 5308–5313
Pardridge W, Moeller T, Mietus L, Oldendorf W (1980) Blood-brain barrier transport and brain sequestration of steroid hormones. Am J Physiol Endocrinol Metab E96–E102
Pederson C, Prange JA (1979) Induction of maternal behavior in virgin rats after intracerebroventricular administration of oxytocin. PNAS 76(12):6661–6665
Pessiglione M, Seymour B, Flandin G, Dolan R, Frith D (2006) Dopaminedependent prediction errors underpin reward-seeking behaviour in humans. Nature 442:1042–1045
Preuschoff K, Bossaerts P, Quartz S (2006) Neural differentiation of expected reward and risk in human subcortical structures. Neuron 381–390
Putman P, Antypa N, Crysovergi P, van der Does J, Does W (2009) Exogenous cortisol acutely influences motivated decision making in healthy young men. Psychopharmacology 298:257–263
Rapport M, Green A, Page I (1948) Serum vasoconstrictor (serotonin) IV. Isolation and characterization. J Biol Chem 176(3):1243–1251
Reddy DS, Jian K (2010) The testosterone-derived neurosteroid androstanediol is a positive allosteric modulator of GABAA receptors. J Pharmacol Exper Therapeut 334(3):1031–1041
Rilling JK et al. (2012) Effects of intranasal oxytocin and vasopressin on cooperative behavior and associated brain activity in men. Psychoneuroendocrinol 37(4):447–461
Roberti JW (2004) A review of behavioral and biological correlates of sensation seeking. J Res Pers 38(3):256–279
Rogers RD (2011) The roles of dopamine and serotonin in decision making: evidence from pharmacological experiments in humans. Neuropsychopharmacol 36(1):114–132
Ronay R, von Hippel W (2010) The presence of an attractive woman elevates testosterone and physical risk taking in young men. Soc Psychol Pers Sci 1(57):57–64
Rupprecht R, Holsboer F (1999) Neuroactive steroids: mechanisms of action and neuropsychopharmacological perspectives. Trends Neurosci 22(9):410–416
Rydén G, Sjöholm I (1969) Half-life of oxytocin in blood of pregnant and non-pregnant women. Acta Endocrinol 61(3):425–431
Santen R, Bardin C (1973) Episodic analysis, clinical interpretation, physiologic mechanisms. J Clin Invest 52:2617–2628
Santern R, Bardin C (1973) Episodic luteinizing hormone secretion in man. J Clin Investig 52:2617–2628
Sapienza P, Zingales L, Maestripieri D (2009) Gender differences in financial risk aversion and career choices are affected by testosterone. PNAS 106(36):15268–15273
Sapolsky R, Romero M, Munck A (2010) How do glucocorticoids influence stress responses? Integrating permissive, suppressive, stimulatory, and preparative actions. Endocr Rev 2(1):55–89
Scarná A, McTavish S, Cowen P, Goodwin G, Rogers R (2005) The effects of a branched chain amino acid mixture supplemented with tryptophan on biochemical indices of neurotransmitter function and decision-making. Psychopharmacology 179:761–768
Schipper BC (2015) Sex Hormones and Competitive Bidding. Manage Sci 61:249–266
Schultz W (1998) Predictive reward signal of dopamine neurons. J Neurophysiol 25(1):236–245

Schultz W (2004) Neural coding of basic reward terms of animal learning theory, game theory, microeconomics and behavioural ecology. Curr Opin Neurobiol 14:139–147

Schwarz S, Pohl P (1992) Steroid hormones and steroid hormone binding globulins in cerebrospinal fluid studied in individuals with intact and with disturbed blood-cerebrospinal fluid barrier. Neuroendocrinology 55(2):174–182

Senior C, Lau A, Butler MJ (2007) The effects of the menstrual cycle on social decision making. Int J Psychophysiol 63(2):186–191

Sevy S, Hassoun Y, Bechara A, Yechiam E, Napolitano B, Burdick K et al (2006) Emotion-based decision-making in healthy subjects: short-term effects of reducing dopamine levels. Psychopharmacology 188:228–235

Shalev II (2011) Vasopressin needs an audience: neuropeptide elicited stress responses are contingent upon perceived social evaluative threats. Horm Behav 60:121–127

Smith V, Suchanek G, Williams A (1988) Bubbles, crashes, and endogenous expectations in experimental spot asset markets. Econometrica 56(6):1119–1151

Stanczyk F, Clarke N (2010) Advantages and challenges of mass spectrometry assays for steroid hormones. J Steroid Biochem Mol Biol 121:491–495

Striepens N et al. (2013) Elevated cerebrospinal fluid and blood concentrations of oxytocin following its intranasal administration in humans. Sci Rep 3:3440

Swerdloff R, Wang C, Cunningham G, Dobs A, Iranmanesh A, Matsumoto A et al (2000) Long-term pharmacokinetics of transdermal testosterone gel in hypogonadal men. J Clin Endocrinol Metab 85(12):4500–4510

Szczypka M, Zhou Q, Palmiter R (1998) Dopamine-stimulated sexual behavior is testosterone dependent in mice. Behav Neurosci 112(5):1229–1235

Takahashi T, Shinada M (2010) Stress hormones predict hyperbolic time-discount rates six months later in adults. Neuro Endocrinol Lett 31(5):616–721

Takahashi T, Sakaguchi K, Oki M, Homma S, Hasegawa T (2006) Testosterone levels and discounting delayed monetary gains and losses in male humans. Neuro Endocrinol Lett 27 (4):439–444

The Concise Encyclopedia of Economics (n.d.) Retrieved 27 Jan 2013, from Library of Economics and Liberty: http://www.econlib.org/library/Enc/bios/Marshall.html

Thilers PP, Macdonald SW, Herlitz A (2006) The association between endogenous free testosterone and cognitive performance: a population-based study in 35 to 90 year-old men and women. Psychoneuroendocrinology 31(5):565–576

Thompson R, Gupta S, Miller K, Mills S, Orr S (2004) The effects of vasopressin on human facial responses related to social communication. Psychoendocrinology 29:35–48

Tobler P, Fiorillio C, Schultz W (2005) Adaptive coding of reward value by dopamine neurons. Science 307(5715):1642–1645

Tuiten A, van Honk J, Koppeschaar H, Bernaards C, Thijssen J, Verbaten R (2000) Time course of effects of testosterone administration on sexual arousal in women. Arch Gen Psychiatry 57:149–153

Uzefovsky F, Shalev I, Israel S, Knafo A, Ebstein R (2012) Vasopressin selectively impairs emotion recognition in men. Psychoneuroendocrinology 37:570–576

van den Bos R, Harteveld M, Stoop H (2009) Stress and decision-making in humans: performance is related to cortisol reactivity, albeit differently in men and women. Psychoneuroendocrinology 34:1449–1458

van Honk J, Peper JS, Schutter DJ (2005) Testosterone reduces unconscious fear but not consciously experienced anxiety: implications for the disorders of fear and anxiety. Biol Psychiatry 58:218–225

van Rooij K et al. (2012) Pharmacokinetics of three doses of sublingual testosterone in healthy premenopausal women. Psychoneuroendocrinol 37(6):773–781

Vercoe M, Zak P (2010) Inductive modeling using causal studies in neuroeconomics: brains on drugs. J Econ Method 2:123–137

Wibral M, Dohmen T, Klingmüller D, Weber B, Falk A (2012) Testosterone administration reduces lying in men. PloS one 7(10):e46774

Wingfield JC, Hegner RE, Dufty AM, Ball GF (1990) The challenge hypothesis: theoretical implications for patterns of testosterone secretion, mating systems, and breeding strategies. Am Nat 829–846

Wood R, Rilling J, Sanfey A, Bhagwager Z, Rogers R (2006) Effects of tryptophan depletion on the performance of an iterated prisoner's dilemma game in healthy adults. Neuropsychopharmacology 31:1075–1084

Yao S, Zhao W (2014) Oxytocin makes females, but not males, less forgiving following betrayal of trust. Int J Neuropsychopharmacol 17:1785–1792

Young L, Wang Z (2004) The neurobiology of pair bonding. Nat Neurosci 7:1048–1054

Young L, Nilsen R, Waymire G, MacGregor G, Insel T (1999) Increased affiliative response to vasopressin in mice expressing the V1a receptor from a monogamous vole. Nature 400:766–768

Youssef F, Dookeeram K (2012) Stress alters personal moral decision making. Psychoneuroendocrinology 37:491–498

Zak P, Matzner W, Matzner W (2004) The neurobiology of trust. Ann NY Acad Sci 1032:224–227

Zak P, Borja K, Matzner W, Kurtzban R (2005a) The neuroeconomics of distrust: sex differences in behavior and physiology. Am Econ Rev Papers Proc 95(2):360–363

Zak PJ, Stanton AA, Ahmadi S (2007) Oxytocin increases generosity in humans. PloS one 2(11): e1128

Zak P (2011) Moral markets. J Econ Behav Organ 77(2):212–233

Zak P (2012) The moral molecule: the source of love and prosperity. Penguin New York

Zak P (2005) Trust: a temporary human attachment facilitated by oxytocin. Behav Brain Sci 28 (3):368–369

Zak P, Kurtzban R, Matzner W (2005b) Oxytocin is associated with human trustworthiness. Horm Behav 48:522–527

Zak P, Kurzban R, Ahmad S, Swerdloff S, Park J (2009) Testosterone administration decreases generosity in the ultimatum game. PLOSone

Zethraeus N et al. (2009) A randomized trial of the effect of estrogen and testosterone on economic behavior. Proc Natl Acad Sci U S A 106(16):6535–6538

Chapter 4
Genes and Human Decision-Making

Martin Reuter and Christian Montag

Abstract The present chapter aims to give a concise overview of the genetics of human decision-making. The focus will be on studies that can be considered within the field of neuroeconomics. Although genetic studies in neuroeconomics are scarce to date, interest in and use of, genetic designs is increasing. This popularity is based on the huge potential of genetic research. The collection of data is more naturalistic as it is not restricted to the laboratory, and genes convey information on brain metabolism relevant to decision-making. The widespread belief, that genetic information is already determined before birth and therefore, is not subject to influence or change over the lifespan, has recently been revised. The young discipline of epigenetics has shown that environmental factors can influence the activity of genes throughout life. Therefore, investigating the interaction between genes and environment on a behavioral, as well as a, molecular level is a promising direction for research. The empirical neuroeconomics studies published to date have adopted a candidate gene approach. This means researchers have identified theory-driven, distinct gene loci relevant for decision-making in economic contexts, from the literature. These pioneer studies are reviewed and discussed below. For those readers who are not familiar with molecular genetics, we recommend the chapter on genetics contained in the methods section of this book as a complementary introduction.

M. Reuter (✉)
Department of Psychology, University of Bonn, Bonn, Germany
e-mail: martin.reuter@uni-bonn-diff.de

M. Reuter
Center for Economics and Neuroscience (CENs), University of Bonn, Bonn, Germany

C. Montag
Department of Molecular Psychology, Institute for Psychology and Education, Ulm University, Ulm, Germany
e-mail: christian.montag@uni-ulm.de

C. Montag
Key Laboratory for NeuroInformation, Center for Information in Medicine, School of Life Science and Technology, University of Electronic Science and Technology of China, Chengdu, People's Republic of China

M. Reuter and C. Montag (eds.), *Neuroeconomics*, Studies in Neuroscience, Psychology and Behavioral Economics, DOI 10.1007/978-3-642-35923-1_4

4.1 The Molecular Genetic Basis of Human Decision-making

Every day we make numerous decisions that influence both our current actions and which can also have strong implications for our future. The choice between cheese cake and apple pie for dessert is obviously not a big thing, but the decision whether or not to invest in high-risk stocks promising enormous profit, may lead to wealth or to financial ruin. From the media and from our own observations, we are well aware that there is great variability in the way people make their decisions. There is, in fact, much greater variability in decision-making behavior between—rather than within—people. This means that individuals typically show a rather stable pattern of behavior—referred to as personality. Some people are notoriously prone to making risky decisions, whereas others tend to procrastinate over a decision and make their choices after considering the alternatives very carefully. Impulsivity and risk-taking on the one hand, and anxiety and reflectiveness on the other, seem to represent two opposing aspects of personality in decision-making (Zaleskiewicz 2001). However, often our decisions do not only influence our own fortune, but also that of others. Man, as a social individual, takes decisions that also have consequences for others and in the same way our own decisions are influenced by the opinions and actions of our fellow human beings (Kaplan and Miller 1987). Therefore, personality traits like cooperativeness, characterizing a person's attitude towards prosocial behavior, are also of great interest to the investigation of human decision-making. It is presumed that the genetic underpinnings of such personality traits build the biological basis for stable patterns in decision-making that seem to be rather invariant across situations and, therefore, allow the prediction of a decision even in rather specific situations. Of course these predictions will not be perfect because we know, e.g., from laboratory experiments, that instructions can mold performance (Wickelgren 1977; see the discussion of the *speed-accuracy-tradeoff* in simple decisions in Chap. 14 of this book) and that states (e.g., current mood) also have a strong impact on decision-making. However, personality traits are highly heritable (up to 50 %; e.g., Tellegen et al. 1988). For the reasons outlined above, genes related to personality prove excellent candidate genes for human decision-making.

Even before neuroeconomics was established as an autonomous scientific discipline, neuroscientists had used genetic approaches for investigating human decision-making. This work was in part triggered by an interest in psychiatric and neurological diseases in which decision-making in patients is strongly impaired. Those patients with brain lesions in clearly defined neural structures, e.g., of the ventromedial prefrontal cortex, offered insights into brain regions relevant for decision-making. A famous paradigm used in such studies is the IOWA Gambling Task, which measures the ability of a person to prefer constant small rewards under low risk of losing money, compared to seldom occurring high rewards under high risk of losing money (IGT; Bechara et al. 2000). The ventromedial prefrontal cortex is essential in impulsive decision-making, with patients with lesions in this area demonstrating a preference for smaller immediate rewards, rather than for larger, delayed rewards. The same pattern of behavior is observed in individuals with a drug addiction, a disorder marked by a lack of impulse control (e.g., Ert et al. 2013; van der Plas et al. 2009). To date there exist numerous genetic association studies in different patient samples and in samples of healthy subjects, that try to explain differences in IGT performance by means of candidate gene polymorphisms. The serotonin transporter linked polymorphic region (5-HTTLPR), an insertion/deletion polymorphism marked by the presence or absence of 43 base pairs in the promoter region of the serotonin transporter gene, has been most extensively studied in this context. Subjects with at least a deletion on one chromosome (genotypes SS and SL) were referred to as S-allele carriers and showed a threefold decrease in mRNA expression compared to subjects with no deletion on either chromosome (genotype LL) (Lesch et al. 1996). In a sample of $N = 885$ healthy Chinese college students, subjects homozygous for the S-allele had significantly lower IGT scores in the first 40 trials than L-allele carriers (He et al. 2010). The IGT task comprises 5 blocks with 20 trials each. The first two blocks (trials 1–40) indicate decisions under uncertainty, whereas blocks 3–5 represent decisions under risk. The main effect of 5-HTTLPR on IGT performance could not be replicated in healthy Caucasian samples (Lage et al. 2011; Stoltenberg and Vandever 2010). In patients with obsessive compulsive disorders, da Rocha et al. (2008) reported significantly lower IGT scores in S-allele carriers, however, in contrast to the study by He et al. (2010), only in the third, fourth, and fifth blocks, i.e., those blocks related to impulsive decision-making.

The IGT has strong parallels to research questions in neuroeconomics, where decisions under ambiguity and under risk are of scientific interest. Neuroscientists also have a long tradition in differentiating subcomponents of the decision-making process by means of biological variables. Cognitive processes related to information processing, memory, and executive control are the prevailing dependent variables. The best supported gene locus with respect to working memory capacity is the COMT Val158Met polymorphism, located in exon 158 of the catecholamine-*O*-methyltransferase (COMT) gene. COMT degrades monoamines (e.g., dopamine) in the synaptic cleft. The catabolizing activity of Met-allele carriers is about three- to fourfold lower than that of Val-allele carriers (Lachman et al. 1996). The Met-allele is associated with higher working memory capacity and better executive control abilities more generally (Goldberg and Weinberger 2008). However—and of interest for neuroeconomics—the Val-allele is related to higher positive emotionality (PE), as compared to the Met-allele (Montag et al. 2012; Reuter and Hennig 2005; Wacker et al. 2012). PE comprises personality traits like extraversion and novelty seeking, as well as other characteristics related to risk-taking and reward-seeking (e.g., Krebs et al. 2009; Smillie 2013).

These examples of genetic association studies have direct implications for the decision processes under investigation in neuroeconomics. Moreover, the gene markers outlined above are among the best studied polymorphisms in neuroeconomic research. It must be noted that we identified thousands of genetic association studies from different disciplines in the literature relating candidate gene loci to various phenotypes. However, for most of these polymorphisms the proof of functionality is missing. The 5-HTTLPR and the COMT Val158met polymorphisms related to altered mRNA expression and differences in metabolic enzyme activity, respectively, are laudable exceptions.

4.2 Molecular Genetics in Neuroeconomics

4.2.1 Ultimatum Game (UG)

Behavioral economists have developed paradigms—so called games—that permit the investigation of human decision-making under experimentally controlled conditions (for an in depth overview of games in neuroeconomic research see Chap. 2 by Civai and Hawes in this book). One of the most prominent paradigms in neuroeconomics is the Ultimatum Game (UG), which provides a framework for the study of prosocial behavior in two players interacting anonymously with each other; Player 1 (the proposer) has to split an endowment with player 2 (the responder). Player 2 can either accept or reject the offer from player 1. If player 2 accepts the offer, the money is split as proposed by player 1. Should player 2 reject the offer, both players receive nothing (Camerer 2003). Thus, the UG represents a classic "take it or leave it" situation. The responder has to either accept the offer of the proposer in its present form or must reject it outright. There is no opportunity for negotiation. The UG played in the laboratory perfectly parallels real life situations. Often we are confronted with such "take it or leave it" situations. For example, if we apply for a job in public service where income brackets are fixed, we have to either accept the salary or leave it and look for a job on the free market. The Greek debt cut is a prime example of a real life UG. A group of banks and investors in Greek government debt had the choice between exchanging their debt for new bonds worth about 70 % less than the original, or encounter the total breakdown of the Greek finance system, resulting in the complete loss of their invested money.

There is a general agreement in the literature that responder behavior in the UG represents fairness preference. If an offer is judged as unfair we are likely to dispense with the offer altogether rather than be satisfied with a small proportion of the pie. Punishing a proposer for an unfair offer (i.e., a low offer that is greater than zero) is not rational and contradicts economists' view on man as a *homo economicus*. It is suggested that the *homo economicus* is merely guided by self-interest and always tries to maximize his profit. In order to do so, he only makes strictly rational decisions (Persky 1995). However, the behavior of the proposer is far more complex than that of the responder. It is a mixture of fairness preference on the one hand and strategic consideration on the other hand: to be a social human being capable of taking the perspective of the responder (theory of mind) and strategic considerations (maximize own profit while minimizing the risk of being punished for an unfair offer).

An initial twin study from Sweden showed that more than 40 % of the variation in subjects' rejection behavior in the UG, is explained by additive genetic effects (Wallace et al. 2007). These data underline the strong genetic basis for the etiology of fairness preferences. However, the heritability estimates for proposer behavior were negligible in this study. To date, two existing molecular genetic studies have tried to identify candidate genes for behavior in the UG. The first study was published by Zhong et al. (2010), who found that the dopamine D4 receptor (DRD4) gene explains at least a small proportion of variance in the UG responder behavior. The DRD4 gene consists of 3400 base pairs (bp) and is located at chromosome 11p15.5. In exon III of this gene a highly polymorphic variable number of tandem repeats (VNTR) polymorphism has been identified, which is characterized by a repetitive sequence of 48 bp (between 2 and 11 repeats) (van Tol et al. 1992). Three alleles are most common; the 2-repeat, the 4-repeat, and the 7-repeat; the ancestral 4-repeat allele is most frequently occurring across all ethnicities. In Caucasians, the 7-repeat is more frequent than the 2-repeat allele. However, in Asians the 7-repeat allele is extremely rare and is typically not considered in Asian genetic association studies.

Besides reported associations between the DRD4 exon III polymorphism and various phenotypes related to decision-making behavior, like impulsivity, novelty seeking, gambling behavior and attention-deficit hyperactivity disorder (ADHD), the functionality of this polymorphism has been further demonstrated (Strobel et al. 1999; Ebstein et al. 1996; Eisenegger et al. 2010; Nikolaidis and Gray 2010). The VNTR region of the DRD4 gene encodes a portion of the third intracellular loop region of the transcribed receptor protein that spans the nerve cell membrane and mediates interaction with second messenger proteins. The 2-repeat allele shows a 50 % reduction in the production of cyclic adenosine monophosphate (cAMP), as compared with the 4-repeat and 7-repeat alleles (Asghari et al. 1995). Regarding decision-making behavior in the UG, Zhong et al. (2010) reported that carriers of the 4/4 genotype stated a 25 % higher minimum acceptable offer in the role of the second mover, as compared with carriers of the 2/4 and 2/2 genotypes. Notably, these results came from a Chinese sample, where the 7-repeat allele is very rare and was therefore not considered in the analysis. The authors did not find an association between the DRD4 exon III polymorphism and the UG proposer behavior. This is in line with the fact that there are no heritability estimates for UG proposer behavior available in the literature to date. Although Zhong et al. reported a significant association between the DRD4 gene and fairness preference as assessed by the UG, the proportion of explained variance is rather small. This is typical for quantitative traits and underlines the necessity of identifying further genetic variants that influence behavior in the UG. With this aim in mind, a second genetic association study focusing on the dopaminergic system's influence on behavior in the UG was devised (Reuter et al. 2013). The DRD2 receptor gene has been linked to various facets of prosocial behaviors like cooperation, attachment style, mentoring, paternal parenting, and positive emotionality, to name but a few (Walter et al. 2011; Gillath et al. 2008; Shanahan et al. 2007; Lucht et al. 2006; Reuter et al. 2006). Two polymorphisms, for which functionality has been established, are most investigated in genetic association studies; the DRD2/ANKK1-Taq Ia (rs1800497) and the DRD2 C957T (rs6277) polymorphisms. The DRD2/ANKK1-Taq Ia polymorphism is a restriction fragment polymorphism on chromosome 11 at q22–q23 (Noble 2003). Three genotypes of the dopamine DRD2/ANNK1-Taq Ia locus can be differentiated: The A1A1 genotype (also referred to as TT genotype), the A1A2 genotype (also referred to as TC genotype), and the A2A2 genotype (CC genotype). Due to the small prevalence of the A1A1 genotype (3 % in healthy Caucasians), A1A1 and A1A2 subjects are commonly grouped as A1 + subjects, whereas A2A2 subjects are referred to as A1− subjects. The prevalence of at least one A1 allele (A1 + group) leads to a reduction of up to 30 % in D_2 receptor density (e.g., Pohjalainen et al. 1998). The direct impact of the DRD2/ANKK1-Taq Ia polymorphism on D_2 receptor density has recently been questioned, since this single nucleotide polymorphism (SNP) is located about 10 kb downstream of the DRD2 gene, within a protein-coding region of the adjacent ANKK1 gene (Neville et al. 2004). Zhang et al. (2007) investigated 23 SNPs within the DRD2 gene and found a decreased expression of the short splice variant of the D2 receptor compared to the long splice variant, caused by two intronic SNPs (rs2283265 and rs1076560). Interestingly, in the study by Zhang et al. (2007) the minor alleles of the two SNPs show strong linkage disequilibrium with the A1 allele of the DRD2/ANKK1-Taq Ia polymorphism ($D' = 0.855$). These data indicate that, due to linkage, the DRD2/ANKK1-Taq Ia polymorphism is indeed a marker for dopamine receptor density.

The above-mentioned study by Reuter et al. (2013) replicates the findings by Zhong et al. (2010). In this study, the 4/4 genotype was associated with significantly higher minimum acceptable offers (20 % higher) as compared to carriers without the 4/4 genotype and, similar to the Zhong et al. study, the DRD4 exon III polymorphism was not related to the proposer behavior. In addition, Reuter et al. detected an association between a haplotype block, spanning 15 kb of the DRD2/ANKK1 region, consisting of the rs18000497 (also known as DRD2/ANKK1-Taq Ia) and rs2283265 SNPs, and the first mover offer in the UG. Carriers of at least one TT haplotype offered significantly less money in the UG (first mover-proposals) than carriers without a TT haplotype. This haplotype effect explains about 6 % of the variance in proposer behavior. The TT haplotype indicates that a subject has, at least on one chromosome, the minor alleles of both gene variants. Both minor alleles have been associated with lower DRD2 receptor density or decreased relative expression of DRD2s mRNA respectively (Pohjalainen et al. 1998; Zhang et al. 2007). On the other hand the second-mover-behavior was not related to genetic variations in the DRD2/ANKK1 region. One may wonder why there is a positive molecular genetic association with proposer behavior in the UG, when heritability estimates from a twin study suggest zero heritability for this distinct behavior (Reuter et al. 2013; Wallace et al. 2007). This is indeed a contradiction. However, the same standards apply to both quantitative genetic studies (twin studies) and molecular genetic studies: only an independent replication can reduce the probability of false positive or false negative findings. While the DRD4 results were successfully replicated, it remains to be clarified whether replication of the twin and the DRD2-proposer data is possible.

The DRD4 exon III effect is unambiguously related to fairness preference (responder behavior, i.e., size of minimum acceptable offer), whereas the DRD2/ANKK1 effect is far more complicated to explain. As described above, the proposer behavior constitutes a mixture of fairness preference and strategic consideration. Future experiments investigating the UG and the related dictator game (DG) in a within-subject design could disentangle these two facets of proposer behavior. In the DG two anonymous players interact with each other. However, the "interaction" is limited to the first mover. He has to split a certain amount of money between himself and the other player. In contrast to the payoff in the UG, which is dependent on the acceptance/rejection of the first mover's proposal by the second mover, the first mover in the DG (for a full description of the dictator game, please see Chap. 2 by Civai and Hawes in this book), also known as "the dictator", makes a proposal that is implemented independently of the second mover, i.e., the second mover cannot reject the dictator's offer. The identification of distinct gene loci related to the proposals in the UG and in the dictator game would help to clarify this issue.

4.2.2 *Dictator Game (DG)*

As mentioned above, the decision-making process of the dictator in the DG is assumed to reflect prosocial behavior (generosity; altruism; the dictator has not to fear punishment for an unfair offer) and is therefore more straightforward to interpret than the ambiguous first mover behavior in the UG. Whereas genetic association studies in the UG concentrate on dopaminergic gene loci, so far all available studies on the dictator game report associations with the classic prosocial hormones, vasopressin, and oxytocin. Vasopressin and oxytocin are phylogenetically very old neuropeptides, consisting of nine amino acids (they are thus referred to as nonapeptides). The two hormones differ only in two amino acids. Invertebrates also possess antecedent forms of oxytocin and vasopressin, which indicates the evolutionary importance of these two hormones. Their relevance for prosocial behavior was impressively demonstrated when differences in social behaviors and related biological systems between prairie (Microtus ochrogaster) and meadow voles (Microtus pennsylvanicus) were discovered (Lim et al. 2004). These two kinds of voles, although from the same species, are marked by extreme differences in pair bonding and fostering of offspring. Male prairie voles are monogamous, are engaged in parental care and exhibit a stable social structure, whereas the closely related meadow voles are solitary, polygamous, and do not care about rearing the offspring. Lim and colleagues discovered that the arginine vasopressin 1a receptor (AVPR1a) mRNA expression in the ventral forebrain is significantly higher in the monogamous prairie voles than in the promiscuous meadow voles. Interestingly, the insertion of a viral AVPR1a gene vector into the ventral forebrain of male meadow voles reduced promiscuity and increased preference for an individual partner. This same genetic region was the focus of interest in the first human genetic studies investigating social behavior by means of the dictator game. In a seminal study of 203 university students who played an online version of the DG, Knafo et al. (2008) demonstrated that allocation of funds in the Dictator Game (DG) was in part determined by length of the arginine vasopressin 1a (AVPR1a) RS3 promoter repeat region. The RS3 polymorphism is one of three promoter region microsatellites (the other two are $(GT)_{25}$ and RS1). Microsatellites represent small repeats of DNA bases, also known as simple sequence repeats (SSR). All three polymorphisms in this study were SSRs, i.e., repetitive DNA sequences characterized by a short base-pair motif that is repeated several to many times in tandem (e.g., CACACACA). It has been demonstrated that such SSRs can exert an influence on the translation and transcription of a gene (Beckmann et al. 2007). In the study by Knafo et al. dictators with short versions (308–325 bp) of the AVPR1a RS3 repeat allocated significantly smaller amounts of money to the other player than participants with long versions (327–343 bp). Additionally, the length of the RS3 repeat region was related to the amounts of AVPR1a mRNA expression in hippocampal postmortem specimens. Long AVPR1a RS3 repeats were associated with higher AVPR1a hippocampal messenger RNA levels than short RS3 repeats. This proof of functionality (although demonstrated in brain tissue of only $n = 15$ individuals) shows that the association of the AVPR1a RS3 polymorphism is not correlative in nature, but that this genetic variant has an influence on central nervous vasopressin receptor metabolism. Results indicate that higher mRNA levels of the APPR1a receptor are related to more generous behavior. In a further study using a modified version of the DG, the initial positive association between AVPR1a RS3 polymorphism and generosity was replicated in a sample of 3.5-year-old twins (Avinun et al. 2011). Children were required to allocate sticker charts instead of money, as in adult games. The findings demonstrate that prosocial behaviors develop earlier in life than is usually expected. Preschoolers at the age of 3.5 already behave socially.

A third study by the same group tested the influence of the oxytocin receptor gene on behavior in the DG (Israel et al. 2009). The human OXTR gene is located on chromosome 3p25.3 spanning approximately 19 kbp, and consists of three introns and four exons (Inoue et al. 1994). The rationale for extending the search for prosocial genes to the oxytocin system is straightforward. Independent studies have demonstrated that both vasopressin and oxytocin promote social behaviors in different mammalian species including humans (for an overview see Ebstein et al. 2009). Vasopressin and oxytocin exert overlapping functions, in part caused by mutual effects on their respective receptors. This is possible because both hormones seem to have comparable receptor affinities (e.g., Landgraf and Neumann 2004; Ragnauth et al. 2004; Pedersen and Boccia 2006).

Israel et al. (2009) investigated $N = 203$ Israeli students with the DG and genotyped for 15 single tagging SNPs across the oxytocin receptor gene (OXTR). Three of these SNPs showed an association with the allocations of the dictator, of which rs1042778 exhibited the strongest effect. Carriers of the G allele (genotypes

GG and GT) as compared to the T allele carriers (genotype TT), split their endowments more generously. Noteworthy, the positive effects of rs1042778 were replicated in an independent study of $N = 98$ female participants (Israel et al. 2009). Unfortunately these findings were not replicated in a sample of $N = 684$ Swedish twins (Apicella et al. 2010).

4.2.3 Trust

Trust is the prerequisite for all successful social interactions, no matter whether these take place on a dyadic (lovers, couples, friends) or on a more complex level (groups, companies, societies, countries) (e.g., Zak and Knack 2001). From an evolutionary perspective, trust is adaptive and has promoted the development of the human species (Kümmerli and Brown 2010). Of relevance to neuroeconomics, studies exist that suggest trust is related to the prosperity of societies (Knack and Keefer 1997). Trust behavior and altruism alike can be subsumed under the term "prosocial behaviors". Therefore, it is not surprising that similar biological systems are involved in altruism and trust behavior. A seminal study by Kosfeld et al. (2005) found that administration of nasal oxytocin increases trust in humans and that this association was not related to risk proneness. In a placebo-controlled pharmacological study, they used the Trust Game. This famous economic game is played in the laboratory and represents a one-shot interaction between two anonymous interaction partners. Participants were randomly assigned the roles of either an investor or a trustee. Both start the game with an initial endowment (monetary units, MU). In the first stage of the game, the investor can send any even number of MUs to the trustee. The amount sent by the investor is tripled by the experimenter. Then, the trustee decides how much money to return to the investor. Any back transfer between 0 MU and the maximum amount available to the trustee is feasible. To obtain measures of trust and trustworthiness (i.e., the size of the back transfer) for each participant, the trust game, in some versions, is played under role uncertainty. First, each player makes the decision in the role of the investor and subsequently must decide as a trustee, how many MUs to send back for any possible amount sent by the investor. After all decisions have been made, a randomization device determines which player in a given pair actually had the role of the investor and which player had the role of the trustee. Players' decisions were then implemented and subjects were paid according to their decisions (for a full description of the Trust Game and other economic games, please see Chap. 2 by Civai and Hawes in this book).

One may argue that the data of Kosfeld and colleagues show that administration of oxytocin increases trust, but that this laboratory experiment does not reflect a natural setting in which no drugs are consumed. However, in addition to the main effect of drug (placebo vs. oxytocin), the data also show large variability in the amount of trust in the placebo condition. A genetic approach may be adopted to investigate this variability in behavior under natural conditions (i.e., free of drug intake). A Swedish twin study of more than 2000 twin pairs showed that trust is indeed heritable. Heritability estimates were 0.33 for males and 0.39 for females (Oskarsson et al. 2012). However, it must be noted that trust was assessed via a self-report measure, using one single item ("Generally speaking, would you say that most people can be trusted, or that you can't be too careful in dealing with people?"). If oxytocin does play a role in trust behavior, as demonstrated by Kosfeld et al. (2005), then it is likely that genetic variants of the oxytocin system are related to individual differences in trust behavior. The first study investigating polymorphisms on the OXTR gene was conducted by Apicella et al. (2010). In their study (outlined above), they investigated both the DG and the Trust Game. The authors did not find an association between any of the nine OXTR SNPs under investigation and participant's behavior in the trust game. In another trust study by Krueger et al. (2012), a SNP in intron 3 of OXTR (rs53576), characterized by an adenine (A) to guanine (G) transition, was related to trust behavior but not to trustworthiness, in a sample of $N = 108$ US students. Participants with the homozygous GG genotype showed higher trust behavior than individuals with at least one A-allele (genotypes AG or AA). This finding is in line with another study reporting higher behavioral and dispositional empathy in carriers of the GG genotype as compared to the A-allele carriers (genotypes AA or AG; Rodrigues et al. 2009). Unfortunately there exists no test of functionality for the rs53576 SNP so far. In a trust study of $N = 100$ healthy Caucasian subjects, the present authors tested nine SNPs—mainly in the promoter region of the OXTR gene—for an association with trust behavior (Reuter et al. 2009). Haplotype analyses (a mathematical method from bioinformatics used in genetics to look for a putative linkage between several SNPs located in close proximity on the same chromosome) revealed a 6-SNP haplotype block or a 3-SNP haplotype block, dependent on the mathematical algorithm used to estimate the haplotypes. The smaller haplotype was included within the larger haplotype. The 6-SNP haplotype was obtained by a more liberal haplotype estimation method. Both haplotypes were significantly related to trust and explained between 6 and 10 % of the observed variance. It turned out that the SNP rs2268498, located in the promoter of OXTR, is the driving force in both haplotype blocks. In order to prove the functionality of these genetic variations, we conducted mRNA expression analyses in hippocampus specimens of pharmaco-therapy resistant temporal lobe epilepsy patients, from whom we also obtained blood samples for genotyping. Results showed that carriers of the haplotype that had been associated with high trust in the laboratory experiment showed a 50 % reduction in OXTR mRNA expression, as compared to the haplotypes that were related to low trust (Reuter et al. 2016). In line with the behavioral data, mRNA expression analyses showed that rs2268498 is the SNP that essentially influences functionality.

Tabak et al. (2014) used an iterated Prisoner's Dilemma in order to investigate behavioral and affective responses following a betrayal of trust. Participants were told to interact with an anonymous partner. In reality they played with/against a computer program that initially exhibited a cooperative tit-for-tat strategy and then began to defect over a series of consecutive trials. The dependent measures were the time point at which participants started to betray even though the computer was cooperative, and when they started to retaliate to unfair decisions made by the computer. Emotional responses to betrayal and retaliation were also assessed. Three different haplotypes on the OXTR were identified that were related either to faster retaliation after betrayal, or with high or low post-betrayal satisfaction, respectively.

4.2.4 Altruism

Many neuroeconomic studies using the dictator game interpret the first mover's (i.e., the dictator's) decision as a marker for generosity or altruism. Therefore, the findings reported in the above section on the dictator game can also be viewed as molecular genetic association studies on altruism. However, the question arises of whether the decision-making in the DG is really altruistic behavior. There exist in the literature countless definitions of altruism. One definition, encompassing many different facets of the concept, defines altruism as the selfless concern for the welfare of others. However, there is a great debate in the literature as to whether true altruism really exists (Fehr and Fischbacher 2003). Pure altruism is giving without regard to reward or the benefits of recognition and need. People who doubt the existence of pure altruism argue that helping others is intrinsically rewarding for altruistic persons and therefore they are pursuing their personal interest, rather than the interests of others. In other words helping others makes them feel good, e.g., by enhancing their self-esteem. This line of argument overcomes the apparent incompatibility with economic concepts like the *homo economicus,* postulating that humans are selfish, rational beings, motivated merely through self-interest (Ng and Tseng 2008). In order to disentangle genetic and environmental (e.g., upbringing, education, etc.) influences contributing to the widely acknowledged variability in altruistic behavior, twin studies are considered the gold standard method. These behavioral genetic studies mostly rely on self-report data: A twin study by Rushton et al. (1986) on 563 pairs of monozygous (MZ) and dizygous (DZ) twins using an altruism scale, reported that 50 % of the variance in altruism was due to genes and the remaining 50 % due to environmental factors. Noteworthy, the total environmental variance came from nonshared environmental sources (e.g., epigenetic factors, peer influences, etc.), not from shared ones. A second study by Matthews et al. (1981) found 72 % heritability for a self-report adjective checklist measure of empathy in 114 MZ and 116 DZ middle-aged male twins. In an additional twin study of 322 pairs of twins, Rushton (2004) replicated the strong genetic effects on prosocial behavior. He reported heritability estimates of 0.40 for females and of 0.50 for males for social responsibility. In contrast to the study from 1986, shared environmental factors accounted for about 23 % of this variance, whereas in the previous studies, environmental effects were exclusively due to nonshared environmental factors.

Two independent groups have investigated the role of the DRD4 Exon 3 polymorphism on self-report measures of altruism and arrived at similar results: Bachner-Melman et al. (2005) observed significantly higher altruism scores in carriers of the 4-repeat allele and Anacker et al. (2013) found significantly higher altruism scores in carriers of the 4/4 genotype. Mertins et al. (2011) used a repeated public goods game (for a detailed description of the public goods game please see Chap. 2 in this book by Civai and Hawes) in order to study genetic effects on altruistic behavior and found evidence for the role of a functional repeat polymorphism in the promoter region of the monoamine oxidase A (MAO-A) gene. In the public goods game participants have to decide whether to give money to a community pool or to take their money for themselves, while also profiting from the donations of the other community members. The money in the community pool is multiplied by the experimenter and shared equally among the community members, irrespective of whether they have contributed. Noteworthy, the return to the individual is always smaller than their original contribution. Therefore, investing in the community pool is considered an altruistic behavior. In addition, the public goods game was not conducted as a single shot game, but in a dynamic setting of increasing information about the other players' contributions. The findings show that male participants carrying the low activity alleles were significantly less altruistic (i.e., gave less money to the community pool) than carriers of the high activity alleles. Interestingly the genetic effects became smaller with increasing information about the other players' behavior in the public goods game. However, in female participants carrying two low activity alleles of the MAO-A polymorphism, the opposite behavior was observed. In females, the low activity alleles were related to higher altruistic decisions. Gender differences with respect to the MAO-A promoter repeat polymorphism must be interpreted with caution, given the fact that the MAO-A gene is located on the X-chromosome. This means that men have only one allele at the polymorphic region (as they possess only one X-chromosome), whereas women have a genotype with an allele on each of their two X-chromosomes. The polymorphism, which is located 1.2 kb upstream of the MAO-A coding sequences, consists of a 30 bp repeated sequence present in 3, 3.5, 4, or 5 copies. Functional studies have shown that the 3.5 and 4 copy alleles (i.e., the high activity alleles) are transcribed 2–10 times more efficiently than the 3 or 5 copy alleles (i.e., the low activity alleles) (Sabol et al. 1998). However, it is not yet proven how strong the functional impact of the second allele is in women compared to the one allele in men. At least the location of the MAO-A polymorphism on the X-chromosome could be a potential explanation for contradictory results based on gender.

In a laboratory study with $N = 101$ healthy Caucasian participants, we investigated altruism in a donation experiment (Reuter et al. 2011). Altruism was not assessed by a classic economic game, but by the amount of money donated to a poor child in a developing country, after participants earned money through participation in two straining computer experiments. The focus was on the dopaminergic system, because dopamine has proven to be—in addition to the prominent neuropeptides oxytocin and vasopressin—crucial for prosocial behaviors like cooperation, attachment style, mentoring, paternal parenting, and positive emotionality, to name but a few (Lucht et al. 2006; Reuter et al. 2006; Shanahan et al. 2007; Gillath et al. 2008; Walter et al. 2011). The functional Val158Met polymorphism was significantly associated with the size of the donations. Carriers of at least one Val-allele donated about twice as much money, compared to those participants without a Val-allele (genotype Met/Met). The functional COMT SNP together with the personality trait cooperativeness, as assessed by the Temperament and Character Inventory (TCI; Cloninger et al. 1993), explained about 15 % of the variance in donation size. The Val-allele, which is characterized by high altruism and also by high catabolic activity with respect to the degradation of dopamine, is the ancestral allele at the polymorphic region in codon 158 on the COMT gene. Today we have equal allele frequencies for the Val- and the Met-allele in Caucasian populations, indicating that the allele frequency of the Met-allele must have dramatically increased over the last 100,000 years. Therefore, it can be concluded that altruism (as represented by the Val-allele) is less adaptive than selfishness (represented by the absence of the Val-allele; i.e., the presence of the Met/Met genotype). However, from the perspective of evolutionary psychology, altruism was the prerequisite for mankind's' success and increases inclusive fitness. Eldakar and Wilson (2008) contributed an interesting model to this debate. Their model identifies a strategy called "selfish punisher" that involves behaving selfishly in first-order interactions and "altruistically" in second-order interactions by punishing other selfish individuals. The act of punishing other selfish individuals can be considered a form of second-order altruism. The idea behind this is that selfish punishers limit selfish behavior and thereby enable altruists to coexist in a stable equilibrium.

Strobel et al. (2011) investigated a related construct named altruistic punishment (for more details, please see Chap. 12 on altruistic punishment by Alexander Strobel, in this book). Altruistic punishment refers to a behavior in which individuals punish others for unfair actions (e.g., free-riding, defection, noncooperation) at a cost to themselves and for the sake of others or the community. According to this definition, altruistic punishment indeed meets the criterion for altruism. The authors conducted a combined fMRI-genetics study. Besides the major finding that brain regions related to cognitive and affective functioning (DLPFC, ACC, insula) were essentially involved in altruistic punishment, they reported an allele load effect (Met/Met > Val/Met > Val/Val) with respect to the BOLD-response (contrast punishment vs. nonpunishment) in the cingulate gyrus, the nucleus accumbens and the insula. Strobel et al. discuss this finding as suggestive in favor of the hypothesis that the evolutionarily younger, Met-allele, is favored in societies where reciprocal behavior—in this case punishment of unfair behavior—is the social norm. Given that, the Met/Met genotype could be related to less altruism (Reuter et al. 2013), but in the study by Strobel et al. to more pronounced brain activity during altruistic punishment, it cannot be taken for granted that altruism and altruistic punishment are highly correlated, although both constructs represent prosocial behavior. Future studies should test this hypothesis in a within-subject design.

4.3 Summary and Perspectives

The present chapter on genes and human decision-making gives an overview of prominent genetic studies that have been published in the field of neuroeconomics to date. It becomes apparent that all studies adopt a candidate gene approach, i.e., they derived their hypotheses for testing specific gene loci from the general literature on genes and human decision-making. Besides the oxytocin and the vasopressin receptor genes, dopaminergic gene loci have been a particular focus of interest. Positive associations between these polymorphisms and economic constructs like trust, altruism, generosity and reciprocity, support results from quantitative genetic studies (twin studies) demonstrating that human decision-making in economic contexts is substantially influenced by genetic factors. However, the explained variance in the phenotypes under investigation is rather small, as is common in genetic association studies. New techniques in molecular genetics like genome wide association studies (GWAS; for an overview please see the methods chapter on genetics in this book), provide the hope of unraveling new gene loci associated with economic decision-making.

As outlined in the introduction to this chapter, the genetic approach in neuroeconomics offers the possibility of assessing biological variables in ecologically valid situations. This is a clear advantage over MRI studies, which provide a rather artificial—and simultaneously, extreme—laboratory setting. Time will tell whether interest in this new and promising genetic approach will increase in the field of neuroeconomics.

References

Anacker K, Enge S, Reif A et al (2013) Dopamine D4 receptor gene variation impacts self-reported altruism. Mol Psychiatry 18(4):402–403

Apicella CL, Cesarini D, Johannesson M et al (2010) No association between oxytocin receptor (OXTR) gene polymorphisms and experimentally elicited social preferences. PLoS One 5(6): e11153. doi:10.1371/journal.pone.0011153

Asghari V, Sanyal S, Buchwaldt S et al (1995) Modulation of intracellular cyclic AMP levels by different human dopamine D4 receptor variants. J Neurochem 65(3):1157–1165

Avinun R, Israel S, Shalev I et al (2011) AVPR1A variant associated with preschoolers' lower altruistic behavior. PLoS One 6(9):e25274. doi:10.1371/journal.pone.0025274

Bachner-Melman R, Gritsenko I, Nemanov L et al (2005) Dopaminergic polymorphisms associated with self-report measures of human altruism: a fresh phenotype for the dopamine D4 receptor. Mol Psychiatry 10(4):333–335

Bechara A, Tranel D, Damasio H (2000) Characterization of the decision-making deficit of patients with ventromedial prefrontal cortex lesions. Brain 123(11):2189–2202

Beckmann J, Estivill X, Antonarakis S (2007) Copy number variants and genetic traits: closer to the resolution of phenotypic to genotypic variability. Nat Rev Genet 8(8):639–646

Camerer CF (2003) Behavioral game theory: experiments on strategic interaction. Princeton University Press, Princeton

Cloninger CR, Svrakic DM, Przybeck TR (1993) A psychobiological model of temperament and character. Arch Gen Psychiatry 50(12):975–990

da Rocha FF, Malloy-Diniz L, Lage NV et al (2008) Decision-making impairment is related to serotonin transporter promoter polymorphism in a sample of patients with obsessive-compulsive disorder. Behav Brain Res 195(1):159–163

Ebstein RP, Novick O, Umansky R et al (1996) Dopamine D4 receptor (D4DR) exon III polymorphism associated with the human personality trait of novelty seeking. Nat Genet 12 (1):78–80

Ebstein RP, Israel S, Lerer E et al (2009) Arginine vasopressin and oxytocin modulate human social behavior. Ann N Y Acad Sci 1167:87–102

Eisenegger C, Knoch D, Ebstein RP et al (2010) Dopamine receptor D4 polymorphism predicts the effect of L-DOPA on gambling behavior. Biol Psychiatry 67(8):702–706

Eldakar OT, Wilson DS (2008) Selfishness as second-order altruism. Proc Natl Acad Sci USA 105 (19):6982–6986

Ert E, Yechiam E, Arshavsky O (2013) Smokers' decision making: more than mere risk taking. PLoS One 8(7):e68064. doi:10.1371/journal.pone.0068064

Fehr E, Fischbacher U (2003) The nature of human altruism. Nature 425(6960):785–791

Gillath O, Shaver PR, Baek JM et al (2008) Genetic correlates of adult attachment style. Pers Soc Psychol Bull 34(10):1396–1405

Goldberg TE, Weinberger DR (2008) Genes and the parsing of cognitive processes. Trends Cogn Sci 8(7):325–335

He Q, Xue G, Chen C et al (2010) Serotonin transporter gene-linked polymorphic region (5-HTTLPR) influences decision making under ambiguity and risk in a large Chinese sample. Neuropharmacology 59(6):518–526

Inoue T, Kimura T, Azuma C et al (1994) Structural organization of the human oxytocin receptor gene. J Biol Chem 269(51):32451–32456

Israel S, Lerer E, Shalev I et al (2009) The oxytocin receptor (OXTR) contributes to prosocial fund allocations in the dictator game and the social value orientations task. PLoS One 4(5):e5535. doi:10.1371/journal.pone.0005535

Kaplan MF, Miller CE (1987) Group decision making and normative versus informational influence: effects of type of issue and assigned decision rule. J Pers Soc Psychol 53(2):306–313

Knack S, Keefer P (1997) Does social capital have an economic payoff? A cross-country investigation. Q J Econ 112(4):1251–1288

Knafo A, Israel S, Darvasi A et al (2008) Individual differences in allocation of funds in the dictator game associated with length of the arginine vasopressin 1a receptor RS3 promoter region and correlation between RS3 length and hippocampal mRNA. Genes Brain Behav 7 (3):266–275

Kosfeld M, Heinrichs M, Zak PJ et al (2005) Oxytocin increases trust in humans. Nature 435 (7042):673–676

Krebs RM, Schott BH, Düzel E (2009) Personality traits are differentially associated with patterns of reward and novelty processing in the human substantia nigra/ventral tegmental area. Biol Psychiatry 65(2):103–110

Krueger F, Parasuraman R, Iyengar V et al (2012) Oxytocin receptor genetic variation promotes human trust behavior. Front Hum Neurosci 6:4

Kümmerli R, Brown SP (2010) Molecular and regulatory properties of a public good shape the evolution of cooperation. Proc Natl Acad Sci USA 107(44):18921–18926

Lachman HM, Papolos DF, Saito T et al (1996) Human catechol-*O*-methyltransferase pharmacogenetics: description of a functional polymorphism and its potential application to neuropsychiatric disorders. Pharmacogenetics 6(3):243–250

Lage GM, Malloy-Diniz LF, Matos LO et al (2011) Impulsivity and the 5-HTTLPR polymorphism in a non-clinical sample. PLoS One 6(2):e16927. doi:10.1371/journal.pone.0016927

Landgraf R, Neumann ID (2004) Vasopressin and oxytocin release within the brain: a dynamic concept of multiple and variable modes of neuropeptide communication. Front Neuroendocrinol 25(3–4):150–176

Lesch KP, Bengel D, Heils A et al (1996) Association of anxiety-related traits with a polymorphism in the serotonin transporter gene regulatory region. Science 274(5292):1527–1531

Lim MM, Wang Z, Olazábal DE, Ren X, Terwilliger EF, Young LJ (2004) Enhanced partner preference in a promiscuous species by manipulating the expression of a single gene. Nature 429(6993):754–757

Lucht M, Barnow S, Schroeder W et al (2006) Negative perceived paternal parenting is associated with dopamine D2 receptor exon 8 and GABA(A) alpha 6 receptor variants: an explorative study. Am J Med Genet B Neuropsychiatr Genet 141B(2):167–172

Matthews KA, Batson CD, Horn J et al (1981) " Principles in his nature which interest him in the fortune of others…": the heritability of empathic concern for others. J Pers 49(3):237–247

Mertins V, Schote AB, Hoffeld W et al (2011) Genetic susceptibility for individual cooperation preferences: the role of monoamine oxidase A gene (MAOA) in the voluntary provision of public goods. PLoS One 6(6):e20959. doi:10.1371/journal.pone.0020959

Montag C, Jurkiewicz M, Reuter M (2012) The role of the catechol-*O*-methyltransferase (COMT) gene in personality and related psychopathological disorders. CNS Neurol Disord: Drug Targets 11(3):236–250

Neville MJ, Johnstone EC, Walton RT (2004) Identification and characterization of ANKK1: a novel kinase gene closely linked to DRD2 on chromosome band 11q23.1. Hum Mutat 23 (6):540–545

Ng ICL, Tseng LM (2008) Learning to be sociable: the evolution of homo economicus. Am J Econ Soc 67(2):265–286

Nikolaidis A, Gray JR (2010) ADHD and the American Journal of Med Genet B DRD4 exon III 7-repeat polymorphism: an international meta-analysis. Soc Cogn Affect Neurosci 5(2–3):188–193

Noble EP (2003) D2 dopamine receptor gene in psychiatric and neurologic disorders and its phenotypes. Am J Med Genet B Neuropsychiatr Genet 116B(1):103–125

Oskarsson S, Dawes C, Johannesson M et al (2012) The genetic origins of the relationship between psychological traits and social trust. Twin Res Hum Genet 15(1):21–33

Pedersen CA, Boccia ML (2006) Vasopressin interactions with oxytocin in the control of female sexual behavior. Neuroscience 139(3):843–851

Persky J (1995) Retrospectives: the ethology of homo economicus. J Econ Perspect 9(2):221–231

Pohjalainen T, Rinne JO, Någren K et al (1998) The A1 allele of the human D2 dopamine receptor gene predicts low D2 receptor availability in healthy volunteers. Mol Psychiatry 3(3):256–260

Ragnauth AK, Goodwillie A, Brewer C et al (2004) Vasopressin stimulates ventromedial hypothalamic neurons via oxytocin receptors in oxytocin gene knockout male and female mice. Neuroendocrinology 80(2):92–99

Reuter M, Hennig J (2005) Association of the functional catechol-*O*-methyltransferase VAL158MET polymorphism with the personality trait of extraversion. NeuroReport 16 (10):1135–1138

Reuter M, Schmitz A, Corr P et al (2006) Molecular genetics support Gray's personality theory: the interaction of COMT and DRD2 polymorphisms predicts the behavioural approach system. Int J Neuropsychopharmacol 9(2):155–166

Reuter M, Montag C, Altmann S (2009) Genetically determined differences in human trust behavior: the role of the oxytocin receptor gene. In: NeuroPsychoEconomics, conference proceedings

Reuter M, Frenzel C, Walter NT et al (2011) Investigating the genetic basis of altruism: the role of the COMT Val158Met polymorphism. Soc Cogn Affect Neurosci 6(5):662–668

Reuter M, Felten A, Penz S et al (2013) The influence of dopaminergic gene variants on decision making in the ultimatum game. Front Hum Neurosci 7:242

Reuter M, Montag C, Altmann S, Bendlow F, Elger C, Kirsch P, Becker A, Schoch-McGovern S, Simon M, Weber B, Felten A (2016) Functional characterization of an oxytocin receptor gene variant (rs2268498) previously associated with socialcognition by expression analysis in vitro and in human brain biopsy. Soc Neurosci. Jul 27:1–8.

Rodrigues SM, Saslow LR, Garcia N et al (2009) Oxytocin receptor genetic variation relates to empathy and stress reactivity in humans. Proc Natl Acad Sci USA 106(50):21437–21441

Rushton JP (2004) Genetic and environmental contributions to pro-social attitudes: a twin study of social responsibility. Proc Biol Sci 271(1557):2583–2585

Rushton JP, Fulker DW, Neale MC et al (1986) Altruism and aggression: the heritability of individual differences. J Pers Soc Psychol 50(6):1192–1198

Sabol SZ, Hu S, Hamer D (1998) A functional polymorphism in the monoamine oxidase A gene promoter. Hum Genet 103(3):273–279

Shanahan MJ, Erickson LD, Vaisey S et al (2007) Helping relationships and genetic propensities: a combinatoric study of DRD2, mentoring, and educational continuation. Twin Res Hum Genet 10(2):285–298

Smillie LD (2013) Extraversion and reward processing. Curr Dir Psychol Sci 22(3):167–172

Stoltenberg SF, Vandever JM (2010) Gender moderates the association between 5-HTTLPR and decision-making under ambiguity but not under risk. Neuropharmacology 58(2):423–428

Strobel A, Wehr A, Michel A et al (1999) Association between the dopamine D4 receptor (DRD4) exon III polymorphism and measures of Novelty Seeking in a German population. Mol Psychiatry 4(4):378–384

Strobel A, Zimmermann J, Schmitz A et al (2011) Beyond revenge: neural and genetic bases of altruistic punishment. Neuroimage 54(1):671–680

Tabak BA, McCullough ME, Carver CS et al (2014) Variation in oxytocin receptor gene (OXTR) polymorphisms is associated with emotional and behavioral reactions to betrayal. Soc Cogn Affect Neurosci 9(6):810–816

Tellegen A, Lykken DT, Bouchard TJ Jr et al (1988) Personality similarity in twins reared apart and together. J Pers Soc Psychol 54(6):1031–1039

van der Plas EA, Crone EA, van den Wildenberg WP et al (2009) Executive control deficits in substance-dependent individuals: a comparison of alcohol, cocaine, and methamphetamine and of men and women. J Clin Exp Neuropsychol 31(6):706–719

Van Tol HH, Wu CM, Guan HC et al (1992) Multiple dopamine D4 receptor variants in the human population. Nature 358(6382):149–152

Wacker J, Mueller EM, Hennig J et al (2012) How to consistently link extraversion and intelligence to the catechol-*O*-methyltransferase (COMT) gene: on defining and measuring psychological phenotypes in neurogenetic research. J Pers Soc Psychol 102(2):427–444

Wallace B, Cesarini D, Lichtenstein P et al (2007) Heritability of ultimatum responder behavior. Proc Natl Acad Sci USA 104(40):15631–15634

Walter NT, Markett SA, Montag C et al (2011) A genetic contribution to cooperation: dopamine-relevant genes are associated with social facilitation. Soc Neurosci 6(3):289–301

Wickelgren WA (1977) Speed-accuracy tradeoff and information processing dynamics. Acta Psychol 41(1):67–85

Zak PJ, Knack S (2001) Trust and growth. Econ J 111(470):295–321

Zaleskiewicz T (2001) Beyond risk seeking and risk aversion: personality and the dual nature of economic risk taking. Eur J Pers 15:105–122

Zhang Y, Bertolino A, Fazio L et al (2007) Polymorphisms in human dopamine D2 receptor gene affect gene expression, splicing, and neuronal activity during working memory. Proc Natl Acad Sci USA 104(51):20552–20557

Zhong S, Israel S, Shalev I et al (2010) Dopamine D4 receptor gene associated with fairness preference in ultimatum game. PLoS One 5(11):e13765. doi:10.1371/journal.pone.0013765

Rodrigues SM, Saslow LR, [illegible] (2009) Oxytocin [illegible] variation relates to empathy and stress reactivity [illegible] Proc Natl Acad [illegible]:21437–21441
Rushton JP (2004) Genetic and [illegible] contributions [illegible] twin study of social responsibility. Proc Biol Sci 27[illegible]
Rushton JP, Fulker DW, Neale [illegible] (1986) Altruism [illegible] heritability of individual differences [illegible] Psychol 50(6):1192–1198
Sabol SZ, Hu S, Hamer D (19[illegible]) [illegible] of polymorphi[illegible] monoamine oxidase A gene promoter. Hum Genet 103(3):273–276
Shanahan MJ, Erickson LD, [illegible] Helping relat[illegible] a combinatorial study of [illegible] and educational [illegible] Res Hum Genet 11(2):293–298
Smillie LD (2013) Extraversion [illegible] processing. Curr [illegible] 22(3):167–172
Stallenberg [illegible] Yudkovsky [illegible] reduces the [illegible] between [illegible] and [illegible] making mode [illegible] Neuro[illegible]
Strobel A, Wehr A, Michel A, [illegible] Association between the [illegible] D4 receptor (DRD4) exon III polymorphism and measures of Novelty Seeking [illegible] population. Mol Psychiatry 4(4):378–384
Strobel A, Zimmermann J, Schmitz A, [illegible] (2011) Beyond [illegible] and genetic bases of altruistic punishment. Neuroimage [illegible]
Tabor [illegible] McCullough ML, [illegible] (2012) Variation [illegible] gene (OXTR) polymorphisms [illegible] and behavior [illegible] Soc Cogn Affect Neurosci [illegible]
Tellegen A, Lykken DT, Bouchard [illegible] (1988) Personality [illegible] twins reared apart and together. J Pers Soc Psychol 54(6):1031–1039
van der Zwaluw CS, [illegible] (20[illegible]) [illegible] substance-dependent individuals [illegible] alcohol [illegible] and [illegible] Psychiatry 17(3):[illegible]
Wang [illegible] HB, Wu GM, [illegible] (199[illegible]) Multiple [illegible] in the human population. Nature [illegible]
Wacker J, Mueller EM, [illegible] (20[illegible]) How to consistently [illegible] and intelligence in [illegible] catechol-O-methyltransferase (COMT) [illegible] measuring psychological phenomena [illegible] research [illegible]
Wallace B, Cesarini D, Lichtenstein P, [illegible] (2007) Heritability of [illegible] behavior. Proc Natl Acad Sci USA 104[illegible]
Walter NT, Markett SA, Montag C, Reuter M (2011) A genetic [illegible] cooperation: [illegible] in the [illegible] associated with [illegible]:[illegible]–301
Wachtel [illegible] WA (19[illegible]) [illegible] and [illegible] Acta Psychol [illegible]
Zak PJ, Knack S (2001) Trust and growth. Econ J 111(470):[illegible]
Zak [illegible] (20[illegible]) Beyond [illegible] half of the neuronal [illegible] Hum [illegible]:[illegible]
Zhong S, [illegible] A, [illegible] (20[illegible]) [illegible] polymorphisms [illegible] receptor [illegible] gene expression [illegible] Natl Acad Sci USA 10[illegible]
Zhong S, Israel S, Xue H, [illegible] (20[illegible]) A neurochemical approach to [illegible] with fairness preference in [illegible]

Chapter 5
Monoamines and Decision-Making Under Risks

Hidehiko Takahashi

Abstract Past neuroeconomics studies using neurophysiology methods (mainly fMRI) have revealed the neural basis of "boundedly rational" or "irrational" decision-making that violates normative economics theory. It is expected that the field of neuroeconomics will be merged with neurotransmitter research and clinical neuroscience. Here, we provide an overview of recent molecular neuroimaging studies to understand how central monoamine transmission is related to "irrational" decision-making. Empirical evidence suggests that central dopamine transmission might be related to distortion of subjective reward probability and noradrenaline and serotonin transmission might influence aversive emotional reaction to financial loss. Positron emission tomography (PET) is a powerful tool to understand the neurochemical basis of decision-making in vivo in human. This approach seems to be a promising direction to understand the neurobiology of impaired decision-making in neuropsychiatric disorders and may help to develop novel pharmacotherapy for them.

5.1 Introduction

In normative economics theory, decision-makers are assumed to be "rational" and purely self-interested. However, we are not always rational, and sometimes show "boundedly rational" or "irrational" decision-making. Laboratory and field evidence from behavioral economics has shown that decision-makers systematically depart from normative theory (Camerer and Loewenstein 2004; Camerer and Fehr 2006; Takahashi et al. 2012b). Because behavioral economics deals with the effects of cognitive and emotional factors on economic decisions, not surprisingly, it has been merged with neuroscientific studies about cognition and emotions, and this interdisciplinary approach is called neuroeconomics.

H. Takahashi (✉)
Department of Psychiatry, Kyoto University Graduate School of Medicine,
54 Shogoin, Kawara-Cho, Sakyo-Ku, Kyoto 606-8507, Japan
e-mail: hidehiko@kuhp.kyoto-u.ac.jp

M. Reuter and C. Montag (eds.), *Neuroeconomics*, Studies in Neuroscience, Psychology and Behavioral Economics, DOI 10.1007/978-3-642-35923-1_5

Over the last decade, neuroeconomics utilizing neurophysiology methods (fMRI or EEG) has been advanced, revealing the neural basis of "boundedly rational" or "irrational" decision-making that violates normative theory. Past neuroeconomics studies have demonstrated that, in addition to cortical regions such as the prefrontal cortex (PFC), subcortical emotion-related brain areas are involved in "irrational" decision-making (Sanfey et al. 2003; Singer et al. 2006; Takahashi et al. 2009; Yamada et al. 2012). The next question is how neurotransmission modulate these central processes (Trepel et al. 2005; Rangel et al. 2008). In this chapter, we provide an overview of recent research to understand the neurochemical basis of "irrational" decision-making under risks.

5.2 Nonlinear Probability Weighting

One type of systematic departure from normative economic theory is that subjective weights on probabilities appear to be nonlinear. Decision-makers often overestimate low probabilities (e.g., playing lotteries) and underestimate high probabilities. A leading alternative to normative theory (expected utility theory) is the prospect theory (Tversky and Kahneman 1992). One of the important components of the prospect theory is nonlinear probability weighting, where objective probabilities, p, are transformed nonlinearly into decision weights $w(p)$ by a weighting function (Fig. 5.1).

Experimental studies suggest that the weighting function is regressive, asymmetric, and inverse S-shaped, crossing the diagonal from above at an inflection point (around 1/3) where $p = w(p)$. Although several functions have been proposed to express nonlinear probability weighting, the one-parameter function derived axiomatically by Prelec (1998), $w(p) = \exp\{-(\ln(1/p))^{\alpha}\}$ with $0 < \alpha < 1$, is widely used. In an inverse S-shaped nonlinear weighting function, low probabilities are overweighted and moderate to high probabilities are underweighted. The function neatly explains the typically observed pattern of risk-seeking for low probability gain and risk-aversion toward high-probability gain.

Paulus and Frank (2006) investigated the neural substrates that are related to nonlinear probability transformation using fMRI with a certainty equivalent procedure. During this procedure, a gamble's certainty equivalent, the amount of sure payoff at which a player is indifferent between the sure payoff and the gamble, was determined. The authors found that differential anterior cingulate activation during estimation of high probabilities relative to low probabilities was positively correlated with Prelec's nonlinearity parameter α across subjects. Another fMRI study with risks of negative outcomes (electric shocks) found similar nonlinear response in brain regions including the caudate/subgenual anterior cingulate (Berns et al. 2008). However, it was reported that the dorsolateral PFC was involved in overweighting low probabilities and underweighting high probabilities, and that the ventral frontal regions showed the opposite pattern (Tobler et al. 2008). More recently, Hsu et al. (2009) reported that the degree of nonlinearity in the neural

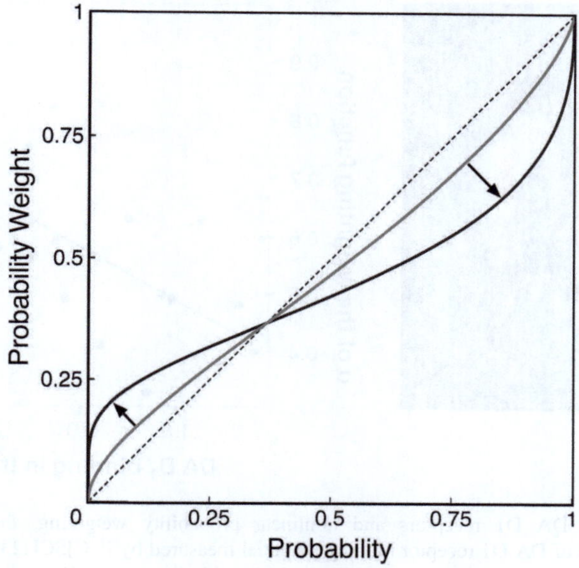

Fig. 5.1 Hypothesized model showing the contribution of central DA transmission to nonlinear probability weighting. A smaller value of α (closer to 0) means a more nonlinear inflected weighting function and a higher value (closer to 1) means a more linear weighting function. At $\alpha = 1$ the function is linear (*dashed line*). DA transmission might play a central role in distorting probability weighting function nonlinearly. Excessive DA transmission might cause exaggerated overestimation of low probability and underestimation of moderate to high probabilities (*black arrow*)

response to anticipated reward in the striatum reflected the nonlinearity parameter as estimated behaviorally. The discrepancies regarding the loci of activation are thought to stem from differences in the task (probability range, context, etc.) and analysis of parameter estimation. However, it seems to be promising to investigate the role of the dopamine (DA) system in nonlinear probability weighting, because DA is linked to risk-seeking behavior (Leyton et al. 2002) and excessive DA release was observed in pathological gambling in Parkinson's disease patients (Steeves et al. 2009). Trepel et al. (2005) hypothesized in an insightful review that DA transmission in the striatum might be involved in shaping probability weighting. Taking advantage of in vivo molecular neuroimaging [positron emission tomography (PET)], we investigated the relationship between central DA transmission and nonlinear probability weighting.

Certainty equivalents were determined outside the PET scanner, and we estimated probability weighting using the Prelec's one-parameter function. There was positive correlation between striatal D1 receptor binding measured by [^{11}C]SCH23390 PET and the nonlinearity parameter α of weighting function (Fig. 5.2) (Takahashi et al. 2010a). No correlation was found between D2 receptor binding measured by [^{11}C]raclopride PET and nonlinearity parameter α. That is, subjects with lower striatal D1 receptor binding tend to show more pronounced

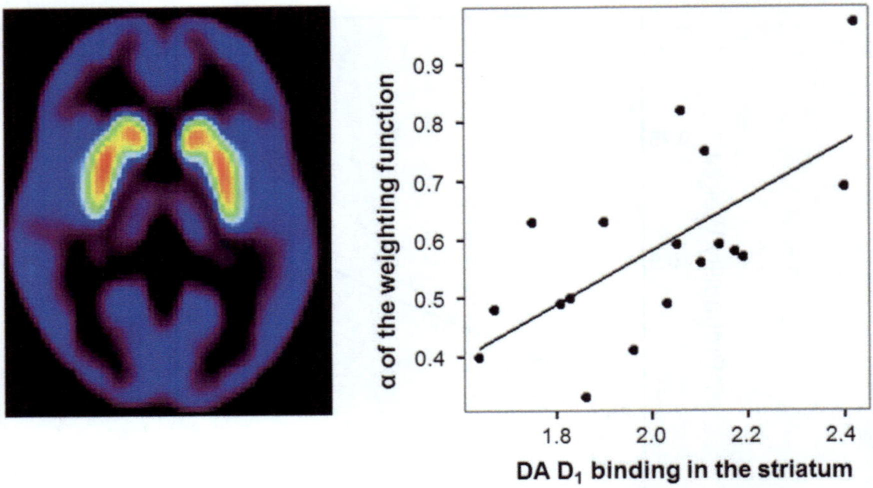

Fig. 5.2 Striatal DA D1 receptors and nonlinear probability weighting. *Left panel* shows parametric image of DA D1 receptor binding potential measured by $[^{11}C]$SCH23390. *Right panel* shows positive correlation between striatal D1 receptor binding and nonlinearity parameter α of weighting function

overestimation of low probabilities and underestimation of high probabilities. $[^{11}C]$SCH23390 is a selective radioligand for D1 receptors, but it also has some affinity for serotonin (5-HT) 2A receptors. 5HT2A receptor density in the striatum is negligible compared to D1 receptor density. However, 5HT2A receptor density is never negligible in extrastriatal regions (Ekelund et al. 2007). Future studies with a more selective radioligand are recommended to test the role of extrastriatal (cortical) D1 receptors in nonlinear weighting.

Mis-estimation of probabilities, especially of low probabilities, might be related to some problematic behaviors in neuropsychiatric disorders. Clinical studies have reported the emergence of pathological gambling in Parkinson's disease patients taking DA agonist medication (Gallagher et al. 2007; Dagher and Robbins 2009), and such patients showed increased DA release in the ventral striatum measured by $[^{11}C]$raclopride PET during gambling (Steeves et al. 2009). Although pathological gambling is a heterogeneous disorder and cannot be solely attributed to misestimating probability, these clinical observations can lead to the hypothesis that excessive DA transmission might cause distortion of subjective probability weights for gains (positive outcomes) (Fig. 5.1) (Takahashi 2012, 2013). However, nonlinear probability weighting is a combination of risk-seeking (overestimation of low probability) and risk-aversion (underestimation of high probability). In fact, a recent study reported that pathological gamblers demonstrated an overall shift towards risk, rather than excessive distortion of nonlinear probability weighting in decision-making under risks (Ligneul et al. 2012). From a psychological point of view, the overweighting of low probability gains may reflect the hope of winning, and it is reasonable to link DA tone and an overall shift towards risk.

Underweighting of high-probability gains may reflect the fear of losing a "near sure thing". Thus, the shape of weighting function, especially in the high-probability portion, should be determined by multiple neurotransmitters other than DA (Takahashi et al. 2010b), such as 5-HT (Takahashi et al. 2005) and norepinephrine (NE) (Onur et al. 2009), which are also known to modulate the emotional reaction of fear. Furthermore, the role of neurotransmitters in nonlinear probability weighting for losses (negative outcomes) should be tested as well.

5.3 Loss Aversion

Pain derived from losing a certain amount of money appears to be greater than the pleasure derived from gaining the equivalent amount. Imagine having a chance to participate in a coin-flip game of chance. Using a fair coin, if the outcome is heads, you will win \$100, and if the outcome is tails, you will lose \$100. Are you willing to participate in this gamble? Typically, most people would say "no". Well, how about the following gamble? If the winning prize is increased to \$200, while the potential loss remains \$100. In this case, some people would say "yes". This means that, typically, losses have at least twice the impact of equivalent gains, a tendency called loss aversion (Tversky and Kahneman 1992). Loss aversion is a robust phenomenon, and many laboratory and field studies have found evidence in monkeys for food rewards, and in humans for financial outcomes, features of consumer goods, food rewards, game show winnings, and apartment sales (Camerer and Loewenstein 2004; Chen et al. 2006; Knutson et al. 2007). In prospect theory, this is expressed by a value function of losses that is steeper than that of gains (Fig. 5.3).

An fMRI study has shown that the PFC and striatum are involved in loss aversion (Tom et al. 2007). On the other hand, brain lesion studies have reported that amygdala lesion patients showed diminished loss aversion (De Martino et al. 2010). Sokol-Hessner et al. (2009) have shown that physiological arousal response (skin conductance response) to losses was greater than to equivalent gains on average. This finding indicates that losses are more emotionally laden and salient than equivalent gains. The study also reported that individuals with greater arousal response to losses versus gains tend to be more loss-aversive. More recently, the same research team, using fMRI, reported a correlation between behavioral loss aversion and amygdala activation in response to losses relative to gains (Sokol-Hessner et al. 2012).

It is known that 5-HT plays an important role in modulating emotional response or affective state, and enhancing central 5-HT transmission by selective serotonin reuptake inhibitors (SSRIs) decreases amygdala activation in response to aversive stimuli (Takahashi et al. 2005). Although there are no PET studies investigating the relationship between 5-HT transmission and loss aversion directly, empirical evidence suggest that central 5-HT tone might be associated with loss aversion. Enhancing 5-HT transmission by tryptophan load reduced the "reflection effect"

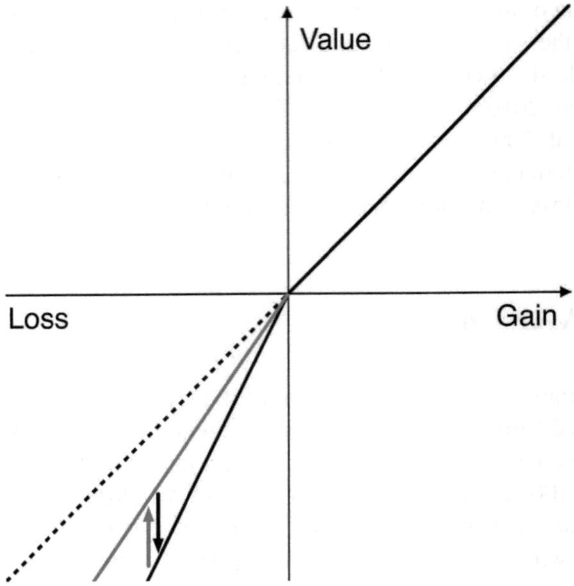

Fig. 5.3 Hypothesized model showing the contribution of central 5HT and NE transmission to loss aversion. 5-HT and NE might contribute to shaping the slope of value function for loss. 5-HT might ease the slope of value function for loss (loss tolerance: *gray*), and NE might intensify the slope (loss aversion: *black*). The value function is usually assumed to be a power function $v(x) = x^{\sigma}$, but we used common simplifying assumptions that σ is 1 for both value functions in gain and loss domains. The ratio (loss/gain) of the slope of linear functions was indicated as λ

(Murphy et al. 2009). "Reflection effect" refers to the fact that decision-makers tend to prefer the sure $50 gain to a 50/50 gamble to win $100 or no gain at all, showing risk-aversion. However, decision-makers tend to prefer a 50/50 gamble to lose $100 or no loss at all to the sure $50 loss, showing risk-seeking. Loss aversion can partly account for "reflection effect" and "framing effect". De Martino et al. (2006) reported that susceptibility to the framing effect was associated with amygdala activation. Then they reported that genetic variation in the promoter region of the 5-HT transporter gene (5-HTTLPR) predicted the susceptibility to the framing effect. Homozygosity for the minor s-allele showed greater amygdala activation during decision-making and stronger framing effect than l-allele carriers (Roiser et al. 2009). More recently, large-sample behavioral economics studies in an Asian sample also showed that homozygosity for the s-allele showed higher loss aversion than l-allele carriers (He et al. 2010). It is difficult to estimate pre-and post-synaptic (and net) 5-HT transmission exactly by genetic variation in 5-HTTLPR (Shioe et al. 2003), but 5-HT neurotransmission seems to attenuate the aversion to financial loss (Fig. 5.3) (Takahashi 2012, 2013).

In addition to 5-HT, a line of evidence suggests that norepinephrine (NE) might be involved in loss aversion. The role of NE in arousal is well established (Berridge and Waterhouse 2003), and it was reported that physiological arousal response was associated with behavioral loss aversion (Sokol-Hessner et al. 2009). Blocking central NE transmission by propranolol attenuated the sensitivity to the magnitude of possible losses at gambles (Rogers et al. 2004). Lack of appropriate PET radioligand has prevented us from investigating the role of central NE transmission in cognition, emotion, and decision-making in vivo. However, (S,S)-18F-FMeNER-D2 has recently been developed as a radioligand for the measurement of NE transporter (NET) for PET (Schou et al. 2004; Arakawa et al. 2008). (S,S)-18F-FMeNER-D2 is a reboxetine analog and has high affinity and high selectivity for NET. We utilized PET scans with (S,S)-[^{18}F]FMeNER-D_2 to investigate the relationship between NET in the brain and loss aversion. A NET-rich region available to PET imaging with this ligand is the thalamus. The amygdala and PFC are also innervated by NE, but the relatively low expression of NET prevented reliable measurement of their NET binding.

Loss aversion parameters were determined outside the PET scanner using a 50:50 mixed gamble (gain–loss). This parameter λ is similar to the parameter in prospect theory but makes the common simplifying assumptions of a linear rather than curvilinear value function (Fig. 5.3), and identical decision weights for a 0.5 probability of a gain or loss. The finding was that there was a negative correlation between λ and NET binding in the thalamus (Fig. 5.4) (Takahashi et al. 2013). In other words, individuals with low thalamic NET are likely to show pronounced loss aversion. Although NE has been implicated in arousal, previous studies also suggest that NE affects processing of salient information (Berridge and Waterhouse 2003). Neurons of the locus coeruleus (LC), the major source of NE in the brain, are physically evoked by salient or emotional stimuli (Aston-Jones et al. 1994), and phasic LC activation also increases NE release in target sites (Berridge and

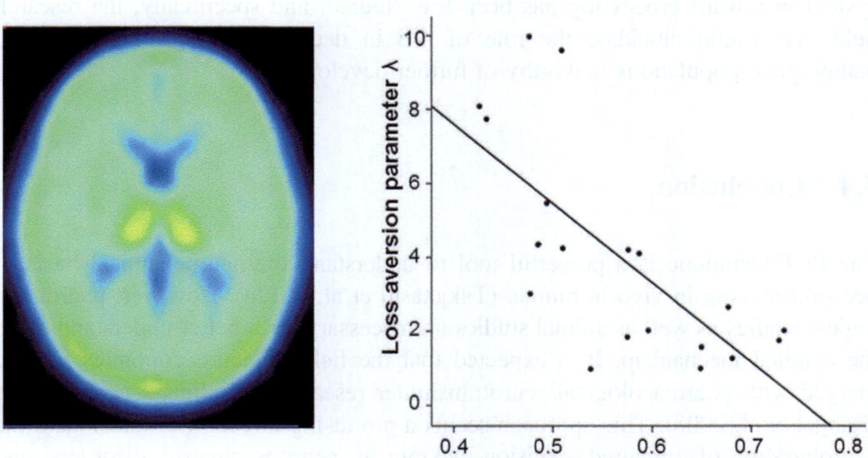

Fig. 5.4 NET in the thalamus and loss aversion. *Left panel* shows average of spatially normalized summed PET image of (S,S)-[^{18}F]FMeNER-D_2. *Right panel* shows negative correlation between NET binding in the thalamus and loss aversion parameter λ

Waterhouse 2003). Enhancing NE tone by NE reuptake inhibitor improves detection of emotional stimuli (De Martino et al. 2008), and blocking central NE transmission by propranolol predominantly impairs processing of negatively emotional stimuli (Cahill et al. 1994). Thus, PET findings suggest that individuals with low NET in the thalamus might show enhanced effect of NE released by salient stimuli due to low reuptake, and consequently show exaggerated emotional or arousal response to losses relative to gains. Thalamic NET might be an indirect mediator of the relationship between NE transmission and loss aversion. Similarly to 5-HT systems, Rasch et al. (2009) reported that a genetic variation of ADRA2B, the gene encoding the α2b-adrenergic receptor, predicted the amygdala responsivity to negative stimuli. Future studies with a more appropriate radioligand for measuring NET in the amygdala and PFC, which are implicated in loss aversion, are recommended. For the present, it is not unreasonable to suppose that central NE tone plays a role in shaping the slope of the value function in the loss domain (Fig. 5.3) (Takahashi 2012, 2013).

In a clinical setting, NET blocker, atomoxetine, is widely used in the pharmacotherapy of Attention-deficit hyperactivity disorder (ADHD). ADHD patients are known to show impulsive and reckless decision-making and have high comorbidity rates of drug addiction and gamble addiction (pathological gambling) (Pattij and Vanderschuren 2008; Breyer et al. 2009). Our finding suggests that atomoxetine might shift ADHD patients' decision-making from reckless (less loss-aversive) to more cautious (more loss-aversive) by blocking NET. Although pathological gambling is a heterogeneous disorder with various social and biological backgrounds, one can make a prediction that NET inhibitors might be beneficial for a subgroup of pathological gambling who show diminished aversive responses to financial losses (Takahashi 2013). Compared to the DA system, the role of the NE system in reward processing has been less studied, and specifically, the research field that would elucidate the role of NE in decision-making in normal and pathological populations is worthy of further development.

5.4 Conclusion

The PET technique is a powerful tool to understand the neurochemical basis of decision-making in vivo in human (Takahashi et al. 2012a). However, pharmacological studies as well as animal studies are necessary for a better understanding of the detailed mechanism. It is expected that the field of neuroeconomics will be merged with pharmacological/neurotransmitter research and clinical neuroscience (Rangel et al. 2008). This approach seems a promising direction to understand the neurobiology of impaired decision-making in neuropsychiatric disorders and develop novel pharmacotherapy for them (Takahashi 2012; Takahashi et al. 2012b; Takahashi 2013).

Acknowledgments A part of this study is the result of "Integrated Research on Neuropsychiatric Disorders" carried out under the Strategic Research Program for Brain Sciences by the Ministry of Education, Culture, Sports, Science and Technology of Japan (MEXT), a Grant-in-Aid for Scientific Research on Innovative Areas: Prediction and Decision Making (23120009), a Grant-in-Aid for Young Scientist A (23680045), a research grant from Takeda Science Foundation, a research grant from Brain Science Foundation, a research grant from Casio Science Foundation and a research grant from Senshin Medical Research Foundation.

References

Arakawa R, Okumura M, Ito H, Seki C, Takahashi H, Takano H, Nakao R, Suzuki K, Okubo Y, Halldin C, Suhara T (2008) Quantitative analysis of norepinephrine transporter in the human brain using PET with (S, S)-18F-FMeNER-D2. J Nucl Med 49:1270–1276

Aston-Jones G, Rajkowski J, Kubiak P, Alexinsky T (1994) Locus coeruleus neurons in monkey are selectively activated by attended cues in a vigilance task. J Neurosci 14(7):4467–4480

Berns GS, Capra CM, Chappelow J, Moore S, Noussair C (2008) Nonlinear neurobiological probability weighting functions for aversive outcomes. Neuroimage 39:2047–2057

Berridge CW, Waterhouse BD (2003) The locus coeruleus-noradrenergic system: modulation of behavioral state and state-dependent cognitive processes. Brain Res Rev 42:33–84

Breyer JL, Botzet AM, Winters KC, Stinchfield RD, August G, Realmuto G (2009) Young adult gambling behaviors and their relationship with the persistence of ADHD. J Gambl Stud 25:227–238

Cahill L, Prins B, Weber M, McGaugh JL (1994) Beta-adrenergic activation and memory for emotional events. Nature 371:702–704

Camerer CF, Fehr E (2006) When does "economic man" dominate social behavior? Science 311:47–52

Camerer C, Loewenstein G (2004) Behavioral economics: past, present, future. In: Camerer C, Loewenstein G, Rabin M (eds) Advance in behavioral economics. Princeton University Press, Princeton, pp 3–51

Chen MK, Lakshminarayanan V, Santos LR (2006) How basic are behavioral biases? Evidence from capuchin monkey trading behavior. J Polit Econ 114:517–537

Dagher A, Robbins TW (2009) Personality, addiction, dopamine: insights from Parkinson's disease. Neuron 61:502–510

De Martino B, Kumaran D, Seymour B, Dolan RJ (2006) Frames, biases, and rational decision-making in the human brain. Science 313:684–687

De Martino B, Strange BA, Dolan RJ (2008) Noradrenergic neuromodulation of human attention for emotional and neutral stimuli. Psychopharmacology 197:127–136

De Martino B, Camerer CF, Adolphs R (2010) Amygdala damage eliminates monetary loss aversion. Proc Natl Acad Sci USA 107:3788–3792

Ekelund J, Slifstein M, Narendran R, Guillin O, Belani H, Guo N, Hwang Y, Hwang D, Abi-Dargham A, Laruelle M (2007) In vivo DA D 1 receptor selectivity of NNC 112 and SCH 23390. Mol Imaging Biol 9:117–125

Gallagher DA, O'Sullivan SS, Evans AH, Lees AJ, Schrag A (2007) Pathological gambling in Parkinson's disease: risk factors and differences from dopamine dysregulation. An analysis of published case series. Mov Disord 22:1757–1763

He Q, Xue G, Chen C, Lu Z, Dong Q, Lei X, Ding N, Li J, Li H, Chen C (2010) Serotonin transporter gene-linked polymorphic region (5-HTTLPR) influences decision making under ambiguity and risk in a large Chinese sample. Neuropharmacology 59:518–526

Hsu M, Krajbich I, Zhao C, Camerer C (2009) Neural response to reward anticipation under risk is nonlinear in probabilities. J Neurosci 29:2231–2237

Knutson B, Rick S, Wimmer GE, Prelec D, Loewenstein G (2007) Neural predictors of purchases. Neuron 53:147–156

Leyton M, Boileau I, Benkelfat C, Diksic M, Baker G, Dagher A (2002) Amphetamine-induced increases in extracellular dopamine, drug wanting, and novelty seeking: a PET/[11C] raclopride study in healthy men. Neuropsychopharmacology 27:1027

Ligneul R, Sescousse G, Barbalat G, Domenech P, Dreher J (2012) Shifted risk preferences in pathological gambling. Psychol Med (Epub ahead of print)

Murphy SE, Longhitano C, Ayres RE, Cowen PJ, Harmer CJ, Rogers RD (2009) The role of serotonin in nonnormative risky choice: the effects of tryptophan supplements on the "reflection effect" in healthy adult volunteers. J Cogn Neurosci 21:1709–1719

Onur OA, Walter H, Schlaepfer TE, Rehme AK, Schmidt C, Keysers C, Maier W, Hurlemann R (2009) Noradrenergic enhancement of amygdala responses to fear. Soc Cogn Affect Neurosci 4:119–126

Pattij T, Vanderschuren LJ (2008) The neuropharmacology of impulsive behaviour. Trends Pharmacol Sci 29:192–199

Paulus M, Frank L (2006) Anterior cingulate activity modulates nonlinear decision weight function of uncertain prospects. Neuroimage 30:668–677

Prelec D (1998) The probability weighting function. Econometrica 66:497–527

Rangel A, Camerer C, Montague PR (2008) A framework for studying the neurobiology of value-based decision making. Nat Rev Neurosci 9:545–556

Rasch B, Spalek K, Buholzer S, Luechinger R, Boesiger P, Papassotiropoulos A, de Quervain DJ (2009) A genetic variation of the noradrenergic system is related to differential amygdala activation during encoding of emotional memories. Proc Natl Acad Sci USA 106:19191–19196

Rogers R, Lancaster M, Wakeley J, Bhagwagar Z (2004) Effects of beta-adrenoceptor blockade on components of human decision-making. Psychopharmacology 172:157–164

Roiser JP, de Martino B, Tan GC, Kumaran D, Seymour B, Wood NW, Dolan RJ (2009) A genetically mediated bias in decision making driven by failure of amygdala control. J Neurosci 29:5985–5991

Sanfey AG, Rilling JK, Aronson JA, Nystrom LE, Cohen JD (2003) The neural basis of economic decision-making in the ultimatum game. Science 300:1755–1758

Schou M, Halldin C, Sóvágó J, Pike VW, Hall H, Gulyás B, Mozley PD, Dobson D, Shchukin E, Innis RB (2004) PET evaluation of novel radiofluorinated reboxetine analogs as norepinephrine transporter probes in the monkey brain. Synapse 53:57–67

Shioe K, Ichimiya T, Suhara T, Takano A, Sudo Y, Yasuno F, Hirano M, Shinohara M, Kagami M, Okubo Y, Nankai M, Kanba S (2003) No association between genotype of the promoter region of serotonin transporter gene and serotonin transporter binding in human brain measured by PET. Synapse 48:184–188

Singer T, Seymour B, O'Doherty JP, Stephan KE, Dolan RJ, Frith CD (2006) Empathic neural responses are modulated by the perceived fairness of others. Nature 439:466–469

Sokol-Hessner P, Hsu M, Curley NG, Delgado MR, Camerer CF, Phelps EA (2009) Thinking like a trader selectively reduces individuals' loss aversion. Proc Natl Acad Sci USA 106: 5035–5040

Sokol-Hessner P, Camerer CF, Phelps EA (2012) Emotion regulation reduces loss aversion and decreases amygdala responses to losses. Soc Cogn Affect Neurosci (Epub ahead of print)

Steeves TD, Miyasaki J, Zurowski M, Lang AE, Pellecchia G, Van Eimeren T, Rusjan P, Houle S, Strafella AP (2009) Increased striatal dopamine release in Parkinsonian patients with pathological gambling: a [11C] raclopride PET study. Brain 132:1376–1385

Takahashi H (2012) Monoamines and assessment of risks. Curr Opin Neurobiol 22:1062–1067

Takahashi H (2013) Molecular neuroimaging of emotional decision-making. Neurosci Res (in press)

Takahashi H, Yahata N, Koeda M, Takano A, Asai K, Suhara T, Okubo Y (2005) Effects of dopaminergic and serotonergic manipulation on emotional processing: a pharmacological fMRI study. Neuroimage 27:991–1001

Takahashi H, Kato M, Matsuura M, Mobbs D, Suhara T, Okubo Y (2009) When your gain is my pain and your pain is my gain: neural correlates of envy and schadenfreude. Science 323: 937–939

Takahashi H, Matsui H, Camerer C, Takano H, Kodaka F, Ideno T, Okubo S, Takemura K, Arakawa R, Eguchi Y, Murai T, Okubo Y, Kato M, Ito H, Suhara T (2010a) Dopamine D receptors and nonlinear probability weighting in risky choice. J Neurosci 30:16567–16572

Takahashi H, Takano H, Kodaka F, Arakawa R, Yamada M, Otsuka T, Hirano Y, Kikyo H, Okubo Y, Kato M, Obata T, Ito H, Suhara T (2010b) Contribution of dopamine D1 and D2 receptors to amygdala activity in human. J Neurosci 30:3043–3047

Takahashi H, Takano H, Camerer CF, Ideno T, Okubo S, Matsui H, Tamari Y, Takemura K, Arakawa R, Kodaka F, Yamada M, Eguchi Y, Murai T, Okubo Y, Kato M, Ito H, Suhara T (2012a) Honesty mediates the relationship between serotonin and reaction to unfairness. Proc Natl Acad Sci USA 109:4281–4284

Takahashi H, Yamada M, Suhara T (2012b) Functional significance of central D1 receptors in cognition: beyond working memory. J Cereb Blood Flow Metab 32:1248–1258

Takahashi H, Fujie S, Camerer C, Arakawa R, Takano H, Kodaka F, Matsui H, Ideno T, Okubo S, Takemura K, Yamada M, Eguchi Y, Murai T, Okubo Y, Kato M, Ito H, Suhara T (2013) Norepinephrine in the brain is associated with aversion to financial loss. Mol Psychiatry 18:3–4

Tobler PN, Christopoulos GI, O'Doherty JP, Dolan RJ, Schultz W (2008) Neuronal distortions of reward probability without choice. J Neurosci 28:11703–11711

Tom S, Fox C, Trepel C, Poldrack R (2007) The neural basis of loss aversion in decision-making under risk. Science 315:515–518

Trepel C, Fox C, Poldrack R (2005) Prospect theory on the brain? Toward a cognitive neuroscience of decision under risk. Cogn Brain Res 23:34–50

Tversky A, Kahneman D (1992) Advances in prospect theory: cumulative representation of uncertainty. J Risk Uncertain 5:297–323

Yamada M, Camerer CF, Fujie S, Kato M, Matsuda T, Takano H, Ito H, Suhara T, Takahashi H (2012) Neural circuits in the brain that are activated when mitigating criminal sentences. Nat Commun 3:759

Part III
Environmental/Situational Factors Influencing Human Decision Making

Chapter 6
Decision-Making Under Uncertainty

Dominik R. Bach

Abstract All decision-making takes place under uncertainty, even in controlled laboratory circumstances. The Bayesian brain hypothesis, a widely accepted theoretical framework of brain function, prescribes that the brain uses probability distributions to store parameter values, rather than point estimates, and is thus able to use uncertainty on various parameters. This allows for investigating value-based decision-making under natural circumstances when information needs to be extracted from noisy input, and it may also impact on decisions based on propositional information. In this chapter, I present experimental approaches to neural representations of uncertainty in value-based decision-making.

6.1 Introduction

Economists often distinguish risk from uncertainty. Risk quantifies the uncertainness of outcomes to be realised from a decision. Uncertainty denotes all imprecision on propositional aspects of a decision-making situation. Following this distinction, many economic theories treat decision-making under uncertainty as a special case. However, for biological agents, all decision-making takes place under uncertainty.

Imagine a controlled laboratory experiment where an agent can make a simple choice between two options. Option A is to win either €10, or €20, depending on whether we draw a black or a white ball from an urn with 50 black and 50 white balls; option B is a sure amount of €15. B. There is no uncertainty in this abstract presentation of the decision problem, other than the risk associated with the outcome of the urn draw, if option A is chosen. To analyse this situation, classic economic theories will rely on the propositional information given to the decision-maker. Consequently, experimenters will try to convince agents that the

D.R. Bach (✉)
University of Zurich, Zurich, Switzerland
e-mail: dominik.bach@bli.uzh.ch

D.R. Bach
University College London, London, UK

M. Reuter and C. Montag (eds.), *Neuroeconomics*, Studies in Neuroscience, Psychology and Behavioral Economics, DOI 10.1007/978-3-642-35923-1_6

abstract, propositional information given to them is accurate, such that classic theory is applicable.

However, human's brains are equipped to solve far more complex problems, extract information from noisy input, and infer the structure of the world. Moreover, typical adults will have multiple experiences with situations similar to the laboratory choice. For example, why should they believe that the urn used by the experimenter really contains equal numbers of black and white balls?

In the next section, I will review a probabilistic framework for how the brain retrieves and stores information, to form the basis for a model of economic decision-making. The neural implementation of these algorithms has been reviewed both on a microscopic/neuronal (Pouget et al. 2013) and on a macroscopic level (Bach and Dolan 2012). Here, I will give an overview of the different kinds of uncertainty that are experimentally manipulated in neuroeconomic investigations.

6.2 A Probabilistic Model for Economic Decisions

Abstract models of value-based decision-making often assume that a decision-making agent receives propositional information in the form of particular numbers, such as values, utilities, probabilities, risk attitudes and so on, and a problem structure that links these numbers. The agent uses this information, for example, to maximise utility. The figures that define the decision problem either arise from current information (for example, the possible outcomes of the urn draw) or from stored values (for example, risk attitudes). Of course, under everyday circumstances, most information is not provided in the form of abstract, propositional information. This begs the question: how does a biological agent extracts the relevant economic information from a continuous stream of sensory input? How does it store and update this information to yield stable preferences and attitudes? This question is relevant also for controlled laboratory experiments because decision-making usually relies on attitudes or preferences acquired previously in real-world contexts—such as a risk preference in our initial example.

It turns out that biological agents use two fairly general principles of processing information: probabilistic computing, and hierarchical causal structures (Dayan and Hinton 1996; Dayan et al. 1995; Friston 2010). Our sensors provide information that is both very noisy and redundant (Kording and Wolpert 2006; Vilares and Kording 2011). This is a challenge for any system, biological or man-made. In order to make inference on the outside world given rich and noisy input, the task is to combine information from various noisy sources. As an example, imagine you are in price negotiations on a bustling street market when the trader announces his final price, and a bystander steps forward to grab the opportunity and buy the item. You have to make a quick decision whether or not to express your interest, but your sensory system is not quite certain whether what the trader said was "thirteen" or "thirty". Although we are never consciously aware of this, our sensory system will combine the auditory information with visual information from observing the traders mouth, and with previous information on the course of the price negotiations. This integration of redundant information reduces noise. In order to make the best possible inference, it is necessary to take the level of uncertainty on each information source into account: more reliable information should be given more weight. The auditory information should count less when it is noisy; but if it is dark, on the contrary, then the visual information should be discounted. A principled framework for making inference under these circumstances is Bayesian Decision Theory (Orban and Wolpert 2011; Trommershauser et al. 2011).

In Bayesian statistics, probabilities represent degrees of belief. This makes it possible to represent uncertainty on parameter values (Fig. 6.1). In an inference problem, a parameter value is given but unknown to us—it needs to be inferred. Classical statistics represents an estimate of a true parameter value as point estimate. For example, if we want to know whether the urn in the initial example is fair, we can draw many times and calculate the probability of drawing white from the sample mean. This is one single number, hence it is called a "point estimate" (Fig. 6.1). Bayesian statistics allows directly expressing our beliefs in particular

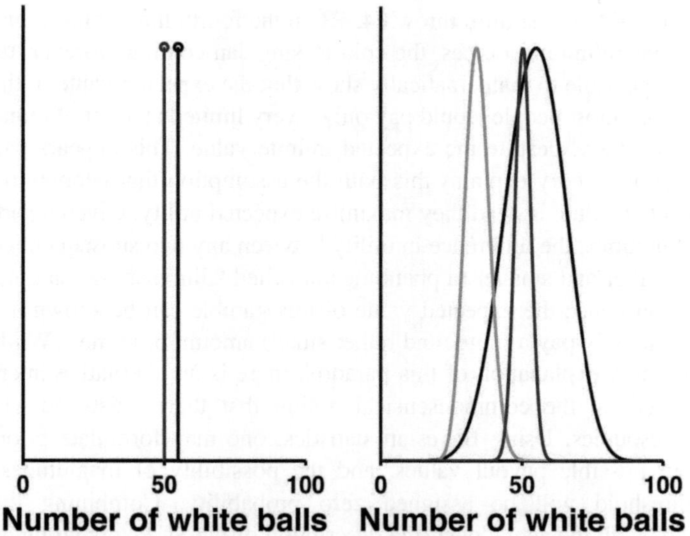

Fig. 6.1 Illustration of point estimation for a parameter value (*left*) and probabilistic representation of parameter values as in Bayesian statistics (*right*). In the initial example, an agent is told that the urn contains 50 black and 50 white balls. He then samples 20 balls and receives 11 white balls. A point estimate of the number of white balls is 55 (*black, left*). The agent has to decide whether to follow this estimate, or the propositional information (*dark grey, left*). A probabilistic estimate expresses the uncertainty associated with the limited number of samples (*black, right*). The agent may also combine the sampling information with a strong belief that the propositional information on the urn is correct (*dark grey, right*) or with his private belief that the experimenter will actually fob him off with an urn that just contains 30 white balls (*light grey, right*). The dispersion of these distributions corresponds to the uncertainty on the number of balls

parameter values, and updating them from experience. For example, from our knowledge, we may start with a strong prior assumption that the urn is fair, and update this after each draw. This will give us a probability distribution over different probabilities of drawing white. Importantly, we are now also able to express skewed priors—for example, because we think that the experimenter will fob us off with a biased urn. We may therefore express an initial belief that the urn contains only 30 white balls, and update this with experience. It appears from sensorimotor research that neural systems compute with probability distributions (Pouget et al. 2013), and use them to express prior beliefs (Brouwer and Knill 2009; Kording et al. 2004; Kording and Wolpert 2004; Tassinari et al. 2006), some of which may be in-built or acquired early during development (Parise et al. 2014). This has been proposed as a general computing principle of the brain (Friston 2009, 2010; Knill and Pouget 2004).

The principle of probabilistic computations provides a particular view on economic decision-making. Take, for example, the famous St. Petersburg paradox (Bernoulli 1738; Camerer 1995). This describes a gamble against a casino with unlimited resources. The player tosses a fair coin until heads appear for the first time. If heads appear on the first throw, he gets €1. If heads appear on the second throw, he gets €2, on the third throw €4, €8 on the fourth throw and so on. Because the casino has infinite resources, the coin tossing can continue forever, until heads appear. It is possible to mathematically show that the expected value of this gamble is infinite. Yet, most people would pay only a very limited amount of money to play this kind of lottery, despite the expected infinite value. This appears paradoxical. Expected utility theory explains this with the assumption that people do not maximise expected value; instead they maximise expected utility. Given a particular set of utility functions, the difference in utility between any two subsequent coin tosses becomes smaller and smaller (a phenomenon called "diminishing marginal utility", Fig. 6.2), and hence the expected value of this gamble can be shown to be finite. Hence people only pay a finite (and rather small) amount of money. While this is a widely accepted explanation of this paradox, there is an alternative interpretation. This is based on the commonsensical notion that there exists no casino with unlimited resources. Using Bayesian statistics, one may formulate prior assumptions about possible payout values, and the possibility of magnitudes above a certain threshold will be assigned zero probability. Combining these prior assumptions with the actual data (the description of the St. Petersburg gamble) will yield a probability distribution for every possible gamble outcome—this distribution describes how likely the outcome will have this or that monetary value. The mean of this distribution will have finite value, different from the abstract information about the problem. Thus, in a Bayesian framework this paradoxon can be explained without invoking non-linear utility functions.

Experimental research has focused on how the brain stores and manipulates probability distributions, and their associated uncertainty. In the following sections, I will review experimental approaches from economic and noneconomic contexts.

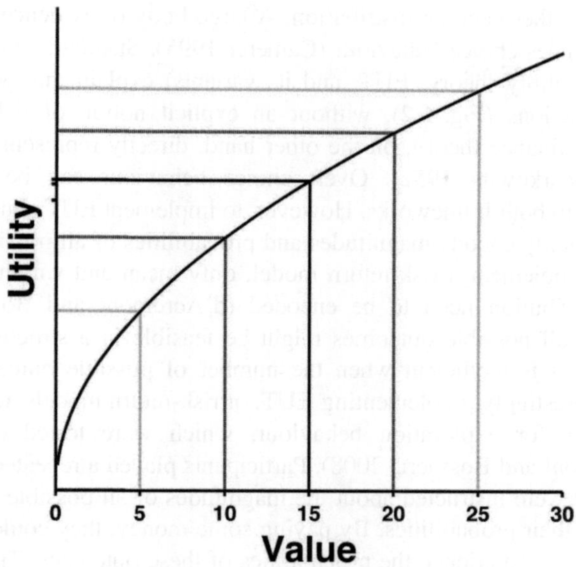

Fig. 6.2 Example of a non-linear utility function which transforms objective value—here in €—into utility. The *lines* indicate values with equal intervals of €5. However, because the utility function is non-linear, the corresponding utilities have unequal spacing—the intervals become smaller with increasing utility. This phenomenon is termed "diminishing marginal utility". Ticks on the utility axis indicate expected utilities of three different situations. The utility corresponding to a value of €15 (*light grey line*) is larger than the expected utility corresponding to a 50 % chance of obtaining €20 and a 50 % change of obtaining €10 (*medium grey lines*)—this is the initial example in the text. The expected utility of a 50 % chance of obtaining €20 and a 50 % change of obtaining €10 (*dark grey lines*) is even smaller. Given a choice between any of these three situations, a utility-maximising decision-making agent would choose the situation with higher utility, and this corresponds to the situation with lower outcome uncertainty (risk). This explains how risk preference (in this case, risk aversion) can be explained with a non-linear utility function

6.3 Uncertainty of Decision Outcomes

Uncertainty of decision outcomes is arguably the most investigated type of imprecision in value-based decision-making. Given the irreversibility of time, there appears an epistemic difference between uncertainty of future outcomes which are unknowable, and uncertainty on present and past events which are in principle knowable. Most empirical research is on future outcomes, but it is not clear whether decision-making systems make this distinction, for example, whether they represent the risk of a past lottery different from the risk of a lottery to be played out in the future. I will first consider economic approaches, and then briefly discuss the non-economic aspects.

(a) **Risk in expected utility and finance theory**: The dispersion of predicted utility or value is usually termed risk in economic (von Neumann and Morgenstern 1944) and finance theory (Markowitz 1952), and quantified as variance of the outcome distribution. A large body of evidence indicates that risk influences choice behaviour (Camerer 1995). Standard economic models (expected utility theory, EUT, and its variants) explain this with non-linear utility functions (Fig. 6.2), without an explicit notion of risk. Risk-return models in finance theory, on the other hand, directly represent risk in calculations (Markowitz 1952). Overt choice behaviour can be equally well described in both frameworks. However, to implement EUT neurally, an agent must explicitly encode magnitudes and probabilities of all possible outcomes, while to implement a risk-return model, only mean and variance of the outcome distribution need to be encoded (d'Acremont and Bossaerts 2008). Encoding all possible outcomes might be feasible in a structured economic gamble, but is inefficient when the number of possible outcomes becomes large. Interestingly, implementing EUT, or risk-return models, makes different predictions for exploration behaviour, which were tested experimentally (d'Acremont and Bossaerts 2008). Participants played a repeated lottery. Each time, they were instructed about the magnitudes of all possible outcomes, but not about their probabilities. By paying some money, they could sample from the lottery to experience the probabilities of these outcomes. Then, they were given a chance to bet on that lottery in order to gain money. In some trials, outcome probabilities and magnitudes changed from the previous trial. In other trials, participants were informed that outcome magnitudes had changed from the previous trial, but that outcome probabilities were held constant. Had participants encoded all the outcome probabilities from the previous trial, there would have been nothing to gain from sampling the new lottery (because the magnitudes, the only thing that changed, were stated explicitly). However, in about half of these trials, participants did sample the probabilities, which suggest that they did not encode all individual outcome probabilities. To be risk-sensitive nevertheless, they would have encoded mean and variance of the outcome distribution. This connects neuroeconomics to reinforcement learning theory: Biological agents can learn with experience to predict the mean of an outcome distribution. To do so, they appear to use algorithms that compute a difference between predicted outcome and actual outcome, termed prediction error (Mackintosh 1983; Pearce and Hall 1980; Rescorla and Wagner 1972; Sutton and Barto 1998). When the outcome distribution has a high dispersion, prediction errors are expected and do not imply that the prediction is wrong; when the outcome dispersion is low, prediction errors are more informative. Hence, the dispersion of the outcome distribution is important for reinforcement learning and has been termed "expected uncertainty" (Yu and Dayan 2005). To acquire a notion of outcome dispersion, update algorithms have been suggested that rely on "risk prediction errors" (d'Acremont et al. 2009; Preuschoff et al. 2006).

(b) **Uncertainty on future states**: A simplifying model of the world is to describe it as a series of discrete states that are connected by transition probabilities (a Markov chain). If the future state is one with economic value, then this model captures a typical economic gamble, but this framework is more general. One can, for example, quantify uncertainty over future states as entropy (Aron et al. 2004).

(c) **Motor uncertainty**: Value-based decisions in biological environments usually involve motor actions. Sensorimotor control theory is concerned with the planning and implementation of uncertain motor action under uncertain sensory guidance (Orban and Wolpert 2011). For example, when catching a ball, if we observe longer, we will have more precise information on the ball trajectory—but if we initiate the movement earlier, the movement will be more precise. Choosing the optimal observation interval requires integrating the uncertainty of the future movement with the uncertainty of past observations (Battaglia and Schrater 2007; Faisal and Wolpert 2009). More generally, when the result of a decision is implemented with an uncertain motor action, normative accounts mandate that motor uncertainty is factored in for the decision. This may be relevant for economic experiments done in non-human species.

How does the brain represent uncertainty of decision outcomes? Phasic firing in dopaminergic midbrain neurons is consistent with prediction error signals during reinforcement learning (Schultz et al. 1997); and these signals are scaled by outcome uncertainty (Tobler et al. 2005), consistent with a reinforcement learning account of expected uncertainty. Neurons in orbitofrontal cortex carry outcome uncertainty signals in their firing rate, independent of value signals (O'Neill and Schultz 2010), although see Ogawa et al. (2013). Third, BOLD fMRI studies have implicated more than ten distinct brain regions as representing outcome uncertainty in various different experimental situations (Abler et al. 2009; Dreher et al. 2006; Fitzgerald et al. 2010; Mohr et al. 2010; Preuschoff et al. 2006; Rolls et al. 2008; Symmonds et al. 2010, 2011; Tobler et al. 2007). These heterogeneous findings may justify speculating that neural encoding of outcome uncertainty depends on the process by which the outcome distribution is estimated (Fig. 6.3).

6.4 Uncertainty on Decision Circumstances

Most decision circumstances are known to the decision-maker only with some imprecision. In the context of value-based decision-making, uncertainty of gamble probabilities has been investigated most. A related concept from reinforcement learning is volatility (Behrens et al. 2007; Mathys et al. 2011). Another is the presence of rule violations indicating reinforcement rule change, and thus, uncertainty on the rules; this situation has been termed "unexpected uncertainty" (Yu and Dayan 2005). Because the impact of such uncertainty on value-based decision-making is less well known, I will focus here on uncertain gamble probabilities and the related economic concept of ambiguity (Ellsberg 1961).

Imagine a lottery in which outcome probabilities are not explicitly stated—instead they can take several particular values. This situation is often termed "ambiguity" and leads to the famous Ellsberg paradox (Ellsberg 1961). Let's say

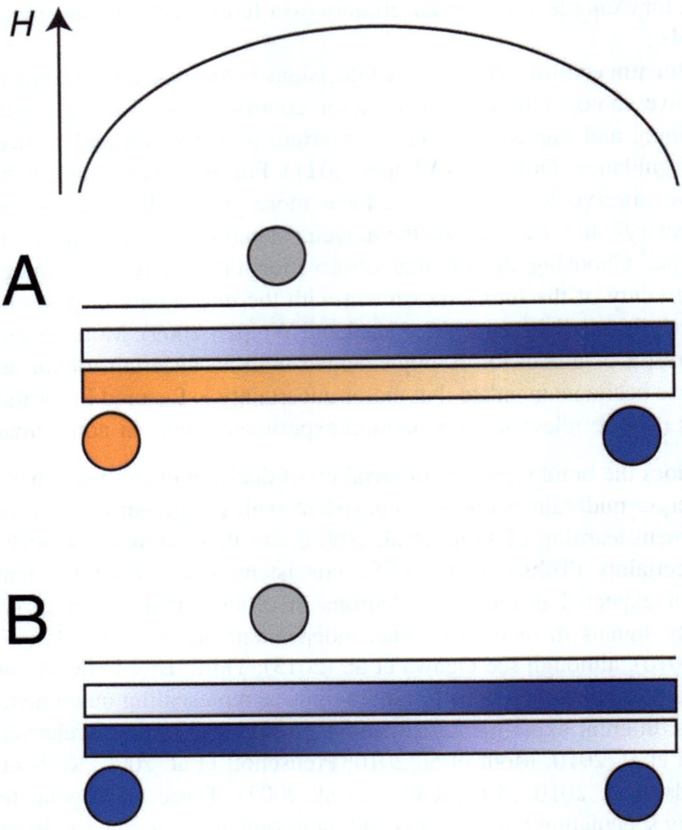

Fig. 6.3 Illustration of a gamble involving uncertainty on outcome probabilities (Bach et al. 2011). *A* A *grey* "bowling" ball appears on the screen. The agent is instructed that it may have come from one of two players (*orange*, *blue*) with dispersion indicated by the *colour gradients* below the *grey* ball. *Orange* and *blue* balls have different probabilities of winning from the gamble. The uncertainty (Shannon entropy *H*) associated with the ball position is indicated on *top*. *B* A corresponding gamble with no uncertainty. The *grey* ball may have come from either of two *blue* players, such that the outcome probabilities correspond to "*blue*" and are known

our initial urn A contains 100 balls, half of which are black, and the other half white. Another urn B contains 100 balls with unknown proportion of black and white. Would you rather bet on black in urn A or B? Some people may say A, so under assumptions of EUT, one can infer that they think the probability of winning from black in urn A (which has $p = 0.5$) is larger than of winning from black in urn B (which has therefore $p < 0.5$). Further, would you rather bet on white in urn A or B? Again, the same people may say A, so that the probability of winning white from urn B is again $p < 0.5$. The proportion of black and white balls in urn B hence does not add up to 1, which constitutes the paradox. Following this thought experiment, empirical investigations have demonstrated that most people avoid ambiguity, even when this does not maximise utility; for example, if one has to pay extra money to bet on urn A (Becker and Brownson 1964; Curley et al. 1986; Keren and Gerritsen 1999; MacCrimmon and Larson 1979; Pulford and Colman 2008; Slovic and Tversky 1974; Yates and Zukowski 1976).

How can we resolve the paradox? Urn A constitutes a single-stage lottery. By contrast, urn B can be understood as a two-stage lottery. There is a bet on the distribution of balls in urn B, termed second-order distribution (Bach et al. 2011; Klibanoff et al. 2005), and a bet on the outcomes of a draw from this distribution. EUT posits that a rational decision-maker should collapse the two stages of this lottery (Camerer 1999). In the absence of further information, one should assume a uniform second-order distribution (that is, all 100 possible compositions in urn B have probability $p = 1/100$). For each possible urn composition, one should multiply the ensuing first-order probabilities of a black or white draw, multiply them with its second-order probability and finally add up all probabilities for black and white across all urn composition in order to make a decision. It turns out that in this case, the expected outcome is the same for both urns.

The Ellsberg paradox however is only a paradox if we assume this reduction of the two-stage bet to a single-stage one. If we assume that people treat the lottery as a full two-stage bet, we can invoke a number of reasons to explain ambiguity aversion. A very simple reason would be that decision-making agents have a prior belief that the number of balls in urn B is not uniformly distributed, and unfavourable urn compositions are more likely. Indeed, restricting the range of possible urn composition reduces ambiguity aversion (Keren and Gerritsen 1999; Larson 1980).

However, ambiguous and non-ambiguous gambles differ across other factors as well. For example, by making a choice between urn A and urn B, people reveal their knowledge and belief about gambles and probabilities, something that might have additional (positive or negative) utility. When people are asked to make their choice publicly in a group, ambiguity aversion is larger than when they make the same choice, but write it down on a piece of paper that is only later to be read by the experimenter (Curley et al. 1986). In line with this, instructing individuals that their choices are going to be evaluated increases ambiguity aversion (Muthukrishnan et al. 2009). Also, when people gamble on getting one of two movies, where the experimenter asks which of the two they prefer, they avoid gambles of type B. However, if the experimenter does not know which movie they prefer, there is no ambiguity avoidance (Trautmann et al. 2008). In this latter case, the experimenter cannot judge people's choices because he does not know what they want to obtain from the gamble. This is a purely social factor that may explain ambiguity aversion but has no relation to uncertainty. Interestingly also, ambiguity aversion (Chow and Sarin 2002) —and also brain responses to ambiguous gambles (Bach et al. 2009)—depend on the fact that something is hidden from the observer rather than completely unknowable.

In summary, ambiguity is a good example how the categorical contrast certainty versus. uncertainty can be imbued with conceptual confounds. At the same time, this type of gamble allows for an elegant manipulation of uncertainty. We can simply use the entropy of the second-order probability distribution (i.e. the probabilities that a certain outcome probability will be realised) to quantify rule uncertainty. In a repeated gamble on aversive outcomes, we presented a "bowling ball game" gamble to participants. A grey ball would appear somewhere on the screen—it could either come from player type 1 or player type 2. Balls from the two player types represented gambles with different outcome probabilities—but the same outcome magnitudes. Now if the ball was close to player type 1, it was more likely to have come from player 1, and if it appeared right in the middle of the screen, it was equally likely to have come from either of the players. Hence, the latter situation involves more uncertainty than the former, and we can realise many ball positions with different rule uncertainties. Indeed, it turned out that gambles are avoided to a degree that depends on the amount of rule uncertainty (Bach et al. 2011). This avoidance of uncertain situations with higher uncertainty appeared to be due to overweighting of the more unfavourable possibility when rule uncertainty was high (Bach et al. 2011). A neural representation of this uncertainty was found in the posterior cingulate cortex. In contrast, neural responses to the contrast ambiguity versus non-ambiguity—possibly not reflecting uncertainty—are consistently reported in a different brain area, namely in the posterior parietal cortex (Bach et al. 2009, 2011; Huettel et al. 2006).

6.5 Sensory Uncertainty

Value-based decisions are often based on abstract, propositional information—here, sensory uncertainty is often irrelevant. In many biological situations, however, information has to be extracted from low-level physical information. This includes a plethora of neuroeconomic experiments done on non-human animals (Camerer 1995). The influence of sensory uncertainty on value-based decisions is illustrated by the certainty effect. When faced with a choice between €4 with $p = 0.2$ and €3 with $p = 0.25$, an agent might choose the €4 lottery. But when given the choice between €4 with $p = 0.8$ and €3 with $p = 1$, he may flip his preference and go for the certain €3 option. According to EUT, these are incompatible choices (von Neumann and Morgenstern 1944), and this discrepancy has been termed the certainty effect. However, sometimes the reverse is observed: a decreased propensity to choose the certain option. It has been demonstrated that a certainty effect can be reversed by manipulating the imprecision on outcome magnitude information (Shafir et al. 2008). Noisy sensory information will also occur in economic transactions that are made under time pressure (think of a stock market), and in economic decisions involving nonmonetary goods.

6.6 Summary and Conclusions

In this chapter, I have discussed how a probabilistic account of the brain can be integrated with economic accounts of decision-making. It becomes clear that in natural environments, the brain must extract all information from noisy sensory data, and take account of uncertainty on all levels of a neural hierarchy. In this process, information sources are combined with each other and with prior knowledge. This can also be relevant for situations in which propositional information is provided as basis for a decision—because even this abstract information may be combined with prior assumptions. Experimental approaches to neural representation of uncertainty in economic variables have focused on the concepts of risk and ambiguity. Sensory and motor uncertainty is less often investigated but may be crucial to understand value-based behaviour in non-human species. It is likely that neuroeconomic investigation of decision-making under uncertainty is going to be a fruitful approach to study the brain, and human decisions.

Acknowledgments I would like to thank Quentin Huys, Marc Guitart Masip, Deborah Talmi and Matthias Staib, for helpful comments on a first draft of this manuscript.

References

Abler B, Herrnberger B, Gron G, Spitzer M (2009) From uncertainty to reward: BOLD characteristics differentiate signaling pathways. BMC. Neurosci. 10:154

Aron AR, Shohamy D, Clark J, Myers C, Gluck MA, Poldrack RA (2004) Human midbrain sensitivity to cognitive feedback and uncertainty during classification learning. J Neurophysiol 92:1144–1152

Bach DR, Dolan RJ (2012) Knowing how much you don't know: a neural organization of uncertainty estimates. Nat Rev Neurosci 13:572–586

Bach DR, Seymour B, Dolan RJ (2009) Neural activity associated with the passive prediction of ambiguity and risk for aversive events. J Neurosci 29:1648–1656

Bach DR, Hulme O, Penny WD, Dolan RJ (2011) The known unknowns: neural representation of second-order uncertainty, and ambiguity. J Neurosci 31:4811–4820

Battaglia PW, Schrater PR (2007) Humans trade off viewing time and movement duration to improve visuomotor accuracy in a fast reaching task. J Neurosci 27:6984–6994

Becker SW, Brownson FO (1964) What price ambiguity? Or the role of ambiguity in decision making. J Polit Econ 72:62–73

Behrens TE, Woolrich MW, Walton ME, Rushworth MF (2007) Learning the value of information in an uncertain world. Nat Neurosci 10:1214–1221

Bernoulli D (1738) Specimen theoriae novae de mensura sortis. Commentarii Academiae Scientiarum Imperialis Petropolitanae 5:175–192

Brouwer AM, Knill DC (2009) Humans use visual and remembered information about object location to plan pointing movements. J Vis 9:24–19

Camerer C (1995) Individual decision making. In: Kagel JH, Roth AE (eds) Handbook of experimental economics. Princeton University Press, Princeton, pp 587–704

Camerer C (1999) Ambiguity aversion and non-additive probability: experimental evidence, models and applications. In: Luini L (ed) Uncertain decisions: bridging theory and experiments. Kluwer, Boston, pp 53–80

Chow CC, Sarin RK (2002) Known, unknown, and unknowable certainties. Theor Decis 52: 127–138

Curley SP, Yates F, Abrams RA (1986) Psychological sources of ambiguity avoidance. Organ Behav Hum Decis Process 38:230–256

d'Acremont M, Lu ZL, Li X, Van der LM, Bechara A (2009) Neural correlates of risk prediction error during reinforcement learning in humans. NeuroImage 47:1929–1939

d'Acremont M, Bossaerts P (2008) Neurobiological studies of risk assessment: a comparison of expected utility and mean-variance approaches. Cogn Affect Behav Neurosci 8:363–374
Dayan P, Hinton GE (1996) Varieties of Helmholtz machine. Neural Netw 9:1385–1403
Dayan P, Hinton GE, Neal RM, Zemel RS (1995) The Helmholtz machine. Neural Comput 7: 889–904
Dreher JC, Kohn P, Berman KF (2006) Neural coding of distinct statistical properties of reward information in humans. Cereb Cortex 16:561–573
Ellsberg D (1961) Risk, ambiguity, and the savage axioms. Quart J Econ 75:643–669
Faisal AA, Wolpert DM (2009) Near optimal combination of sensory and motor uncertainty in time during a naturalistic perception-action task. J Neurophysiol 101:1901–1912
Fitzgerald TH, Seymour B, Bach DR, Dolan RJ (2010) Differentiable neural substrates for learned and described value and risk. Curr Biol 20:1823–1829
Friston K (2009) The free-energy principle: a rough guide to the brain? Trends Cogn Sci 13: 293–301
Friston K (2010) The free-energy principle: a unified brain theory? Nat Rev Neurosci 11:127–138
Huettel SA, Stowe CJ, Gordon EM, Warner BT, Platt ML (2006) Neural signatures of economic preferences for risk and ambiguity. Neuron 49:765–775
Keren G, Gerritsen LEM (1999) On the robustness and possible accounts of ambiguity aversion. Acta Psychol 103:149–172
Klibanoff P, Marinacci M, Mukerji S (2005) A smooth model of decision making under ambiguity. Econometrica 73:1849–1892
Knill DC, Pouget A (2004) The Bayesian brain: the role of uncertainty in neural coding and computation. Trends Neurosci 27:712–719
Kording KP, Wolpert DM (2004) Bayesian integration in sensorimotor learning. Nature 427:244–247
Kording KP, Wolpert DM (2006) Bayesian decision theory in sensorimotor control. Trends in cognitive sciences 10:319–326
Kording KP, Ku SP, Wolpert DM (2004) Bayesian integration in force estimation. J Neurophysiol 92:3161–3165
Larson JR (1980) Exploring the external validity of a subjectively weighted utility model of decision making. Organ Behav Hum Performance 26:293–304
MacCrimmon KR, Larson S (1979) Utility theory: axioms versus 'paradoxes'. In: Allais M, Hagen O (eds) Expected utility hypotheses and the allais paradox. D. Reidel, Dordrecht, pp 333–410
Mackintosh NJ (1983) Conditioning and associative learning. Oxford University Press, Oxford
Markowitz H (1952) Portfolio selection. J Finan 7:77–91
Mathys C, Daunizeau J, Friston KJ, Stephan KE (2011) A Bayesian foundation for individual learning under uncertainty. Front Hum Neurosci 5:39
Mohr PN, Biele G, Krugel LK, Li SC, Heekeren HR (2010) Neural foundations of risk-return trade-off in investment decisions. NeuroImage 49:2556–2563
Muthukrishnan AV, Wathieu L, Xu AJ (2009) Ambiguity aversion and the preference for established brands. Manage Sci 55:1933–1941
Ogawa M, van der Meer MA, Esber GR, Cerri DH, Stalnaker TA, Schoenbaum G (2013) Risk-responsive orbitofrontal neurons track acquired salience. Neuron 77:251–258
O'Neill M, Schultz W (2010) Coding of reward risk by orbitofrontal neurons is mostly distinct from coding of reward value. Neuron 68:789–800
Orban G, Wolpert DM (2011) Representations of uncertainty in sensorimotor control. Curr Opin Neurobiol 21:629–635
Parise CV, Knorre K, Ernst MO (2014) Natural auditory scene statistics shapes human spatial hearing. Proc Natl Acad Sci USA 111:6104–6108
Pearce JM, Hall G (1980) A model for Pavlovian learning—variations in the effectiveness of conditioned but not of unconditioned stimuli. Psychol Rev 87:532–552
Pouget A, Beck JM, Ma WJ, Latham PE (2013) Probabilistic brains: knowns and unknowns. Nat Neurosci 16:1170–1178

Preuschoff K, Bossaerts P, Quartz SR (2006) Neural differentiation of expected reward and risk in human subcortical structures. Neuron 51:381–390

Pulford BD, Colman AM (2008) Size doesn't really matter. ambiguity aversion in Ellsberg urns with few balls. Exp Psychol 55:31–37

Rescorla RA, Wagner AR (1972) A theory of Pavlovian conditioning: variations in the effectiveness of reinforcement and nonreinforcement. In: Black AH, Prokasy WF (eds) Classical conditioning II: current research and theory. Appleton-Century-Crofts, New York, pp 64–99

Rolls ET, McCabe C, Redoute J (2008) Expected value, reward outcome, and temporal difference error representations in a probabilistic decision task. Cereb Cortex 18:652–663

Schultz W, Dayan P, Montague PR (1997) A neural substrate of prediction and reward. Science 275:1593–1599

Shafir S, Reich T, Tsur E, Erev I, Lotem A (2008) Perceptual accuracy and conflicting effects of certainty on risk-taking behaviour. Nature 453:917–920

Slovic P, Tversky A (1974) Who accepts Savage's axiom? Behav Sci 19:368–373

Sutton RS, Barto AG (1998) Reinforcement learning: an introduction. MIT Press, Cambridge

Symmonds M, Bossaerts P, Dolan RJ (2010) A behavioral and neural evaluation of prospective decision-making under risk. J Neurosci 30:14380–14389

Symmonds M, Wright ND, Bach DR, Dolan RJ (2011) Deconstructing risk: separable encoding of variance and skewness in the brain. NeuroImage

Tassinari H, Hudson TE, Landy MS (2006) Combining priors and noisy visual cues in a rapid pointing task. J Neurosci 26:10154–10163

Tobler PN, Fiorillo CD, Schultz W (2005) Adaptive coding of reward value by dopamine neurons. Science 307:1642–1645

Tobler PN, O'Doherty JP, Dolan RJ, Schultz W (2007) Reward value coding distinct from risk attitude-related uncertainty coding in human reward systems. J Neurophysiol 97:1621–1632

Trautmann ST, Vieider FM, Wakker PP (2008) Causes of ambiguity aversion: known versus unknown preferences. J Risk Uncertainty 36:225–243

Trommershauser J, Kording K, Landy MS (2011) Sensory cue integration. Oxford University Press, Oxford

Vilares I, Kording K (2011) Bayesian models: the structure of the world, uncertainty, behavior, and the brain. Ann NY Acad Sci 1224:22–39

von Neumann J, Morgenstern O (1944) Theory of games and economic behavior. Princeton University Press, Princeton

Yates JF, Zukowski LG (1976) Characterization of ambiguity in decision-making. Behav Sci 21:19–25

Yu AJ, Dayan P (2005) Uncertainty, neuromodulation, and attention. Neuron 46:681–692

Chapter 7
Emotion Regulation and Economic Decision-Making

Renata M. Heilman, Andrei C. Miu and Daniel Houser

Abstract Emotion plays an important role in human social and economic decision-making. Only in the last few decades has this view been accepted in mainstream research by economists and psychologists studying decisional processes. Many studies now provide compelling evidence that the experience of various different emotions influence changes in decisional outcomes. Emerging research on emotion regulation, however, highlights that humans typically make efforts to control emotion experiences. This leaves open the possibility that decision effects attributed to acute emotions may be affected by regulatory strategies. If so, this raises the additional possibility that different regulation strategies could have different implications for economic decisions. Researchers have recently begun to study these possibilities and discovered that emotion regulation can indeed modulate effects that task-related and incidental emotions have on decisions. In this chapter, we provide a review of the empirical studies that have investigated the effects of regulatory strategies on social and economic decisions. We present an overview of the concept of emotion regulation by referring to the different types of regulatory strategies and their cognitive and behavioral effects. We proceed to review empirical studies relevant for various types of decision-making, such as risk and ambiguity, susceptibility to framing, and economic bargaining. We conclude by discussing practical implications of this rapidly evolving research topic.

R.M. Heilman · A.C. Miu
Department of Psychology, Babes-Bolyai University,
37 Republicii, 400015 Cluj-Napoca, CJ, Romania

D. Houser (✉)
Department of Economics and Interdisciplinary Center for Economic Science,
George Mason University, 4400 University Drive, MSN 1B2,
Fairfax, VA 22030, USA
e-mail: dhouser@gmu.edu

M. Reuter and C. Montag (eds.), *Neuroeconomics*, Studies in Neuroscience, Psychology and Behavioral Economics, DOI 10.1007/978-3-642-35923-1_7

7.1 Introduction[1]

Emotion plays an important role in human social and economic decision-making (see, e.g., Elster 1998; Loewenstein 2000; Peters et al. 2006). People evaluate objective features of alternatives in a subjective way (Edwards 1962; Kahneman and Tversky 1979), and emotions influence these subjective evaluations (Loewenstein and O'Donoghue 2004; Naqvi et al. 2006; Slovic et al. 2007). Emerging research on emotion regulation (henceforth, ER), however, highlights that humans typically make efforts to control emotion experiences (Gross 2002). This leaves open the possibility that decision effects attributed to acute emotions may be affected by ER strategies. If so, this raises the additional possibility that different regulation strategies could have different implications for economic decisions. Only very recently, however, have scholars begun to investigate these possibilities, and in doing so have discovered that ER can indeed mediate effects that task-related and incidental emotions have on decisions (Crockett et al. 2008; Kahneman and Frederick 2007; Miu and Crişan 2011; Heilman et al. in preparation).

The intrinsic role of emotion in decision holds greater importance when the value of prospects (i.e., actions with uncertain rewards) is computed in "emotion-cognition brain hubs" (Pessoa 2008) thought to include midbrain dopaminergic regions and their targets (e.g., ventral and dorsal striatum, ventromedial and ventrolateral prefrontal cortex, anterior cingulate cortex). For this reason, neuroeconomists have emphasized that the interaction of emotion and decision-making is profitably studied in environments that include risk (where the decision-maker has perfect information regarding the stochastic relationship between actions and outcomes) and uncertainty (where the decision-maker does not have full information about the stochastic environment; see, e.g., Rangel et al. 2008).

When a human anticipates or experiences an emotion, s/he will often use strategies to control that experience. It follows that ER, a concept subsuming the processes controlling which emotions we have, when we have them, and how we experience and express them (Gross 2002), could be crucial to decision-making as well as other cognitive processes (e.g., memory; Richards and Gross 1999, 2000). This chapter discusses the impact of ER on a variety of behaviors of particular interest to economists, including environments that involve strategic and non-strategic risk. Despite the fact that research in ER is relatively new, our chapter cannot begin to cover the rapidly expanding literature in this area. Our approach has been to include reviews of key findings connected to economic decision-making.

Economists, and especially cognitive economists, will be especially interested in understanding ER. A reason is that ER can be trained, manipulated, and controlled. Moreover, different ER strategies generate predictable systematic differences in people's decisions in the face of the same emotion. An implication is that one might be able to design decision support tools, or more generally economic institutions, that encourage specific types of ER. This would seem to carry particular value when

[1]This section draws in part from Heilman et al. (2010, pp. 257–258).

decisions must be made in high-stress situations that create acute emotional responses.

The remainder of this chapter is organized as follows. In the next section we discuss in greater detail the concept of emotion regulation. Following this is a broad overview of ER's impact on cognition and decision. The next section proceeds to explain why and how ER can confound our inferences about the impact of emotion on decision. The subsequent three sections discuss what we know about the impact of ER in three specific contexts: choice under risk, choice under different frames, and choices in bargaining environments. The final section is a concluding discussion.

7.2 What Is Emotion Regulation?

Obviously, people are not at the whim of their emotions. They can use a number of regulation strategies designed to alter their emotional reactions. One of the most influential current approaches in the study of emotion and emotion regulation (ER) is the process model of emotions (Gross 1998a, b, 2002). The starting point of Gross's process model of ER is a concept of emotion that is shared with a number of prior theorists (Ekman 1972; Frijda 1988). According to this model, it is the way that we interpret or appraise external or internal events that mediates the emotional impact that such events have on us. Such appraisals can be very rapid and do not necessarily require our conscious awareness. An emotional response is triggered whenever a situation is interpreted as presenting important opportunities or challenges. In the following phase, a biologically based emotion program is started and its nature depends upon how the situation has been appraised (Lazarus 1991). As soon as an emotional program has been triggered, changes at the behavioral, subjective, or physiological levels prepare the individual for an adaptive response to the situation that s/he has perceived (Ekman 1992). Although these response tendencies help people respond quickly and adaptively, they can not enforce any given action (Frijda 1988) due to the fact that emotional response tendencies can be modulated (e.g., exaggerated, diminished, or even entirely inhibited) before they are expressed as observable behaviors, emotion self-reports, or physiological changes (Gross and Munoz 1995).

ER is a construct that subsumes all the actions that people take in order to control which emotions they have, when they have them and how they experience or express those emotions (Gross 2002). Although the line of research on ER was initially related to the developmental literature (Campos et al. 1983, 1989; Thompson 1990, 1991), the adult literature on ER has rapidly expanded in the last decade (e.g., Gross and Levenson 1993; Izard 1990; Ochsner and Gross 2005; Gross and Thompson 2007).

Because emotion unfolds over time, regulatory strategies can be distinguished in terms of when they have their primary impact on the emotion-generative process (Gross and John 2003). Generally, it is assumed that an emotion increases in intensity over time and that the sooner one starts downregulating that emotion, the more effective that intervention would be. Therefore, according to a generic timing hypothesis of ER, early interventions would be more effective than late interventions (Sheppes and Gross 2011). More specifically, the process model of ER distinguishes between antecedent-focused ER and consequence-focused ER strategies. The antecedent-focused strategies occur before appraisals give rise to full-blown emotional response tendencies and involve modifying the inputs to the emotional system—that is, changing the external or internal environment, whereas response-focused ER occurs after the responses are generated and refers to altering the behavioral manifestations associated with the emotional response (Gross and Munoz 1995).

Within these two broad classes of ER, more fine-grained distinctions can be made (Frijda 1988; Gross 1998b; Gross and Thompson 2007). On one hand, antecedent-focused ER includes *situation selection*, in which one avoids people or situations that might trigger an emotional reaction; *situation modification*, in which one modifies the environment so as to alter its emotional impact; *attention deployment*, associated with turning the attention away or toward something in order to influence emotions; and *cognitive change*, that means the re-evaluation of the situation one is in or one's capacity to manage the situation to the purpose of altering the emotional response. On the other hand, response-focused ER includes strategies that intensify, diminish, prolong, or curtail ongoing emotional experience, expression, or physiological responding (Gross 1998b). In the context of the generic timing hypothesis, it is thought that antecedent-focused ER strategies are more effective in changing an emotional outcome than response-focused strategies due to the fact that the former ER strategies come into act much sooner than the latter, when the emotional response has not yet reached it's highest intensity levels.

A recent alternative to the generic timing hypothesis is the process-specific timing hypothesis (Sheppes and Gross 2011). This hypothesis has the following main ideas: (a) time of the initiation of a regulatory process is essential for some of the ER strategies and not so important for others; and (b) higher levels of emotional intensity reduce the effectiveness of late ER strategies. Based on this perspective, it is possible to further distinguish between the costs and benefits of different ER strategies among the antecedent-focused category. More specifically, recent investigations were aimed at comparing distraction, as an early selection strategy and reappraisal, as a late selection strategy (Sheppes and Gross 2011; McRae et al. 2010; Ochsner and Gross 2005). So far, the main results indicate that early selection regulation strategies are relatively uninfluenced by the intensity of the emotion, whereas the effectiveness of late selection strategies is reduced for intense emotions, although the impact of intense emotions is moderated by the relative strength of the ER strategy that is being used (Sheppes and Gross 2011). In the following section we will present empirical results most relevant to economics that compare the effectiveness of distraction, reappraisal, and suppression.

7.3 Cognitive and Behavioral Effects of Emotion Regulation

Until recently, most of the research done in this area was focused on comparing the cost and benefits of antecedent and response-focused ER. One form of antecedent-focused ER that has received particular attention is cognitive reappraisal (Gross 2002; John and Gross 2007; Ochsner and Gross 2007; Siemer et al. 2007). This strategy implies changing the situation's meaning so that it alters its emotional impact. From the response-focused ER strategies category, expressive suppression is the most studied, which involves inhibiting ongoing emotion-expressive behaviors (Gross 1998a, b; John and Gross 2007; Ochsner and Gross 2007).

Individual differences in ER (e.g., the habitual use of expressive suppression or cognitive reappraisal) have been shown to impact affect, social functioning, well-being, cognitive functioning, and physiological activation (John and Gross 2004). Habitual suppressors experience more negative than positive emotions, they have poorer interpersonal functioning, and report reduced well-being in comparison to habitual reappraisers (Gross and John 2003). Suppressing negative emotions is associated with decreased emotion expression and unchanged emotional experience, whereas reappraisal decreases both the emotional experience and expression of negative emotions (Gross 1998b). In contrast, suppression of positive emotions diminishes both their expression and experience (Gross 1998b; Gross and Levenson 1997; Stepper and Strack 1993; Strack et al. 1988). Neuroimaging data indicates that suppression increases activation in the emotion-generative brain regions, such as the insula and the amygdala, whereas reappraisal is associated with increased early prefrontal cortex responses, frequently activated in executive cognitive control tasks (Goldin et al. 2008).

In a study that investigated the cognitive demands of reappraisal and suppression, Richards and Gross (2000) concluded that suppression, as opposed to reappraisal, resulted in memory impairment for social information presented while participants were regulating their emotions. Academic performance was shown to be decreased in suppressors, but not cognitive reappraisal, through a mechanism that involved working memory decrements (Johns et al. 2008). The cognitive costs of suppression are accompanied by social and physiological costs as well, for it was experimentally shown that interacting with a partner who used suppression was more stressful than interacting with a partner who used reappraisal, as indexed by increases in blood pressure (Butler et al. 2003). Moreover, a series of studies (Gross and Levenson 1993, 1997) indicated that suppression produces a mixed pattern of physiological responses, including decreased somatic activity and heart rate, but increased signs of sympathetic activation of the electrodermal and cardiovascular systems.

Motivated by the theoretical support of the process-specific timing hypothesis, several studies also contrasted reappraisal and distraction. Distraction is an early selection regulatory strategy that involves diverting attention away from an emotional situation by loading working memory with independent neutral contents

(Van Dillen and Koole 2007). In consequence, distraction acts like a very powerful filter that prevents the processing of an emotional stimulus.

The upside effects of using distraction compared to reappraisal include down-regulating both high-and low-intensity negative emotions (Sheppes and Meiran 2007), and minimal cognitive resources requirements even under high levels of emotional intensity (Sheppes et al. 2009; Sheppes and Meiran 2008). In contrast, reappraisal can successfully reduce only the impact of low-intensity negative emotions, whereas the reappraisal of high intensity negative affect might result in counterintentional increases in negative emotional states and a resulting inability of further regulating these escalating emotional states (Sheppes and Meiran 2007). Nevertheless, when reappraisal was given more time to operate, it successfully reduced high-intensity negative emotions (Sheppes et al. 2009; Sheppes and Meiran 2007, 2008).

The use of distraction to downregulate negative emotions might also have some downside effects. For instance, participants who were instructed to distract themselves while watching a sad movie demonstrated impaired memory for the emotional details of the movie; meanwhile, reappraisers had intact memory both for the emotional and neutral details of the movie (Sheppes and Meiran 2007, 2008). Another possible problem with distraction might stem from its long-term use: even though distraction is highly effective in reducing participants' sadness on the short run, it does not change the way participants evaluate and respond to negative emotional experiences in the long run (Kross and Ayduk 2008).

7.4 Emotion Regulation: Confound in the Effect of Emotions on Economic Decision-Making?

Because ER is widespread in our daily lives, it is possible that it might actually mediate the involvement of emotion in economic decision-making. Most of the previous studies on emotion and decision-making have not controlled for ER. Therefore, effects on economic decision-making, ranging from "coloring" the content of thoughts to interfering with information processing, which have been previously attributed to acute emotions might actually be mediated by ER strategies such as cognitive reappraisal or expressive suppression.

The important role of ER in decision-making is supported by at least four lines of evidence: (1) emotions are frequently regulated, in a spontaneous or incidental manner; the ubiquity of ER in situations that trigger emotions makes difficult the isolation of the direct and specific effects of emotion; (2) the distinct effects of specific emotions on decision-making are explained by differences in the underlying pattern of appraisals, particularly on the certainty and control dimensions (Lerner and Keltner 2000; also see the next paragraph); by effectively downregulating emotion experience, ER contributes to an increased sense of emotional control that might influence decision-making; (3) recent neuropsychological studies indicated that certain brain lesions (e.g., ventromedial prefrontal cortex) have detrimental effects on both economic behavior (e.g., bargaining behavior) and emotion regulation (e.g., Koenigs and Tranel 2007); at the same time, pharmacological manipulations of serotonin signaling, which very likely affect prefrontal functioning, influence both inequity aversion in economic bargaining, and ER (e.g., Crockett, Clark, Tabibnia, Lieberman, and Robbins 2008); and (4) both ER, and decision-making dimensions that are critically influenced by emotions (e.g., risk-taking, susceptibility to framing, bargaining behavior) depend on similar emotion-cognition brain hubs (Pessoa 2008), such as increased functional coupling prefrontal-amygdala circuits (e.g., Goldin et al. 2008; De Martino et al. 2006). A recent review documented the common neural mechanisms that underlie ER and decision-making, by focusing on the involvement of ventrolateral, medial, dorsomedial, and dorsolateral prefrontal cortex in both ER strategies and reversal learning (i.e., the capacity to alter choice behavior when the value of response options change) (see Mitchell 2011).

There are at least two mechanisms by which ER can influence economic decision-making. One, an "emotional" route, stems from differences between reappraisal and suppression in their effectiveness in mitigating the experience of negative and positive emotions (Gross 2002). The second, a "nonemotional" route, stems from differences in the level of effort (cognitive load) required to implement reappraisal or suppression, which could perhaps be related to differences between their respective contributions to ego depletion (Baumeister 2003; Richards and Gross 1999). For instance, in comparison to cognitive reappraisal that diminishes emotion at an early stage and without the need of sustained effort over time, expressive suppression instead involves increased efforts to actively inhibit prepotent emotional responses (Gross and Thompson 2007). Richards and Gross (1999, 2000) have invoked the "nonemotional route" (i.e., differences in computational resources taken away by ER from online information processing) for explaining why expressive suppression, but not cognitive reappraisal, impairs declarative memory. In one of our studies (Heilman et al. 2010), we tested the influence of two ER strategies (i.e., cognitive reappraisal and expressive suppression) on the effects of negative and positive emotions on economic decision-making under uncertainty and risk. The regulation of negative affect allowed us to contrast cognitive reappraisal, which effectively reduced the experience of emotion, and the ineffective expressive suppression. Cognitive reappraisal, but not expressive suppression reduced the effect of negative emotions on economic decision-making. The regulation of positive affect offered a situation in which both cognitive reappraisal and expressive suppression are effective in reducing the experience of emotion, so the only difference that remained was in the cognitive load associated with each of these ER strategies. In this condition, both reappraisal and suppression influenced the effects of positive emotions on decision-making. Therefore, this study suggests that ER impacts economic decision-making by its effects on reducing the experience of emotions (the emotional route), rather than ego depletion.

7.5 Emotion Regulation and Economic Risk-Taking

Risk is defined as the probability, size or subjective "negative utility" of a potential loss (Vlek and Stallen 1980). Although scientists have distinguished decision-making under uncertainty or risk based on the completeness of information about the prospects (i.e., probabilities of costs and benefits associated with the alternatives), most decisions involve some degree of risk in the sense that action-outcome associations are probabilistic (Rangel et al. 2008). Various measures of economic risk-taking exist, ranging from questionnaires such as the Domain-Specific Risk-Taking questionnaire (Blais and Weber 2006) to behavioral tasks such as the Balloon Analogue Risk Task (BART) (Lejuez et al. 2002) or gambling tasks that allow the estimation of utility functions that drive choices (e.g., Tom et al. 2007). These and other similar measures are the workhorses of a growing literature that investigates the effects of emotion, and more recently emotion regulation on economic decision under risk.

Studies first focused on the effects of a single emotion on economic risk-taking, or contrasted emotions of different valence (i.e., pleasant vs. unpleasant). For instance, Gaul (1977) was probably the first to report that anxiety biased attention to losses in a task that involved choosing between different lotteries (see also Lauriola and Levin 2001; Miu et al. 2008). It was suggested that valence is the crucial dimension that determines the direction of the effects of emotion on economic decision-making. Therefore, distinct emotions of the same valence (e.g., sadness, anger, fear) were assumed to have similar effects of decision-making. Based on the cognitive-appraisal theories of emotion, Lerner and Keltner (2000) argued that emotions of the same valence, which differ in their underlying appraisals of the target stimulus, would exert different effects on decision-making. Particularly, the appraisal dimensions of certainty and control would influence risk perception because they map onto the factors of "unknown risk" and "dread risk" that determine risk assessments (see Slovic et al. 1986). Therefore, fear and anger would influence risk-taking in opposing directions because the former is supported by appraisals of low certainty over an outcome and low control over a situation, whereas the latter is supported by appraisals of high certainty and high individual control (Lerner and Keltner 2000; see also Smith and Ellsworth 1985). In a similar vein, Raghunathan and Pham (1999) argued that emotions of the same valence, but differing in their underlying appraisal pattern, would have distinct effects on decision-making because they may activate different implicit goals. They tested the effects of anxiety and sadness on performance in a task that involved choosing between a gamble associated with a lower risk and a lower reward, and one associated with a higher risk and a higher reward. They hypothesized that anxiety would bias preferences toward low-risk/low-reward options because the appraisal pattern underlying this emotion may activate an implicit goal of uncertainty reduction and risk avoidance. In contrast, sadness would bias preferences toward high-risk/high-reward options because this emotion activates an implicit goal of reward substitution (since sadness is based on the perceived loss or absence of a cherished object or person). Their results confirmed these predictions, and further showed that anxiety and sadness influenced economic risk-taking only when the participants evaluated the gambles from their own perspective, and not when they evaluated the gambles as if they were making the decision for someone else (Raghunathan and Pham 1999). Therefore, anxiety and sadness seem to impact risk-taking by a consciously controlled monitoring of feelings, when people use a "What would I feel better about…?" heuristic to assess the implications of decisions that would affect themselves—presumably, feelings are less relevant when one makes decisions on behalf of someone else. These results, together with those of Lerner and Keltner (2000, 2001), indicated that appraisal dimensions and their motivational influences explain the distinct effects that emotions of the same valence may have on economic risk-taking, and emphasized the need to examine the effects of specific emotions on decision-making.

By modulating emotional experience, ER strategies contribute to an increased sense of emotional control that might influence risk-taking. Considering that people typically make efforts to control their emotions (Gross 2002), it is thus possible that ER dampens effects of acute emotions on decision making. This possibility was extensively tested by our group in a couple of studies that directly investigated the influence of two of the most commonly used ER strategies on the effects of emotions on economic risk-taking (Heilman et al. 2010). The first experiment compared the influence of cognitive reappraisal and expressive suppression that the participants were instructed to use in order to regulate the fear and disgust induced by two film clips. In line with Lerner and Keltner's approach (Lerner and Keltner 2000), we were interested in fear and disgust because they fall at opposite ends of the certainty dimension of appraisal. After viewing the emotion-inducing films, the participants played BART, a measure of risk-taking in which financial rewards are earned by pumping balloons with variable explosion thresholds, presented on a computer screen. Risk-taking is defined in terms of mean pumps per unexploded balloon. The main finding from this study was that the use of cognitive reappraisal, which effectively downregulated fear and disgust, also mitigated against the risk aversion associated with these negative emotions. In a second study, we tested whether this effect also held for the incidental use of cognitive reappraisal in a natural situation that triggered positive and negative affect. The results confirmed that cognitive reappraisal effectively reduced the negative affect induced by learning the previously overestimated results of an exam, and consequently decreased risk aversion in a BART that was played immediately after learning the exam result. In summary, these studies showed that by downregulating the experience of negative emotions, cognitive reappraisal reduces economic risk aversion. Based on previous functional neuroimaging studies on cognitive reappraisal (Goldin et al. 2008) and BART performance (Rao et al. 2008), we suggested that the effects of cognitive reappraisal on economic risk-taking might be supported by increased functional coupling between frontal and mesolimbic brain regions (Heilman et al. 2010).

Several recent studies have investigated emotion regulation strategies that are similar to cognitive reappraisal. Two of these studies focused on the regulation of reward expectancy, by relaxing imagery (Delgado et al. 2008; Martin and Delgado 2011). This approach is based on the assumption that ER of reward-related emotions such as anticipation or excitement could prevent approach-related behaviors associated with such positive emotions from becoming maladaptive (e.g., drug seeking, compulsive gambling) in the long-term. Down-regulating the emotional response to a conditioned stimulus that predicted a monetary reward, by thinking of an image that helped calming down, effectively decreased physiological arousal (i.e., skin conductance responses) linked to the anticipation of rewards (Delgado et al. 2008). This imagery-focused regulation strategy also attenuated the cerebral activation in the striatum that was triggered by reward-predicting conditioned stimuli. Building on this study, Martin and Delgado (2011) examined whether a similar imagery-focused strategy during the presentation of a cue that preceded a financial decision-making phase affected economic risk-taking. Briefly, participants were presented with one of two cues: the image of a slot machine, which predicted that a monetary decision between a safe and a risky lottery followed, or the image of a stamp machine, which predicted a non-monetary choice between stamps. During the presentation of the cues, each participant was prompted to imagine a calming (e.g., a sunny day in the park) or exciting scene (e.g., a roller coaster ride), and think of the same image each time the word "relax" was presented above the picture. In the control condition, the participants just thought about the decision that followed, either financial or non-financial. The main findings from this study (Martin and Delgado 2011) were that the imagery-focused strategy that downregulated the response to the anticipation of monetary gambles reduced the number of risky choices in the decision phase, and attenuated the activation of the striatum to risky choices. Therefore, regulating the emotional response by means of relaxing imagery modulated the neural response (i.e., striatum activation) to the anticipation of monetary rewards, and promoted goal-directed behaviors (i.e., less risky choices).

A third study (Sokol-Hessner et al. 2009) investigated a different regulation strategy that involved taking the perspective of a trader who would treat each monetary decision, even those associated with losses, as one of many decisions, which will be summed together to produce a portfolio. This type of perspective taking resembles cognitive reappraisal in that it involves an intentional reinterpretation of the stimulus. Using an economic game in which participants made forced choices between a binary gamble (i.e., mixed valence gambles with positive or negative outcomes, and gain-only gambles with positive or zero outcomes) and a guaranteed amount (i.e., zero or positive amounts), Sokol-Hessner et al. (2009) found that taking the perspective of a trader reduces behavioral loss aversion, and the physiological arousal to losses.

7.6 Emotion Regulation and Susceptibility to Framing

For centuries, economists referred to the normative models when judging whether a decision is rational or not. Although the definition of rationality has been largely debated, there is a general agreement that rational choices should, among others, satisfy an invariance requirement (Tversky and Kahneman 1981). According to the invariance principle, the preference between options should be independent of their description (Tversky and Kahneman 1986). In other words, when the options of the same situation are presented in different frames, this presentation should not influence people's preference for one option or another. However, an extensive body of evidence piles against the rationality of the decision-maker by proving that people do not act according to this principle and they are, in fact, predisposed to persistent decisional biases.

One of the most studied violations of the invariance principle is the framing effect, where extensionally equivalent descriptions lead to different choices by altering the relative salience of different aspects of the problem (Kahneman 2003). This effect was first demonstrated using the *Asian Disease Problem* (Tversky and Kahneman 1981), which showed that people display risk aversion when alternatives are framed as gains, and risk seeking when objectively equivalent alternatives are framed as losses (Tversky and Kahneman 1981). Ever since the first appearance of this classic problem, hundreds of studies have been published that provide further support for this general decisional bias.

Among the task-related aspects that were invoked to explain this decisional preference, in the last decade scholars have turned their attention to the role that emotions might play in the evaluation of the options (Kahneman 2003) and even more recently ER became part of the explanation. A study conducted by De Martino et al. (2006) supported the role of emotions in the framing effect, by showing that this decisional bias was associated with increased amygdala activity and was negatively predicted by orbital and medial prefrontal cortex activity. Thus, the authors conclude that the framing effect might be the consequence of an affect heuristic by which individuals incorporate emotional information in their decisional process (De Martino et al. 2006). However, this interpretation was recently challenged by another study (Talmi et al. 2010). These recent results showed that patients with selective amygdala lesions resulting from Urbach-Wiethe disease display the same framing effect as neurologically intact controls, with the sole difference that the patient group manifested a higher frequency of risk-taking in both frames. Other studies provide further support for the ER-decisional processes interaction by implying that ER impairments due to ventromedial prefrontal cortex lesions (Koenings and Tranel 2007), dysfunctional serotonin signaling (Crokett et al. 2008), or a common genetic polymorphism of the human serotonin transporter gene (Crisan et al. 2009; Roiser et al. 2009) might account for irrational economic decisions.

Miu and Crişan (2011) conducted the first study that directly tested the effects of ER on the susceptibility to framing in a gambling task (see De Martino et al. 2006). These results indicated that the use of cognitive reappraisal, but not expressive suppression during the gambling task reduced the susceptibility to framing (i.e., it increased decision invariance). Building on these results, we further investigated the effects of individual differences in ER on risk attitudes in framing problems that addressed aspects related to health, financial, and nature issues (Heilman and Miclea 2016). In accordance with the large majority of studies that have observed a framing effect, we also found that participants showed a significant preference for risky choices in the loss frame and a reduced preference for these choices in the gain frame, for all the three problem domains. More importantly, our results indicated a major impact of ER on risk preference, with ER strategies accounting for up to 46.9 % of the total variance. By analyzing each category of framing problems, we found that regulatory strategies were more relevant for domains related to human life, such as financial or health related issues, than nature-related aspects (Heilman and Miclea 2016). To the best of our knowledge, this is the first study to have investigated the complex interaction between ER and framing effects in problems that mimic various real life situations.

7.7 Emotion Regulation and Economic Bargaining

Cooperation between genetically unrelated people has evolved as an adaptive mechanism for the survival of the species, since many objectives are achieved more efficiently if people cooperate. However, successful cooperation requires complicated decisions on how resources should be divided among collaborators (Van den Bergh and Dewitte 2006). For this purpose, fairness norms are particularly important. The Ultimatum Game (UG) illustrates the tension between self-interest and reciprocity and equity motives in a social decision situation (Guth et al. 1982).

The standard form of the UG involves two players. One of them, known as the proposer, has to make a monetary offer to the second player, the responder, concerning an amount of money that the two must split between them. The responder can either accept or reject the offer. If the offer is accepted, then the money is split as proposed. But if the responder rejects the offer, then neither player receives anything. Both players are fully aware of the rules and consequences of the game. The UG is typically played with real money, provided by the experimenter. Based on the two major economic assumptions regarding human nature, namely, the decision-maker's rationality and his/her self-regarding preferences (Camerer and Fehr 2006), the normative solution for this decision-making task would be for the proposer to offer as little money as possible and for the responder to accept any nonzero offer. Nevertheless, the large majority of the proposers offer about 50 % of the pot to the responders, and responders accept roughly only half of the unfair offers, defined as 30 % or less of the total amount of money (Camerer 2003).

Numerous scholars advanced theories in attempt to explain these violations of the economic predictions. Some of the most influential theories of UG behavior focus on inequity aversion (Bergh 2008; Bolton and Ockenfels 2000; Fehr and Schmidt 1999). An UG player displays inequity aversion if s/he dislikes and sanctions outcomes that are inequitable, particularly when they are to his/her disadvantage (Fehr and Schmidt 1999). The original models of inequity aversion and UG behavior neglected "non-consequentialist" reasons for reciprocally fair action (Falk et al. 2003, 2008; Falk and Fischbacher 2006). They have thus been modified to include essential aspects such as the consequences of actions and the underlying intentions (Falk and Fischbacher 2006). Nonetheless, even these expanded models fail to a certain degree to describe and predict decision-making accurately. The impact of a number of variables (e.g., property rights, social sharing, emotions, and biological influences) has been investigated and we will review some of the major findings in the following paragraphs.

When facing an unfair offer, the cognitive goal of gaining money and the affective goal of resisting unfairness are in conflict (Sanfey et al. 2003). As a consequence, negative emotional reactions to ultimatum offers might prevail over self-interest motives and result in higher rejection rates. For instance, previous studies found that the probability of rejection is positively correlated with the intensity of self-reported negative emotions (Pillutla and Murnighan 1996; Bosman et al. 2001; Oechssler et al. 2008). In particular, incidental sadness results in lower acceptance rates (Harle and Sanfey 2007), whereas another study found that clinical depression is associated with increased responders' rationality, by accepting more frequently unfair offers (Harle et al. 2010). Furthermore, Xiao and Houser (2005) found that rejection rates fall when responders can express their negative emotions directly to proposers. Other scholars have investigated various biological correlates of behavioral responses in the UG. For instance, emotional responses at the physiological level have also been associated with offer acceptance, indicating that the rejection of unfair offers from human proposers triggers a higher skin conductance response, compared to the electrodermal activation in response to unfair offers randomly generated by a computer (van't Wout et al. 2006). High rejection of unfair offers is also associated with lower serotonin levels (Crockett et al. 2008; Emanuele et al. 2008), lower levels of serum omega-3 fatty acids (Emanuele et al. 2009) and it also seems to have a genetic component (Wallace et al. 2007). So far, the studies revealed that sex hormones have mixed effects for men and women. While high levels of testosterone or estrogen have no effects on women's acceptance rates (Zethraeus et al. 2009), men who reject low offers have significantly higher testosterone levels than those who accept (Burnham 2007), and lower 2D:4D digit ratio, known as an index of prenatal exposure to androgens (Van den Bergh and Dewitte 2006). Neuroimaging studies highlighted several brain areas relevant for responders' behavior. A study conducted by Sanfey et al. (2003) indicated that rejection of unfair offers is associated with activation in the anterior insula, which scholars have linked to the experience of anger and disgust, Koenigs and Tranel (2007) found that patients with ventromedial prefrontal cortex lesions rejected more unfair offers compared to the control group, whereas another research group showed that temporary inhibition of the right dorsolateral prefrontal cortex reduces responders' willingness to reject their partners' intentionally unfair offers Knoch et al. (2006, 2008).

While extensive research has been conducted on responders' behavior, affective aspects of the proposers' behavior have been almost entirely disregarded (Cappelletti et al. 2008). For instance, whether or not a proposer is relying on his/her personal emotions affects his/her offer to the responder (Stephen and Pham 2008). Other factors were also investigated in relation to proposers' offers. More specifically, a higher frequency of unfair offers was correlated with low levels of serum omega-3 fatty acids (Emanuele et al. 2009) and low serotonin levels (Emanuele et al. 2008).

Since numerous studies in behavioral economics and neuroeconomics demonstrate that emotions have an impact on UG behavior, it is reasonable to assume that ER might also influence decision-making. Even though this line of research is still in its infancy, scholars have already acknowledged that ER might mediate the effects that task-related and incidental emotions have on UG and other types of decision-making (Kahneman and Frederick 2007; Crockett et al. 2008). For instance, one study found that individual differences in ER explain 55 % of the variance in negotiators' profit in a simulated negotiation (Yurtsever 2008). However, only recently was directly addressed a possible mediating effect of ER on the complex relation between emotion and decision-making. In a recent study we used an experimental manipulation that allowed us to test the impact of ER strategies on ultimatum decisions under an anxiety condition. We discovered that habitual suppressors accept more unfair offers than non-habitual suppressors. In addition, the habitual use of expressive suppression and anxiety seem to interact, with the result that habitual suppressors make more unfair offers when they are anxious, and accept more unfair offers when they are not anxious (Heilman et al. in preparation). Therefore, efficient ER strategies might be the key to enabling one to accept unfair offers (Crockett et al. 2008).

7.8 Conclusion

Humans routinely use ER strategies to alter their emotional reactions. This chapter described and discussed several of these strategies. We reviewed a substantial literature showing that the effect of any particular emotion on economic decision-making varies systematically with the type of ER it triggers. For example, we reviewed research showing that ER moderates the impact of fear and disgust on risky decisions, as well as the impact of anxiety on decisions in bargaining. The import of these results to economists derives in part from the fact that ER can be controlled, and even trained. This may have implications for the design of economic institutions to promote improved economic decision-making and so enhance social welfare.

There are a large number of economic environments where ER might be expected to be especially important, but where to the best of our knowledge research has not yet been conducted. Consider, for example, situations that involve temptation. These are important because tempting desires can interfere with the ability to attain one's long-term goals. Consequently, the ability to resist temptation is a necessary skill. Unfortunately, self-control is difficult. It seems of particular interest, and of especially high social value, to understand how different ER strategies might impact the ability to successfully implement self-control in order to avoid succumbing to temptation.

Temptation is one among many economic domains where continued research in ER is sure to be profitably conducted. All such research is fundamentally interdisciplinary in nature, bringing together leading scholars in economics, psychology, and neuroscience. We look forward to the results of these synergistic interactions, and share an optimistic excitement for the many discoveries that lie ahead.

Acknowledgments Writing this paper was supported by grant grant number PCE-ID-2012-4-0498 from the Romanian National Authority for Scientific Research, CNCS–UEFISCDI to A.C.M. and by CNCSIS—UEFISCDI, project number PNII—IDEI, 2440/2008.

References

Baumeister RF (2003) Ego depletion and self-regulation failure: a resource model of self-control. Alcohol Clin Exp Res 27(2):281–284

Bergh A (2008) A critical note on the theory of inequity aversion. J Soc Econ 37:1789–1796

Blais AR, Weber EU (2006) A domain-specific risk-taking (DOSPERT) scale for adult populations. Judgm Decis Making 1:33–47

Bolton G, Ockenfels A (2000) A theory of equity, reciprocity and competition. Am Econ Rev 90:166–193

Bosman R, Sonnemans J, Zeelenberg M (2001) Emotions, rejections, and cooling off in the ultimatum game. Unpublished working paper. University of Amsterdam

Burnham TC (2007) High-testosterone men reject low ultimatum game offers. Proc Biol Sci 274 (1623):2327–2330

Butler EA, Egloff B, Wilhelm FH, Smith NC, Erickson EA, Gross JJ (2003) The social consequences of expressive suppression. Emotion 3(1):48–67

Camerer CF (2003) Behavioral game theory—experiments in strategic interaction. Princeton University Press, Princeton

Camerer CF, Fehr E (2006) When does "economic man" dominate social behavior? Science 311 (5757):47–52

Campos JJ, Barrett KC, Lamb ME, Goldsmith MH, Stenberg C (1983) Socioemotional development. In: Haith M, Campos JJ (eds) Handbook of child psychology: vol. 2. Infancy and developmental psychology. Wiley, New York, pp 783–915

Campos JJ, Campos RG, Barrett KC (1989) Emergent themes in the study of emotional development and emotion regulation. Dev Psychol 25:394–402

Cappelletti D, Guth W, Ploner M (2008) Being of two minds: an ultimatum experiment investigating affective processes. Unpublished manuscript. Jena Economic Research Papers

Crisan LG, Pana S, Vulturar R, Heilman RM, Szekely R, Druga B et al (2009) Genetic contributions of the serotonin transporter to social learning of fear and economic decision making. Soc Cogn Affect Neurosci 4(4):399–408

Crockett MJ, Clark L, Tabibnia G, Lieberman MD, Robbins TW (2008) Serotonin modulates behavioral reactions to unfairness. Science 320:1739

De Martino B, Kumaran D, Seymour B, Dolan RJ (2006) Frames, biases, and rational decision-making in the human brain. Science 313(5787):684–687

Delgado MR, Gillis MM, Phelps EA (2008) Regulating the expectation of reward via cognitive strategies. Nat Neurosci 11(8):880–881

Edwards W (1962) Subjective probabilities inferred from decisions. Psychol Rev 69:109–135

Ekman P (1972) Universal and cultural differences in facial expression of emotion. In: Cole J (ed) Nebraska symposium in motivation. University of Nebraska Press, Lincoln, pp 207–283

Ekman P (1992) An argument for basic emotions. Cogn Emot 6:169–200

Elster J (1998) Emotions and economic theory. J Econ Lit Am Econ Assoc 36(1):47–74

Emanuele E, Brondino N, Bertona M, Re S, Geroldi D (2008) Relationship between platelet serotonin content and rejections of unfair offers in the ultimatum game. Neurosci Lett 437 (2):158–161
Emanuele E, Brondino N, Re S, Bertona M, Geroldi D (2009) Serum omega-3 fatty acids are associated with ultimatum bargaining behavior. Physiol Behav 96(1):180–183
Falk A, Fischbacher U (2006) A theory of reciprocity. Game Econ Behav 54:293–315
Falk A, Fehr E, Fischbacher U (2003) On the nature of fair behavior. Econ Inq 41:20–26
Falk A, Fehr E, Fischbacher U (2008) Testing theories of fairness-intentions matter. Game Econ Behav 62:287–303
Fehr E, Schmidt KM (1999) A theory of fairness, competition and cooperation. Q J Econ 14:815–848
Frijda NH (1988) The laws of emotion. Am Psychol 43(5):349–358
Gaul M (1977) Influences of anxiety level on evaluation of dimensional importance in risky tasks. Pol Psychol Bull 8:165–170
Goldin PR, McRae K, Ramel W, Gross JJ (2008) The neural bases of emotion regulation: reappraisal and suppression of negative emotion. Biol Psychiatry 63(6):577–586
Gross JJ (1998a) Antecedent- and response-focused emotion regulation: divergent consequences for experience, expression, and physiology. J Pers Soc Psychol 74(1):224–237
Gross JJ (1998b) The emerging field of emotion regulation: an integrative review. Rev Gen Psychol 2:271–299
Gross JJ (2002) Emotion regulation: affective, cognitive, and social consequences. Psychophysiology 39(3):281–291
Gross JJ, John OP (2003) Individual differences in two emotion regulation processes: implications for affect, relationships, and well-being. J Pers Soc Psychol 85(2):348–362
Gross JJ, Levenson RW (1993) Emotional suppression: physiology, self-report, and expressive behavior. J Pers Soc Psychol 64(6):970–986
Gross JJ, Levenson RW (1997) Hiding feelings: the acute effects of inhibiting negative and positive emotion. J Abnorm Psychol 106(1):95–103
Gross JJ, Munoz RF (1995) Emotion regulation and mental health. Clin Psychol Sci Pract 2:151–164
Gross JJ, Thompson RA (2007) Emotion regulation: conceptual foundations. In: Gross JJ (ed) Handbook of emotion regulation. The Guilforg Press, New York, pp 3–26
Guth W, Schmittberger R, Schwarze B (1982) An experimental analysis of ultimatum bargaining. J Econ Behav Organ 75:367–388
Harle KM, Sanfey AG (2007) Incidental sadness biases social economic decisions in the ultimatum game. Emotion 7(4):876–881
Harle KM, Allen JJ, Sanfey AG (2010) The impact of depression on social economic decision making. J Abnorm Psychol 119(2):440–446
Heilman RM, Miclea M (2016) Risk seeking preferences: an investigation of framing effects across decisional domains. Cogn, Brain, Behav: Interdisc J 20:1–17
Heilman RM, Miu AC, Houser D Emotion regulation modulates inequity aversion in an ultimatum game. (in preparation)
Heilman RM, Crisan LG, Houser D, Miclea M, Miu AC (2010) Emotion regulation and decision making under risk and uncertainty. Emotion 10(2):257–265
Izard CE (1990) Facial expressions and the regulation of emotions. J Pers Soc Psychol 58 (3):487–498
John OP, Gross JJ (2004) Healthy and unhealthy emotion regulation: personality processes, individual differences, and life span development. J Pers 72(6):1301–1333
John OP, Gross JJ (2007) Individual differences in emotion regulation. In: Gross JJ (ed) Handbook of emotion regulation. The Guilford Press, New York, pp 351–372
Johns M, Inzlicht M, Schmader T (2008) Stereotype threat and executive resource depletion: examining the influence of emotion regulation. J Exp Psychol Gen 137(4):691–705
Kahneman D (2003) A perspective on judgment and choice: mapping bounded rationality. Am Psychol 58(9):697–720

Kahneman D, Frederick S (2007) Frames and brains: elicitation and control of response tendencies. Trends Cogn Sci 11(2):45–46
Kahneman D, Tversky A (1979) Prospect theory: an analysis of decision under risk. Econometrica 47(2):263–292
Knoch D, Pascual-Leone A, Meyer K, Treyer V, Fehr E (2006) Diminishing reciprocal fairness by disrupting the right prefrontal cortex. Science 314(5800):829–832
Knoch D, Nitsche MA, Fischbacher U, Eisenegger C, Pascual-Leone A, Fehr E (2008) Studying the neurobiology of social interaction with transcranial direct current stimulation—the example of punishing unfairness. Cereb Cortex 18(9):1987–1990
Koenigs M, Tranel D (2007) Irrational economic decision-making after ventromedial prefrontal damage: evidence from the ultimatum game. J Neurosci 27(4):951–956
Kross E, Ayduk O (2008) Facilitating adaptive emotional analysis: distinguishing distanced-analysis of depressive experiences from immersed-analysis and distraction. Pers Soc Psychol Bull 34(7):924–938
Lauriola M, Levin IP (2001) Personality traits and risky decision making in a controlled experimental task: an exploratory study. Personality Individ Differ 31:215–226
Lazarus RS (1991) Emotion and adaptation. Oxford University Press, Oxford
Lejuez CW, Read JP, Kahler CW, Richards JB, Ramsey SE, Stuart GL et al (2002) Evaluation of a behavioral measure of risk taking: the balloon analogue risk task (BART). J Exp Psychol Appl 8(2):75–84
Lerner JS, Keltner D (2000) Beyond valence: toward a model of emotion-specific influences on judgement and choice. Cogn Emot 14(4):473–493
Lerner JS, Keltner D (2001) Fear, anger, and risk. J Pers Soc Psychol 81(1):146–159
Loewenstein G (2000) Emotions in economic theory and economic behavior. Am Econ Rev Pap Proc 90:426–432
Loewenstein G, O'Donoghue T (2004) Animal spirits: affective and deliberative processes in economic behavior (Electronic version). Retrieved 9 Jan 2008 from http://cbdr.cmu.edu/seminar/Loewenstein.pdf
Martin LN, Delgado MR (2011) The influence of emotion regulation on decision making under risk. J Cogn Neurosci 23:2569–2581
McRae K, Hughes B, Chopra S, Gabrieli JJD, Gross JJ, Ochsner KN (2010) The neural correlates of cognitive reappraisal and distraction: an fMRI study of emotion regulation. J Cogn Neurosci 22:248–262
Mitchell DG (2011) The nexus between decision making and emotion regulation: a review of convergent neurocognitive substrates. Behav Brain Res 217(1):215–231
Miu AC, Crişan LG (2011) Cognitive reappraisal reduces the susceptibility to the framing effect in economic decision making. Personality Individ Differ 51(4):478–482
Miu AC, Miclea M, Houser D (2008) Anxiety and decision-making: toward a neuroeconomics perspective. Adv Health Econ Health Serv Res 20:55–84
Naqvi N, Shiv B, Bechara A (2006) The role of emotion in decision making. a cognitive neuroscience perspective. Curr Dir Psychol Sci 15(5):260–264
Ochsner KN, Gross JJ (2005) The cognitive control of emotion. Trends Cogn Sci 9(5):242–249
Ochsner KN, Gross JJ (2007) The neural architecture of emotion regulation. In: Gross JJ (ed) Handbook of emotion regulation. The Guilford Press, New York, pp 87–109
Oechssler J, Roider A, Schmitz P (2008) Cooling-off in negotiations—does it work? Unpublished working paper. University of Bonn
Pessoa L (2008) On the relationship between emotion and cognition. Nat Rev Neurosci 9 (2):148–158
Peters E, Västfjäll D, Gärling T, Slovic P (2006) Affect and decision making: A "hot" topic. J Behav Decis Making 19:79–85
Pillutla MM, Murnighan JK (1996) Unfairness, anger, and spite: emotional rejections of ultimatum offers. Organ Behav Hum Decis Process 68:208–224
Raghunathan R, Pham MT (1999) All negative moods are not equal: motivational influences of anxiety and sadness on decision making. Organ Behav Hum Decis Process 79(1):56–77

Rangel A, Camerer C, Montague PR (2008) A framework for studying the neurobiology of value-based decision making. Nat Rev Neurosci 9(7):545–556

Rao H, Korczykowski M, Pluta J, Hoang A, Detre JA (2008) Neural correlates of voluntary and involuntary risk taking in the human brain: an fMRI study of the balloon analog risk task (BART). Neuroimage 42(2):902–910

Richards JM, Gross JJ (1999) Composure at any cost? The cognitive consequences of emotion suppression. Pers Soc Psychol Bull 25(8):1033–1044

Richards JM, Gross JJ (2000) Emotion regulation and memory: the cognitive costs of keeping one's cool. J Pers Soc Psychol 79(3):410–424

Roiser JP, de Martino B, Tan GC, Kumaran D, Seymour B, Wood NW et al (2009) A genetically mediated bias in decision making driven by failure of amygdala control. J Neurosci 29 (18):5985–5991

Sanfey AG, Rilling JK, Aronson JA, Nystrom LE, Cohen JD (2003) The neural basis of economic decision-making in the ultimatum game. Science 300(5626):1755–1758

Sheppes G, Gross JJ (2011) Is timing everything? Temporal considerations in emotion regulation. Pers Soc Psychol Rev 15:319–331

Sheppes G, Meiran N (2007) Better late than never? On the dynamics of online regulation of sadness using distraction and cognitive reappraisal. Pers Soc Psychol Bull 33(11):1518–1532

Sheppes G, Meiran N (2008) Divergent cognitive costs for online forms of reappraisal and distraction. Emotion 8(6):870–874

Sheppes G, Catran E, Meiran N (2009) Reappraisal (but not distraction) is going to make you sweat: physiological evidence for self-control effort. Int J Psychophysiol 71(2):91–96

Siemer M, Mauss I, Gross JJ (2007) Same situation—different emotions: how appraisals shape our emotions. Emotion 7:592–600

Slovic P, Fischhoff B, Lichtenstein S (1986) The psychometric study of risk perception. In: Covello VT, Menkes J, Mumpower J (eds) Risk evaluation and measurement. Plenum, New York, pp 3–24

Slovic P, Finucane ML, Peters E, MacGregor DG (2007) The affect heuristic. Eur J Oper Res 177:1333–1352

Smith CA, Ellsworth PC (1985) Patterns of cognitive appraisal in emotion. J Pers Soc Psychol 48 (4):813–838

Sokol-Hessner P, Hsu M, Curley NG, Delgado MR, Camerer CF, Phelps EA (2009) Thinking like a trader selectively reduces individuals' loss aversion. Proc Natl Acad Sci USA 106(13):5035–5040

Stephen AT, Pham MT (2008) On feelings as a heuristic for making offers in ultimatum negotiations. Psychol Sci 19(10):1051–1058

Stepper S, Strack F (1993) Proprioceptive determinants of emotional and nonemotional feelings. J Pers Soc Psychol 64:211–220

Strack F, Martin LL, Stepper S (1988) Inhibiting and facilitating conditions of the human smile: a nonobtrusive test of the facial feed-back hypothesis. J Pers Soc Psychol 54:768–777

Talmi D, Hurlemann R, Patin A, Dolan RJ (2010) Framing effect following bilateral amygdala lesion. Neuropsychologia 48(6):1823–1827

Thompson RA (1990) Emotion and self-regulation. In: Thompson RA (ed) Socioemotional development. Nebraska symposium on motivation, vol 36. University of Nebraska Press, Lincoln, pp 367–467

Thompson RA (1991) Emotional regulation and emotional development. Educ Psychol Rev 3:269–307

Tom SM, Fox CR, Trepel C, Poldrack RA (2007) The neural basis of loss aversion in decision-making under risk. Science 315(5811):515–518

Tversky A, Kahneman D (1981) The framing of decisions and the psychology of choice. Science 211(4481):453–458

Tversky A, Kahneman D (1986) Rational choice and the framing of decisions. J Bus 59 (4):251–278

Van den Bergh B, Dewitte S (2006) Digit ratio (2D:4D) moderates the impact of sexual cues on men's decisions in ultimatum games. Proc Biol Sci 273(1597):2091–2095

Van Dillen LF, Koole SL (2007) Clearing the mind: a working memory model of distraction from negative mood. Emotion 7(4):715–723

van't Wout M, Kahn RS, Sanfey AG, Aleman A (2006) Affective state and decision-making in the ultimatum game. Exp Brain Res 169(4):564–568

Vlek C, Stallen PJ (1980) Rational and personal aspects of risk. Acta Psychol 45:273–300

Wallace B, Cesarini D, Lichtenstein P, Johannesson M (2007) Heritability of ultimatum game responder behavior. Proc Natl Acad Sci USA 104(40):15631–15634

Xiao E, Houser D (2005) Emotion expression in human punishment behavior. Proc Natl Acad Sci USA 102(20):7398–7401

Yurtsever G (2008) Negotiators' profit predicted by cognitive reappraisal, suppression of emotions, misrepresentation of information, and tolerance of ambiguity. Percept Mot Skills 106:590–608

Zethraeus N, Kocoska-Maras L, Ellingsen T, von Schoultz B, Hirschberg AL, Johannesson M (2009) A randomized trial of the effect of estrogen and testosterone on economic behavior. Proc Natl Acad Sci USA 106(16):6535–6538

Chapter 8
How the Experience of Time Shapes Decision-Making

Marc Wittmann and Martin P. Paulus

Abstract We present an outline of a model for how the subjective experience of time influences decision-making. First, an individual's time perspective determines how strongly attention is directed to time. A stronger emphasis on the present perspective at the expense of the future perspective—as seen in impulsive individuals—leads to a stronger focus on the passage of time in waiting situations. This in turn causes longer estimates of duration. In intertemporal decisions, a relative overestimation of duration can lead to the perception of delayed rewards lying too far in the future. As a consequence, the value of a future commodity is discounted and more immediate but less valuable rewards are preferred. We present empirical evidence on the relationship between time perception and intertemporal decision-making and discuss these findings within the respective psychological and neural models.

8.1 Introduction

Time plays a pivotal role in decision-making. Several temporal aspects can be identified which are relevant for different processing stages in decision-making (Ariely and Zakay 2001; Klapproth 2008). For example, time can be a scarce or abundant resource when making decisions: how much time do I have before I must choose an option? Time is a commodity and subject matter of decisions: how many

M. Wittmann (✉)
Institute for Frontier Areas of Psychology and Mental Health,
Wilhelmstr. 3a, 79098 Freiburg, Germany
e-mail: wittmann@igpp.de

M.P. Paulus
Laureate Institute for Brain Research, Tulsa, Oklahoma, USA

M.P. Paulus
Psychiatry Service, VA San Diego Healthcare System La Jolla, San Diego, USA

M.P. Paulus
Department of Psychiatry, University of California, San Diego, La Jolla, CA, USA

M. Reuter and C. Montag (eds.), *Neuroeconomics*, Studies in Neuroscience, Psychology and Behavioral Economics, DOI 10.1007/978-3-642-35923-1_8

days of vacation do I take? In both examples, subjective estimates of duration influence our decisions. Duration and the passage of time are either experienced at present ("at the moment I have plenty of time") or they are anticipated as future intervals ("tomorrow I won't have enough time"). Moreover, future intervals are anticipated in relation to the momentary experience of time, i.e., as waiting time until the occurrence of an event or as waiting time until the end of an ongoing event.

In a classification of subjective time at least three interrelated dimensions can be discerned, which are important for decision-making (Wittmann 2009a): (1) the *time perspectives* of past, present, and future; (2) an individual's *time estimation* abilities as measured by the accuracy in estimating clock time; (3) *time awareness* as the subjective impression of time passing relatively fast or slow. Typically, we flexibly switch between *time perspectives*, either focusing on the past to evaluate present options or imaging the future to create alternative goal states. Decisions thus are based on the tripartite structure of temporal experience with past, present, and future (Zimbardo and Boyd 1999). Choices are also related to the explicit *estimation of duration* and the *awareness of the passage of time* experienced just now and as anticipated for future time intervals. Do we stand in line at the post office or do we move on to run some other errands? At what point in time do we complain when the waiter does not bring the ordered meal? When purchasing goods on the Internet do we pay an extra sum in order to receive the product earlier?

In this chapter, we show how the three interrelated dimensions of subjective time are subject to modifications and how this influences our decisions. In short, we propose that a more pronounced present perspective at the expense of the future perspective leads to a stronger focus on the momentary passage of time, which results in longer duration estimates. A relative overestimation of duration, the feeling of having to wait too long, consequently leads to choices with short-term outcomes over those with long-term consequences (Wittmann and Paulus 2008, 2009). Such shortsightedness in decision-making can lead to negative consequences in the long run, subjective time strongly affecting life achievements (Mischel et al. 1989). Thereafter, the ability to delay gratification is to a certain extent dependent upon the temporal perspective as well as the consequent experience of duration.

In this chapter, we will discuss models of time perception and show how experienced time is subject to alterations. Investigations with impulsive individuals have led the way for understanding the relation between time perception and decision-making. As a paradigmatic example of how modifications of time occur and how these modifications influence decision-making, we will specifically highlight research on addiction. Individuals who are substance dependent show a profound change in subjective time, which is associated with maladaptive decision-making. The empirical findings on additive behavior will be related to models of time perception. Finally, we will summarize results from neuroimaging studies showing how similar brain systems are involved in the estimation of duration and the anticipation of future intervals in intertemporal decision-making.

8.2 Modifications in the Experience of Time

An individual's *time perspective* dynamically changes according to transient situational demands. Present choices are made for future outcomes, reasoning that depends upon past experiences. However, a dominance regarding one of the temporal perspectives can manifest itself as a consequence of life events. Sudden critical life events, i.e., unexpected unemployment or a life-threatening illness (van Laarhoven et al. 2011) can dramatically shorten the future perspective of a person and lead to the dominance of the present perspective as only short-term plans become relevant (Carstensen 2006). In addition, the relative dominance of the present perspective at the expense of the future perspective can be seen as a personality trait in impulsive individuals (Zimbardo and Boyd 1999). Impulsive behavior is defined as reacting to the immediate situation without thinking about future consequences. That is, impulsivity is conceptualized as extreme temporal shortsightedness.[1] For example, a stronger focus on the present at the expense of the future is strongly related to impulsive behavior such as gambling, having unprotected sex, risky driving, or using drugs (Zimbardo et al. 1997; Keough et al. 1999).

In cognitive models of *time estimation* and, related, of *time awareness*, prospective and retrospective time is distinguished (Zakay and Block 1997, 2004). Prospective time estimation is concerned with the perception of duration or the passage of time as presently experienced. An observer directs more or less attention to the passage of time while a particular duration is judged. If more attention is directed to time during an explicit duration estimation task, a temporal interval is experienced as lasting comparatively longer than when attention is distracted from time (Conti 2001). In retrospective time, when we judge the passage of time of an interval that already has passed, duration is reconstructed from memory. The more contextual changes and events experienced during that time span (which are stored in memory and later retrieved), the longer that interval appears to have lasted in retrospect (Flaherty et al. 2005; Bailey and Areni 2006). Similarly, the more event markers defining the duration between the present moment and the past event are remembered, the longer retrospective time (Zauberman et al. 2010). Whereas prospective time perception only applies to intervals up to a few minutes—an assumed prospective timing mechanism has an upper limit of temporal integration—retrospective time estimation can refer to memory contents of an individual's life time (Wittmann and Lehnhoff 2005).

In prospective time two factors lead to modifications in the estimation of duration as explained within the framework of pacemaker-accumulator models

[1]In this context one has to differentiate between an impulsive present orientation and present-mindedness as trained through meditation techniques. The former is associated with a strong urge to act in the present moment, whereas the latter is associated with an observational state associated with more self-control. This is important because meditation is now being considered as a way of treating drug-addicted individuals. The value for meditation is that it can disconnect experience from impulsive action leading to more self-controlled behavior.

(Gibbon et al. 1984; Treisman 1963; Zakay and Block 1997). In these models a pacemaker produces a regular series of pulses that are fed into an accumulator. The number of pulses that has been recorded for a given time span represents experienced duration. However, pulses are only accumulated when attention is actually directed to time. Therefore, and since the number of collected pulses represents duration, more attention to time leads to longer duration estimates. Additionally, an increased arousal level is related to a higher pacemaker rate which in turn leads to greater accumulation of pulses during a given time period—thus causing longer estimates of duration (Burle and Casini 2001; Droit-Volet and Meck 2007; Wittmann and Paulus 2008).

Modifications of subjective duration caused by emotional states are interpreted within the framework of the pacemaker-accumulator models (Droit-Volet 2009; Noulhiane et al. 2007; Wittmann 2009b). First, and related to attention, patients in psychological distress such as those with depression (Bschor et al. 2004) or cancer patients with high levels of anxiety (Wittmann et al. 2006) perceive a slowing of the pace of time and overestimate duration. The presence of emotional distress in these individuals diverts attentional resources away from ongoing thoughts and actions to the passage of time (Twenge et al. 2003). Second, and related to transient arousal states in healthy subjects, stimuli with emotional content are typically judged to last longer than more neutral stimuli (Droit-Volet and Gil 2009). These findings are explained by heightened arousal levels of subjects judging the duration of stimuli with emotional content. For example, participants who watched footage of the 9/11 events perceived the duration of the film to last longer than a more neutral film with the same duration (Anderson et al. 2007). In another study, young healthy male college students first had to rate photographs of sexually attractive female lingerie models. In a following session, subjects anticipated future intervals to last longer than subjects who had previously rated less exiting photographs (Kim and Zauberman 2012).

Studies with impulsive individuals provide the empirical basis for conceptual explanations of how subjective time influences decision-making. Experimentally, analyzing performance in intertemporal decision tasks can reveal the temporal shortsightedness in impulsive individuals. Participants have to choose between smaller rewards that can be consumed earlier and rewards that can be obtained only later. Most of people have the tendency to discount future rewards, the subjective value of future rewards decreases considerably with increasing waiting time (Ainslie 1975; Prelec and Loewenstein 1997). Given the choice, individuals are more likely to prefer an immediate reward over a later reward (even when it has greater value). This tendency becomes stronger when waiting times for the delayed reward increase. The discounting function is steeper at shorter delays (the rate of discounting is higher) and becomes flatter as the delay of the reward increases (the rate of discounting is lower). This specific hyperbolic discounting pattern is discussed as stemming from the fact that people's estimates of duration do not correspond to objective time in a linear way, but that objective time is perceived logarithmically (Takahashi 2005; Wackermann 2007; Zauberman et al. 2008). Thereafter, hyperbolic discounting of delayed rewards is dependent upon the psychophysical characteristics of subjective time. The diminishing sensitivity to longer time intervals, as delays are perceived logarithmically, thus contributes to the degree of hyperbolic discounting (Kim and Zauberman 2009).

The propensity to even more strongly discount delayed rewards (an even stronger preference for immediate consumption) is one of the key findings in studies with impulsive individuals (Kirby et al. 1999; Barkley et al. 2001). Behavior of impulsive individuals in delay discounting tasks has been related to alterations in subjective time (Wittmann and Paulus 2008, 2009): Impulsive individuals more strongly attend to time due to their increased focus on the present moment at the expense of the future perspective. Stronger attention to the passage of time leads to an overestimation of duration. A general overestimation of duration in turn causes the impression that delayed events lie too far in the future—as a consequence value of future commodities is discounted and more immediate rewards are preferred. For example, children and adults with attention deficit hyperactivity disorder (Rubia et al. 2009) and subjects who are transiently more impulsive through sleep deprivation (Reynolds and Schiffbauer 2004) overestimate duration in the multiple seconds range and they discount more strongly future rewards. That is, a relative overestimation of duration caused by a stronger present time orientation may underlie the shortsighted choices in intertemporal decision-making.

8.3 An Excursus: Time Experience in Addiction

Findings in addiction research shed light on core questions of neuroeconomics. Within the context of conceptualizations related to impulsivity and self-control, an addicted individual shows strong preference changes which are based on alterations in the representation of "value" and "decision utility" (Monterosso et al. 2012). Specifically, a drug addict's altered sense of time could be one reason for a stronger decay of value of temporally delayed rewards. Drug-dependent individuals might opt more often for smaller and immediate rewards over delayed but higher rewards because they anticipate the temporal delay to last longer than do people who have no addiction (Wittmann and Paulus 2008, 2009).

The conceptualization related to an altered experience of time in individuals with addiction has at least three component processes related to trait impulsivity: First, these individuals show a stronger emphasis on the present time perspective (as opposed to the future perspective). Second, attention is directed more strongly to the momentary passing of time. Third, allocating attentional resources to the here and now leads to a subjective slowing of time and an overestimation of future time intervals. Resulting from this experience of time, drug-dependent individuals will discount delayed rewards more strongly as these rewards lie subjectively too far in the future.

Several studies suggest that chronic drug users exhibit a stronger present time perspective. First of all, individuals who have a stronger focus on the present time and less focus on the future perspective are also more likely to use alcohol, tobacco, and other drugs (Keough et al. 1999). Moreover, alcohol-dependent persons have a less extensive future perspective as compared to social drinkers (Smart 1968). Similarly, drug-dependent patients enrolled in a drug treatment program are less motivated for the future than control subjects (Lavelle et al. 1991); heroin addicts are less likely to predict events far into the future as they have shortened time horizons (Alvos et al. 1993). Evidence suggests that a trait-related focus on the present time is strongly related to trait impulsivity. For example, heroin addicts, who score significantly lower on a future orientation scale, also show more impulsive behavior in decision-making. These individuals were more likely to play from a deck of cards that contained immediate gains but that resulted in large, delayed punishers and overall net losses (Petry et al. 1998). A similar pattern of impulsive behavior emerged in a study with substance abusing pathological gamblers (Petry 2001). Thus, a shortened future time perspective, that is, a poor perception of events in the future, may explain drug-addicted individuals' persistent drug use despite future adverse consequences. A shortened time horizon could increase the likelihood of opting for behavior associated with an immediate benefit (the pleasurable state induced by the drug) over behavior that leads to delayed rewards (long-term health).

There is some consensus that drug users show an altered temporal horizon of risks and benefits which is consistent with a steeper delay discounting function, i.e., the more a reward is delayed in time the stronger the value of the reward decreases for drug using individuals as compared to controls (Kirby and Petry 2004; Madden et al. 1999). This tendency to discount values of delayed rewards more rapidly than comparison subjects has been identified in different types of substance users (Vuchinich and Simpson 1998; Reynolds 2004). Future rewards are more strongly discounted as a function of the number of illicit drugs used and the age onset of first taking substances (alcohol, nicotine, marijuana) in a sample of college students (Kollins 2003).

According to the pacemaker-accumulator models of time perception, drug-dependent individuals who may experience distress because habitual impulsive acts of drug taking cannot be instantiated allocate more attentional resources to the passage of time and, additionally, are in a state of increased physiological arousal. Both factors in combination lead to an overestimation of experienced duration. For example, smokers who felt craving for a cigarette but were asked to wait through a certain time interval before they were again allowed to smoke experienced time to pass more slowly (Sayette et al. 2005).

However, it is noteworthy that there are only few studies concerning the estimation of duration in patients with drug addiction. In one study, a significant relative overestimation of an interval in the multiple-second range in patients with methamphetamine or cocaine dependence as compared to matched controls was attributable to increased self-reported impulsivity (Wittmann et al. 2007a). Moreover, patients with heroin addiction showed stronger under-reproductions of 12-second intervals as compared to controls, a behavior which varied as a function of withdrawal duration (Aleksandrov 2005). A more pronounced under-reproduction of duration, essentially an indication that subjective duration has elapsed faster, is a typical behavior seen in impulsive individuals (Wittmann et al. 2011). Future research on addiction will have to combine the assessments of subjective time and of behavioral tests in one study in order to more precisely understand the relationships between time perspective, duration estimation, and intertemporal decision-making.

8.4 The Neural Correlates of Subjective Time

Individual state- and trait-related modulations of the time perspectives are associated with subjective estimates of duration, which in turn affects temporal decision-making. The dynamical relation within the dimensions of subjective time determines time-related preferences and behavioral choices. If intertemporal decisions are so strongly dependent upon the perception and anticipation of duration, then research in the neurosciences should reveal activation of similar underlying brain structures for time perception and for intertemporal decision-making. However, the neural basis underlying subjective time and temporal preferences in decision-making remains controversial. That is, a large number of brain areas and systems have been identified as underlying the experience of duration as well as governing behavior in intertemporal decision-making. For reviews of functional neuroimaging studies on the perception of duration, see Lewis and Miall (2003); Wiener et al. (2010); for neuroimaging reviews on intertemporal decision-making, see Monterosso and Luo (2010) and Carter et al. (2010).

Here we want to focus on two brain areas, which have repeatedly been shown to be coactivated in functional neuroimaging studies, and which might specifically be related to the processing of time: the striatum and the insular cortex. Functional neuroimaging studies of time perception show how the striatum and the insular cortex are activated during the perception of shorter duration in the millisecond-to-a-few-seconds range (e.g., Livesey et al. 2007; for a review of studies see Wittmann 2009b) as well as during the perception of multiple-second intervals (Wittmann et al. 2011). Also related to intertemporal decisions, both the striatum and the insula are often activated as related to different component processes (McClure et al. 2007; Wittmann et al. 2007b). Tanaka et al. (2004) provided the most compelling evidence that the insular cortex and the striatum code for the selection of immediate and delayed rewards. Ventroanterior regions of these two brain areas were more activated for immediate choices, dorsoposterior regions were more activated when subjects learned to choose delayed rewards. In another study with choices between immediate and delayed rewards in the multiple-second range, the anterior insula and the striatum were both activated when subjects chose either the immediate (smaller) or the delayed (larger) option (Wittmann et al. 2010a).

The insular cortex and the striatum are strongly related to expectation and anticipation. That is, these two areas are part of brain systems whose functioning is inherently related to time. The striatum is not only involved in reward evaluations guiding goal-directed behavior as it tracks perceived value (Gregorios-Pippas et al.

2009; Kable and Glimscher 2007); it is strongly involved in reward expectation (Hassani et al. 2001). Similarly, the insular cortex plays a fundamental role in the neural decision-making system, integrating visceral sensations and emotional states to modulate decisions (Craig 2002). That is, the insula integrates interoceptive information as a basis for judging prospective reward value. More specifically, the anterior insula, together with the striatum, has been associated with the anticipation of rewards and the expectation and evaluation of upcoming events (Lovero et al. 2009). In this line, the anterior insula has been discussed as generating a predictive model, which provides an individual with a signal of how he or she will feel (Paulus and Stein 2006).

The close connection between the insular cortex and the integration of ascending body signals on the one hand (Craig 2002) and findings of modulations of insular cortex activation and physiological changes of the body affecting time perception on the other hand (Wittmann et al. 2010; Meissner and Wittmann 2011) have led to the hypothesis that the accumulation of physiological changes in the body is the basis for the experience of duration (Craig 2009). Similarly, body states and visceral factors are strongly involved in decision-making. Drive states such as hunger, thirst, sexual desire, or the craving for drugs lead to an intense desire that can dominate the decisions we make (Loewenstein 1996). But emotion- and body-related signals, changes in visceral states, are to some extent inherently involved in all of our decisions, not only when self-control is compromised by overwhelming desires (Damasio 1994; Reimann and Bechara 2010). Complex decisions related to time and experiences of duration are governed by emotions and feeling states attributable to processes associated with the striatum and the insula.

8.5 Summary

In order to anticipate upcoming events and to adjust to environmental demands an individual has to adequately process temporal information. How long do I have to wait for something to happen or to end? Do I have enough time to prepare for an exam? Decisions are made between possible outcomes that are perhaps minutes apart: should I wait for the elevator, do I take the stairs? Human decision-making in the context of economic planning can encompass the anticipation of outcomes that lie years or even decades ahead: should I start now to save for retirement? The human brain constantly generates predictions about present versus future outcomes: should I now watch TV or go running for long-term physical fitness? The way time is perceived and how subjective time is modified is an important factor for understanding how decisions pertaining to the temporal properties of outcomes are made.

Knowledge of how a present bias causes the discounting of delayed benefits can lead to supportive measures of self-control. Humans intend to be self-controlled but when faced with concrete options in the here and now they often give into impulsive acts. For example, due to temporal shortsightedness one might not want to deduct a considerable amount of money for a saving option the bank is offering from now on. However, one might now agree and sign a contract on a plan with monthly payments that starts in 1 year. If a commitment to act is shifted from the present to the future more abstract reasoning sets in; when events are temporally distant, they become psychologically more abstract and people are more inclined to act according to their more self-controlled reasoning (Trope and Liberman 2003).

The ultimate test concerning our outline of a model on the relationship between subjective time and decision-making would be through conducting intervention studies. Would it be possible to influence maladaptive decision-making through modifications of state- and trait-related experiences of time? Intervention programs could be developed that manipulate the temporal delay of rewards and restructure the temporal perspective on events and the experience of time in order to promote desired decision-making.

Acknowledgments The empirical work leading to the theoretical account presented here was supported by grants from the Max Kade Foundation, New York, the National Institute of Drug Abuse, Bethesda, and the Kavli Institute for Brain and Mind, San Diego. Discussions on the topic with Dan Zakay and Alejandro Vasquez were instrumental for developing the ideas presented here.

References

Ainslie G (1975) Specious reward—behavioral theory of impulsiveness and impulse control. Psychol Bull 82:463–496

Aleksandrov SG (2005) Dynamics of assessments of time intervals by patients with heroin addiction. Neurosci Behav Physiol 35:371–374

Alvos L, Gregson RA, Ross MW (1993) Future time perspective in current and previous injecting drug users. Drug Alcohol Depend 31:193–197

Anderson MJ, Reis-Costa K, Misanin JR (2007) Effects of September 11th terrorism stress on estimated duration. Percept Mot Skills 104:799–802

Ariely D, Zakay D (2001) A timely account of the role of duration in decision-making. Acta Psychol 108:187–207

Bailey N, Areni CS (2006) Background music as a quasi clock in retrospective duration judgments. Percept Mot Skills 102:435–444

Barkley RA, Edwards G, Laneri M, Fletcher K, Metevia L (2001) Executive functioning, temporal discounting, and sense of time in adolescents with attention deficit hyperactivity disorder (ADHD) and oppositional defiant disorder (ODD). J Abnorm Child Psychol 29:541–556

Bschor T, Ising M, Bauer M, Lewitzka U, Skerstupeit M, Müller-Oerlinghausen B, Baethge C (2004) Time experience and time judgment in major depression, mania and healthy subjects. A controlled study of 93 subjects. Acta Psychiatr Scand 109:222–229

Burle B, Casini L (2001) Dissociation between activation and attention effects in time estimation: implications for internal clock models. J Exp Psychol Hum Percept Perform 27:195–205

Carstensen LL (2006) The influence of a sense of time on human development. Science 312:1913–1915

Carter RMK, Meyer JR, Huettel SA (2010) Functional neuroimaging of intertemporal choice models: a review. J Neurosci, Psychol, Econ 3:27–45

Conti R (2001) Time flies: investigating the connection between intrinsic motivation and the experience of time. J Pers 69:1–26

Craig AD (2002) How do you feel? Interoception: the sense of the physiological condition of the body. Nat Rev Neurosci 3:655–666

Craig AD (2009) Emotional moments across time: a possible neural basis for time perception in the anterior insula. Philoso Trans Royal Soc B 364:1933–1942

Damasio AR (1994) Descartes' error: emotion, reason, and the human brain. Grosset-Putnam, New York

Droit-Volet S, Gil S (2009) The time-emotion paradox. Philos Trans R Soc B 364:1943–1954

Droit-Volet S, Meck WH (2007) How emotions colour our perception of time. Trends Cogn Sci 11:504–513

Flaherty MG, Freidin B, Sautu R (2005) Variation in the perceived passage of time: a cross-national study. Soc Psychol Q 68:400–410

Gibbon J, Church RM, Meck WH (1984) Scalar timing in memory. In: Gibbon J, Allan L (eds) Timing and time perception. Annals of the New York Academy of Sciences: vol 423. New York Academy of Sciences, New York, pp 52–77

Gregorios-Pippas L, Tobler PN, Schultz W (2009) Short-term temporal discounting of reward value in human ventral striatum. J Neurophysiol 101:1507–1523

Kable JW, Glimcher PW (2007) The neural correlates of subjective value during intertemporal choice. Nat Neurosci 10:1625–1633

Keough K, Zimbardo PG, Boyd JN (1999) Who's smoking, drinking, and using drugs? Time perspective as a predictor of substance use. Basic Appl Soc Psychol 21:149–164

Kim BK, Zauberman G (2009) Perception of anticipatory time in temporal discounting. J Neurosci Psychol Econo 2:91–101

Kim BK, Zauberman G (2012) Can victoria's secret change the future? A subjective time perception account on sexual cue effects on impatience. J Exp Psychol: Gen 10.1037/a0028954

Kirby KN, Petry NM (2004) Heroin and cocaine abusers have higher discount rates for delayed rewards than alcoholics or non-drug-using controls. Addiction 99:461–471

Kirby KN, Petry NM, Bickel WK (1999) Heroin addicts have higher discount rates for delayed rewards than non-drug-using controls. J Exp Psychol Gen 128:78–87

Klapproth F (2008) Time and decision-making in humans. Cogn Affect Behav Neurosci 8:509–524

Kollins SH (2003) Delay discounting is associated with substance use in college students. Addict Behav 28:1167–1173

Lavelle T, Hammersley R, Forsyth A (1991) Personality as an explanation of drug use. J Drug Issues 21:593–604

Lewis PA, Miall RC (2003) Distinct systems for automatic and cognitively controlled time measurement: evidence from neuroimaging. Curr Opin Neurobiol 13:250–255

Livesey AC, Wall MB, Smith AT (2007) Time perception: manipulation of task difficulty dissociates clock functions from other cognitive demands. Neuropsychologia 45:321–331

Loewenstein G (1996) Out of control: visceral influences on behaviour. Organ Behav Hum Decis Process 65:272–292

Lovero KL, Simmons AN, Aron JL, Paulus MP (2009) Anterior insular cortex anticipates impending stimulus significance. NeuroImage 15:976–983

Madden G, Bickel W, Jacobs E (1999) Discounting of delayed rewards in opioid-dependent outpatients: exponential or hyperbolic discounting functions? Exp Clin Psychopharmacol 7:284–293

McClure SM, Ericson KM, Laibson DI, Loewenstein G, Cohen JD (2007) Time discounting for primary rewards. J Neurosci 27:5796–5804

Meissner K, Wittmann M (2011) Body signals, cardiac awareness, and the perception of time. Biol Psychol 86:289–297

Mischel W, Shoda Y, Rodriguez ML (1989) Delay of gratification in children. Science 244:933–938

Monterosso J, Luo S (2010) An argument against dual valuation system competition: cognitive capacities supporting future orientation mediate rather than compete with visceral motivations. J Neurosci Psychol Econo 3:1–14

Monterosso J, Piray P, Luo S (2012) Neuroeconomics and the study of addiction. Biol Psychiatry 72:107–112

Noulhiane M, Mella N, Samson S, Ragot R, Pouthas V (2007) How emotional auditory stimuli modulate time perception. Emotion 7:697–704

Paulus M, Stein M (2006) An insular view of anxiety. Biol Psychiatry 60:383–387

Petry NM (2001) Substance abuse, pathological gambling, and impulsiveness. Drug Alcohol Depend 63:29–38

Petry NM, Bickel W, Arnett M (1998) Shortened time horizons and insensitivity to future consequences in heroin addicts. Addiction 93:729–738

Prelec D, Loewenstein G (1997) Beyond time discounting. Mark Lett 8:97–108

Reimann M, Bechara A (2010) The somatic marker framework as a neurological theory of decision-making: Review, conceptual comparisons, and future neuroeconomics research. J Econ Psychol 31:767–776

Reynolds B (2004) Do high rates of cigarette consumption increase delay discounting? A cross-sectional comparison of adolescent smokers and young-adult smokers and nonsmokers. Behav Processes 67:545–549

Reynolds B, Schiffbauer R (2004) Measuring state changes in human delay discounting: an experiential discounting task. Behav Processes 67:343–356

Rubia K, Halari R, Christakou A, Taylor E (2009) Impulsiveness as a timing disturbance: neurocognitive abnormalities in attention-deficit hyperactivity disorder during temporal processes and normalization with methylphenidate. Philos Trans R Soc B 364:1919–1931

Sayette MA, Loewenstein G, Kirchner TR, Travis T (2005) Effects of smoking urge on temporal cognition. Psychol Addict Behav 19:88–93

Smart RG (1968) Future time perspectives in alcoholics and social drinkers. J Abnorm Psychol 73:81–83

Takahashi T (2005) Loss of self-control in intertemporal choice may be attributable to logarithmic time-perception. Med Hypotheses 65:691–693

Tanaka SC, Doya K, Okada G, Ueda K, Okamoto Y, Yamawaki S (2004) Prediction of immediate and future rewards differentially recruits cortico-basal ganglia loops. Nat Neurosci 7:887–893

Treisman M (1963) Temporal discrimination and the difference interval: Implications for a model of the "internal clock". Psychol Monogr 77:1–31

Trope Y, Liberman N (2003) Temporal construal. Psychol Rev 110:403–421

Twenge JM, Catanese KR, Baumeister RF (2003) Social exclusion and the deconstructed state: Time perception, meaninglessness, lethargy, lack of emotion, and self-awareness. J Pers Soc Psychol 85:409–423

van Laarhoven HWM, Schilderman J, Verhagen CA, Prins JB (2011) Time perception of cancer patients without evidence of disease and advanced cancer patients in a palliative, end-of-life setting. Cancer Nurs 34:453–463

Vuchinich RE, Simpson CA (1998) Hyperbolic temporal discounting in social drinkers and problem drinkers. Exp Clin Psychopharmacol 6:292–305

Wackermann J (2007) Inner and outer horizons of time experience. Span J Psychol 10:20–32

Wiener M, Turkeltaub P, Coslett HB (2010) The image of time: a voxel-wise meta-analysis. NeuroImage 49:1728–1740

Wittmann M (2009a) The subjective flow of time. In: Birks HJ (ed) The Encyclopedia of Time. SAGE Publications, Thousand Oaks, pp 1322–1324

Wittmann M (2009b) The inner sense of time. Philosophical Trans Royal Soc B 364:1955–1967

Wittmann M, Lehnhoff S (2005) Age effects in perception of time. Psychol Rep 97:921–935

Wittmann M, Paulus MP (2008) Decision-making, impulsivity and time perception. Trends in Cogn Sci 12:7–12

Wittmann M, Paulus MP (2009) Temporal horizons in decision-making. J Neurosci Psychol Econ 2:1–11

Wittmann M, Vollmer T, Schweiger C, Hiddemann W (2006) The relation between the experience of time and psychological distress in patients with hematological malignancies. Palliat Support Care 4:357–363

Wittmann M, Leland D, Churan J, Paulus MP (2007a) Impaired time perception and motor timing in stimulant-dependent subjects. Drug Alcohol Depend 90:183–192
Wittmann M, Leland DS, Paulus MP (2007b) Time and decision-making: differential contribution of the posterior insular cortex and the striatum during a delay discounting task. Exp Brain Res 179:643–653
Wittmann M, Lovero KL, Lane SD, Paulus MP (2010a) Now or later? Striatum and insula activation to immediate versus delayed rewards. J Neurosci Psychol Econo 3:15–26
Wittmann M, Simmons AN, Aron J, Paulus MP (2010b) Accumulation of neural activity in the posterior insula encodes the passage of time. Neuropsychologia 48:3110–3120
Wittmann M, Simmons AN, Flagan T, Lane SD, Wackermann J, Paulus MP (2011) Neural substrates of time perception and impulsivity. Brain Res 1406:43–58
Zakay D, Block RA (1997) Temporal cognition. Curr Dir Psychol Sci 6:12–16
Zakay D, Block RA (2004) Prospective and retrospective duration judgments: an executive-control perspective. Acta Neurobiol Exp 64:319–328
Zauberman G, Kyu Kim B, Malkoc SA, Bettman JR (2008) Discounting time and time discounting: subjective time perception and intertemporal preferences. J Mark Res 46:543–556
Zauberman G, Levav J, Diehl K, Bhargave R (2010) 1995 feels so close yet so far: the effect of event markers on subjective feelings of elapsed time. Psychol Sci 21:133–139
Zimbardo PG, Boyd JN (1999) Putting time in perspective: a valid, reliable individual-difference metric. J Pers Soc Psychol 77:1271–1288
Zimbardo PG, Keough K, Boyd JN (1997) Present time perspective as a predictor of risky driving. Personality Individ Differ 23:1007–1023

Chapter 9
Framing Effects: Behavioral Dynamics and Neural Basis

Xiao-Tian Wang, Lilin Rao and Hongming Zheng

Abstract Research on framing effects has been one of few multidisciplinary endeavors joined by psychologists, economists, political scientists, and management and marketing researchers. Framing effects epitomize the power of linguistic subtlety in regulating decision-making, showing that different ways of framing, phrasing, or presenting virtually identical choice options systematically affect risk preference, evaluation of experience, and persuasiveness of messages. Given its central role in the studies of decision biases, the framing effect has been used as an experimental probe for understanding general mechanisms of human judgment and decision-making. Researchers have proposed various models explaining the framing effect. However, it was not until recently that research of framing effects started to focus more on psychological mechanisms above and beyond phenomenology. We conducted a meta-analysis of neural correlates of framing effects. The topographic convergences from a total of 26 foci found in the fMRI studies of framing effects revealed two key brain areas underlying framing effects: the left anterior cingulate cortex (ACC) and the right inferior frontal gyrus (IFG). Together with behavioral findings, these results suggest that valence framing as a secondary cue becomes most salient and effective when primary contextual or social cues are absent or incongruent. The processing of choice problems under these conditions call for an ambiguity-reducing and conflict-monitoring function, which would result in the ACC activation. Second, the right IFG activation suggests that the nature of valence framing is both semantic and hedonic, involving not only verbatim linguistic analysis, but also interpretation of its affective tones and metaphorical implications.

X.-T. Wang (✉)
Psychology Department, University of South Dakota,
Vermillion, SD 57069, USA
e-mail: xtwang@used.edu

L. Rao
Institute of Psychology, Chinese Academy of Sciences,
Beijing, China

H. Zheng
Southwest Jiaotong University, Chengdu, China

M. Reuter and C. Montag (eds.), *Neuroeconomics*, Studies in Neuroscience, Psychology and Behavioral Economics, DOI 10.1007/978-3-642-35923-1_9

9.1 Behavioral Studies of Framing Effects

9.1.1 Framing and Framing Effects: Definition and Typology

Since the seminal work by Kahneman and Tversky (1979), Tversky and Kahneman (1981), framing effects have been widely studied by researchers from across the social sciences for over 30 years. Research on framing effects has been one of few multidisciplinary endeavors joined by psychologists, economists, political scientists, and management and marketing researchers. Framing effects epitomize the power of linguistic subtlety in regulating decision-making, oftentimes without awareness of the decision-maker. Different ways of framing, phrasing or presenting virtually identical choice options systematically affect risk preference, evaluation of experience, products or job candidates, and persuasiveness of arguments in negotiation and communication. A meta-analysis has shown that among some primary predictors of risky choice, framing condition was the top predictor (partial $r = 0.44$) followed by the value of risky payoff (partial $r = 0.14$), and type of good at stake (partial $r = 0.13$) while gain or loss condition and probability of payoff were not significant in predicting choice preference (Kühberger et al. 1999, p. 213).

Tversky and Kahneman (1981) used the term 'decision frame' broadly to refer to "the decision-maker's conception of the acts, outcomes, and contingencies associated with a particular choice" (p. 453). Later definitions of framing effects, despite the differences in their connotations and coverage, pinpoint two typical features: equivalency in choice outcomes and opposing valences in presentations of the choice outcomes. That is, framing effects occur when frames that cast "the same critical information in either a positive or a negative light" cause individuals to have different choice preferences (Chong and Druckman 2007; Druckman 2001; Levin et al. 1998, p. 150).

Consider the well-known Asian disease problem demonstrated first by Tversky and Kahneman (1981). In the cover story of the problem, the respondents were asked to imagine that "the US is preparing for the outbreak of an unusual Asian disease, which is expected to kill 600 people. Two alternative programs to combat the disease have been proposed." The outcomes of the disease-combating plans were then framed (phrased) differently. In the positive frame the respondents were told: "If Plan A is adopted, 200 people will be saved. If Plan B is adopted, there is a one-third probability that all 600 people will be saved, and two-thirds probability that none of them will be saved." Given a binary choice between the two alternative plans, the majority of the respondents (72 %) were risk averse, preferring the sure option (Plan A) over its gamble equivalent (Plan B). However, when the same outcomes were 'negatively framed' in terms of lives lost ("If Plan A is adopted, 400 people will die. If Plan B is adopted, there is a one-third probability that none of them will die, and two-thirds probability that all 600 people will die."), the majority of the respondents (78 %) were risk-taking, preferring the gamble option (Plan B) over its sure-thing equivalent. This classic framing effect has been reliably replicated with different samples and across cultures and disciplines (e.g., see Kühberger 1998; Kühberger et al. 1999).

Such risk preference reversal due to valence framing of virtually equivalent choice outcomes raises radical doubts about basic assumptions in economic theories of rationality. How can decision agents make consistent and rational choices when empirical evidence of choice preference appears to be so malleable, so vulnerable to framing effects? The classic framing effect is thus viewed as an irrational decision bias and a cognitive illusion because it violates the invariance axiom of expected utility theory, which requires a rational decision-maker to have a consistent preference order among identical choice prospects independent of the way the prospects are presented or framed.

As illustrated in the Asian disease problem, equivalent choice outcomes can be framed with either positive valence or negative valence (e.g., lives saved vs. lives killed, survival rate vs. mortality rate, gain vs. loss, opportunity vs. threat, success vs. failure, benefits vs. costs). A cumulative body of evidence has shown a wide range of behavioral consequences due to framing of decision problems (Chong and Druckman 2007; Kühberger 1998; Levin et al. 1998). Based on their meta-analysis of different kinds of framing effects, Levin et al. (1998) proposed a well-adopted typology. They proposed three types of framing effects: (1) Risky choice framing, as illustrated by the Asian disease problem where framing of choice outcomes affects risk preference (risk averse or risk seeking) of the decision-maker; (2) Attribute framing, where framing of attributes or characteristics of an object or event affects the evaluation of the object (e.g., a sample of ground beef was rated as better tasting when it was labeled as '75 % lean' rather than '25 % fat', see Levin and Gaeth 1988); and (3) Goal framing, where framing of either the positive consequences of performing an act or the negative consequences of not performing the act affects implicit goals an individual adopts and persuasiveness of a message. A well-known example of goal framing effects has been documented by Meyerowitz and Chaiken (1987), showing that women were more apt to engage in breast self-examination when presented with information stressing the negative consequences of not engaging in this action than when presented with information stressing the positive consequences of engaging in this action.

As stated in the title of the paper by Levin et al. (1998), all frames are not created equally. Similarly, not all framing effects are the same. Some involve irrational preference reversals (e.g., the opposite choice preferences under different framings) while others involve a shift in choice preference (e.g., making one option even more attractive than another). A choice shift differs from a choice reversal in that the proportion of risky choices differs across framing conditions but is not significantly greater than 0.5 under one framing condition and significantly less than 0.5 in an alternative framing condition. Wang (1996a) makes the distinction between *bidirectional* framing effects which involve a reversal in risk preference and *unidirectional* framing effects which involve a preference shift (e.g., from risk seeking to even more risk seeking under negative framing, or from risk averse to even more risk averse under positive framing).

9.1.2 Theories of Framing Effects

Given its central role in the studies of decision biases, the framing effect has been used as an experimental probe for understanding general mechanisms of human judgment and decision-making. Researchers have proposed various models explaining framing effects.

1. Prospect Theory

An initial and prevalent explanation of framing effects is based on the prospect theory's (Kahneman and Tversky 1979) S-shaped value function which is separated by a zero point (status quo) with a gain region above the status quo and a loss region below the status quo. Tversky and Kahneman (1981) theorized that framing manipulation determines whether outcomes are encoded as gains or losses and that this encoding determines which portion of the S-shaped value function would contribute to the risk preference of the decision-maker. For gains implicated in the positive framing condition, the subjective value function is concave and promotes risk aversion, whereas for losses implicated in the negative framing condition, the value function is convex and promotes risk seeking. However, this prospect of theoretical account of framing effects is limited to risky choice framing. In case of attribute framing where the presence of risk is not essential, valence of framing (e.g., success rate vs. failure rate) may either evoke favorable or unfavorable associations in memory (Levin and Gaeth 1988) or evoke different reference points (goal or minimum requirement rather than status quo, Wang and Johnson 2012).

2. Fuzzy-Trace Theory

Another famous account of framing effects is derived from fuzzy-trace theory proposed by Reyna and Brainerd (1991, 1995, 2011). The theory posits that people form two types of mental representations about a past event, called verbatim and gist traces. Gist traces are fuzzy representations of a past event (e.g., its bottom-line meaning), whereas verbatim traces are detailed representations of a past event. Although people (adults) are capable of processing both verbatim and gist information, they prefer to reason with gist traces rather than verbatim traces. In the case of encoding choice outcomes of a sure thing and a gamble of equal expected value in the Asian disease problem, the gist translations boil down to choosing between (a) saving some people, (b) saving some people or saving no one under the positive (gain) frame, and choosing between (c) some people die, and (d) some people die or none die under the negative (loss) framing. Reyna and Brainerd (1991, 1995) showed that relational gist of quantities (e.g., some, all, none) was sufficient to replicate the classic framing effect. Thus, framing effects can be seen as a result of translating expected choice outcomes to gist representations of quantities with positive or negative connotations. These nonnumerical framing effects demonstrated that numbers, which are essential to predictions of prospect theory and all other utility theories, were not necessary to observe the framing effect. Moreover, according to prospect theory or other expected utility theories, the outcome of zero in the gamble option literally contributes nothing to predictions of choice behavior because the value of zero, based on the S-shaped value function, is also set at zero. However, when the information about the zero outcomes was removed from the classic Asian disease problem, framing effects disappeared (see Reyna and Brainerd 2011).

3. Asymmetry in Dual Processing

Another approach to theorizing framing effects emphasizes the interplay between two distinct processing systems or asymmetry in affective or cognitive processing. Dunegan (1993), for example, proposed that framing may act as a catalyst for different modes of cognitive processing. Characteristics of controlled cognitive modes were found when information was negatively framed; characteristics of more automatic processing were found when information was positively framed, suggesting that positive framing may be perceived as a default with a lower cognitive processing load. Kuo et al. (2009) employed eye tracking to measure cognitive effort in a framing study. Their results suggest an asymmetry in cognitive effort, as indicated by eye fixation and eye-movement time per word, due to positive and negative frames. More effort was observed in the negative framing condition than in the positive framing condition. However, a study by Whitney et al. (2008) found that working memory load slightly reduced risk-seeking tendency but not framing effects, suggesting that valence framings did not have differential effects on cognitive processing effort.

Another account of framing effects proposed by McElroy and Seta (2003, 2004) focuses on the asymmetry between analytic processing and holistic processing. This model assumes that analytic processing is insensitive to the influence of framing, whereas holistic processing is more susceptible to framing. By behaviorally inducing selective hemispheric activation, framing effects were found when the right hemisphere was selectively activated, whereas they were not observed when the left hemisphere was selectively activated.

4. Ambiguity-Ambivalence Hypothesis

There are other limits and constraints to framing effects in addition to hemispheric dominance. Meta-analyses (Levin et al. 1998; Kühberger 1998) show that the effect of framing is overall significant but not as 'pervasive' or 'robust' as previously believed. Many researchers have explored the premise and moderators of framing effects. Some examples of the moderators are sex of the decision-maker (Fagley and Miller 1990, 1997), cognitive ability (Stanovich and West 1998), personal involvement (Levin et al. 1998, p. 160), reflection on and rationale for the decision (Takemura 1994; Fagley and Miller 1987; Miller and Fagley 1991; Sieck and Yates 1997), personal knowledge about risks involved (e.g., Bohm and Lind 1992; Levin and Chapman 1990), perceived ambiguity of the values presented in the problem descriptions (Kühberger 1995), task context (e.g., medical vs. statistical, Bless et al. 1998), need for cognition (LeBoeuf and Shafir 2003), perceived interdependence between individuals at risk (Bloomfield et al. 2006; Wang et al. 2001), and social group size and composition (Bloomfield 2006; Shimizu and Udagawa 2011a; Wang 1996a, b; Zhang and Miao 2008).

The aforementioned studies focusing on the variables that can enhance, limit, or even eliminate a framing effect have taken into consideration individual as well as situational factors that determine the magnitude as well as process of framing effects. This line of research calls for an overarching theory of framing effects that addresses these ecological, social, and dispositional constraints on and premises of framing effects. This led to the development of an ambiguity-ambivalence (AA) hypothesis of framing effects and decision biases (Wang 2008). The AA hypothesis assumes that (1) decision cues are selected and used in accordance to their priorities. (2) Cue priority reflects ecological and social validity of a cue in predicting specific risks, beyond the stated values and probabilities of outcomes in a choice task. Primary cues are valid ecological, social, and life-history variables (such as kith-and-kin relations, social group size, sex, age, health, socioeconomic status, and mating/reproductive cues, etc.). Secondary cues in decision-making are mainly communicational, such as verbal framing, facial expression, tone of voice, etc. (3) Primary cues determine the settings of decision reference points (e.g., goals and bottom lines) and anchor decision preference while secondary cues (and individual dispositional factors) fine tune the settings of reference points and choice preference. (4) Decision biases, such as framing effects, tend to occur as a result of secondary cue use when primary cues are either absent in risk communication (i.e., an ambiguity condition) or when primary cues elicit conflicting preferences (i.e., an ambivalence condition).

From this perspective, framing effects would occur as a result of ambiguity and ambivalence in decision cues and preferences. Similar ideas can be seen in fuzzy-trace theory (Reyna and Brainerd 1991) and the discussion on the use of probabilistic mental models (Gigerenzer et al. 1991) to deal with informational ambiguity as a premise of framing effects (Chang et al. 2002; Kühberger 1995).

As predicted from the AA hypothesis, framing effects occurred when making life–death decisions in evolutionarily novel and socially unfamiliar large group contexts but disappeared when the same problems were scaled down to a small group context with a handful of people (Bloomfield 2006; Shimizu and Udagawa 2011a; Zhang and Miao 2008; Wang and Johnston 1995; Wang 1996a, b). Moreover, work experience in large organizations reduced sensitivity to framing manipulation in large group contexts (Shimizu and Udagawa 2011b). Interestingly, framing effects in a small group reappeared when the small group included both strangers and kin relatives, thus creating ambivalence in risk preference between a "we all live or die together" risk-seeking preference for kith-and-kin and a more risk-averse preference for strangers (Wang et al. 2001).

9.2 A Neuroscience Approach to Understanding Framing Effects

The discussion so far summarizes behavioral studies of framing effects. These studies leave open the possibility for more theoretical development and better understanding of underlying processes that produce or inhibit framing effects. It was not until recently that research on framing effects started to focus more on the underlying psychological mechanisms above and beyond phenomenology, owing to rapid development in neuroimaging technology. Neuroscience and neuroimaging can help forward our understanding of framing effects in at least four different yet coherent ways.

First, the neuroscience approach would allow us to evaluate framing theories by mapping the framing-related brain activations against the key brain areas implicated by different theories of framing effects. Second, many hypothetical mechanisms and post-hoc explanations based on behavioral effects can be better understood, verified, or disproved by using the brain imaging technique. Third, neuroimaging studies often shed light into puzzling or conflicting findings from behavioral studies of framing effects. Fourth, brain imaging studies capture online neural activities of the entire brain during the decision-making. This brain map of activation or deactivation may highlight uncharted brain areas and provide insights and new leads for future investigations.

In the following sections of the chapter, we illustrate the use and usefulness of the neuroscience approach to framing effects by generating predictions about neural correlates of framing effects based on different theoretical viewpoints, and by conducting a meta-analysis of the fMRI studies of framing effects. We end the chapter with a summary and conclusions regarding the status quo of behavioral and neuroimaging studies of framing effects.

9.2.1 Contrasting and Evaluating Theories of Framing Effects

As illustrated in Table 9.1, theories of framing effects make overlapping but distinctive predictions about brain regions underpinning the proposed functions. The predictions presented in Table 9.1 are by no means meant to be thorough or systematic.

Table 9.1 is only a sketchy outline to exemplify how theories of framing effects can be tested with neuroimaging studies. For a more specific example, consider an fMRI study of framing effects by Gonzalez et al. (2005). The authors proposed a cognitive–affective tradeoff model which can be classified as one of the 'asymmetry in dual processing models' as listed in Table 9.1.

The cognitive–affective tradeoff model assumes that the framing effect occurs due to a tradeoff between the cognitive effort required to calculate expected values

Table 9.1 Theories of framing effects and predicted brain activations

Theory	Assumptions	Behavioral effects	Predicted key brain areas associated with valence framing
Prospect theory (Kahneman and Tversky 1979)	S-shaped value function separated by a zero point (status quo) indicates risk aversion in gains or under positive framing and risk seeking in losses or under negative framing	Risk averse in a positive frame and risk seeking in a negative frame	Brain areas associated with gains/rewards and losses/punishments (e.g., the nucleus accumbens, the ventral striatum, and the orbital frontal cortex, see, e.g., Breiter et al. 2001; O'Doherty et al. 2001; Tom et al. 2007) are expected to be differentially activated by valance framing
Fuzzy-trace theory (Reyna and Brainerd 1991, 1995, 2011)	Framing effects are a result of translating expected choice outcomes to gist representations of quantities (e.g., some, many, all, none)	Classic framing effects can be replicated or reversed by replacing numerical information by relational gist of quantities	Activation levels in the prefrontal cortex, the site of working memory and the regions of the parietal lobe for numerical processing, including the intraparietal sulcus and the inferior parietal lobule (Honey et al. 2002; Owen 1997; Piazza et al. 2004; Pinel et al. 2004) would reflect the appearance and disappearance of framing effects
Asymmetry in dual processing models (Dunegan 1993; Kuo et al. 2009; McElroy and Seta 2003, 2004)	Valence framing catalyzes different modes of processing. Framing effects are strengthened by holistic processing and attenuated by analytical processing	More cognitive effort was observed in the negative framing condition than in the positive framing condition. Selective right hemisphere activation promoted framing effects	Framing effects are expected to be associated with right hemispheric activation and differential activations in the brain areas related to loss aversion (e.g., the ventral striatum) and a tradeoff between cognitive/analytical processing in the dorsal lateral prefrontal cortex (dlPFC) and affective/holistic processing in the ventral medial prefrontal cortex (vmPFC) (e.g., Cendri et al. 2012; Montague and Berns 2002; Sokol-Hessner et al. 2012; Tom et al. 2007)
Ambiguity-Ambivalence Hypothesis (Wang 2008)	Framing effects tend to occur when primary choice cues are either absent or incongruent, thus inducing ambiguity or ambivalence in choice preference and making secondary cues of framing more salient and decisive	Framing effects occurred in large anonymous group contexts when primary social cues were absent, but disappeared in small face-to-face group contexts. However, framing effects reappeared when the same life–death problem was presented in a small group context involving both strangers and kin	Ambiguity and ambivalence in risk preference may be reflected in differential activations in the brain regions associated with linguistic processing and the regions responsible for conflict monitoring and resolution (e.g., anterior cingulate cortex (ACC), Botvinick et al. 2004; Holroyd et al. 2004)

of an alternative and the affective value of the alternative. The model predicts that decision-makers prefer choice options that are cognitively less effortful to process or/and affectively more pleasant. In a positive frame, the compromise between maximizing hedonic feeling and minimizing cognitive effort is easier to achieve. For instance, selecting the option in which "200 people will be saved" feels more positive in an emotional sense and is less effortful (i.e., no calculations are necessary). In contrast, such a compromise is more difficult to attain in the negative frame. Although the option in which "400 people will die" requires little calculation, the relatively bad outcome makes it less attractive due to a stronger feeling of displeasure. Thus when selecting among options presented in a negative frame, individuals are more willing to undertake the cognitive effort demanded to assess the more risky option in order to get the hedonically less unpleasant outcome.

Previous studies have shown that individuals take longer to make decisions when the options are framed as losses rather than gains (Payne et al. 1993). However, it is still unknown whether it is a result of a greater cognitive effort in the negative than in the positive frame or a result of a larger affective cost. To answer these questions derived from the cognitive-affective tradeoff model, the authors conducted an fMRI study which revealed that the cognitive effort required to select a sure gain was considerably lower than the cognitive effort required to choose a risky gain. The fMRI results, although not directly correlated with the behavioral framing effects, showed significantly higher activation levels in the frontal and parietal lobes when making risky rather than certain choices under positive frames. This finding is consistent with their theoretical view that the sure gain is a default choice over its gamble equivalent when considering cognitive processing effort.

9.2.2 Neuroimaging Tests of Alternative Behavioral Accounts of Framing Effects

Behavioral studies of framing effects have yielded interesting results that are often open to alternative explanations. These alternative accounts of behavioral findings can be further evaluated in neuroimaging studies. A recent fMRI study by Zheng et al. (2010) on group- size-dependent framing effects illustrates such a value of the neuroimaging approach.

In a series of studies (Wang 1996a, b; Wang et al. 2001), we examined the appearance and disappearance of framing effects when the size of the group (the total number of lives at stake) was systematically manipulated. The same life–death problem was framed either in terms of lives saved or in terms of lives lost. The framing effect was evident, but it occurred only when the problem was presented in a large, anonymous, and thus ambiguous group context involving 600 lives or more. The framing effect disappeared when the size of the endangered group was within a two-digit number (<100), and the majority of the participants unambiguously preferred the gamble option under both the saving lives and losing lives framing conditions.

These findings suggest that the small size of a social group signals a higher interdependence between group members and evokes a kith-and-kin rationality that guides a live-or-die together risk preference. In contrast, risk preference of a decision-maker becomes erratic when prioritized group cues are absent in a large anonymous group context. When risk preference is ambiguous, secondary cues such as verbal framing are attended and used to direct choices. However, an alternative hypothesis of group size-dependent framing effects based on a utility theory would posit that people are more competent in calculating small numbers than large numbers when evaluating risky outcomes, and thus are less ambiguous in their choice preference and less susceptible to framing manipulation.

To evaluate these rival accounts of group size-dependent framing effects, Zheng et al. (2010) conducted an fMRI study. The results of this study, as shown in Table 9.2, help to evaluate the two alternative hypotheses.

Group size difference was captured by activation in the middle frontal gyrus. Verbal framing in the large group context was associated with activation of the right inferior frontal gyrus while the same valence framing in the small group context was associated with activation of different brain structures, including the insula and an area in the parietal lobe. These differential activations were all located in the right hemisphere. Note that the right inferior frontal gyrus includes the homologue of the Broca's area.

These results support the group size account rather than the numerical size account of framing effects. First, behavioral framing effects only occurred in the large group context and were associated with the right IFG activation. In contrast, the disappearance of the framing effect in the small group context was associated with activation of different brain structures (e.g., the right insular). Second, the group size-dependent framing effect was restricted in the right hemisphere, thus was unlikely an explicit numerical processing effect. If numerical processing played a major role in determining appearance or disappearance of framing effects, the left hemisphere and the brain structures related to numerical processing should be

Table 9.2 Brain activation by framing in large and small group contexts (from Zheng et al. 2010)

Brain Area Experimental Condition	Hemisphere	Cluster size (voxels)	Z max	MNI coordinates		
				x	*y*	*z*
IFG P600 minus N600	Right	9	3.35	33	29	−8
Insula P6 minus N6	Right	12	3.61	33	−13	13
Parietal lobe P6 minus N6	Right	17	3.42	33	−31	52
MFG 600 minus 6	Right	23	3.70	24	32	-8

Note: *IFG* Inferior frontal gyrus, *MFG* Middle frontal gyrus. *P* denotes positive frame, and *N* denotes negative frame. The number (600 or 6) represents the number of people at stake. The montreal neurological institute (MNI) coordinates are used to map images. The activation under positive framing was larger than that under negative framing condition in both 600 and 6 group size conditions. The differences were all detected at the level of uncorrected $p < 0.001$

differentially activated (e.g., Piazza et al. 2004; Pinel et al. 2004). Additional behavioral results further exclude the 'large-number' account of framing effects. Wang (1996a) demonstrated that framing effects occurred in large groups of 6000 as well as 600 people, and disappeared in small groups of 60 as well as 6 people. The tenfold difference between the two large groups and between the two small groups did not make a difference in choice preference. Further evidence comes from the Wang, et al. (2001) study, where the classic framing effect occurred in the context of 6 billion human lives but disappeared in the context of 6 billion ET (extraterrestrial) lives. Thus, the framing effect is not likely a large number effect, but is human group size sensitive.

Our findings overall are consistent with the predictions of the AA hypothesis: distinct brain areas are recruited for solving ambiguity or ambivalence caused by the lack of primary social or relational cues in a large anonymous group context. In contrast, framing effects diminished in a small group context while the insula and parietal lobe in the right hemisphere were distinctively activated, suggesting an important role of emotion in switching choice preference from an indecisive mode to a more consistent risk-taking inclination.

The brain imaging findings are interesting in that they suggest that the framing effect is both linguistic and implicit. The affective component of valence framing may direct a holistic, right hemispheric process while the cognitive connotation of framing activates implicit linguistic processing in the right hemisphere in addition to activation of the common linguistic processing regions in the left hemisphere. Previous studies show that the right IFG is involved in response inhibition and impulse control (Aron et al. 2003, 2004; Asahi et al. 2004). Thus, the higher activation found in the right IFG under positive framing suggests a greater control effort induced by using positive framing than negative framing in making hypothetical life–death choices.

9.2.3 Gaining New Insights from Brain Imaging Studies of Framing Effects

The findings from neuroimaging studies of framing effects often provide new insights and new research directions. In this section, we report several new findings derived from a current analysis of an fMRI study of framing effects, in which we examined brain activations when making hypothetical risky choices.

The risky choice problems, adopted from those used in the study by Wang (1996a), involved monetary investment (either 600 or 60,000 Chinese Yuan) and property (either 6 or 600 precious oil paintings) at stake. The structure of the choice problems used in this study was identical to that of the Asian disease problem. The participants were asked to make a binary choice between a sure outcome (framed either as saving one-third of a total amount or losing two-thirds of the total amount) and a gamble of equal expected value. Twenty-one participants (8 males) took part in this study. The procedure was the same as that used in Zheng et al. (2010).

The behavioral effects of valence framing were significant in the painting scenario (risk-seeking choices increased from 47.8 % under positive framing to 57.6 % under negative framing), $\chi^2(1) = 7.617$, $p = 0.0058$. The framing effect was not significant in the money investment scenario (the risk-seeking choice = 55.7 % under positive framing and 62.2 % under negative framing; $\chi^2(1) = 0.246$, $p = 0.620$).

The results of the fMRI data analysis are shown in Table 9.3 and Fig. 9.1. The left inferior frontal gyrus and the left superior frontal gyrus were identified to have a higher activation level under negative framing than positive framing while the left medial temporal cortex showed a higher activation under the positive framing than negative framing. These results are consistent with some previous findings. The inferior frontal gyrus is also found to be associated with framing effects in our previous study (Zheng et al. 2010). In addition, the left medial temporal cortex and the left superior frontal gyrus identified in the present analysis are geographically close to two of the activation regions reported by De Martino et al.' (2006): the left medial temporal gyrus and the left superior frontal sulcus, respectively.

Human neuropsychological studies have highlighted the importance of both frontal cortex and the medial temporal cortex, including the hippocampus in memory encoding and retrieval (e.g., Cohen and Eichenbaum 1993; Tulving et al. 1994). Previous studies also suggest that effortful and strategic memory processes are mainly mediated by the frontal lobes, whereas automatic associative memory processes are mainly mediated by the medial temporal lobes and hippocampus (MTL/H) (e.g., Moscovitch 1994). Thus, our results suggest a differential activation pattern by framing where the negative framing involves more elaborative encoding in the frontal lobe and the positive framing elicits more holistic encoding in the medial temporal cortex.

Table 9.3 Brain regions associated with framing effects

Cluster size	Region	BAs	Peak *T* value	MNI coordinates		
Frame: positive > negative						
122	Left medial temporal cortex	20	4.61	−39	−21	−21
Frame: negative > positive						
120	Left superior frontal gyrus	6	−2.85	−15	−21	75
41	Left inferior frontal gyrus		−2.89	−57	21	9

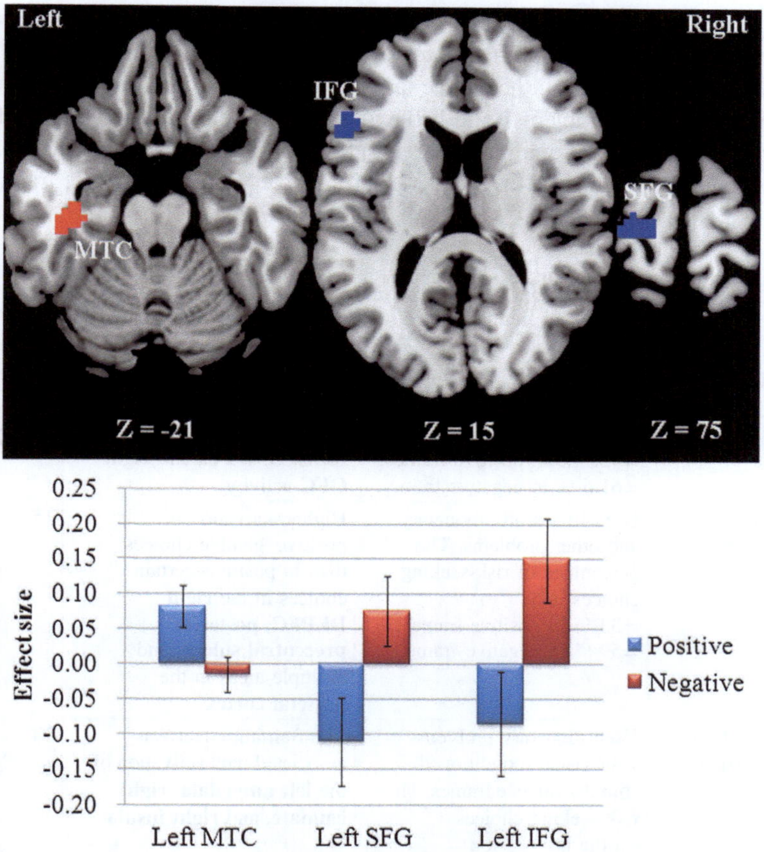

Fig. 9.1 Brain regions showing the effect of frame. *Abbreviation IFG* Inferior frontal gyrus; *MTC* Medial temporal cortex; *SFG* Superior frontal gyrus. *Error bars* denote standard errors. The statistical significance was determined by Monte Carlo simulations to obtain a P-corrected value <0.05, after correcting for whole brain comparisons. The corrected threshold corresponds to a P-uncorrected value <0.005 with a minimum cluster size of 1080 mm^3 (a gray mask with 55,342 voxels was used)

9.2.4 *A Meta-Analysis of Neuroimaging Studies of Framing Effects—Toward a Better Understanding of Framing Effects*

In the field of neuroscience, there have not been many neuroimaging studies on framing effect so far. We carried out a systematic search using the keywords 'framing effect' and 'MRI' to identify relevant studies included in the PubMed and PsycINFO databases; the search was conducted in December 2012, and no time span was specified for date of publication. Our inclusion criteria were that: (1) the studies presented coordinate-based analyses of the data; (2) all or most of the brain was imaged; and (3) participants were asked to choose in different frames. The fMRI studies obtained from this search are summarized below in Table 9.4.

As shown in Table 9.4, the choice tasks involved both life–death and monetary and other types of problems. The behavioral framing effects were all significant, either bidirectional or unidirectional. Non-verbal valence framing had similar behavioral effects. Guitart-Masip et al. (2010) adopted a unique valence framing by

Table 9.4 Functional MRI studies of framing effects

Study	Behavioral framing effects inside the scanner	Main neural correlates	*n*	Foci
De Martino et al. (2006)	Used monetary problems. The percentage of risk-seeking choices =42.9 % in positive frame =61.6 % in negative frame	The framing effect was associated with **amygdala** activity; but reduced by **ACC**, **OMPFC**, and **right OFC** activity	20	6
Gonzalez et al. (2005)	Used life–death, monetary, and other problems. The percentage of risk-seeking choices =33 % in positive frame =59 % in negative frame	Higher activation in positive–gamble choices than in positive–certain choices in the **right DLPFC**, **posterior precentral sulcus**, and multiple areas in the **parietal cortex**	15	7
Guitart-Masip et al. (2010)	Used monetary problems with visual (conditioned stimuli) valence frames. The risk-seeking choices (in the last session) =40 % in positive frame =50 % in negative frame	The framing effect was associated with activation of the **left amygdala, right caudate,** and **right insula**	24	3
Roiser et al. (2009)	Used monetary problems. Framing susceptible and in-susceptible participants differed in the size of the alleles at 5-HTTLPR. The risk-seeking choices =43.3 % in positive frame =56.7 % in negative frame	Framing effects-prone individuals had higher activation in the **left amygdala,** whereas insusceptibility to framing was associated with **ACC** activation.	30	4
Zheng et al. (2010)	Used life–death problems. The risk-seeking choices =52 % in positive frame =65 % in negative frame	The framing effect was correlated with activation in the **right IFG**. The reduction of the framing effect was associated with activation in the **right insula** and an area in the **right parietal cortex**	22	3

Note: *ACC* Anterior cingular cortex; *OMPFC* Orbital and medial prefrontal cortex; *DLPFC* Dorsal lateral prefrontal cortex; *5-HTTLPR* serotonin transporter-linked polymorphic region; *OFC* The orbitofrontal cortex; *IFG* inferior frontal gyrus associating visual stimuli (different texture patterns printed on a card) with gains or losses before using them as valence frames for framing choice outcomes. When the sure-thing outcome of a choice is either presented on a 'gain' card (positive frame) or on a 'loss' card (negative frame), classic monetary framing effects occurred.

Brain regions which were reported to contribute to the appearance of framing effects include the amygdala (De Martino et al. 2006; Guitart-Masip et al. 2010), the insula, and the right IFG (Zheng et al. 2010).

Some brain regions inhibit or reduce framing effects. De Martino et al. (2006) found that greater activity in the orbital and medial prefrontal cortex (OMPFC) predicted a reduced susceptibility to the framing effect across participants. The anterior cingulate cortex (ACC) was also implicated in reduced framing effects (De Martino et al. 2006; Roiser et al. 2009).

The ACC is thought to reduce framing effects by acting as a teaching signal (Botvinick 2007). Because ACC activation was greater when participants' choices were incongruent with frame effects (De Martino et al. 2006), it may modulate the motivational influence of the amygdala on choice. This possibility is supported by the greater coupling between the ACC and the amygdala in participants who were less susceptible to the frame (Roiser et al. 2009). A similar role might be attributed to the orbital frontal cortex (OFC), a region which correlates with resistance to monetary framing effects (De Martino et al. 2006), and with which the ACC has strong reciprocal connectivity (Kringelbach and Rolls 2004).

Interestingly, researchers discovered that susceptibility to framing is related to genetic variation in the serotonin transporter gene (the 5-HTTLPR). Individuals homozygous for the short allele at the 5-HTTLPR rather than individuals homozygous for the long allele were found to be more susceptible to valence framing when making hypothetical monetary risky choices. This susceptibility to framing effects is associated with altered amygdala activity and lack of prefrontal regulatory control (Roiser et al. 2009). However, this finding has been challenged by a follow-up study. Talmi, Hurlemann, Patin, and Dolan (2010) reported that two patients with Urbach-Wiethe disease, a rare condition associated with congenital, complete bilateral amygdala degeneration, exhibited an intact framing effect. However, choice preference in these patients did show a qualitatively distinct pattern compared to controls, as evidenced by a significantly increased risk-seeking preference. These findings suggest that the amygdala does exert an overall influence on risk-taking but may not play a causal role in framing effects.

We then conducted our meta-analysis by entering the coordinates of all the foci that were reported in the above fMRI studies of framing effects to get their topographic convergences. The goal of coordinate-based meta-analysis of neuroimaging data is to identify brain areas where the reported foci of activation converge across published experiments. In this meta-analysis of neural correlates of framing effects, we adopted a widely used technique for coordinate-based meta-analyses of neuroimaging data, called activation likelihood estimation (ALE). ALE assesses the overlap between foci based on modeling them as probability distributions centered at their respective coordinates. ALE maps are then obtained by computing the union of activation probabilities for each voxel. To differentiate true convergence of foci from random clustering (i.e., noise), a permutation test is applied: to obtain an ALE null distribution, the same number of foci as in the real analysis are randomly redistributed throughout the brain (see Eickhoff et al. 2009; Turkeltaub et al. 2002). In other words, the meta-analysis determines if the clustering is significantly higher than expected under the null distribution of a random spatial association of results from the considered experiments.

Our analysis was implemented using GingerALE Version 2.1.1 (available at http://brainmap.org/ale). The meta-analysis was performed using the Montreal Neurological Institute (MNI) stereotactic coordinates derived from the studies listed in Table 9.4. Coordinates published in Talairach space were transformed to the Montreal Neurological Institute (MNI) template according to the Lancaster transform (icbm2tal) in GingerALE. Statistical significance was determined using a permutation test of randomly generated foci. No assumptions were made concerning the distribution or spatial separation of these random foci; however, clusters of activity were required to exceed 200 mm^3 in volume. The test was corrected for multiple comparisons using the false discovery rate (FDR) method (Genovese et al. 2002). Anatomical labels of final cluster locations are provided by the Talairach Daemon.

Initially, a total of 23 foci were analyzed. The ALE meta-analysis showed that high ALE values were observed in the left ACC and the right IFG (see Table 9.5; Fig. 9.2), indicating these two brain regions are the key neural correlates of framing effects.

We then added into the ALE meta-analysis the coordinates of the three foci associated with framing effects as reported in the previous section of this chapter. The brain localization results for framing effects based on the 26 foci remained the same.

Neurobiological studies of the right IFG have shown some converging evidence for its unique role in 'semantic selection' by which competing activated concepts are sorted out, inhibiting competing concepts while selecting one concept for action, particularly when the concept to be selected involves atypical usage. For instance, when people are given a common noun (e.g., cake) and asked to produce a typical use ('bake' or 'eat'), the left IFG is strongly active; but when asked to produce an unusual use (e.g. 'sell') of such nouns, the right IFG is more strongly active (see Jung-Beeman 2005 for a review). Both the left and the right IFG play a

Table 9.5 Neural correlates of framing effects derived from ALE meta-analysis

Cluster	Region	BA	Talairach coordinates			ALE ($\times 10^{-2}$)	Volume (mm^3)
			x	*y*	*z*		
1	Left ACC	24	−4	36	−6	1.16	408 (880)
2	Right IFG	47	32	30	−8	0.90	216 (216)

Note: *ACC* Anterior cingulate cortex; *IFG* Inferior frontal gyrus, *BA* Brodmann area. The values in parentheses are the results after adding four foci from the study reported in this chapter in the prior section

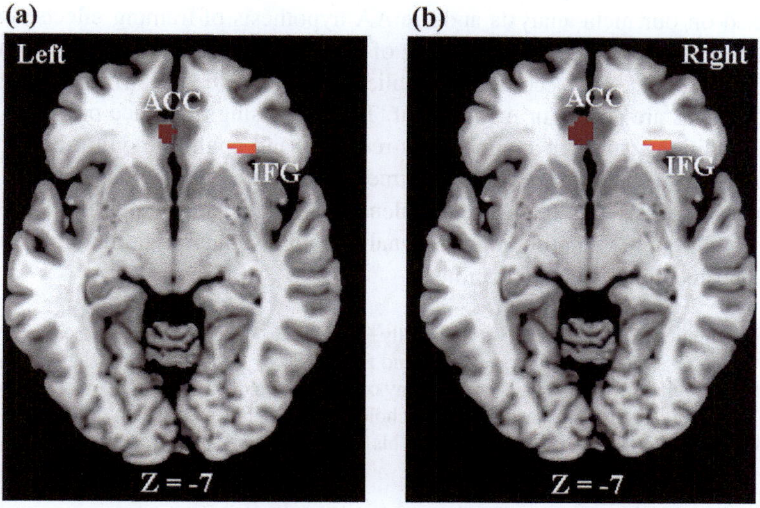

Fig. 9.2 Meta-analysis results of neural correlates of framing effects. **a** Topographic convergences from the foci reported in the studies in Table 9.4. **b** Topographic convergences from the foci reported in the studies in Table 9.4 plus additional foci reported in this chapter

unique role in understanding the figurative meaning of a metaphor beyond semantic analysis (Rapp et al. 2004). In addition, the right IFG is active in response inhibition and selection. The go/no-go task, which taps the ability to inhibit prepotent response tendency (for instance, stop pressing a button when a red signal appears), has consistently activated the lateral prefrontal cortex, particularly the right IFG (Aron et al. 2003; Asahi et al. 2004; Chikazoe et al. 2007). Some researchers consider the right IFG the most important prefrontal structure that exerts inhibition and cognitive control over subcortical structures of the brain (Aron et al. 2004). The same area is also implicated in risk aversion: higher risk aversion is correlated with higher activity at the right IFG (Christopoulos et al. 2009).

Coupled with the results of our meta-analysis, these findings suggest that the right IFG is a unique neural correlate of framing effects. It plays an integrative role in evaluating risk-related cues and in regulating risk preference and choice selection based on both semantic and affective meanings imbedded in choice problems.

Researchers have suggested a variety of ways to interpret the neuroimaging findings of the framing effect. De Martino et al. (2006) suggested that the framing effect is driven by an affect heuristic underwritten by an emotional system. From the perspective of a dual-system framework, Kahneman and Frederick (2006) interpreted these findings as evidence that different frames evoke distinct emotional responses that different individuals can suppress to various degrees. Similarly, Gonzalez et al. (2005) proposed a model which tries to incorporate a tradeoff between the cognitive effort required to calculate expected values of an alternative and the affective value of the alternative to explain the choice process underlying the framing effect.

Based on our meta-analysis and the AA hypothesis of framing effects, we propose a new view on neural mechanisms of framing effects. First, valence framing as a secondary choice cue becomes most salient and effective when primary contextual or social cues are absent or incongruent. The processing of choice problems under these conditions calls for an ambiguity-reducing and conflict-monitoring function, which would result in the ACC recruitment and activation. Second, the right IFG activation suggests that the nature of valence framing is both semantic and hedonic, involving not only verbatim linguistic analysis but also interpretation of its affective tones and metaphorical implications.

Acknowledgments This work was partially supported by National Science Foundation (NSF) Grant SES-1123341, and a research fund from the Key Laboratory of Behavioral Science, Institute of Psychology of the Chinese Academy of Sciences to the first author, and a research fund (No. Y2CQ043005) from the Institute of Psychology of the Chinese Academy of Sciences to the second author. We thank Steven Garelik for his helpful comments on an earlier version of this paper.

References

Aron AR, Fletcher PC, Bullmore ET, Sahakian BJ, Robbins TW (2003) Stop-signal inhibition disrupted by damage to right inferior frontal gyrus in humans. Nat Neurosci 6:115–116

Aron AR, Robbins T, Poldrack R (2004) Inhibition and the right inferior frontal cortex. Trends Cogn Sci 8:170–177

Asahi S, Okamoto Y, Okada G, Yamawaki S, Yokota N (2004) Negative correlation between right prefrontal activity during response inhibition and impulsiveness: an fMRI study. Eur Arch Psychiatry Clin Neurosci 254:245–251

Bless H, Betsch T, Franzen A (1998) Framing the framing effect: the impact of context cues on solutions to the "Asian disease" problem. Euro J Soc Psychol 28:287–291

Bloomfield AN (2006) Group size and the framing effect: treats to human beings and animals. Mem Cogn 34:929–937

Bloomfield AN, Sager JA, Bartels DM, Medin DL (2006) Caring about framing effects. Mind Soc 5:123–138

Bohm P, Lind H (1992) A note on the robustness of a classical framing result. J Econ Psychol 13:355–361

Botvinick MM (2007) Conflict monitoring and decision making: Reconciling two perspectives on anterior cingulate function. Cogn Affect Behav Neurosci 7:356–366

Botvinick MM, Cohen JD, Carter CS (2004) Conflict monitoring and anterior cingulate cortex: an update. Trends in Cognitive Sciences 8:539–546

Breiter HC, Aharon H, Kahneman D, Dale A, Shizgal P (2001) Functional imaging of neural responses to expectancy and experience of monetary gains and losses. Neuron 30:619–639

Cendri A, Hutcherson CA, Plassmann H, Gross JJ, Rangel A (2012) Cognitive regulation during decision making shifts behavioral control between ventromedial and dorsolateral prefrontal value systems. J Neurosci 32:13543–13554

Chang CJ, Yen SH, Duh RR (2002) An empirical investigation of competing theories to explain the framing effect in accounting-related decision. Behav Res Acc 14:35–64

Chikazoe J, Konishi S, Asari T, Jimura K, Miyashita Y (2007) Activation of right inferior frontal gyrus during response inhibition across response modalities. J Cogn Neurosci Arch 19:69–80

Chong D, Druckman J (2007) Framing theory. Annu Rev Polit Sci 10:103–126

Christopoulos GI, Tobler PN, Bossaerts P, Dolan RJ, Schultz W (2009) Neural correlates of value, risk, and risk aversion contributing to decision making under risk. J Neurosci 29:12574–12583
Cohen NJ, Eichenbaum HE (1993) Memory, amnesia, and the hippocampal system. MIT Press, Cambridge, MA
De Martino B, Kumaran D, Seymour B, Dolan RJ (2006) Frames, biases, and rational decision-making in the human brain. Science 313:684–687
Druckman JN (2001) On the limits of framing effects: Who can frame? J Polit 63:1041–1066
Dunegan KJ (1993) Framing, cognitive modes, and image theory: toward an understanding of a glass half full. J Appl Psychol 78:491–503
Eickhoff SB, Laird AR, Grefkes C, Wang LE, Zilles K, Fox PT (2009) Coordinate-based activation likelihood estimation meta-analysis of neuroimaging data: a random-effects approach based on empirical estimates of spatial uncertainty. Hum Brain Mapp 30:2907–2926
Fagley NS, Miller PM (1987) The effects of decision framing on choice of risky vs. certain options. Organ Behav Hum Decis Process 39:264–277
Fagley NS, Miller PM (1990) The effect of framing on choice: interactions with risk-taking propensity, cognitive style, and sex. Pers Soc Psychol Bull 16:496–510
Fagley NS, Miller PM (1997) Framing effects and arenas of choice: your money or your life? Organ Behav Hum Decis Process 71:355–373
Genovese CR, Lazar NA, Nichols TE (2002) Thresholding of statistical maps in functional neuroimaging using the false discovery rate. NeuroImage 15:870–878
Gigerenzer G, Hoffrage U, Kleinbölting H (1991) Probabilistic mental models: a Brunswikian theory of confidence. Psychol Rev 98:506–528
Gonzalez C, Dana J, Koshino H, Just M (2005) The framing effect and risky decisions: examining cognitive functions with fMRI. J Econ Psychol 26:1–20
Guitart-Masip M, Talmi D, Dolan R (2010) Conditioned associations and economic decision biases. NeuroImage 53:206–214
Holroyd CB, Nieuwenhuis S, Mars RB, Coles MGH (2004) Anterior cingulate cortex, selection for action, and error processing. In Posner MI (ed) Cognitive neuroscience of attention. Guilford Press, New York, pp 219–231
Honey GD, Fu CH, Kim J et al (2002) Effects of verbal working memory load on corticocortical connectivity modeled by path analysis of functional magnetic resonance imaging data. NeuroImage 17:573–582
Jung-Beeman M (2005) Bilateral brain processes for comprehending natural language. Trends Cogn Sci 9:512–518
Kahneman D, Frederick S (2006) Frames and brains: elicitation and control of response tendencies. Trends Cogn Sci 11:45–46
Kahneman D, Tversky A (1979) Prospect theory. Econometrica 47:263–292
Kringelbach ML, Rolls ET (2004) The functional neuroanatomy of the human orbitofrontal cortex: evidence from neuroimaging and neuropsychology. Prog Neurobiol 72:341–372
Kühberger A (1995) The framing of decisions: a new look at old problems. Organ Behav Hum Decis Process 62:230–240
Kühberger A (1998) The influence of framing on risky decisions: a meta-analysis. Organ Behav Hum Decis Process 75:23–55
Kühberger A, Schulte-Mecklenbeck M, Perner J (1999) The effects of framing, reflection, probability, and payoff on risk preference in choice tasks. Organ Behav Hum Decis Process 78:204–231
Kuo FY, Hsu CW, Day RF (2009) An exploratory study of cognitive effort involved in decision under Framing-an application of the eye-tracking technology. Decis Support Syst 48:81–91
LeBoeuf RA, Shafir E (2003) Deep thoughts and shallow frames: on the susceptibility to framing effects. J Behav Decis Mak 16:77–92
Levin IP, Chapman DP (1990) Risk taking, frame of reference, and characterization of victim groups in AIDS treatment decisions. J Exp Soc Psychol 26:421–434
Levin IP, Gaeth GJ (1988) Framing of attribute information before and after consuming the product. J Consum Res 15:374–378

Levin IP, Schneider S, Gaeth GJ (1998) All frames are not created equal: a typology and critical analysis of framing effects. Organ Behav Hum Decis Process 76:149–188
McElroy T, Seta J (2003) Framing effects: an analytic-holistic perspective. J Exp Soc Psychol 39:610–617
McElroy T, Seta JJ (2004) On the other hand Am I rational? Hemispheric activation and the farming effect. Brain Cogn 55:572–580
Meyerowitz BE, Chaiken S (1987) The effect of message framing on breast self-examination attitudes, intentions, and behavior. J Pers Soc Psychol 52:500–510
Miller PM, Fagley NS (1991) The effects of framing, problem variations, and providing rationale on choice. Pers Soc Psychol Bull 17:517–522
Montague PR, Berns GS (2002) Neural economics and the biological substrates of valuation. Neuron 36:265–284
Moscovitch M (1994) Cognitive resources and dual-task interference effects at retrieval in normal people: the role of the frontal lobes and medial temporal cortex. Neuropsychology 8:524
O'Doherty J, Kringelbach ML, Rolls ET, Hornak J, Andrews C (2001) Abstract reward and punishment representations in the human orbitofrontal cortex. Nat Neurosci 4:95–102
Owen AM (1997) The functional organization of working memory processes within human lateral frontal cortex: the contribution of functional neuroimaging. Euro J Neurosci 9:1329–1339
Payne JW, Bettman JR, Johnson EJ (1993) The adaptive decision maker. Cambridge University Press, New York
Piazza M, Izard V, Pinel P, Le Bihan D, Dehaene S (2004) Tuning curves for approximate numerosity in the human intraparietal sulcus. Neuron 44:547–555
Pinel P, Piazza M, Le Bihan D, Dehaene S (2004) Distributed and overlapping cerebral representations of number, size, and luminance during comparative judgments. Neuron 41:983–993
Rapp AM, Leube DT, Erb M, Grodd W, Kircher TTJ (2004) Neural correlates of metaphor processing. Cogn Brain Res 20:395–402
Reyna VF, Brainerd CJ (1991) Fuzzy-trace theory and framing effects in choice: Gist extraction, truncation, and conversion. J Behav Decis Mak 4:249–262
Reyna VF, Brainerd CJ (1995) Fuzzy-trace theory: an interim synthesis. Learn Individ Differ 7:1–75
Reyna VF, Brainerd CJ (2011) Dual processes in decision making and developmental neuroscience: a fuzzy-trace model. Dev Rev 31:180–206
Roiser JP, De Martino B, Tan GCY, Kumaran D, Seymour B, Wood NW, Dolan R (2009) A genetically mediated bias in decision making driven by failure of amygdala control. J Neurosci 29:5985–5991
Shimizu K, Udagawa D (2011a) A re-examination of the effect of contextual group size on people's attitude to risk. Judgment and Decision Making 6:156–162
Shimizu K, Udagawa D (2011b) How can group experience influence the cue priority? A re-examination of the ambiguity-ambivalence hypothesis. Front Evol Psychol 2:1–9
Sieck W, Yates JF (1997) Exposition effects on decision making: choice and confidence in choice. Organ Behav Hum Decis Process 70:207–219
Sokol-Hessner P, Hutcherson C, Hare T, Rangel A (2012) Decision value computations adjust to the available decision time. Eur J Neurosci 35:1065–1074
Stanovich KE, West RF (1998) Individual differences in framing and conjunction effects. Think Reason 4:289–317
Takemura K (1994) Influence of elaboration on the framing of decision. J Psychol 128:33–39
Talmi D, Hurlemann R, Patin A, Dolan R (2010) Framing effect following bilateral amygdala lesion. Neuropsychologia 48:1823–1827
Tom SM, Fox CR, Trepel C, Poldrack RA (2007) The neural basis of loss aversion in decision making under risk. Science 315:515–518
Tulving E, Kapur S, Craik FIM, Markowitsch HJ, Houle S (1994) Hemispheric encoding/retrieval asymmetry in episodic memory: positron emission tomography findings. Proc Natl Acad Sci USA 91:2016–2020

Turkeltaub PE, Eden GF, Jones KM, Zeffiro TA (2002) Meta-analysis of the functional neuroanatomy of single-word reading: method and validation. Neuroimage 16:765–780
Tversky A, Kahneman D (1981) The framing of decisions and the psychology of choice. Science 211:453–458
Wang XT (1996a) Framing effects: Dynamics and task domains. Organ Behav Hum Decis Process 68:145–157
Wang XT (1996b) Domain-specific rationality in human choices: violations of utility axioms and social contexts. Cognition 60:31–63
Wang XT (2008) Risk communication and risky choice in context: ambiguity and ambivalence hypothesis. Ann N Y Acad Sci 1128:78–89
Wang XT, Johnson GJ (2012) A tri-reference point theory of decision making under risk. J Exp Psychol Gen 141:743–756
Wang XT, Johnston V (1995) Perceived social context and risk preference: a re-examination of framing effects in a life-death decision problem. Journal of Behavioral Decision Making 8:279–293
Wang XT, Simons F, Brédart S (2001) Social cues and verbal framing in risky choice. J Behav Decis Mak 14:1–15
Whitney P, Rinehart CA, Hinson JM (2008) Framing effects under cognitive load: The role of working memory in risky decisions. Psychon Bull Rev 15:1179–1184
Zhang Y, Miao D (2008) Social cues and framing effects in risky decisions among Chinese military students. Asian J Soc Psychol 11:241–246
Zheng H, Wang XT, Zhu L (2010) Framing effects: behavioral dynamics and neural basis. Neuropsychologia 48:3198–3204

Chapter 10
The Influence of Costs, Benefits and Their Interaction on the Economic Behaviour of Consumers

Luca Panzone and Deborah Talmi

> *"Tis the sharpness of our mind that gives the edge to our pains and pleasures."*
>
> —Michel de Montaigne

Abstract Recent neuroscientific research on economic behaviour of consumers explores how individuals translate information into value in their brain, and what mechanisms underlie this process. The typical aim of this research is to establish how single attributes are valued and combined into a single utility, neglects findings in multi-attribute utility theory on how utility is achieved when both costs and benefits are involved. This chapter argues that it is important to consider how the marginal utility of costs and benefits changes in the respective presence of one another. This point is discussed by reviewing behavioural and brain imaging data that illuminate this interplay, with a focus on the implications on econometric models of consumer behaviour.

10.1 Introduction

Any gain in utility comes at the expense of an initial loss, a concept recognised by the adage 'No pain, no gain'. For instance, consumers accept the payment of a price (a cost) because they expect a gain in hedonic satisfaction through product quality (a consequent benefit). Similarly, current policies targeting behavioural change impose a trade-off between effort (an immaterial cost) and a reward associated with that effort, e.g. a health improvement (a private benefit), or an environmental improvement (a public benefit) (Steg and Vlek 2009). How the brain estimates the overall utility of a person facing the cost and the reward prospect and converts them

---

L. Panzone
School of Social Science, Sustainable Consumption Institute, Manchester, UK
e-mail: luca.panzone@manchester.ac.uk

D. Talmi (✉)
School of Psychological Science, University of Manchester, Manchester, UK
e-mail: deborah.talmi@manchester.ac.uk

M. Reuter and C. Montag (eds.), *Neuroeconomics*, Studies in Neuroscience, Psychology and Behavioral Economics, DOI 10.1007/978-3-642-35923-1_10

into a single value is not yet fully understood (Camerer et al. forthcoming; Levy and Glimcher 2012; Louie and Glimcher 2012; Vlaev et al. 2011).

There is a growing literature on the neural representation of value, i.e. how individual consumers determine the value of choice options. Yet some of the recent discoveries on brain valuation activity have not been integrated into a coherent economic theoretical framework. The aim of this chapter is to understand how individuals convert costs and benefits into utility in their decisions to consume, particularly focusing on understanding how these combine into a final valuation of utility. Costs and benefits are defined as anticipated consequences of the acceptance of a choice, whereby they respectively increase (benefits) or decrease (costs) the subjective value of the choice in the decision-making process. This definition hinges on the nature of these phenomena, since benefits (such as rewards) induce positive feelings of pleasure (Schultz 2000), while costs (such as effort or pain) are an aversive feature associated with consumption (Kivetz 2003).

In the most basic model of optimal decision-making, an individual chooses by introspectively estimating the expected amount of satisfaction (e.g. happiness, utility) that a good can deliver upon consumption. This process entails the evaluation of the intrinsic quality provided by each of the characteristics within a good (Lancaster 1966), which are then integrated into a single value for each option in the choice set. The resulting choice then corresponds to the option delivering the highest overall value (Vlaev et al. 2011; Kahneman et al. 1997) or the option with the highest probability of a positive outcome at consumption. Because quality is estimated before consumption, these stages are limited by cognitive bounds that reduce consumer efficacy: consumers use heuristics to determine quality, and contextual stimuli influence their final decisions (Vlaev et al. 2011; Kahneman 2003). It is important to keep in mind that heuristics and external stimuli span across attributes, reinforcing the links between them.

This simple model of behaviour assumes independence of product attributes. However, when goods provide 'mixed prospects', i.e. those associated both with costs and with benefits, valuation is more complex than what has been described so far. For instance, the purchase of an energy-efficient appliance requires the trade-off between high environmental quality together with prospective savings (two benefits) and a price premium (cost). The extra money consumers are willing to pay for the 'green' (i.e. energy-efficient) choice depends both on the expected benefits from energy efficiency and on the utility of money. Consumers with high utility from money (e.g. low-income households) would switch to energy-efficient appliances only when deriving high levels of utility from energy conservation. Instead, environmentally motivated consumers are likely to switch even at a very high premium. Individuals face such dilemmas frequently in non-purchasing decisions as well: for example, when they decide whether to go to the gym regularly (effort cost) to lose weight (benefit), or decide to walk to the recycling bin instead of throwing their paper into the regular bin (effort cost) to improve the quality of the environment (benefit).

Neuroscientific research typically explores the representation of reward and punishment separately (see e.g. Schultz 2006; Montague and King-Casas 2007;

Seymour et al. 2007), and how they are integrated during goal-directed behaviour is relatively unexplored (Phillips et al. 2007; Walton et al. 2007). Goal-directed behaviour engenders conflict when trading the prospect of an appetitive gain against an equal prospect of an aversive cost. This conflict is represented empirically by the dynamic interaction of costs and benefits in the brain, a concept that Multi-Attribute Utility Theory (MAUT) describes through the use of an interaction term. This point received empirical support in the pattern of activation of the ventral striatum and the subgenual cingulated gyrus in the brain. Talmi et al. (2009) reported a positive correlation between BOLD signal and reward when participants are offered money (a benefit) to accept mild electric stimulations (low-cost condition); the correlation was attenuated when the electric stimulation was painful (high-cost condition) (see also Talmi and Pine 2012). This decreasing correlation is indicative of a nonlinear relation between costs and benefits.

The remainder of this chapter will proceed as follows. The next section reviews the representation of costs and benefits in economic modelling, with a focus on utility theory, and describes how costs and benefits interact in the final determination of utility. The analysis focuses on decisions between outcomes with a limited number of attributes (one cost and one benefit) in a single context. The reason for this rather narrow focus is that the integration of many attributes into a single utility may exceed cognitive bounds: evidence (Vlaev et al. 2011) suggests that in complex multi-attribute choices consumers may not be able to determine the overall utility, and the simple model described in this section might not hold. The model presented here will, therefore, be more valid when only two characteristics are used. Section 10.3 describes the same interaction between costs and benefits in a neuroscientific model of decision-making. The microeconomic implications of the model described in Sect. 10.3 are discussed in Sect. 10.4 to extend current microeconometric models of choice, examining its influence on estimable parameters of consumer choice through a simple simulation exercise. Finally, Sect. 10.5 concludes by summarising the key issues emerging from the chapter.

10.2 The Economic Perspective of Cost and Benefits

Economic models of choice and one-person decision problems are an important part of economic theory. According to the Hicksian model of utility maximisation (Hicks 1939), individual consumers value goods for the utility they derive upon consumption. For instance, a consumer derives consumption utility from good i of characteristic X_i in the form $U_i = f(X_i)$. More precisely, the total value of a good corresponds to the sum (in utility terms) of the values assigned to the quantity of each attribute the good provides (Lancaster 1966). Specifically, the consumer compares two or more options i (where $i = 1, 2, \ldots, I$) that differ in their cost C_i and benefits B_i in terms of their consumption utility $U_i = f(C_i, B_i)$, where the matrix $[C_i, B_i]$ corresponds to X_i. Given income y and price p, the consumer then chooses by solving the utility maximisation problem.

$$\max_{C,B} U_i = U(C_i, B_i) \quad \text{s.t. } p_i \leq y_j \tag{10.1}$$

The marginal utility of costs and benefits is expected to be negative and positive, respectively.

This general economic model of utility maximisation assumes that consumers determine the value of a good in three steps. First, they gather information on each attribute constituting the good with certainty, e.g. what is the environmental impact of refrigerators rated A+ over those rated A. Second, they 'convert' this information into a measure of utility, e.g. assuming goods do not differ in any other dimension, they determine the utility derived from the lower carbon footprint associated to A+ refrigerators and the net present value of prospective savings, and the disutility of the price premium required by an A+-rated refrigerator. Third, they sum the utility of each attribute to determine the total value of each option. The choice then needs to fit within a budget constraint and requires the trade-off between costs and benefits, which can be observed through the Marginal Rate of Substitution (MRS), $\mathrm{MRS} = \frac{\partial U_i}{\partial B_i} \Big/ \frac{\partial U_i}{\partial C_i}$, a measure of how much utility from one unit of benefit is needed to compensate the utility of one unit of cost.

The general utility function of Eq. (10.1) can take different functional forms. Psychological and neuroeconomic models of internal valuation generally base their analysis of decision-making on a von Neumann–Morgenstern utility function (von Neumann and Morgenstern 1947), which models utility under risk. This function, known as **expected utility function**, assumes that rational consumers estimate the utility of costs and benefits perfectly, with certain knowledge of the probability of occurrence of an outcome. In this model, the subjective value U of an action corresponds to the sum of the utility (u) of each outcome, evaluated separately, multiplied by the corresponding probability π. The value of a mixed outcome based on one beneficial (B) and one costly (C) attribute corresponds to

$$U = U(C_i, B_i) = \pi_i^{\mathrm{C}} \cdot u(C_i) + \pi_i^{\mathrm{B}} \cdot u(B_i) \tag{10.2}$$

A positive U resulting from Eq. (10.2) favours a decision to act, and in a choice set the alternative with the greatest U is always preferred.

In Eq. (10.2), when costs and benefits are exclusive of each other $\pi_i^{\mathrm{C}} + \pi_i^{\mathrm{B}} = 1$. More generally, when costs and benefits are not the only two possible outcomes that exists but rather two characteristics of only one outcome in the space $\pi_i^{\mathrm{C}} + \pi_i^{\mathrm{B}}$ can differ from 1. For instance, total probability is below 100 % if an A+ refrigerator costs an extra £100 with $\pi_i^{\mathrm{C}} = 0.99$ and the subjective belief of an environmental improvement is $\pi_i^{\mathrm{B}} = 0.001$. Likewise, if a consumer believes that the refrigerator certainly helps the environment ($\pi_i^{\mathrm{B}} = 1$) and is certainly expensive ($\pi_i^{\mathrm{C}} = 1$), total probability is above 100 %. Similarly, optimistic consumers might overestimate the probability of a gain much more than they underestimate the complementary probability, and pessimistic consumers might do the same for losses.

Expected utility theory (EUT), described by Eq. (10.2), is a restrictive model for decision-making under uncertainty because it assumes an objective probability π. Prospect theory (PT) extended this utility function by fitting a subjective probability function that fits empirical data better than EUT (Kahneman and Tversky 1979; Tversky and Kahneman 1992). In particular, PT indicates that accessibility (the way individuals perceive signals), reference-dependence (the reference point used to evaluate a signal) and framing (the way the signal is presented) are crucial determinant of behaviour (Kahneman 2003). Consumers then make decisions based on expected subjective probability rather than objective one, as

$$U = \omega(\pi_i^{\mathrm{C}}) \cdot u(C_i) + \omega(\pi_i^{\mathrm{B}}) \cdot u(B_i) \tag{10.3}$$

where ω are subjective decision weights. Empirically, this utility function is concave in the gain domain and convex in the loss domain with a steeper slope in the latter (i.e. the disutility of a loss is greater than the utility of a gain), a phenomenon known as loss aversion (Kahneman and Tversky 1979).

Equations (10.2) and (10.3) do not define a functional form for the utility of costs, $u(C_i)$, and benefits, $u(B_i)$, but impose a linear integration of these functions into a single value. This restriction complies with the **independence assumption**: the utility derived from the consumption of a good is independent from the utility derived from other goods. This assumption might hold when the consumption of two goods is self-exclusive, e.g. the consumer has budget to purchase either a refrigerator or a washing machine. Its existence is more controversial when considering product characteristics (e.g. energy-efficiency class against a price premium), or costs and benefits of the same action. In fact, consumers pursue a certain number of consumption goals that are satisfied through product characteristics that are rarely self-exclusive (see e.g. Khan et al. 2004; Dhar and Simonson 1999). For instance, the goal of environmental friendliness in a refrigerator requires both a positive utility for high efficiency (a benefit) and positive willingness to pay for it (a cost), making these two characteristics dependent on each other.

Importantly, the independence assumption applied to Eq. (10.3) implicitly assumes that costs reduce (and benefits increase) the subjective value derived from benefits (costs) by a constant amount, regardless of how valuable those benefits (costs) are. However, an increase in costs can increase or decrease not only the utility expected by the consumer, but also the utility derived from a unit increase in benefits; similarly, an increase in benefits can modulate the utility derived from a unit increase in costs. For instance, Kivets (Kivetz 2003) observes that the imposition of an effort requirement (the cost) for a loyalty reward scheme influences the expected utility of the reward (the benefit): small effort leads to expectations of an immediate, certain, but small reward; considerable effort instead causes expectations of a large, even if uncertain, reward. While in a linear model an increase in costs can lead to higher expected benefits, Kivets observes that an increasing effort requirement influences the disutility derived from effort itself as well as the expected marginal utility of the reward, a relation suggesting that these two variables interact in a utility function. This “synergic effect” implies that a change in costs and benefits does not only impact total utility directly (as intercept shifters), but the final joint effect is higher or lower than the sum of both effects, depending on the existing relation between them.

Multi-attribute Utility Theory (MAUT, see e.g. Keeney and Raiffa 1993) provides some theoretical background for the representation of the interaction between costs and benefits in a utility function. MAUT maintains that individual characteristics might have a synergic influence on utility. As a simple example, a consumer might have a negative utility from the price of a piece of cake, and a low but positive utility for its size; nonetheless, utility for size can become very high when the dessert is particularly expensive—an interaction between size and price. Following MAUT, the utility function in (10.3) can then be then generalised as

$$U = \omega(\pi_i^{\mathrm{C}}) \cdot u(C_i) + \omega(\pi_i^{\mathrm{B}}) \cdot u(B_i) + \omega(\pi_i^{\mathrm{C \times B}}) \cdot u(C_i) \cdot u(B_i) \tag{10.4}$$

where $\omega(\pi_i^{\mathrm{C \times B}})$ is the subjective probability that costs and benefits interact. The subjective value U as a function of the utility u of costs (the same applies to benefits) is portrayed in Fig. 10.1, under the assumption of linear functions u: unlike the additive model, an interactive model allows for a shift in both intercept and slope of total utility.

The key difference between Eqs. (10.3) and (10.4) is the ability, unique to interactive models, to allow the sensitivity of benefits to vary as a function of the costs associated with the choice, and vice versa (Talmi and Pine 2012). In particular, the marginal utility of costs and benefits from Eq. (10.4) correspond to[1]

$$\frac{\partial U}{\partial C_i} = \omega(\pi_i^{\mathrm{C}}) \cdot \frac{\partial u(C_i)}{\partial C_i} + \omega(\pi_i^{\mathrm{C \times B}}) \cdot u(B_i) \cdot \frac{\partial u(C_i)}{\partial C_i} \tag{10.5a}$$

and

$$\frac{\partial U}{\partial B_i} = \omega(\pi_i^{\mathrm{B}}) \cdot \frac{\partial u(B_i)}{\partial B_i} + \omega(\pi_i^{\mathrm{C \times B}}) \cdot u(C_i) \cdot \frac{\partial u(B_i)}{\partial B_i} \tag{10.5b}$$

Returning to the choice of an energy-efficient refrigerator, assume two otherwise identical refrigerators, differing only in their efficiency rating (A and A+) and price (£100 higher for the A+ option). According to additive models (Eq. 10.3), the higher price reduces the likelihood of purchasing the efficient option regardless of the value the decision-maker assigns to the environmental characteristic. By contrast, the interactive model (Eq. 10.4) accounts for the dependence between the value consumers may assign to price premium and efficiency rating. This

[1]Equations (10.5a) and (10.5b) could be further generalised to account for loss aversion, which affects the derivative of costs, so that the change in cost influences benefits in a different way compared to how the change in benefits influences costs. However, this point would complicate the argument presented in this article, and it is left for future research.

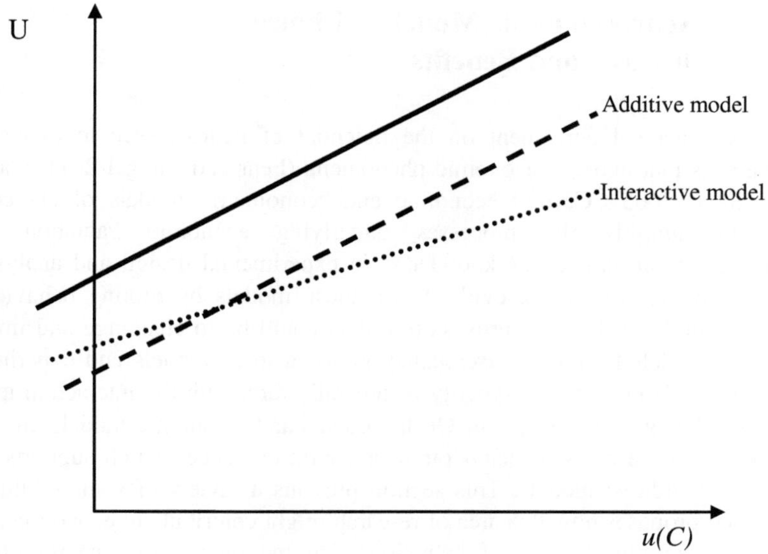

Fig. 10.1 Value according to additive and interactive models

generalisation allows, for instance, for the possibility of a lower disutility from the cost for those consumers who deem energy conservation to be very valuable.

An important feature of an interactive model is the possibility to test whether the assumptions of attribute independence is valid by comparing the additional variance explained by the interactive model and the additive model. In fact, because the interaction term is nested within the additive model, its contribution can be tested using a Student's t-test or a Fisher's F test, or comparable statistics for maximum likelihood and Bayesian models such as likelihood ratio test, or Akaike or Bayesian Information Criteria. Through a model-comparison exercise, earlier research suggested that two-way interactions (the interaction of pairs of attribute) in linear regressions account for 5–15 % of the overall variance of the model, while other interactions (the interaction of three or more attributes) are negligible in their predictive power (see Louviere et al. 2000). However, to our knowledge there is no measure of the statistical importance of interaction terms on models of choice, implying the absence of estimates on the relevance of the interaction between costs and benefits on utility.

As described in the next section, recent behavioural and neuroimaging findings support the challenge to the independence assumption by corroborating the existence of an interaction term. In fact, the presence of mixed outcomes is processed in the brain differently from the simple presence of costs or benefits alone, a point that calls for a reconsideration of the simple additive model of costs and benefits. The next section reviews the existing evidence indicating that interaction terms increase the predictive accuracy of a model and reflect the cognitive process of valuation.

10.3 A Neuroeconomic Model of Choice with Costs and Benefits

There is a general agreement on the potential of neuroscience to explain the mechanisms that underlie economic phenomena (Fehr and Rangel 2011; Camerer et al. 2004, 2005). Current economic and econometric models of choice and behaviour simplify the processes underlying economic valuation tasks. Neuroscience can utilise this knowledge in experimental design and analysis, as well as provide converging evidence for these models by rooting behaviour in biology. Another role for neuroscientific data could be to challenge and improve economic models by noting discrepancy between these models and experimental observations. However, this synergy is generally rare, with the interaction term in Eq. (10.4) being a case in point. On the other hand, economic models are rarely challenged on the basis of behavioural or neural evidence, which supports interactive over additive models. This section presents a review of some of this evidence, and proposes how this area of research might contribute to economic theory and extend current models of behaviour. On the other hand, neurobiological research of cost–benefit analysis often ignores value integration functions available in utility theory, an issue reviewed by Talmi and Pine (2012).

Neuroeconomics supports the usual assumption that consumer choice is the result of a process of valuation in the brain. Therefore, understanding value representation can help understand consumer behaviour. This area of research showed major recent advances (Dayan, Dayan 2012), identifying a key role for the striatum and the ventromedial prefrontal cortex (Levy and Glimcher 2012). Earlier work explored the process of valuation in animals (e.g. Premack 2007). This research identified a network of brain regions involved in the decision to accept a cost, usually physical effort, in order to attain a benefit, usually food (e.g. Floresco et al. 2008). Animals' decision in this situation relies on the dopamine system and the nucleus accumbens (NAc) as well as the anterior cingulate cortex (ACC) and the amygdala (Phillips et al. 2007; Salamone et al. 2007; Floresco et al. 2008). Animal models do not provide strong support for the notion of a single representation of utility in the brain (Roesch and Bryden 2011), and controversy remains on whether costs and benefits share representations (Fehr and Rangel 2011; Rilling and Sanfey Rilling and Sanfey 2011).

Levy and Glimcher (2012) identify one possible mechanism for the decision-making process in the human brain, portrayed in Fig. 10.2. Their model is based on the controversial yet common assumption in neuroscience (see e.g. Vlaev et al. 2011) that the final neural representation of value uses a single 'neural currency'. According to this model, in a first step information from all cortical and all subcortical structures is processed and aggregated into a single value. This first step identifies the expected utility of external stimuli using incoming sensory information and internal signals (e.g. satiety). In a second step, this aggregate utility estimate is combined with the same initial sensory and internal stimuli to make a final choice through a motor control circuitry. This final choice reveals the good

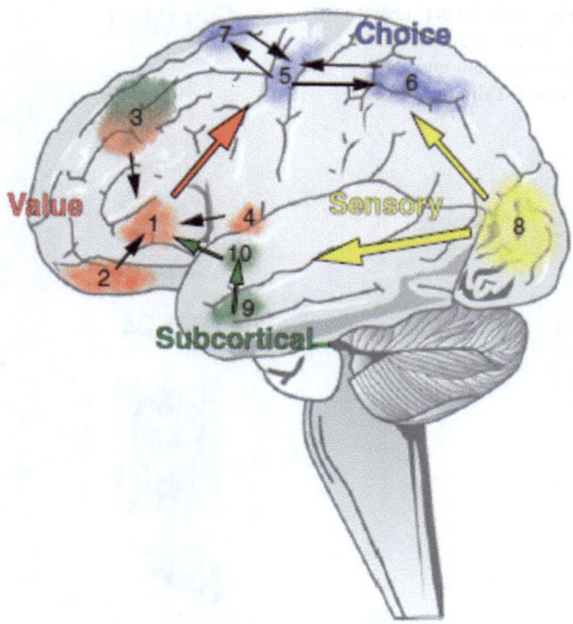

Fig. 10.2 Neuroscientific model of choice. *Source*: Levy and Glimcher (2012). Note: *1* vmPFC; *2* OFC; *3* DLPFC; *4* insula; *5* primary motor cortex (M1); *6* posterior parietal cortex; *7* frontal eye fields; *8* visual cortex; *9* amygdala; *10* striatum

with the highest expected utility. The second step can be more or less optimal: for instance, the food chosen could be determined by product attributes and level of hunger, but the colour of the package and the level of hunger may also have a direct influence on the final choice.

These models identify a general process of choice and value integration, but do not illuminate how mixed prospects are evaluated. The first published empirical work in cognitive neuroscience examining the assumption of a simple additive utility of costs and benefits is Talmi et al. (2009). They tested non-additive valuation by comparing the fit of two models, one additive (as in Eq. 10.3) and one non-additive (as in Eq. 10.4), to the same set of behavioural and neural data. In the experiment, participants were offered choices that incorporated simultaneous rewarding and punishing consequences, namely monetary gain (a benefit) and physical pain (a cost) (Fig. 10.3). The paradigm required participants to 'accept' or 'reject' a series of offers comprising of mixtures of costs and benefits; outcomes were delivered at a specified percentage of times, and levels of monetary reward were manipulated parametrically. The experimental design enabled testing whether pain attenuates the neural representation of reward, providing evidence for a significant interaction between costs and benefits. However, the study did not explore the impact of reward on the neural representation of pain.

The effect of the interaction term on choice can be observed in the drift function portrayed in Fig. 10.4 from Talmi et al. (2009). This figure represents the path leading to a choice for an individual participant. Circles represent empirical data on

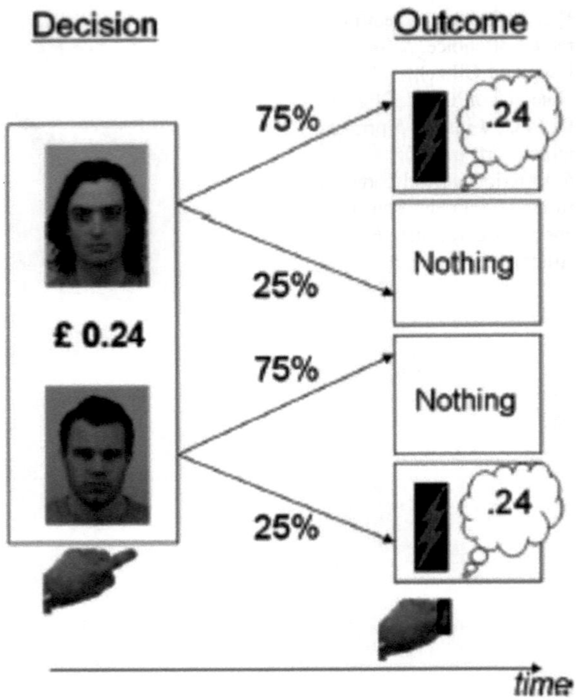

Fig. 10.3 Experimental task requiring the trade-off between pain and money. *Source* Talmi et al. (2009)

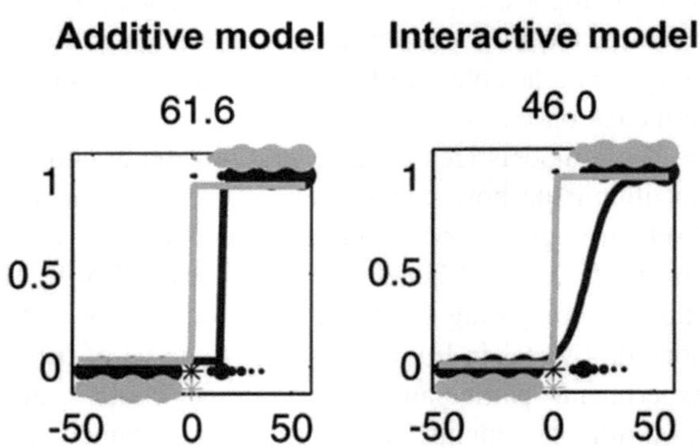

Fig. 10.4 Empirical and modelled choice behaviour for a single participant. *Source* Talmi et al. (2009)

choice frequency (a function of money offered and pain), while line graphs represent modelled choices for pain in the additive (left) and interactive (right) models. Black refers to trials where the stimulation was painful, and gray refers to trials where the stimulation was mild. The interactive model allows for a change in the slope of the sigmoid function caused by pain, while the additive effect of pain is only a rightward horizontal shift. Figure 10.4 clearly shows that the additive model was a better predictor of choice for this individual, a result supported by statistical tests of model fit across the whole sample (see Talmi et al. 2009, for more detail and interesting individual differences).

Talmi et al. (2009) supported the interactive model by demonstrating that the interaction of costs and benefits engaged a brain system implicated in value learning (dopaminergic target structures), namely the ventral striatum, which is thought to be involved in reinforcement learning when processing appetitive (O'Doherty et al. 2004; Tobler et al. 2007) as well as aversive (Iordanova et al. 2006; Hoebel et al. 2007) outcomes. It also engaged the subgenual ACC, which processes optimal decisions as well as appetitive and aversive choice (Walton et al. 2007), and works with the ventral striatum in the valuation of costs and benefits in rats (e.g. Salamone and Correa 2002; Walton et al. 2003; Schweimer et al. 2005).

Some of Talmi et al.'s (2009) results have been independently replicated by Park et al. (2011). Both studies find that the interactive model is superior to an additive model in modelling observed behaviour. However, while Talmi et al. (2009) only modelled a linear utility function u, Park et al. (2011) compared linear and non-linear utility functions, showing that the interaction term contributes significantly to explaining behaviour only when utility is modelled linearly. Nevertheless, their neurobiological data also fit the interactive model better than the additive model regardless of the utility function used. Finally, Prevost et al. (2010) also provide supporting evidence for the behavioural and neural interaction between costs and benefits in a study where participants were invited to exert physical effort in order to gain erotic pleasure. The functional form of the interaction term in Prevost et al. (2010) differed from Talmi et al. (2009) and Park et al. (2011), as did the neural substrate serving this interaction, a discrepancy that deserves further research. Strikingly, only a handful of studies on cost–benefit analysis in cognitive neuroscience engaged in model comparison, but those who tested the interaction of costs and benefits found clear supporting evidence.

The paradigms in the three neuroeconomic studies reviewed in this section were somehow limited: they focused on a single type of cost (pain or effort); and the model-comparison exercise used a limited number of functional forms for the utility function $u(\cdot)$. Furthermore, results are based on experimental evidence, hence limited by the absence of real-world equivalents. Nonetheless, these works open future research on the interaction between costs and benefits. For instance, research can expose participants to a differentiated series of consequential real-world decisions that include both primary reinforcers (either positive, e.g. food; or negative, e.g. pain) and secondary reinforcers (e.g. money), which recruit different brain systems (Dreher 2012), to test for possible differences in their interaction. Regrettably, neuroeconomic experiments require expensive data collection, inevitably slowing the pace of current research. Furthermore, the typical use of time delay between the decision and outcome delivery (Kurniawan et al. 2010; Park et al. 2011), employed to separate the neural signal of valuation from the neural processing of outcomes, alters the value of prospects (Loewenstein 1987), a limit to the practical implementation of Eq. (10.4) that should be addressed in future research.

10.4 Cost, Benefits, and Their Interaction in Microeconomic Models of Consumer Behaviour

The previous section discussed its influence on behavioural and neurobiological data from a neuroscientific perspective. This section discusses the microeconometric side of mixed prospects and how the interaction between costs and benefits can best be incorporated in econometric analysis, with a focus on models of utility such as choice models. In particular, the objective is to understand the implications of neuroscientific research for econometric models of consumer behaviour, which describe how consumers make decisions by trading-off attributes. This extension of modelling can be of interest to several areas of research concerned on consumption, such as for instance environmental public policy and sustainable consumption, which is discussed in a separate box.

To understand the microeconomic problem, imagine a consumer j of demographic D_j considering several options where each option i is characterised by monetary or non-monetary costs C_i and benefits B_i. In a generic utility function as in Eq. (10.1), the utility derived by consumer j from option i corresponds to a deterministic (observable) component V_{ij}, and a random (unobservable) component v_{ij} (see e.g. Burton et al. 2001), as

$$U_{ij}(B_i, C_i; D_j) = V_{ij}(B_i, C_i) + v_{ij}(B_i, C_i; D_j) \tag{10.6}$$

Consumers trade-off between costs and benefits (given personal preferences) and choose the option with the highest utility. In a two-option set, the probability of consumer j choosing option 1 is

$$P[(V_{j1} + v_{j1})] > P[(V_{j2} + v_{j2})] \tag{10.7a}$$

which implies

$$P[(V_{j1} - V_{j2})] > P[(v_{j2} - v_{j1})] \tag{10.7b}$$

Consumers then choose option 1 whenever the difference in deterministic utility is larger than the difference in random utility. The same result applies for larger choice sets.

Assuming random coefficients (consistent with Eq. 10.7b), the utility function of a consumer can be modelled using a Random Parameter Logit model[2] (RPL) (see Berry 1994). In the case of independence of costs and benefits (as in Eq. 10.4), utility corresponds to

[2]The RPL is a general choice model used in applied econometrics. The utility function used to estimate demand parameters can be simplified on the basis of the assumption made by the investigator (e.g. nested logit, conditional logit, simple logit). See Berry (1994).

$$U_{ij} = \alpha_{0,j} \cdot C_{ij} + \alpha_{1,j} \cdot B_{ij} + \varepsilon_{ij} \tag{10.8}$$

where for each of the k coefficients $\alpha_{k,j} = \bar{\alpha}_k + \delta_k \cdot D_j + \sigma_k \cdot e_{k,j}$, and ε_{ij} and $e_{k,j}$ are residuals. Equation (10.8) assumes that the utility $u(\cdot)$ from Eq. (10.4) is linear for both costs and benefits, but more general options can be also applied. In a binary decision (buy versus no-buy), utility can be derived as choice probability and reveals the point with the highest utility as[3]

$$\begin{aligned} U_{ij} &= 1 \quad \text{if} \quad 0 < U_{ij}^* < +\infty \\ U_{ij} &= 0 \quad \text{if} \quad -\infty < U_{ij}^* \leq 0 \end{aligned} \tag{10.9}$$

so that the probability terms $\omega(\cdot)$ in Eq. (10.3) are captured in the coefficients. To relax the assumption of independence of C and B, Eq. (10.8) can be extended to include an interaction term as in Eq. (10.4), to obtain the regression

$$U_{ij} = \alpha_{0,j} \cdot C_{ij} + \alpha_{1,j} \cdot B_{ij} + \alpha_{2,j} \cdot (B_{ij} \cdot C_{ij}) + \varepsilon_{ij} \tag{10.10}$$

where $\alpha_{k,j} = \bar{\alpha}_k + \delta_k \cdot D_j + \sigma_k \cdot e_{k,j}$. Noticeably, interaction terms should always accompany main terms to avoid omitted variable bias, and the possible endogeneity of C or B is treated using usual econometric methods (see Ozer-Balli and Sorensen 2010).

Interactions between independent variables are not infrequent in applied econometric models of utility (both choice models and contingent valuations). However, these models tend to treat costs as a single variable (price), while benefits (quality) are disaggregated into a variety of benefits. For instance, the quality of a refrigerator is not included as a single measure, but as a matrix of variables representing its product characteristics (e.g. energy-efficiency class, size, colour, brand). Thus, the interaction between product characteristics generally corresponds to reward-reward interactions and captures synergies between benefits (e.g. the size of a refrigerator and its efficiency), and makes no reference to the literature presented in the previous section. However, more frequently the difficulty of implementing full factorial designs in experimental tasks justifies the exclusion of interaction terms, particularly in light of their relatively low explanatory role observed in linear models (Dawes and Corrigan 1974; Louviere et al. 2000). However, it remains unclear what consequence their exclusion has on the explanatory power of models of choice.

Importantly, the inclusion of an interaction term allows for a changing marginal utility of costs and benefits. Equation (10.8) assumes consumers have a constant marginal utility from costs and benefits equal to $\frac{\partial U_{ij}}{\partial C_i} = \alpha_{0,j}$ and $\frac{\partial U_{ij}}{\partial B_i} = \alpha_{1,j}$. A nonlinear functional form (e.g. logarithmic) for $u(\cdot)$ in Eq. (10.4) would give a marginal utility changing in its argument (i.e. a marginal utility of costs depending on costs), but with no impact on the independence assumption. In the interactive

[3]Contingent valuation methods allow the inclusion of a scale of probability in Eq. (10.9) to account for hypothetical bias (see Wang 1997).

model of Eq. (10.10), the marginal utility of costs depends on the benefits and that of benefits depends on costs: these equal, respectively, $\frac{\partial U_{ij}}{\partial C_i} = \alpha_{0,j} + \alpha_{2,j} \cdot B_i$ and $\frac{\partial U_{ij}}{\partial B_i} = \alpha_{1,j} + \alpha_{2,j} \cdot C_i$. This difference implies a different MRS: it is constant and equal to $\mathrm{MRS_{B,C}} = \frac{\alpha_{1,j}}{\alpha_{0,j}}$ in an additive model; while in an interactive model it corresponds to $\mathrm{MRS_{B,C}} = \frac{\alpha_{1,j} + \alpha_{2,j} \cdot C_i}{\alpha_{0,j} + \alpha_{2,j} \cdot B_i}$. Importantly, while economic theory expects $\alpha_0 < 0$ and $\alpha_1 > 0$, the sign of the marginal utility and MRS in an interactive model depends also on α_2 and on the level of costs and benefits observed. Note that for purely monetary costs, the absolute value of the MRS measures the marginal willingness to pay for *B*.

Figure 10.5 explores the economic meaning of an interaction term. In the figure, δ refers to the marginal utility of the combination of costs and benefits, and the sum of the δs in a diagonal equals the coefficient α_2 of Eq. (10.10). In particular, a negative interaction term indicates a movement along the δ_B–δ_C axis and a negative synergy between costs and benefits: the marginal utility of a benefit declines with increasing costs and vice versa. On the contrary, a positive interaction term corresponds to a movement along the δ_A–δ_D axis and a positive synergy of costs and benefits: the marginal utility of benefits increases or decreases together with costs. Each individual component δ can be identified by an appropriate experimental design that separately assesses consumer response in each of the four conditions in Fig. 10.5 through, for instance, a choice experiment.

To understand the practical implications of these considerations, take a consumer who is contemplating the trade-off between energy-efficiency (benefit) and the price premium (cost) associated with the label in the market for efficient refrigerators. Coefficients of costs and benefits in both the additive and the interactive models (10.8) and (10.10) indicate a negative contribution of the price premium alone ($\alpha_{0,j} < 0$) and a positive contribution of the energy-efficiency label alone ($\alpha_{1,j} > 0$) (see Ward et al. 2011). The interaction term would indicate the relation between these attributes, which is possibly negative: individuals with high utility for energy efficiency would probably show a low sensitivity to the market premium required (δ_C); similarly, consumers with high utility from money (low-income households) would give low value to technology (δ_B).

Figures 10.6 and 10.7 show the difference in results using the parameters estimated for energy-efficient refrigerators in Ward et al. (2011), i.e. $\alpha_0 = -0.008$

Fig. 10.5 Possible combinations of an interaction between costs and benefits

		Benefits	
		High	Low
Costs	High	δ_A	δ_B
	Low	δ_C	δ_D

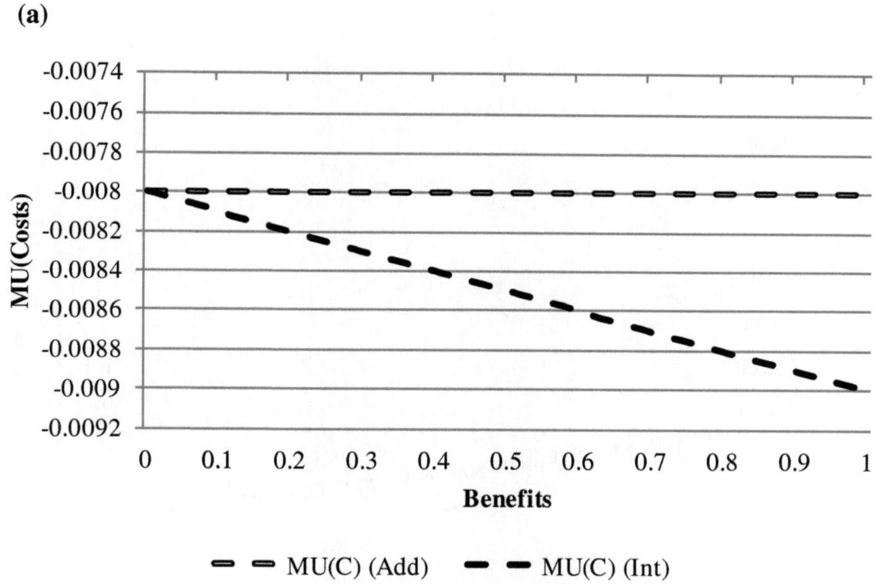

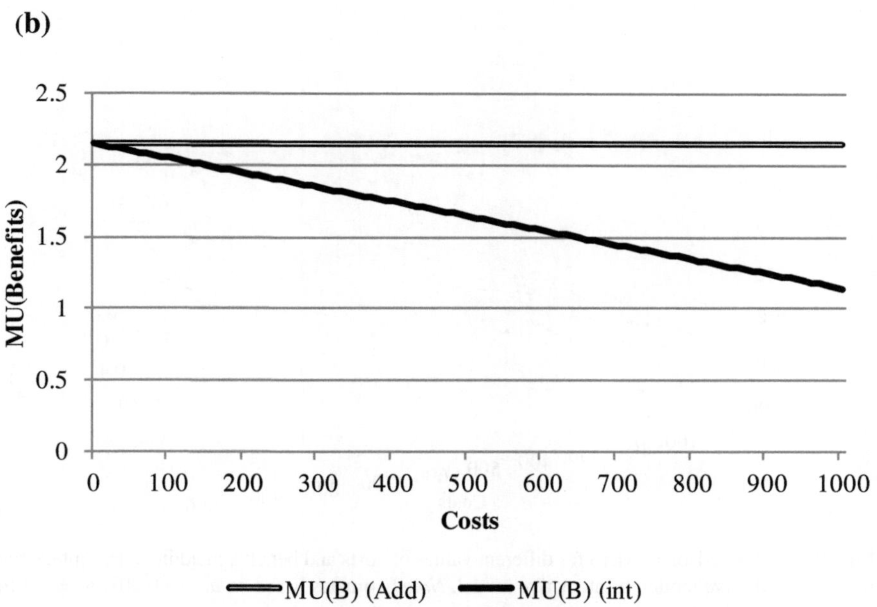

Fig. 10.6 Comparison of marginal utility (MU) using additive and interactive models. **a** Costs. **b** Benefits. *Note* Estimates based on $\alpha_0 = -0.008$, $\alpha_1 = 2.146$, and $\alpha_2 = -0.001$. *Add* additive model; *Int* interactive model

(a)

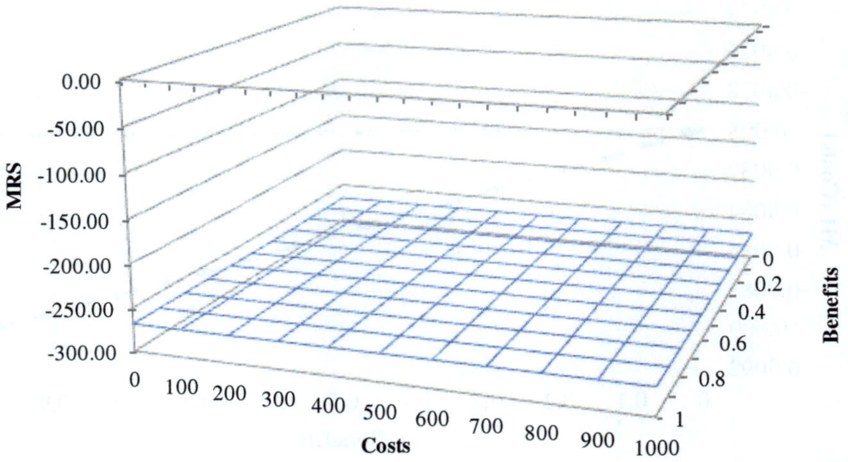

(b)

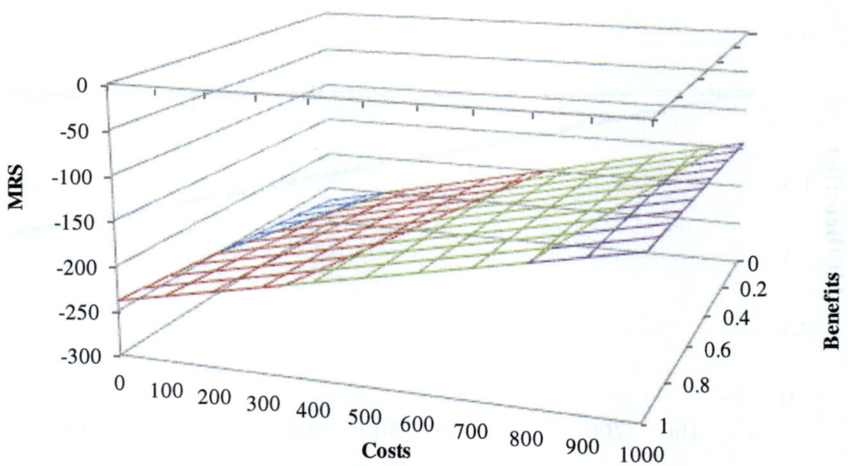

Fig. 10.7 Comparison of MRS for different values of costs and benefits in additive and interactive models. **a** Additive model. **b** Interactive model. *Note* Estimates based on $\alpha_0 = -0.008$, $\alpha_1 = 2.146$, and $\alpha_2 = -0.001$

(for price) and $\alpha_1 = 2.146$ (for the efficiency label). Ward et al. (2011) report no interaction term, so the simulation uses an arbitrary value $\alpha_2 = -0.001$. Intervals chosen for the axis account for the nature of the experimental task in Ward et al. (2011): they use a dummy variable for the label, going from zero to one; and they use a price going just above \$1000. Figures 10.6a and 10.6b show that the additive model imposes a constant marginal utility, while in an interactive model the marginal utility declines as costs and benefits increase. Similarly, Fig. 10.7 shows on a tri-dimensional space how the MRS changes for varying levels of price and label. Again, while this value is constant for an additive model (a flat surface), it varies for the interactive model. A change in the coefficient α_2 could give substantially different figures from those presented here, but would not modify the essence of the argument.

Using an interaction term for costs and benefits comes with two main difficulties in the empirical application. First, consumer models treat C as a single variable (price) and B as a matrix of product characteristics (a proxy for quality). An important reason for this practice is the need to observe the economic trade-off between goods and money, particularly with the objective of measuring willingness to pay through the MRS. However, consumers face a series of non-monetary losses alongside price, such as for example effort, which needs to be included in a matrix of costs C. This point applies for instance to the decision to purchase readily prepared vegetables against the natural product, which require different levels of effort and preparation time once price is considered. The vector C in this case includes as costly characteristics price as well as the effort required (e.g. in time terms) to prepare the vegetables for consumption. This distinction between costs and benefits should be clearly outlined in modelling exercises, to provide a fruitful understanding of results from a behavioural perspective besides the pure econometric viewpoint.

A second limitation is the difficulty in the *ex-ante* classification of product characteristics as costs or benefits, which is not always straightforward in an empirical analysis of market behaviour. For instance, organic can be viewed as a benefit for its impact on food quality and as a cost because of its possible implications on global food security. Similarly, energy-efficiency can be seen as a benefit in terms of energy conservation, but can be seen as a cost whenever the change in standard requires an effortful change in behaviour (e.g. an energy-efficient machine requiring low-temperature wash) or adaptation (e.g. the longer warm-up time of energy-efficient lightbulbs). These differences can be rarely separated into two different coefficients, causing difficulties in the identification of the impact on consumers. This complication implies the definition of a correct interaction between costs and benefits might not always be straightforward to define, and in some instances might require preliminary exploratory research, for instance through a qualitative analysis, to fully understand consumer perception of a specific characteristic.

The Interaction Term and Environmental Public Policy

As described in the main text of this chapter, the inclusion of a term to account for the interaction for costs and benefits can provide a more complete representation of consumer behaviour. Research in behavioural science to date has typically explored the trade-off between private goods, for instance pain or effort versus money or pleasure (Park et al. 2011; Prevost et al. 2010; Talmi et al. 2009). A rather unexplored area of research regards choices requiring a trade-off between public and private goods, a concern to environmental economists and policymakers. For instance, a consumer deciding whether to recycle trades the environmental improvement (a public good) against a loss of time (a private good). Public goods are an important feature of consumption because the individual making a choice or behaviour does not fully pay for, or benefit from, the consequences of his negligence or thoughtfulness, sharing them instead with the society. As an example, a firm discharging toxic waste (a private good) in a river (a public good) causes an environmental damage that primarily affects other individuals using the river (e.g. fishermen, swimmers, kayakers, etc.). Noticeably, shared recipients or victims could be very distant in both space and time.

Environmental public policy aims to preserve the environment as a public good. In particular, to change behaviour in a desirable direction, policies need to increase individuals' intrinsic motivation for the preservation of the public good, changing their implicit attitudes (Beattie and Sale 2011) and intrinsic values (Blankenship et al. 2012). To achieve its objective, public policy uses specific instruments (e.g. a subsidy, a tax, a ban or a penalty) in order to increase the individual benefits from socially desirable actions, or increase the costs of socially undesirable actions of firms and consumers (Gneezy et al. 2011). However, these instruments are often designed without clear consideration of the behavioural implications (Gneezy et al. 2011). For instance, earlier research established that extrinsic rewards such as prizes decrease long-term intrinsic motivation and effort; while intrinsic rewards such as personal satisfaction increase long-term intrinsic motivation (Gneezy et al. 2011; Kaplan and Oudeyer 2007). Similarly, there is generally no attention to the interaction between costs and benefits in the empirical evaluation of a policy instrument.

An understanding of the role of costs and benefits in consumer utility is crucial for a correct use of economic policy instruments to address environmental consumption. In fact, cost–benefit decisions lead to an internal conflict because a person's goal to achieve the benefit may clash with the goal of avoiding the cost (Locke et al. 1994), leading to high physiological arousal (Talmi et al. 2009). Because policy instruments directly influence costs and benefits, the way consumers integrate them into a value will have consequences to the consequent effectiveness of the policy. As an example, a carbon tax on meat decreases the utility from its consumption by increasing the price of purchase against a positive utility for the environmental benefit. However, the marginal utility of the effort in consumers with high preferences for the environmental sustainability benefit would be expected to increase with the price. In other words, an environmentally sensitive consumer may react positively to the tax when this is targeting a good like meat with a high carbon footprint, showing a positive interaction that is only captured through the term $\alpha_{2,i}$ in Eq. (10.10). The extent to which recent findings in neuroscience on private goods can apply to environmental policy requires further research.

10.5 Conclusions

A quote attributed to G.E.P. Box states that "All models are wrong, but some are useful" (a thought brought forward also in Box 1976). In fact, while the large amount of quantitative data available in the current information age stimulates the creative development of behavioural model of choice and decision-making, those do not always incorporate the underlying neurobiological mechanisms. This limitation detracts from the ability of such models to inform policy and research. Particularly, although neuroscience developed a good understanding of how the brain represents pleasure and pain separately, there is relatively little knowledge on how they are integrated into a final utility function, and how they lead to decisions and choices. While the prevalent strategy in the presence of mixed outcomes is to model the impact of costs and benefits independently, cognitive neuroscience indicates that costs and benefits interact. This chapter shows that the potential gain from this area of research to decision-making science is a transformation of the way cost–benefit analysis research is conducted in humans and animals, particularly integrating interdisciplinary knowledge from economics and cognitive neuroscience.

The valuation of mixed prospects is of great interest to cognitive neuroscientists, a point made evident in a large body of work on this topic. Despite the potential synergy of economics and neuroscience in the development of neuroeconomics (Camerer et al. 2004, 2005), neuroscience has seen the greater influence in the field, while economics has so far struggled to incorporate the recent developments on how the brain makes decisions into appropriate modelling. Applied economic research often assumes additive integration of costs and benefits out of convenience, without questioning the validity of this assumption. Instead, economics could benefit from the integration of experimental psychological findings into economic models of human decision-making. The discussion presented in this chapter provides an overview of the benefits from incorporating interaction terms through simulated data. Applying this model can be done in a straightforward manner through the collection of primary data or through the observation of market behaviour, a task left for future research.

More generally, future research should aim at better understanding of the functional form of utility, exploring and improving the validity and robustness of existing models. In particular, more attention should be given to the potential interactions between costs and benefits, and to the influence of perceived risk (outcome probability), as well as to delay (time between decision and outcome) on costs, benefits and their interaction. Neuroscientific research should aim at generalising the current Von Neumann–Morgensten type of utility function while supporting functional needs that comply with economic theory, rooting economic behaviour in biology. We hope that this chapter will prove useful by providing an understanding the status quo of current research, and outlining areas of research that require further exploration. In fact, the overall objective of this chapter is to enhance the understanding of economics, neuroscience, as well as *their* interaction.

Acknowledgments We are grateful to Matthieu Arnoult, Ariane Kehlbacher, Katarzyna Werner for their useful comments on earlier version of this text.

References

Beattie G, Sale L (2011) Shopping to save the planet? Implicit rather than explicit attitudes predict low carbon footprint consumer choice. Int J Environ Cult Econ Soc Sustain 7(4):211–232

Berry ST (1994) Estimating discrete-choice models of product differentiation. Rand J Econ 25 (2):242–262

Blankenship KL, Wegener DT, Murray RA (2012) Circumventing resistance: using values to indirectly change attitudes. J Pers Soc Psychol 103(4):606–621

Box GEP (1976) Science and statistics. J Am Stat Assoc 71(356):791–799

Burton M, Rigby D, Young T, James S (2001) Consumer attitudes to genetically modified organisms in food in the UK. Eur Rev Agric Econ 28(4):479–498

Camerer CF, Loewenstein G, Prelec D (2004) Neuroeconomics: why economics needs brains. Scand J Econ 106(3):555–579

Camerer C, Loewenstein G, Prelec Drazen (2005) Neuroeconomics: how neuroscience can inform economics. J Econ Lit 43(1):9–64

Camerer C, Cohen J, Fehr E, Glimcher P, Laibson D (forthcoming). Neuroeconomics. In: Kagel J, Roth A (eds) Handbook of experimental economics. Princeton University Press, USA

Dawes RM, Corrigan B (1974) Linear models in decision making. Psychol Bull 81(2):95–106

Dayan P (2012) How to set the switches on this thing. Curr Opin Neurobiol

Dhar R, Simonson I (1999) Making complementary choices in consumption episodes: highlighting versus balancing. J Mark Res 36(1):29–44

Dreher JC (2012) Neural coding of computational factors affecting decision making. Prog Brain Res 202:289–320

Fehr E, Rangel A (2011) Neuroeconomic foundations of economic choice—recent advances. J Econ Perspect 25(4):3–30

Floresco SB, Onge JRS, Ghods-Sharifi S, Winstanley CA (2008) Cortico-limbic-striatal circuits subserving different forms of cost-benefit decision making. Cogn Affect Behav Neurosci 8:375–389

Gneezy U, Meier S, Rey-Biel P (2011) When and why incentives (don't) work to modify behavior. J Econ Perspect 25(4):191–210

Hicks JR (1939) The foundations of welfare economics. Econ J 49(196):696–712

Hoebel BG, Avena NM, Rada P (2007) Accumbens dopamine-acetylcholine balance in approach and avoidance. Curr Opin Pharmacol 7:617–627

Iordanova MD, Westbrook RF, Killcross AS (2006) Dopamine activity in the nucleus accumbens modulates blocking in fear conditioning. Eur J Neurosci 24:3265–3270

Kahneman D (2003) Maps of bounded rationality: psychology for behavioral economics. Am Econ Rev 93(5):1449–1475

Kahneman D, Tversky A (1979) Prospect theory: an analysis of decision under risk. Econometrica 47(2):263–292

Kahneman D, Wakker PP, Sarin R (1997) Back to Bentham?—explorations of experienced utility. Quart J Econ 112:375–405

Kaplan F, Oudeyer P-Y (2007) In search of the neural circuits of intrinsic motivation. Front Neurosci 1(1):225–236

Keeney RL, Raiffa H (1993). Decisions with multiple objectives: preferences and value tradeoffs. Cambridge University Press

Khan U, Dhar R, Wertenbroch K (2004) A behavioral decision theory perspective on hedonic and utilitarian choice. In: Ratneshwar S, Mick DG (eds) Inside consumption: consumer motives, goals, and desires. Routledge, Abingdon

Kivetz R (2003) The effects of effort and intrinsic motivation on risky choice. Mark Sci 22: 477–502

Kurniawan IT, Seymour B, Vlaev I, Trommershäuser J, Dolan RJ, Chater Nick (2010) Pain relativity in motor control. Psychol Sci 21(6):840–847

Lancaster KJ (1966) A new approach to consumer theory. J Polit Econ 74(2):132–157

Levy DJ, Glimcher PW (2012) The root of all value: a neural common currency for choice. Curr Opin Neurobiol 22:1–12

Locke EA, Smith KG, Erez M, Chah D-O, Schaffer A (1994) The effects of intraindividual goal conflict on performance. J Manag 20(1):67–91

Loewenstein G (1987) Anticipation and the valuation of delayed consumption. Econ J 97 (387):666–684

Louie K, Glimcher PW (2012) Efficient coding and the neural representation of value. Ann NY Acad Sci 1251(1):13–32

Louviere JJ, Hensher DA, Swait JD (2000) Stated choice methods: analysis and application. Cambridge University Press, Cambridge

Montague PR, King-Casas B (2007) Efficient statistics, common currencies and the problem of reward-harvesting. Trends Cogn Sci 11(12):514–519

O'Doherty J, Dayan P, Schultz J, Deichmann R, Friston K, Dolan RJ (2004) Dissociable roles of ventral and dorsal striatum in instrumental conditioning. Science 304:452–454

Ozer-Balli H, Sorensen BE (2010) Interaction effects in econometrics. CEPR discussion paper no. DP7929, available online at SSRN http://papers.ssrn.com/sol3/papers.cfm?abstract_id=1641004

Park SQ, Kahnt T, Rieskamp J, Heekeren HR (2011) Neurobiology of value integration: when value impacts valuation. J Neurosci 31:9307–9314

Phillips PE, Walton ME, Jhou TC (2007) Calculating utility: preclinical evidence for cost-benefit analysis by mesolimbic dopamine. Psychopharmacology 191:483–495

Premack D (2007) Human and animal cognition: continuity and discontinuity. PNAS 104 (35):13861–13867

Prevost C, Pessiglione M, Metereau E, Clery-Melin ML, Dreher JC (2010) Separate valuation subsystems for delay and effort decision costs. J Neurosci 30:14080–14090

Rilling JK, Sanfey AG (2011) The neuroscience of social decision-making. Annu Rev Psychol 62:23–48

Roesch MR, Bryden DW (2011) Impact of size and delay on neural activity in the rat limbic corticostriatal system. Front Neurosci 5(130):1–13

Salamone JD, Correa M (2002) Motivational views of reinforcement: implications for understanding the behavioral functions of nucleus accumbens dopamine. Behav Brain Res 137(1):3–25

Salamone JD, Correa M, Farrar A, Mingote SM (2007) Effort-related functions of nucleus accumbens dopamine and associated forebrain circuits. Psychopharmacology 191(3):461–482

Schultz W (2000) Multiple reward signals in the brain. Nat Rev Neurosci 1(3):199–208

Schultz W (2006) Behavioral theories and the neurophysiology of reward. Annu Rev Psychol 57:87–115

Schweimer J, Saft S, Hauber W (2005) Involvement of catecholamine neurotransmission in the rat anterior cingulate in effort-related decision making. Behav Neurosci 119(6):1687–1692

Seymour B, Daw N, Dayan P, Singer T, Dolan R (2007) Differential encoding of losses and gains in the human striatum. J Neurosci 27(18):4826–4831

Steg L, Vlek C (2009) Encouraging pro-environmental behaviour: an integrative review and research agenda. J Environ Psychol 29(3):309–317

Talmi D, Pine A (2012) How costs influence decision values for mixed outcomes. Front Neurosci 6(146):1–21. doi:10.3389/fnins.2012.00146, http://journal.frontiersin.org/article/10.3389/fnins.2012.00146

Talmi D, Dayan P, Kiebel SJ, Frith CD, Dolan RJ (2009) How humans integrate the prospects of pain and reward during choice. J Neurosci 29(46):14617–14626

Tobler PN, O'Doherty JP, Dolan RJ, Schultz W (2007) Reward value coding distinct from risk attitude-related uncertainty coding in human reward systems. J Neurophysiol 97:1621–1632

Tversky A, Kahneman D (1992) Advances in prospect theory: cumulative representation of uncertainty. J Risk Uncertainty 5:297–323

Vlaev I, Chater N, Stewart N, Brown GD (2011) Does the brain calculate value? Trends Cogn Sci 15(11):546–554

Von Neumann J, Morgenstern O (1947) Theory of games and economic behavior. Princeton University Press, Princeton

Walton ME, Bannerman DM, Alterescu K, Rushworth MFS (2003) Functional specialization within medial frontal cortex of the anterior cingulate for evaluating effort-related decisions. J Neurosci 23:6475–6479

Walton ME, Rudebeck PH, Bannerman DM, Rushworth MFS (2007) Calculating the cost of acting in frontal cortex. Ann NY Acad Sci 1104(1):340–356

Wang H (1997) Treatment of 'don't-know' responses in contingent valuation surveys: a random valuation model. J Environ Econ Manag 32(2):219–232

Ward DO, Clark CD, Jensen KL, Yen ST, Russell CS (2011) Factors influencing willingness-to-pay for the ENERGY STAR® label. Energy Policy 39(3):1450–1458

Part IV
Decision Making in Social Contexts

Chapter 11
Individual Differences in Decision-Making: A Neural Trait Approach to Study Sources of Behavioral Heterogeneity

Kyle Nash and Daria Knoch

Abstract People display rich heterogeneity in decision-making, but from where do these individual differences originate? And what are the processes that underlie decision-making heterogeneity? In this chapter, we explore a 'neural trait approach' in which neuroscience measures of meaningful dispositional differences are used to illuminate the sources of variability in decision-making. We begin by outlining the neural trait approach, with a focus on two methods: resting state electroencephalography and structural magnetic resonance imaging. Next, we review innovative studies that have used these methodologies to explore time and social preferences in decision-making. We then outline future research considerations and close by discussing certain opportunities and challenges afforded by this research and the neural trait approach in general.

11.1 Introduction

Behavioral heterogeneity in decision-making is ubiquitous. One person struggles with temptation and another effortlessly resists the same enticement. One individual takes a bold risk and another person eschews tempting chance. One person shows charity and another shows little to no regard for others. Given the same choices in the same situation, it is relatively assured that people will not act uniformly and will instead display a rich variety of responses. Economic models of decision-making, however, primarily describe the ways in which the average person will approach a given choice (Sanfey 2007; Smith and Huettel 2010; Wischniewski et al. 2009). Such an approach is uniquely capable of revealing general processes in decision-making and has been immensely powerful in predicting behavior across groups (Camerer 2003). Given the manifest heterogeneity of behavior, though, individual differences have been somewhat underemphasized and represent a 'final

K. Nash · D. Knoch (✉)
Department of Social Psychology and Social Neuroscience, Institute of Psychology, University of Bern, Bern, Switzerland
e-mail: daria.knoch@psy.unibe.ch

M. Reuter and C. Montag (eds.), *Neuroeconomics*, Studies in Neuroscience, Psychology and Behavioral Economics, DOI 10.1007/978-3-642-35923-1_11

frontier' that, if explored, can deepen our understanding of decision-making processes (Appelt et al. 2011; Braver et al. 2010; Heatherton 2011; Kanai and Rees 2011; Levallois et al. 2012; Wischniewski et al. 2009). Key questions include: From where do these (often stark) differences in decision-making originate? And once these sources are discovered, what can be inferred about the psychological processes that lead to such behavioral heterogeneity?

Prior work on individual differences in economic decision-making has often focused on conceptually relevant personality traits. Notable traits include the Big 5 (Becker et al. 2012; Kurzban and Houser 2001; Lu and Argyle 1991), behavioral inhibition and behavioral activation systems (BIS and BAS, respectively; Carver and White 1994; Scheres and Sanfey 2006), Machiavellianism (Christie and Geis 1970; Spitzer et al. 2007; Wilson et al. 1998) and social value orientation (Van Lange 1999, 2000). For example, Scheres and Sanfey (2006) examined whether BIS (trait aversive motivation) and BAS (trait approach motivation) predicted behavior in the Dictator game (DG) and the Ultimatum game (UG). They found that trait levels of approach motivation amongst proposers predicted a strategic shift from low offers in the DG to higher, even-split offers in the UG (Scheres and Sanfey 2006). People scoring high on Machiavellianism prefer exploitative over cooperative strategies (Wilson et al. 1998), such as opportunistic defection in a bargaining game (Gunnthorsdottir et al. 2002). And a recent study by Becker et al. (2012) found that Big 5 traits in general showed small degrees of association with classic economic preferences, including patience, risk, positive and negative reciprocity, trust in others, and altruistic tendencies.

These and many other studies have shown that variance in decision-making behavior can be partially explained by personality trait questionnaires. However, this approach is not without certain limitations. Questionnaire research is somewhat restricted by the degree to which people can access and accurately judge personality processes. For example, in the classic paper by Nisbett and Wilson (1977), people demonstrated that they are sometimes unaware of stimuli that lead to their behavior, unaware of the behavior itself, and unaware of the stimulus → behavior link. Essentially, people may not know why they act in certain ways, which poses a large problem for self-report. Additionally, self-report is vulnerable to bias. People tend to respond in socially desirable ways, may respond randomly, and may change their responses based on their estimation of an experiment's purpose (Edwards 1957; Nichols and Maner 2008). Finally, personality measurements that precede dependent variable measurements could introduce statistical noise. For example, people who fill out questionnaires that make certain preferences more salient may act differently in subsequent economic games. Conversely, people who play an economic game may respond differently on subsequent questionnaires.

As an alternative to the questionnaire, we hold that neuroscience methods uniquely allow for objective measurement of meaningful dispositional differences that could shed light on the sources of variability in decision-making, particularly in terms of key economic preferences (Berkman and Falk 2013; Kanai and Rees 2011; Levallois et al. 2012). This chapter, which details the neural trait approach in neuroeconomics, is composed of four sections. The first outlines the neural trait approach. The second section highlights contemporary studies that have specifically utilized the neural trait approach to explore preferences in decision-making. The third section highlights future research considerations on the neural trait approach to individual differences in decision-making. The final section outlines certain opportunities and challenges afforded by this research and the neural trait approach in general.

11.2 The Neural Trait Approach

We define a neural trait as a quantifiable brain-based characteristic that is stable over time and capable of influencing economic and/or social preferences. In the field of neuroeconomics, most studies that have employed the neural trait approach have focused on the characteristics of brain structure or resting state brain activity. Generally, the neural trait approach involves two steps; (1) indexing task-independent, brain-based differences and (2) examining whether these neural trait indices predict behavior in decision-making or processes directly relevant to decision-making. The first and most salient objective of this approach is to determine sources of behavioral heterogeneity (Braver et al. 2010; Kanai and Rees 2011). As previously stated, economic models of decision-making typically describe aggregate behavior across individuals (Levallois et al. 2012; Scheres and Sanfey 2006; Wischniewski et al. 2009). However, individuals can evidence quite divergent behavior in the same scenarios. Smith and Huettel (2010) note that even in Kahneman and Tversky's (1979) work on decision-making biases, sizable minorities within their sample displayed behavior that directly opposed fundamental decision-making phenomena. Traits, and neural traits in particular, can explain this remarkable amount of variance in decision-making. The second objective of the neural trait approach is to add an additional level of analysis that can supplement and be informed by task-dependent analyses of neural and psychological processes (Berkman and Falk 2013). Based on prior literature it can be stated that neural traits can help researchers infer why differences in decision-making may occur (though interpretation should be cautious, see the *Cautions and Implications* section below). In sum, neural traits associated with certain functions can demonstrate how and suggest why people differ in their preferences.

11.3 Neural Trait Measurement

What makes an effective neural trait measure, then? We suggest that such a measure must meet certain criteria. First, to be able to capture dispositional differences the neural trait must be stable over time. For example, the measure should demonstrate high test–retest reliability. Second, the measure should be unique or specific to that individual, or, in other words, highly capable of recognizing a person based on the measure.

We focus on two methods that thoroughly fulfill these criteria and have been used in neuroeconomic research. The first is *resting state electroencephalographic (rsEEG) activity (for a general description of the EEG method see the chapter of Debener* et al. *in this book)*. Essentially, this method involves recording brain electrical activity on the scalp when the participant is at rest (as little as 2 min of recording is needed). As such, rsEEG reflects baseline patterns of brain activity that are not related to any particular task. Within neural trait research, rsEEG is typically analyzed by computing power values for different frequency bands. Importantly, these frequency-based measures of resting EEG activity are relatively stable in the adult brain, revealing test–retest reliabilities of up to 0.8 over a period of 5 years (Dunki et al. 2000; Näpflin et al. 2007), are heritable (estimates ranging from 70 to 96 % heritability; see de Geus 2010; Smit et al. 2008; van't Ent et al. 2009; Zietsch et al. 2007), and are unique or specific to the individual (i.e., it is possible to predict who the individual is based on the particular pattern of EEG activity—at up to a 99 % recognition rate; Dunki et al. 2000; Näpflin et al. 2007). These findings comprehensively demonstrate that rsEEG can effectively capture dispositional differences in neural functioning. Additionally, rsEEG overcomes a significant limitation of neuroscience research—one that is particularly germane to individual difference research: the limitation of small sample size (Button et al. 2013). Larger sample sizes are needed to detect individual sources of variability, and rsEEG is a relatively inexpensive and efficient neuroscience method.

The second measure that fulfills the criteria for a good neural trait is structural magnetic resonance imaging (MRI) of brain anatomy (*for a general description of the MRI method see the chapters 8c (fMRI), 8d (VBM), and 8e (DTI) in this book*). Neuroanatomical differences can be quantified through measuring and analyzing MRI data (i.e., high resolution images of the brain) with gray-matter approaches (such as *voxel based morphometry*, *VBM*, or vertex-based analyses, for example with the freesurfer image analysis suite) and white-matter approaches (such as diffusion tensor imaging, DTI). Whereas gray-matter approaches typically yield regionally specific estimations of cortical/subcortical volume, thickness, or surface area (Ashburner and Friston 2000), white-matter approaches quantify features of the neural connections or pathways in the brain (Basser 1995). The basic notion behind both gray- and white-matter approaches is that these brain differences reflect different functional or processing capacity (Boyke et al. 2008; DeYoung et al. 2010). Unsurprisingly, cortical volume (i.e., the combination of cortical thickness and surface area) is highly stable in the adult brain (DeYoung et al. 2010; Han et al. 2006), heritable (estimates ranging from 80 % heritability; Thompson et al. 2001; Panizzon et al. 2009) and highly specific to the individual (Mechelli et al. 2005). White-matter outcome variables have also demonstrated excellent test–retest reliability (Benson et al. 2007; Buechel et al. 2004) and are highly heritable too (estimates from 75 to 90 % heritability; Chiang et al. 2009). Much like resting EEG, then, structural MRI measures are ideal to quantify brain-based inter-individual differences (Kanai and Rees 2011). Moreover, structural measures have the added benefit of being able to image cortical and subcortical anatomy with a high degree of spatial precision.

Research utilizing the neural trait approach, either with rsEEG or with structural MRI, carries with it further advantages. First, because resting EEG and structural MRI can be measured separately from behavioral performance, researchers can then measure behavioral performance in more ecologically valid environments—that is, outside of the MRI scanner or without being hooked up to EEG-electrodes. Relatedly, the separation of behavioral measurement from neural measurement permits multiple tasks to be administered to the same set of subjects. Therefore, it is distinctively possible to link superficially different preferences, such as risk, time, and social preferences, to some common neural circuitry and/or common psychological process.

11.4 Neural Trait Research and Decision-Making

In this section, we review a number of studies that applied the neural trait approach to investigate the sources of individual differences in economic decision-making. Note that the following explication is not exhaustive. We instead focus on contemporary research that has utilized the neural trait approach to investigate significant preferences in economic decision-making; namely time preferences and social preferences.

11.4.1 Time Preferences

Time preferences describe the degree to which a person favors rewards or behavioral options as a function of time (Berns et al. 2007). Generally, people overlook or 'discount' future rewards in favor of smaller, sooner rewards, depending on the amount of delay of the future reward. A reduced preference for sooner rewards is critical in forgoing immediate temptations to reach longer term goals (Casey et al. 2011; Mischel et al. 2011). Time preferences are highly variable across individuals (Peters and Büchel 2011). Moreover, such individual differences in time preference are stable and predict a wide range of impulsive, yet consequential, every-day behaviors, like smoking, infidelity, and food consumption (Beck and Triplett 2009; Casey et al. 2011; Mischel et al. 2011; Reimers et al. 2009). However, relatively little is known about the sources of these individual differences (Olson et al. 2009; Peters and Büchel 2011; Shamosh et al. 2008). Thus, the neural trait approach appears well suited to explore such heterogeneity, as was demonstrated in a recent study by Gianotti et al. (2012). In this study, participants first had resting EEG activity measured. After, participants completed a task in which they made a series of decisions between smaller–sooner rewards and later–larger rewards that differed in magnitude and delay. In their analyses, the researchers used a source localization technique to identify the brain regions that generated the rsEEG activity (sLORETA; Pascual-Marqui 2002) and correlated this estimated neural activity with time preferences. Results demonstrated that approximately 14 % of the variance in time preferences was explained by rsEEG source localized to the left DLPFC. Specifically, lower levels of baseline activity in this region were associated with increased delay discounting (see Fig. 11.1). Thus, in using a task-independent measure of baseline neural activity, which allowed quantification of a trait-like neural signature of each individual's brain function at rest, the authors were able to explain a sizable degree of heterogeneity in intertemporal choice. Notably, this is not the only example of using the neural trait approach to predict time preferences. Bjork et al. (2009) employed structural MRI to measure gray matter volume in prefrontal regions to determine whether anatomical differences in the PFC predict behavioral heterogeneity in a similar measure of time preference. Mirroring the resting EEG findings, Bjork et al. (2009) found that a greater degree of delay discounting was associated with reduced PFC cortical volume, including reduced volume in lateral PFC regions. Similarly, Yu (2012) used whole-brain voxel-based morphometric analysis to examine white-matter volume and time preferences. Results demonstrated that delay discounting was negatively associated with lateral PFC subgyral white-matter volume and positively correlated with parahippocampal/hippocampal white-matter volume. Olson et al. (2009) used a similar method and examined the relationship between delay discounting and white-matter integrity using DTI. They found that delay discounting was negatively associated with increased white-matter integrity in several regions, including lateral

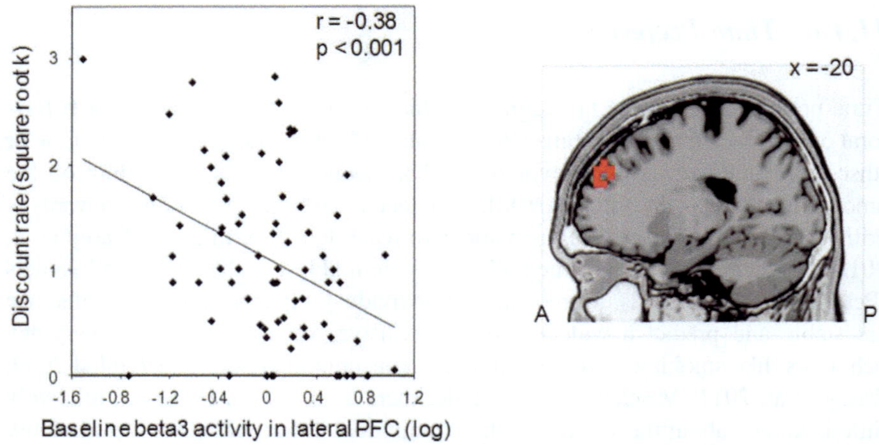

Fig. 11.1 Adapted from Gianotti et al. (2012): *On the left* the graph shows the significant correlation between delay discount rate and beta 3 activity level in the DLPFC assessed by rsEEG. Beta 3 activity is associated with increased cortical activity. *Higher numbers on the Y-axis* indicate higher delay discounting rates and *higher numbers on the X-axis* indicate higher levels of resting brain activity in left DLPFC. *On the right* locations of the voxels that showed negative significant correlations are indicated in red in a sagittal view of the brain. *A* anterior; *P* posterior

PFC regions (Olson et al. 2009). In all four studies, then, higher lateral PFC functioning capacity (increased resting activity, cortical volume, white-matter volume, and white-matter integrity) predicted lower levels of delay discounting. These resting EEG and structural results are further complimentary with studies that have experimentally modulated lateral PFC functioning. For example, reducing activity in the left lateral PFC through *transcranial magnetic stimulation* (*TMS*) causes an increase in delay discounting (Figner et al. 2010). Because this region has been related to self-control in other decision-making processes, these findings thus converge to support speculation that dispositional differences in the lateral PFC might reflect differences in a more general 'self-control capacity' (p. 6, Gianotti et al. 2012).

11.4.2 Social Preferences

Social preferences are predominantly defined as the degree to which an individual positively or negatively accounts for the well-being or material gains and losses of other people (Fehr 2009). A large body of research demonstrates that people are heterogeneous in their social preferences, including preferences for cooperation (Fehr and Gächter 2000), rejection of unfair offers (Güth 1995; Roth 1995), and altruism (Andreoni and Miller 2002). For example, the DG is regarded as a putatively 'pure' measure of prosocial preference (Camerer and Fehr 2002; for an introduction into behavioral economic games see the Chap. 2 by Civai and Hawes in this book). In this game, roughly 80 % of proposers do offer more than nothing to their interaction partner and up to 20 % offer an even-split (Forsythe et al. 1994), indicating a wide range of social preferences. In the UG, modal offers tend toward a fair split, yet the mean offer is typically around 40 %, again indicating significant individual variation (Camerer 2003). We thus turn to neural trait research that has begun to delve into the sources of heterogeneity in social preferences.

11.4.3 Prosocial Behavior and Altruism

Often defined as behavior that benefits another individual or group at a personal cost to the actor (Camerer 2003; Fehr and Fischbacher 2003; Henrich et al. 2006), altruism is a strong other-regarding tendency as a large proportion of people help unrelated strangers (Camerer 2003). However, such altruistic yet costly behavior is characterized by 'enormous individual heterogeneity' (p. 73, Morishima et al. 2012). For example, Andreoni and Miller (2002) found that people could be classified into three very different groups based on behavior in a modified DG; a purely selfish group, a more strategic group, and a fair group. Similarly, Kurzban and Houser (2005) found that 63 % of players in a public goods game could be classified as conditional cooperators, 13 % as altruists, and 20 % as selfish free-riders. Nonetheless, variables such as sex, wealth, age, and education are poor predictors of individual differences in altruistic behavior (Camerer 2003; Henrich et al. 2006). To fill this empirical gap, Morishima et al. (2012) used structural MRI to explore whether anatomical differences in the temporo-parietal junction (TPJ) predicted individual differences in altruism, as measured with dictator and reciprocity games. The TPJ has been repeatedly linked to perspective-taking—the ability to account for other peoples' thoughts, emotions, and goals (Saxe and Kanwisher 2005; Young et al. 2010). Differences in TPJ structure, and thereby differences in perspective-taking, were hypothesized to predict altruism, as measured with a game in which participants allocated money between themselves and an anonymous partner. The researchers found that increased TPJ volume predicted an increased tendency to act altruistically (i.e., assign money to their partner at a personal cost), particularly in situations when the participant was in an advantageous position relative to other individuals (Morishima et al. 2012). Thus, this study not only provided a link between individual differences in brain structure to preferences for altruism, but it also suggests that altruism is dependent upon an individual's capacity to engage in perspective-taking, as evidenced by TPJ volume.

11.4.4 Costly Punishment

The threat of punishment plays an important role in norm compliance. A key reason that the threat of punishment is so effective is that many people are willing to enforce the fairness norm by delivering punishment, even at a cost (Fehr and Gächter 2000). Unsurprisingly, given that such behavior requires sacrifice, preferences for costly punishment behavior are characterized by significant variation (Herrmann et al. 2008). For example, Fischbacher et al. (2013) used a unique approach in which they examined responders' reaction times to decisions made in a modified UG to characterize heterogeneity in costly punishment behavior. They found that strictly selfish respondents and respondents who displayed social preferences showed distinct reaction times to unkind and unfair offers. This empirically demonstrates a large degree of heterogeneity in costly punishment behavior. Findings from the few attempts to explain sources of such behavioral heterogeneity have been mixed, however. Gender, income, wealth, and education have low predictive power that varies strongly according to the idiosyncrasies of the different study designs (Camerer, 2003). On the other hand, up to 40 % of variation in costly punishment behavior can be attributed to genetic components (Wallace et al. 2007), hinting that stable and objective dispositional differences might be uniquely capable of explaining heterogeneity in costly punishment. Knoch et al. (2010) thus utilized the neural trait approach to index such stable and objective differences. In their study, participants first had rsEEG activity measured. After, participants then played as the responder in the UG. As the responder, participants accepted or rejected a proposed division of 20 CHF from 12 anonymous, ostensibly different interaction partners. A rejection ensures that both parties get no money at all. Thus, the responder can punish the proposer, but at a personal cost. Knoch et al. (2010) conducted a whole-brain correlational analysis between source localized resting EEG activity and costly punishment behavior. Results demonstrated that the propensity to punish was predicted by baseline cortical activity in the right DLPFC. That is, higher resting state activity in this area predicted higher levels of costly punishment behavior. Strikingly, this pattern of rsEEG activity explained approximately 50 % of the variance in costly punishment. Because there is considerable evidence that the lateral PFC is linked to self-control processes (Aron et al. 2003; Miller and Cohen 2001) and given that costly punishment conflicts with economic self-interest, the authors surmised that resting state activity in the DLPFC might predict costly punishment because such activity reflects a capacity to implement self-control to overcome self-interested choices. This is further supported by brain stimulation research in which disrupted lateral PFC functioning caused reduced costly punishment behavior (Baumgartner et al. 2011; Knoch et al. 2008; Knoch et al. 2006; van't Wout et al. 2005).

11.4.5 Impartiality and Intergroup Bias

People can and do act with impartiality. This is certainly not always the case, however, as ingroup bias or favoritism may be the default response (Brewer 1999). Thus, behavior in an intergroup context can run the gamut from fierce parochialism to stoic impartiality. In general, personality measures have been relatively poor predictors of heterogeneity in impartiality and intergroup bias (Hewstone et al. 2002). To better explain sources of individual differences in the propensity for impartiality, it would appear that more objective individual markers, as afforded by the neural trait approach, would be more apposite. Recently, Baumgartner et al. (2013a, b) examined whether impartiality may be explained by structural differences in the brain. They used structural MRI and measured impartiality by having participants—in the role of a third-party—view a series of prisoner's dilemma interactions between dyads of ingroup and outgroup members. In this game, the interacting players simultaneously decided whether to cooperate and transfer points or defect and keep points which were later turned in for real money. The 'third-party' participants viewing the interactions could then deliver punishment (or not). Administering punishment was costly, as participants had to give up points to strip the other player of their points. Impartiality was computed as the differences between punishments of ingroup members versus punishment of outgroup members in situations of unilateral defection. Intriguingly, Baumgartner et al. (2013a, b) found that increased dmPFC volume was associated with increased impartiality or fairness (see Fig. 11.2). In addition, Baumgartner and colleagues used the dmPFC volume to split participants into three separate groups (low, medium, and high dmPFC) and examined punishment levels of both ingroup and outgroup defectors. As expected, these three groups demonstrated strong differences in impartiality. Those with low dmPFC volumes demonstrated ingroup favoritism (reduced

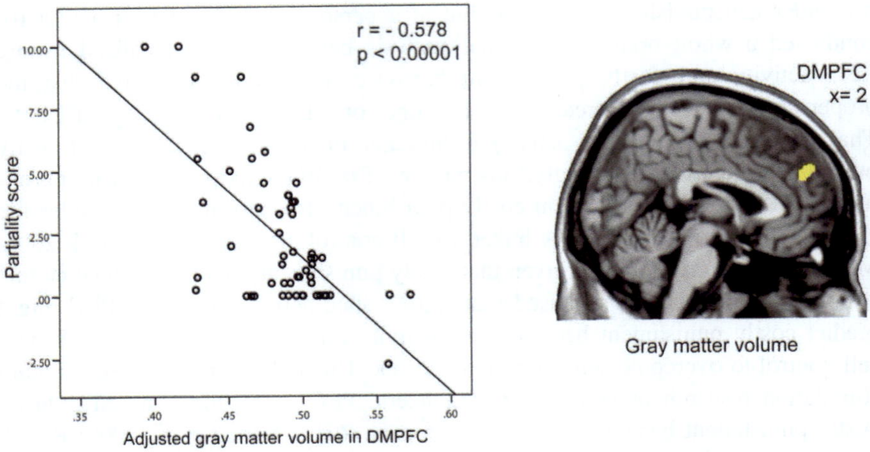

Fig. 11.2 Adapted from Baumgartner et al. (2013a, b): *On the left* the *scatter plot* shows the significant negative correlation between the partiality score and *gray*-matter volume in the dorsomedial prefrontal cortex (DMPFC), which is adjusted for several covariates (e.g., age and brain size, see article for details). *High values* on the partiality score indicate that third-parties strongly differed in the punishment of outgroup and ingroup perpetrators, whereas *low values* indicate that third-parties demonstrated an impartial punishment pattern, i.e., they equally punished outgroup and ingroup perpetrators. *On the right* locations of voxels in the DMPFC that showed significant negative correlations with the partiality score are indicated in *yellow color* in a sagittal view of the brain

punishment of the ingroup) and outgroup prejudice (increased punishment of the outgroup). Those with medium levels of dmPFC volume demonstrated ingroup favoritism but not outgroup prejudice. Finally, those with high levels of dmPFC volume demonstrated equal treatment of ingroup and outgroup defectors (i.e., ingroup perpetrators were punished just as strongly as outgroup perpetrators). Thus, heterogeneity in impartiality was tightly bound to dmPFC volume. These findings point to a potential psychological mechanism as, much like the TPJ, the dmPFC has been repeatedly implicated in social cognition and perspective-taking processes (Adolphs 2003; Van Overwalle 2009). Thus, the ability to overcome ingroup favoritism and outgroup prejudice may depend on the capacity to engage in perspective-taking equally for ingroups and outgroups—a capacity that may be largely determined by dmPFC volume.

11.4.6 Deception

Lying and deception are ubiquitous in social interactions. Typically, people deceive others for personal gain, but the propensity to deceive is highly variable amongst individuals (Gino and Pierce 2009; Kashy and DePaulo 1996). Attempts to explain this heterogeneity have primarily focused on self-reported personality traits (e.g.,

Phillips et al. 2011). However, personality traits tend to have low predictive power of deception (DePaulo 2004). A neural trait approach might prove a better predictive tool. As evidence of this, Baumgartner et al. (2013a, b) recently used source localization of resting EEG to examine whether trait levels of neural activation might explain differences in deceiving others. They used an ecologically valid measure of self-initiated deception with real, monetary consequences. In this task, participants promised whether or not they would return money to real partners in a trust game, but were later given the opportunity to break that promise (i.e., keep the investment). A cluster analysis revealed three groups; (1) dishonest subjects that had uniformly high promise rates but uniformly low return rates; (2) moderately dishonest subjects that had relatively lower promise rates and moderate return rates; and (3) honest subjects that had high promise rates and high return rates. Heterogeneity in deception was thus easily discernible. In their primary analyses, they found that the tendency to deceive was associated with reduced resting state activation in the anterior insula. The higher the neural baseline activation in this area is the lower individuals' propensity to deceive. This resting state activation in the insula accounted for 23 % of the variance in deceptive behavior. Furthermore, they found that the same pattern of baseline anterior insula activation that was associated with reduced deception was also associated with trait levels of negative affect and dispositional tendencies to avoid aversive emotional situations. As such, these findings are a good example of the power of the neural trait approach as they suggest who lies and who remains honest. One might primarily expect that people who are less inclined to deceive have strong prosocial preferences, as opposed to those who deceive, an expectation which future research could explore. This study revealed a unique relationship, however. Honesty was related to more resting activity in a brain area associated with mapping internal bodily states and in representing emotional arousal and conscious feelings (Craig 2009). Thus, people with heightened baseline insula activity may be predisposed towards honesty because their hyperactive emotional system could make a deceptive act too aversive. In this case, the examination of neural traits indicated predisposing psychological processes that may have otherwise been unknown or considered relatively unimportant.

In sum, across these studies, even in the highly complex domain of social preferences, neural traits are demonstrably capable of explaining behavioral heterogeneity in decision-making. In the next section, we explore how the neural trait approach may be adapted for future research.

11.5 The Neural Trait Approach in Future Research

This chapter has explored the neural trait approach to explaining dispositional differences in decision-making with a particular focus on time and social preferences. In addition to explaining behavioral heterogeneity in decision-making, however, we further contend that the neural trait approach holds promise as a key component in fuller research designs.

For example, prior research has primarily emphasized contextual or situational factors that determine decision-making behavior or explain economic preferences. Given the twin advantages of stability and objectivity, neural traits have the ability to powerfully augment contextual research by uncovering brain-based individual differences and examining whether these differences interact with situational variables. Such an approach is not new, as it was classically formulated by Kurt Lewin in his now famous equation describing (B)ehavior as a function of (P)erson × (E)nvironment (Lewin 1946). We argue that modern forms of this 'equation' could include neural trait measures of (P) in combination with experimental manipulations (E) known to impact decision-making behavior (B). For example, subtle or even subliminal cues can shift behavior from selfish to prosocial behavior and vice versa (Bargh et al. 2001; Bateson et al. 2006; Rand et al. 2012). Neural traits associated with differences in self- and other-regarding preferences should moderate reactions to economic and social cues. That is, neural traits associated with self-regarding preferences might predict reactions to monetary cues; whereas neural traits associated with other-regarding preferences might predict reactions to social cues (see also Declerck et al. 2013, for similar reasoning). Future research paradigms that combine neural trait measures and situational manipulations could deepen our understanding of decision-making processes.

Future research may also utilize the neural trait approach as a unique component in fuller analyses of genetic → brain → behavior pathways. In parallel with the neural trait approach, a number of researchers have examined individual differences in decision-making by assessing potential genetic contributions (e.g., Boettiger et al. 2007; Cesarini et al. 2009; Kuhnen and Chiao 2009; for an overview on genes and human decision-making, see also Chap. 4 in this book). The question is, how do these genes then affect behavior? More specifically, what are the mechanisms? The answer, proposed as the *intermediate phenotype model*, is that genes impact behavior through neural mechanisms (Meyer-Lindenberg and Weinberger 2006). The basic notion is that genes influence an individual's behavior through their effects on the brain. To be an effective, brain-based intermediate phenotype, certain criteria have been articulated. Central amongst these criteria are the characteristics of stability and heritability (Gottesman and Gould 2003; Green et al. 2008). As noted previously, resting EEG activity and structural MRI are both highly stable and heritable. Neural traits are thus ideal intermediate phenotypes.

As an example, Gianotti et al. (2012) also employed the intermediate phenotype approach in the aforementioned time preference research. In addition to measuring resting EEG, these researchers genotyped participants on the COMT Val158Met polymorphism, which has been related to dopamine levels in the PFC and delay discounting in prior research (Boettiger et al. 2007; Paloyelis et al. 2010). Results demonstrated that participants with more Val alleles (greater COMT activity and lower dopamine levels in the PFC) exhibited greater delay discounting. This effect was mediated by the baseline activation levels in the left lateral PFC: Higher numbers of Val alleles lead to lower baseline activation which, in turn, biases choices towards greater impatience. More studies such as this could provide a mechanistic understanding of the contribution of specific gene variants to individual

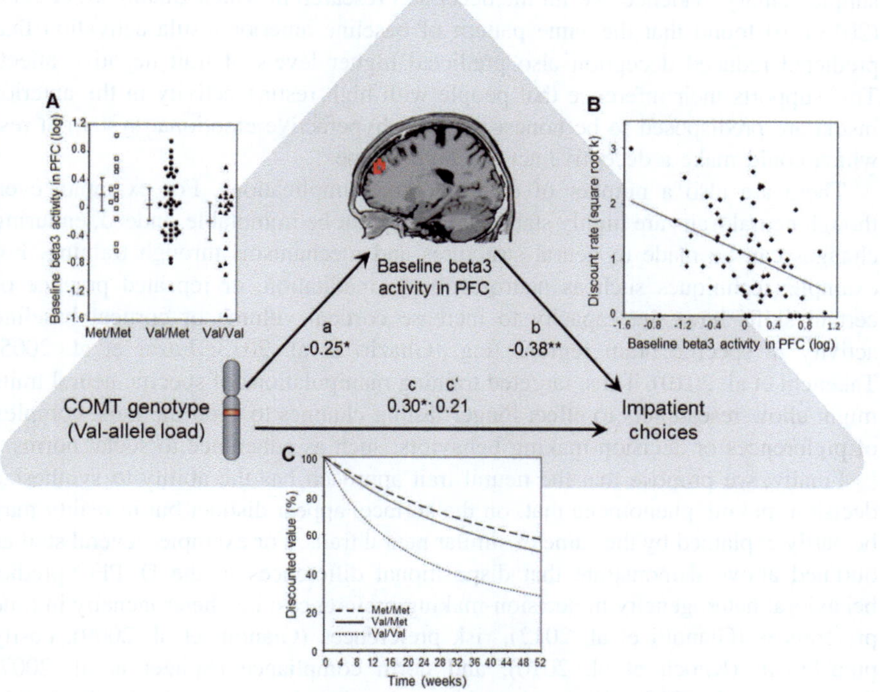

Fig. 11.3 From Gianotti et al. (2012): a demonstration of the intermediate phenotype approach. Pictured is the path diagram that illustrates that baseline EEG beta 3 activity level in the left DLPFC mediates COMT-determined differences in steepness of delay discounting

differences in neural traits and behavior that goes beyond simple biomarker identification (see Fig. 11.3).

11.6 Cautions and Implication

One should be cognizant of potential issues involved in the neural trait approach. Detecting neural traits that relate to certain behaviors can often involve examining activity throughout the entire brain. Thus, the probability of finding a spurious relationship is greatly inflated unless the researcher uses appropriate corrections for multiple testing. One should also be aware that there is a temptation to over-interpret neural trait findings. By relating decision-making behavior to certain brain areas, there is an inclination to 'guess' at the psychological process that mediates the brain—behavior link. Indeed, this potential was often discussed in this chapter as an advantage of the approach. To be clear, however, inferring the psychological mechanism is more appropriate when a brain region has been tightly linked to specific, germane functioning. To address this issue, one can also provide supplementary evidence. Recall the deception research in which Baumgartner et al. (2013a, b) found that the same pattern of baseline anterior insula activation that predicted reduced deception also predicted higher levels of trait negative affect. This supports their inference that people with high resting activity in the anterior insula are predisposed to be honest due to a hyperactive emotional system at rest which could make a deceptive act too bothersome.

There are also a number of exciting broad implications. For example, even though neural traits are highly stable, they may not be immutable. Indeed, enduring changes can be made to neural structures and mechanisms through training. For example, techniques such as neurofeedback, meditation, or repeated practice of certain skills have the capacity to increase cortical volume or cortical baseline activity in specific brain regions (e.g., Ghaziri et al. 2013; Lazar et al. 2005; Takeuchi et al. 2010). Thus, targeted training manipulations of specific neural traits might allow researchers to effect longer lasting changes to even the most complex of preferences or decision-making behaviors, such as adherence to social norms.

Finally, we propose that the neural trait approach has the ability to synthesize decision-making phenomena that, on the surface, appear distinct but in reality may be partly explained by the same or similar neural traits. For example, several studies outlined above demonstrate that dispositional differences in the DLPFC predict behavioral heterogeneity in decision-making preferences; i.e., heterogeneity in time preferences (Gianotti et al. 2012), risk preferences (Gianotti et al. 2009), costly punishment (Knoch et al. 2010), and norm compliance (Spitzer et al. 2007; Steinbeis et al. 2012). Given that this region has been strongly implicated in self-control (Cohen and Lieberman 2010; Coutlee and Huettel 2012; Figner et al., 2010; Heatherton and Wagner 2011; Miller and Cohen 2001), it is a tantalizing possibility that these disparate decision-making domains could be linked to a common neural trait and/or psychological capacity. Future research could take advantage of the neural trait approach to potentially demonstrate a shared basis between risk, time, and social preferences.

References

Adolphs R (2003) Cognitive neuroscience of human social behaviour. Nat Rev Neurosci 4(3):165–178

Andreoni J, Miller J (2002) Giving according to GARP: an experimental test of the consistency of preferences for altruism. Econometrica 70(2):737–753

Appelt KC, Milch KF, Handgraaf MJ, Weber EU (2011) The Decision Making Individual Differences Inventory and guidelines for the study of individual differences in judgment and decision-making research. Judgment Decis Making 6(3):252–262

Aron AR, Fletcher PC, Bullmore T, Sahakian BJ, Robbins TW (2003) Stop-signal inhibition disrupted by damage to right inferior frontal gyrus in humans. Nat Neurosci 6(2):115–116

Ashburner J, Friston KJ (2000) Voxel-based morphometry—the methods. Neuroimage 11(6):805–821

Bargh JA, Lee-Chai A, Barndollar K, Gollwitzer PM, Trötschel R (2001) The automated will: nonconscious activation and pursuit of behavioral goals. J Pers Soc Psychol 81(6):1014–1027

Basser PJ (1995) Inferring microstructural features and the physiological state of tissues from diffusion weighted images. NMR Biomed 8:333–344
Bateson M, Nettle D, Roberts G (2006) Cues of being watched enhance cooperation in a real-world setting. Biol Lett 2(3):412–414
Baumgartner T, Knoch D, Hotz P, Eisenegger C, Fehr E (2011) Dorsolateral and ventromedial prefrontal cortex orchestrate normative choice. Nat Neurosci 14(11):1468–1474
Baumgartner T, Gianotti L, Knoch D (2013a) Who is honest and why: baseline activation in anterior insula predicts inter-individual differences in deceptive behavior. Biol Psychol 94 (1):192–197
Baumgartner T, Schiller B, Hill C, Knoch D (2013b) Impartiality in humans is predicted by brain structure of dorsomedial prefrontal cortex. Neuroimage 81:317–324
Beck RC, Triplett MF (2009) Test–retest reliability of a group-administered paper–pencil measure of delay discounting. Exp Clin Psychopharmacol 17(5):345–355
Becker A, Deckers T, Dohmen T et al (2012) The relationship between economic preferences and psychological personality measures. Annu Rev Econ 4:453–478
Benson RR, Meda SA, Vasudevan S, Kou Z, Koushik GA, Hanks RA, Millis SR, Makke M, Latif Z, Coplin W, Maythaler J, Haacke EM (2007) Global white matter analysis of diffusion tensor images is predictive of injury severity in traumatic brain injury. J Neurotraum 24:466–459
Berkman ET, Falk EB (2013) Beyond brain mapping using neural measures to predict real-world outcomes. Curr Dir Psychol Sci 22(1):45–50
Berns GS, Laibson DI, Loewenstein G (2007) Intertemporal choice—toward an integrative framework. Trends Cogn Sci 11(11):482–488
Bjork JM, Momenan R, Hommer DW (2009) Delay discounting correlates with proportional lateral frontal cortex volumes. Biol Psychiatry 65(8):710–713
Boettiger CA, Mitchell JM, Tavares VC, Robertson M, Joslyn G, D'Esposito M, Fields HL (2007) Immediate reward bias in humans: fronto-parietal networks and a role for the catechol-O-methyltransferase 158Val/Val genotype. J Neurosci 27(52):14383–14391
Boyke J, Driemeyer J, Gaser C, Büchel C, May A (2008) Training-induced brain structure changes in the elderly. J Neurosci 28(28):7031–7035
Braver TS, Cole MW, Yarkoni T (2010) Vive les differences! Individual variation in neural mechanisms of executive control. Curr Opin Neurobiol 20(2):242–250
Brewer MB (1999) The psychology of prejudice: ingroup love and outgroup hate? J Soc Issues 55 (3):429–444
Buechel C, Raedler T, Sommer M, Sach M, Weiller C, Koch MA (2004) White matter asymmetry in the human brain: a diffusion tensor MRI study. Cereb Cortex 14(9):945–951
Button KS, Ioannidis JP, Mokrysz C, Nosek BA, Flint J, Robinson ES, Munafò MR (2013) Power failure: why small sample size undermines the reliability of neuroscience. Nat Rev Neurosci 14 (5):365–376
Camerer C (2003) Behavioral game theory: experiments in strategic interaction. Princeton University Press, Princeton
Camerer CF, Fehr E (2002) Measuring social norms and preferences using experimental games: a guide for social scientists
Carver CS, White TL (1994) Behavioral inhibition, behavioral activation, and affective responses to impending reward and punishment: the BIS/BAS scales. J Pers Soc Psychol 67:319
Casey BJ, Somerville LH, Gotlib IH, Ayduk O, Franklin NT, Askren MK et al (2011) Behavioral and neural correlates of delay of gratification 40 years later. Proc Natl Acad Sci 108 (36):14998–15003
Cesarini D, Dawes CT, Johannesson M, Lichtenstein P, Wallace B (2009) Genetic variation in preferences for giving and risk taking. Q J Econ 124(2):809–842
Chiang MC, Barysheva M, Shattuck DW, Lee AD, Madsen SK, Avedissian C et al (2009) Genetics of brain fiber architecture and intellectual performance. J Neurosci 29(7):2212–2224
Christie R, Geis FL (1970) Studies in Machiavellianism. Academic Press, New York

Cohen JR, Lieberman MD (2010) The common neural basis of exerting self-control in multiple domains. In: Trope Y, Hassin R, Ochsner KN (eds) Self control in society, mind, and brain. Oxford University Press, Oxford, pp 141–162

Coutlee CG, Huettel SA (2012) The functional neuroanatomy of decision making: prefrontal control of thought and action. Brain Res 1428:3–12

Craig AD (2009) How do you feel–now? The anterior insula and human awareness. Nat Rev Neurosci 10(1):59–70

Declerck CH, Boone C, Emonds G (2013) When do people cooperate? The neuroeconomics of prosocial decision making. Brain Cogn 81(1):95–117

de Geus EJ (2010) From genotype to EEG endophenotype: a route for post-genomic understanding of complex psychiatric disease? Genome Med 2(9):1–4

DePaulo BM (2004) The many faces of lies. In: Miller AG (ed) The social psychology of good and evil. Guilford Press, New York, pp 303–326

DeYoung CG, Hirsh JB, Shane MS, Papademetris X, Rajeevan N, Gray JR (2010) Testing predictions from personality neuroscience brain structure and the big five. Psychol Sci 21 (6):820–828

Dunki RM, Schmid GB, Stassen HH (2000) Intraindividual specificity and stability of human EEG: comparing a linear vs a nonlinear approach. Methods Inf Med (Methodik der Information in der Medizin) 39(1):78–82

Edwards AL (1957) The social desirability variable in personality assessment and research. Dryden Press, New York

Fehr E (2009) Social preferences and the brain. In: Glimcher PW, Camerer C, Poldrack RA et al (eds) Neuroeconomics: decision making and the brain. Academic Press, London, pp 215–232

Fehr E, Fischbacher U (2003) The nature of human altruism. Nature 425(6960):785–791

Fehr E, Gächter S (2000) Fairness and retaliation: the economics of reciprocity. J Econ Perspect 14 (3):159–181

Figner B, Knoch D, Johnson EJ, Krosch AR, Lisanby SH, Fehr E, Weber EU (2010) Lateral prefrontal cortex and self-control in intertemporal choice. Nat Neurosci 13(5):538–539

Fischbacher U, Hertwig R, Bruhin A (2013) How to model heterogeneity in costly punishment: insights from responders' response times. J Behav Decis Mak 26(5):462–476

Forsythe R, Horowitz JL, Savin NE, Sefton M (1994) Fairness in simple bargaining experiments. Games Econ Behav 6(3):347–369

Ghaziri J, Tucholka A, Larue V, Blanchette-Sylvestre M, Reyburn G, Gilbert G et al (2013) Neurofeedback training induces changes in white and gray matter. Clin EEG Neurosci 44 (4):265–272

Gianotti LR, Figner B, Ebstein RP, Knoch D (2012) Why some people discount more than others: baseline activation in the dorsal PFC mediates the link between COMT genotype and impatient choice. Front Neurosci 6

Gianotti LR, Knoch D, Faber PL, Lehmann D, Pascual-Marqui RD, Diezi C et al (2009) Tonic activity level in the right prefrontal cortex predicts individuals' risk taking. Psychol Sci 20 (1):33–38

Gino F, Pierce L (2009) Dishonesty in the name of equity. Psychol Sci 20(9):1153–1160

Gottesman II, Gould TD (2003) The endophenotype concept in psychiatry: etymology and strategic intentions. Am J Psychiatry 160(4):636–645

Green AE, Munafò MR, DeYoung CG, Fossella JA, Fan J, Gray JR (2008) Using genetic data in cognitive neuroscience: from growing pains to genuine insights. Nat Rev Neurosci 9(9):710–720

Gunnthorsdottir A, McCabe K, Smith V (2002) Using the Machiavellianism instrument to predict trustworthiness in a bargaining game. J Econ Psychol 23(1):49–66

Güth W (1995) On ultimatum bargaining experiments—A personal review. J Econ Behav Organ 27(3):329–344

Han X, Jovicich J, Salat D, van der Kouwe A, Quinn B, Czanner S et al (2006) Reliability of MRI-derived measurements of human cerebral cortical thickness: the effects of field strength, scanner upgrade and manufacturer. Neuroimage 32(1):180–194

Heatherton TF (2011) Neuroscience of self and self-regulation. Annu Rev Psychol 62:363–390
Heatherton TF, Wagner DD (2011) Cognitive neuroscience of self-regulation failure. Trends Cogn Sci 15(3):132–139
Henrich J, McElreath R, Barr A, Ensminger J, Barrett C, Bolyanatz A et al (2006) Costly punishment across human societies. Science 312(5781):1767–1770
Herrmann B, Thöni C, Gächter S (2008) Antisocial punishment across societies. Science 319 (5868):1362–1367
Hewstone M, Rubin M, Willis H (2002) Intergroup bias. Annu Rev Psychol 53:575–604
Kahneman D, Tversky A (1979) Prospect theory: an analysis of decision under risk. Econometrica: J Econometric Soc 263–291
Kanai R, Rees G (2011) The structural basis of inter-individual differences in human behaviour and cognition. Nat Rev Neurosci 12(4):231–242
Kashy DA, DePaulo BM (1996) Who lies? J Pers Soc Psychol 70(5):1037–1050
Knoch D, Pascual-Leone A, Meyer K, Treyer V, Fehr E (2006) Diminishing reciprocal fairness by disrupting the right prefrontal cortex. Science 314(5800):829–832
Knoch D, Nitsche MA, Fischbacher U, Eisenegger C, Pascual-Leone A, Fehr E (2008) Studying the neurobiology of social interaction with transcranial direct current stimulation—The example of punishing unfairness. Cereb Cortex 18(9):1987–1990
Knoch D, Gianotti LR, Baumgartner T, Fehr E (2010) A neural marker of costly punishment behavior. Psychol Sci 21(3):337–342
Kuhnen CM, Chiao JY (2009) Genetic determinants of financial risk taking. PLoS ONE 4(2): e4362
Kurzban R, Houser D (2001) Individual differences in cooperation in a circular public goods game. Eur J Pers 15(S1):S37–S52
Kurzban R, Houser D (2005) Experiments investigating cooperative types in humans: a complement to evolutionary theory and simulations. Proc Natl Acad Sci USA 102(5):1803–1807
Lazar SW, Kerr CE, Wasserman RH, Gray JR, Greve DN, Treadway MT et al (2005) Meditation experience is associated with increased cortical thickness. Neuroreport 16(17):1893–1897
Levallois C, Clithero JA, Wouters P, Smidts A, Huettel SA (2012) Translating upwards: linking the neural and social sciences via neuroeconomics. Nat Rev Neurosci 13(11):789–797
Lewin K (1946) Behavior as a function of total situation. In: Cartwright D (ed) Field theory in social science: selected theoretical papers. Harper & Row, New York, pp 238–304
Lu L, Argyle M (1991) Happiness and cooperation. Personality Individ Differ 12(10):1019–1030
Mechelli A, Friston KJ, Frackowiak RS, Price CJ (2005) Structural covariance in the human cortex. J Neurosci 25(36):8303–8310
Meyer-Lindenberg A, Weinberger DR (2006) Intermediate phenotypes and genetic mechanisms of psychiatric disorders. Nat Rev Neurosci 7(10):818–827
Miller EK, Cohen JD (2001) An integrative theory of prefrontal cortex function. Annu Rev Neurosci 24(1):167–202
Mischel W, Ayduk O, Berman MG, Casey BJ, Gotlib IH, Jonides J et al (2011) 'Willpower' over the life span: decomposing self-regulation. Soc Cogn Affect Neurosci 6(2):252–256
Morishima Y, Schunk D, Bruhin A, Ruff CC, Fehr E (2012) Linking brain structure and activation in temporoparietal junction to explain the neurobiology of human altruism. Neuron 75(1):73–79
Näpflin M, Wildi M, Sarnthein J (2007) Test–retest reliability of resting EEG spectra validates a statistical signature of persons. Clin Neurophysiol 118(11):2519–2524
Nichols AL, Maner JK (2008) The good-subject effect: investigating participant demand characteristics. J Gen Psychol 135(2):151–166
Nisbett RE, Wilson TD (1977) Telling more than we can know: verbal reports on mental processes. Psychol Rev 84(3):231–259
Olson EA, Collins PF, Hooper CJ, Muetzel R, Lim KO, Luciana M (2009) White matter integrity predicts delay discounting behavior in 9-to 23-year-olds: a diffusion tensor imaging study. J Cogn Neurosci 21(7):1406–1421

Paloyelis Y, Asherson P, Mehta MA, Faraone SV, Kuntsi J (2010) DAT1 and COMT effects on delay discounting and trait impulsivity in male adolescents with attention deficit/hyperactivity disorder and healthy controls. Neuropsychopharmacology 35:2414–2426

Panizzon MS, Fennema-Notestine C, Eyler LT, Jernigan TL, Prom-Wormley E, Neale M et al (2009) Distinct genetic influences on cortical surface area and cortical thickness. Cerebral Cortex 19(11):2728–2735

Pascual-Marqui RD (2002) Standardized low-resolution brain electromagnetic tomography (sLORETA): technical details. Methods Find Exp Clin Pharmacol 24(Suppl D):5–12

Peters J, Büchel C (2011) The neural mechanisms of inter-temporal decision-making: understanding variability. Trends Cogn Sci 15(5):227–239

Phillips MC, Meek SW, Vendemia JMC (2011) Understanding the underlying structure of deceptive behaviors. Personality Individ Differ 50(6):783–789

Rand DG, Greene JD, Nowak MA (2012) Spontaneous giving and calculated greed. Nature 489 (7416):427–430

Reimers S, Maylor EA, Stewart N, Chater N (2009) Associations between a one-shot delay discounting measure and age, income, education and real-world impulsive behavior. Personality Individ Differ 47(8):973–978

Roth A (1995) Bargaining experiments. In: Kagel J, Roth A (eds) Handbook of experimental economics. Princeton University Press, Princeton, pp 254–348

Sanfey AG (2007) Social decision-making: insights from game theory and neuroscience. Science 318(5850):598–602

Saxe R, Kanwisher N (2005) People thinking about thinking people: the role of the temporo-parietal junction in "theory of mind". In: Social neuroscience: key readings. Psychology Press, New York, pp 171–182

Scheres A, Sanfey AG (2006) Individual differences in decision making: drive and reward responsiveness affect strategic bargaining in economic games. Behav Brain Functions 2(1):35

Shamosh NA, DeYoung CG, Green AE, Reis DL, Johnson MR, Conway AR et al (2008) Individual differences in delay discounting relation to intelligence, working memory, and anterior prefrontal cortex. Psychol Sci 19(9):904–911

Smit DJ, Stam CJ, Posthuma D, Boomsma DI, De Geus EJ (2008) Heritability of "small-world" networks in the brain: a graph theoretical analysis of resting-state EEG functional connectivity. Hum Brain Mapp 29(12):1368–1378

Smith DV, Huettel SA (2010) Decision neuroscience: neuroeconomics. Wiley Interdiscip Rev Cogn Sci 1(6):854–871

Spitzer M, Fischbacher U, Herrnberger B, Grön G, Fehr E (2007) The neural signature of social norm compliance. Neuron 56(1):185–196

Steinbeis N, Bernhardt BC, Singer T (2012) Impulse control and underlying functions of the left DLPFC mediate age-related and age-independent individual differences in strategic social behavior. Neuron 73(5):1040–1051

Takeuchi H, Sekiguchi A, Taki Y, Yokoyama S, Yomogida Y, Komuro N et al (2010) Training of working memory impacts structural connectivity. J Neurosci 30(9):3297–3303

Thompson PM, Cannon TD, Narr KL, Van Erp T, Poutanen VP, Huttunen M et al (2001) Genetic influences on brain structure. Nat Neurosci 4(12):1253–1258

Van Lange PA (1999) The pursuit of joint outcomes and equality in outcomes: an integrative model of social value orientation. J Pers Soc Psychol 77:337–349

Van Lange PA (2000) Beyond self-interest: a set of propositions relevant to interpersonal orientations. Eur Rev Soc Psychol 11(1):297–331

Van Overwalle F (2009) Social cognition and the brain: a meta-analysis. Hum Brain Mapp 30 (3):829–858

Van't Ent D, Van Soelen IL, Stam CJ, De Geus EJ, Boomsma DI (2009) Strong resemblance in the amplitude of oscillatory brain activity in monozygotic twins is not caused by "trivial" similarities in the composition of the skull. Human Brain Mapp 30(7):2142–2145

van't Wout M, Kahn RS, Sanfey AG, Aleman A (2005) Repetitive transcranial magnetic stimulation over the right dorsolateral prefrontal cortex affects strategic decision-making. Neuroreport 16(16):1849–1852
Wallace B, Cesarini D, Lichtenstein P, Johannesson M (2007) Heritability of ultimatum game responder behavior. Proc Natl Acad Sci 104(40):15631–15634
Wilson DS, Near DC, Miller RR (1998) Individual differences in Machiavellianism as a mix of cooperative and exploitative strategies. Evol Human Behav 19(3):203–212
Wischniewski J, Windmann S, Juckel G, Brüne M (2009) Rules of social exchange: game theory, individual differences and psychopathology. Neurosci Biobehav Rev 33(3):305–313
Young L, Dodell-Feder D, Saxe R (2010) What gets the attention of the temporo-parietal junction? An fMRI investigation of attention and theory of mind. Neuropsychologia 48(9):2658–2664
Yu R (2012) Regional white matter volumes correlate with delay discounting. PLoS ONE 7(2): e32595
Zietsch BP, Hansen JL, Hansell NK, Geffen GM, Martin NG, Wright MJ (2007) Common and specific genetic influences on EEG power bands delta, theta, alpha, and beta. Biol Psychol 75 (2):154–164

Chapter 12
Altruistic Punishment

Alexander Strobel

Abstract Altruism can be regarded as one crucial challenge in the neuroeconomic study of human behavior. This challenge is at least threefold: First, altruism needs to be defined properly; second, its diverse behavioral expressions require elucidation; and third, its underlying mechanisms need to be delineated in order to understand why individuals behave altruistically. The present chapter therefore aims at providing a working definition of altruism and at focusing on one particular behavioral expression—namely altruistic punishment—and its role in human cooperation. It proceeds with exemplifying several experimental paradigms in the study of altruistic punishment and finally summarizes key findings on its neuroscientific underpinnings. A special emphasis will be on neuroimaging and psychophysiological studies and on an outlook on potential neuromodulatory influences. In doing so, this overview can by no means be exhaustive; rather, it is intended to provide a general impression of the current state of the neuroscientific study of the complex trait of altruism.

12.1 Introduction

12.1.1 A Working Definition of Altruism

Among the diverse definitions of altruism, the present outline favors the view of De Quervain and et al. (2004), who—referring to Sober and Wilson (1998)—distinguish between a biological and a psychological definition of altruism. The *biological definition* regards altruistic behavior as any costly behavior that confers an economic benefit to other individuals, regardless of the motives behind such behavior. The *psychological definition*, in contrast, requires that such behavior is driven by a non-hedonic motive. A bee, which sacrifices itself for the sake of the beehive by stinging an intruder, acts altruistically, but as we cannot know about any motive

A. Strobel (✉)
Department of Psychology, TU Dresden, 01062 Dresden, Germany
e-mail: alexander.strobel@tu-dresden.de

M. Reuter and C. Montag (eds.), *Neuroeconomics*, Studies in Neuroscience, Psychology and Behavioral Economics, DOI 10.1007/978-3-642-35923-1_12

behind this behavior, we have to refer to the biological definition of altruism. A nun, who sacrifices herself for the sake of the poor, acts altruistically as well, and here, we could assume some non-hedonic motive behind this behavior, and thus, could refer to the psychological definition of altruism. Still, however, the nun might expect some reward for her behavior—in her community, in the clergy, or in the "kingdom to come," and as we cannot know for sure about the motive behind her behavior, we need to refer to the biological definition of altruism as well—as costly behavior that confers an economic benefit to other individuals, regardless of the motives behind such behavior. While some of the neuroscientific studies summarized below also set out to elucidate the motives behind altruistic behavior, relying solely on neurobiological data cannot address the question of the motives of altruistic behavior properly. Nevertheless, such a neurobiological approach can provide empirical evidence that can back up discussions on the motives presumably underlying altruistic behavior.

Therefore, throughout the present chapter, the term altruism, if not otherwise specified, is used synonymously with "altruistic behavior." Thus, a biological perspective is taken, thereby abstracting from potential psychological motives behind altruistic behavior—without assuming there might be none.

12.2 Altruistic Punishment

Altruistic behavior can be observed not only in humans, but in several other species as well, and can often be explained based on genetic relatedness or benefits arising in repeated interactions (Bowles and Gintis 2004; Fehr and Gächter 2002). Such accounts, however, cannot explain why humans show altruistic behavior and cooperation even in anonymous, unattended, and/or one-shot interactions with strangers. The strong reciprocity account (Bowles and Gintis 2004) provides an explanation for these situations: It is based on so-called *altruistic punishment*, i.e., *the costly punishment of norm violations without any personal benefit for the punishing individual, but with potential benefit for other individuals.*

Altruistic punishment can be observed across a wide variety of cultures (Henrich et al. 2006; Herrmann et al. 2008), rendering it a plausible mechanism underlying human cooperation throughout different societies. It may have developed by gene-culture coevolution (Gintis 2003), i.e., a possible genetic propensity to internalize norms strengthened these norms and evolutionarily favored individuals who exhibited such a propensity (see also Henrich et al. 2006, p. 1770). Indeed, experimental and simulation studies have shown that cooperation can be maintained even in larger groups and in one-shot interactions, if there is the possibility to punish defectors (e.g., Fehr and Gächter 2002; Boyd et al. 2003; Fehr and Fischbacher 2003). Before discussing the details of these studies, however, it is necessary to outline several experimental paradigms that can be used to study altruistic behavior in general and altruistic punishment in particular.

12.3 Experimental Paradigms for the Study of Altruistic Punishment

Figure 12.1 depicts four economic games, i.e., abstract social situations where interacting individuals exchange some goods, usually money or so-called monetary units (MU). A rather simple game is the *Dictator Game*, where one player A, the "Dictator" receives a certain amount of MU (here: 20) part of which he or she can share at his or her discretion with another player B, the "Recipient." Player B has to accept any assignment made by player A—hence the name of the game. In the Dictator Game, altruism can be operationalized as the amount of MU which player A transfers to player B, as he or she is not bound to share, and sharing means bearing a cost without a benefit—at least in one-shot interactions between two players that do not know each other. In Fig. 12.1, player A would be considered as

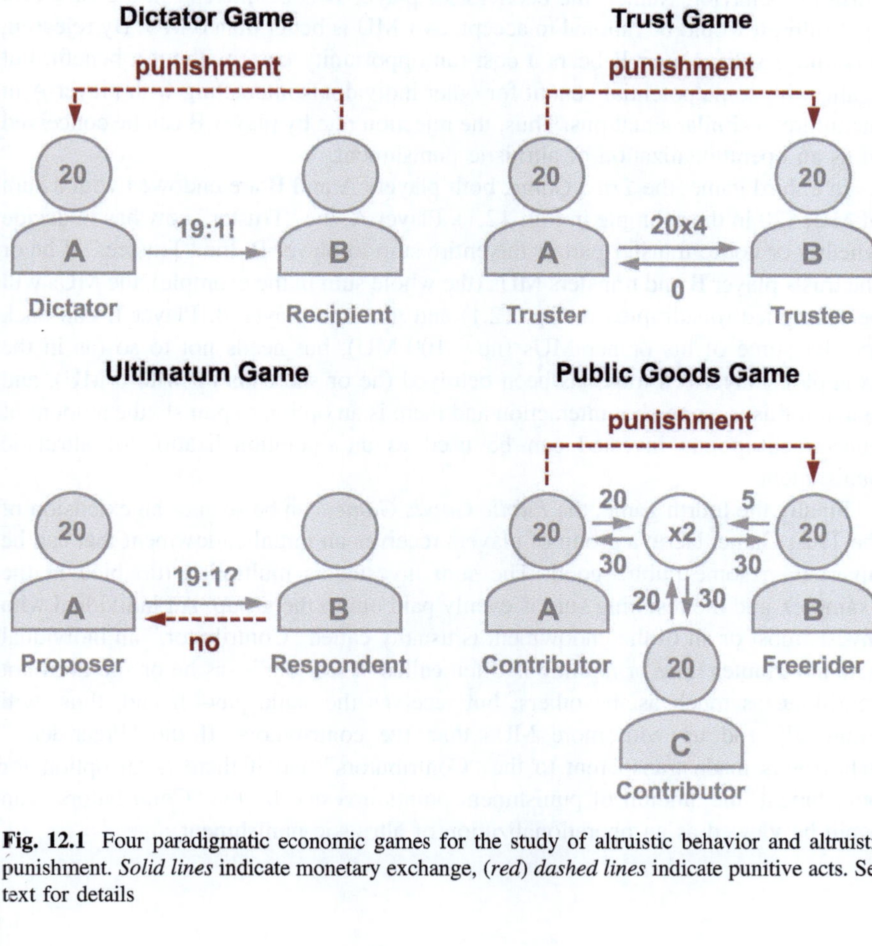

Fig. 12.1 Four paradigmatic economic games for the study of altruistic behavior and altruistic punishment. *Solid lines* indicate monetary exchange, (*red*) *dashed lines* indicate punitive acts. See text for details

less altruistic, as most of the 20 MU is kept, and only 1 MU is shared with player B. In a modified version of this game, player B can be given the opportunity to punish player A, e.g., via assigning MUs from a separate account as punishment points, thereby reducing the financial outcome of player A, but at the same time bearing a cost, as the separate account will eventually been paid out in cash to player B. Hence, punishing unfair assignments in the one-shot Dictator Game can be seen as an operationalization of altruistic punishment, as the punisher bears a cost without a financial benefit—but a potential benefit for other individuals, as player A might behave differently in future exchange situations to avoid punishment.

A related, but slightly more complex game is the *Ultimatum Game*, the name of which derives from the fact that player A acts as "Proposer" of a split of a sum of MU, and player B, the "Respondent" decides whether this offer is acceptable. If player B rejects the offer of player A, both players end up with nothing. Hence, it is in the self-interest of player A to propose a rather fair or at least not too unfair share. This disqualifies the amount of MUs offered by player A as operationalization of altruistic behavior. Rather, the behavior of player B is of interest: in the case of a 19:1 offer, it would be rational to accept, as 1 MU is better than 0 MU. By rejecting this unfair split, player B bears a cost (an opportunity cost) without a benefit, but again, with some potential benefit for other individuals interacting with player A in the future in similar situations. Thus, the rejection rate by player B can be conceived of as an operationalization of altruistic punishment.

In a third game, the *Trust Game*, both players A and B are endowed with a sum of MUs (20 in the example in Fig. 12.1). Player A, the "Truster" now has to decide whether or not to transfer part or the entire sum to player B, the "Trustee." If he or she trusts player B and transfers MUs (the whole sum in the example), the MUs will be multiplied (quadrupled in Fig. 12.1) and given to player B. Player B can back transfer some of his or her MUs (now 100 MU), but needs not to so (as in the example). Player A's trust has been betrayed (he or she ends up with 0 MU), and again, if this is a one-shot interaction and there is an option to punish, the amount of punishment points invested can be used as an operationalization of altruistic punishment.

Finally the fourth game, the *Public Goods Game*, can be seen as an extension of the Trust Game. Here, a group of players receives an initial endowment that can be invested in some public good. The sum invested is multiplied (doubled in the example), and the resulting sum is evenly paid out to the group. An individual who invests most or all of the endowment is usually called "Contributor," an individual who contributes little or nothing is often called "Freerider"—as he or she does not contribute as much as the others, but receives the same pay-off and, thus, will eventually end up with more MUs than the contributors. If the "Freerider's" behavior is made transparent to the "Contributors" and if there is an option for punishment, the amount of punishment points invested by the "Contributors" can again be viewed as an operationalization of altruistic punishment.

12.4 Maintenance of Cooperation by Altruistic Punishment

In the already-mentioned experiment by Fehr and Gächter (2002), groups of participants played several rounds of a Public Goods game, which was designed in a way that it was in the material self-interest of the players to act as "Freeriders." Varying group compositions ensured the one-shot nature of the interaction. In sessions where the individual contribution was made transparent to the others and punishment was possible, cooperation started at a higher level and increased, while in sessions where no punishment was possible, it started at a rather low level and even decreased. Thus, even in situations where freeriding is reasonable, altruistic punishment appears to maintain cooperation. This has also been substantiated by simulation studies, which have shown that altruistic punishment (and the more so punishment of non-punishers) maintains cooperation even in large groups (Boyd et al. 2003; Fehr and Fischbacher 2003). This view has not gone without criticism, either based on other simulation protocols (Ohtsuki et al. 2009) or on a review of the literature on experimental versus field studies on the factors that maintain cooperation (Guala 2012). Altruistic punishment therefore should be viewed as one possible, but by no means the most important or efficient way to ensure cooperation in human societies. Nevertheless, it has been proven to be a particularly useful approach for the study of the neurobiological correlates of altruistic behavior, which will be exemplified in the following.

12.5 The Role of Emotion-Related Brain Areas in Altruistic Punishment

One of the first studies investigating the neural bases of economic decision-making was a functional magnetic resonance imaging study by Sanfey et al. (2003) who scanned their participants while they had to decide whether to accept or to reject fair or unfair monetary offers in a sequential one-shot Ultimatum Game. As mentioned above, the rejection of unfair offers bears a cost without a benefit and can thus be seen as an operationalization of altruistic punishment. The main finding of this study was that right dorsolateral prefrontal cortex (DLPFC), anterior cingulate cortex (ACC), and anterior insula exhibited stronger activation during unfair offers. Moreover, the activity of the anterior insula was positively correlated with rejection rates for unfair offers. This result was interpreted based on the prominent role of the DLPFC in cognitive control processes, of the ACC in monitoring of—both cognitive and affective—conflict, and of the insula in subserving emotional processing via representations of signals of internal states.

This result highlights the importance of emotions in economic decision-making. Indeed, recent evidence supports the view that insular representations of emotional states (Singer et al. 2009) may serve as bias signals in economic decision-making.

This may drive the motivation to reject unfair offers and thereby to punish norm violations (see Montague and Lohrenz 2007). The role of emotions in altruistic punishment is further substantiated by a lesion study by Koenigs and Tranel (2007), who showed that individuals with lesions in the ventromedial/orbitofrontal cortex (OFC) were more likely than control subjects to reject unfair offers in the Ultimatum Game. As OFC damage has been associated with emotional dysregulation and failures in emotion-guided decision-making (e.g., Bechara et al. 2000), this evidence suggests that OFC lesions might impair downregulation of (negative) emotional responses when facing unfair offers, which could lead to economically irrational behavior such as the rejection of unfair, but nonzero offers. Thus, this interpretation fosters the notion that (negative) emotional processes play a prominent role in altruistic punishment, which in turn also suggests that non-altruistic motives can drive costly punishment.

12.6 The Role of Cognition-Related Brain Areas in Altruistic Punishment

Coming back to the study of Sanfey et al. (2003), one result deserves further consideration. While in that study, DLPFC activation was observed during the presentation of unfair offers, it was not directly correlated with rejection rates. However, another line of evidence nevertheless suggests a causal role for DLPFC in the rejection of unfair Ultimatum Game offers. In a study by Knoch et al. (2006), transcranial magnetic stimulation (TMS) was applied over the right and left DLPFC of respondents in an Ultimatum Game. The authors showed that as compared to a sham condition, transient disruption of right, but not left DLPFC activity resulted in reduced rejections of unfair offers, while fairness ratings remained unaffected. Thus, although participants still viewed offers as unfair, they were more likely to accept them and, hence, to exhibit reduced altruistic punishment behavior. Comparable results were obtained by this group in a later study using transcranial direct current stimulation (tDCS; Knoch et al. 2008); again, transient inhibition of the right DLPFC resulted in reduced rejection rates while fairness ratings remained unaffected. In another study, they examined electroencephalogram (EEG) alpha activity at rest, which in the setting of that study was conceived as a trait-like (inverse) indicator of cortical activation. In line with their earlier observations, they found higher right frontal resting alpha activity to be associated with reduced rejection rates during an Ultimatum Game (Knoch et al. 2010). Taken together, these results argue for a causal role of the integrity of especially the right DLPFC, a region that has been implicated in exerting cognitive control, particularly the inhibition of prepotent responses (e.g., Aron et al. 2004).

Interestingly, if (TMS- or tDCS-induced) disruption of the right DLPFC results in reduced rejection rates as shown by Knoch et al. (2006, 2008), then the "prepotent response" when facing unfair Ultimatum Game offers appears to be to *reject* them. This could be viewed as an economically irrational, but socially desirable behavioral act that suggests an important influence of cultural norms on altruistic behavior, which may be interpreted in terms of the above-mentioned gene-culture coevolution. Yet, a study by Crockett et al. (2010) suggests other plausible explanations: In that study, the authors assessed their participants' impulsive choice (via a delay discounting task) and altruistic punishment (operationalized as rejection of unfair Ultimatum Game offers) both during serotonin reduction via tryptophan depletion as experimental condition and during placebo as a control condition. They found that across both conditions, the magnitude of impulsive choice correlated with altruistic punishment, and that increases in impulsive choice after tryptophan depletion correlated with increases in altruistic punishment after tryptophan depletion. Given this experimental evidence, cultural norms need either be so deeply internalized that it requires less self-control to abide by them, or altruistic punishment may be driven by other motives that are associated with less self-control. A study by de Quervain et al. (2004) indeed points in the latter direction, as will be outlined in the following.

12.7 Is Revenge Indeed Sweet? Retribution as a Potential Motive for Altruistic Punishment

The studies summarized so far all used the Ultimatum Game, where the cost of altruistic punishment is to *abstain* from a gain from so far *unavailable* monetary resources. This might involve quite different internal processes than to actually *deploy available* monetary resources for the sake of punishment. Hence, a study by de Quervain et al. (2004) deserves mention, as these authors performed a more direct investigation of the neural processes underlying altruistic punishment using a Trust Game. Their participants—all in the role of a betrayed truster—underwent positron emission tomography while they could use part of their financial reimbursement to punish betraying trustees. Four conditions were realized: First, the betrayal of trust was *i*ntentional and the punishment was *c*ostly (IC), that is, the betrayed player had to invest own resources to punish the betrayer whose financial outcome was reduced; second, the betrayal of trust was *i*ntentional, but the punishment was *f*ree of costs, i.e., no own resources had to be invest to impose punishment with the effect of reducing the trustees financial outcome (IF). Third, the betrayal of trust was *i*ntentional, but the punishment was only *s*ymbolic (IS), that is, it was neither costly for the punisher nor did it reduce the betrayer's financial outcome; and fourth, there was a condition, where a random device had generated the decision not to transfer back any money from the trustee to the truster. If the punishment option was chosen, then both the truster and the trustee—who was not responsible for the decision—had to bear a cost.

At the behavioral level, this setting produced high pay-off reductions imposed on the trustee in the IC condition, even higher reductions in the IF condition, but no or only marginal pay-off reductions in the IS and NC conditions. At the neuronal level, an activity difference was observed in the caudate nucleus, with the IC and IF conditions being accompanied by higher, and the IS and NC conditions by lower caudate nucleus activation as compared to mean activation. Moreover, the amount of MUs invested for punishing in the IC condition correlated with caudate activity in both the IC and the IF condition. The authors suggested that—as the caudate nucleus has been implicated in reward processing (e.g., Delgado et al. 2003)—the motivation to punish defectors could be partly due to feelings of satisfaction when social norm violations are punished and justice is reestablished. Thus, this result points to revenge as a potential motive underlying altruistic punishment in direct one-shot interactions.

However, two issues in this seminal study require further examination. First, in the study by de Quervain et al. (2004) as well as in the other studies referred to so far, participants were directly affected by the unfair behavior of the other players, the offers of whom they could reject or whom they could punish directly (which will be referred to here as *first-party punishment*). In such situations, punishment may be driven by anger and revenge-like motives, thus presumably reflecting conditions where punishment is subjectively beneficial via satisfaction through revenge. To test this interpretation, it would be important to contrast such conditions with others where the punisher is not directly affected by unfair behavior, so that revenge-like motives cannot account for punishment. Several studies have therefore employed *third-party punishment* (with the potential punisher being not directly affected by the unfair behavior of the other player; see, e.g., Fehr and Fischbacher 2004). Here, the punishment of unfair behavior should not be motivated by the satisfaction of revenge.

Second, in the de Quervain et al. (2004) study, caudate activation was stronger in the two conditions where punishment was effective (i.e., IC and IF) and was reduced in the conditions where punishment was ineffective (IS) or was no punishment at all, as the trustee was not responsible for the "betrayal" (NC). Yet, only the IC condition was an operationalization of altruistic–i.e., costly–punishment. This leaves open the question whether the caudate nucleus activation might have been associated with *effective* rather than with *altruistic* punishment.

12.8 Third-Party Punishment

In an own functional magnetic resonance imaging study (Strobel et al. 2011), we therefore addressed these two issues in order to provide a more differentiated basis for a neuroscientific account of the motives underlying altruistic punishment. To this end, we used a variant of the Dictator Game, where player B (the "Recipient") had to accept even very unfair offers, but could exert punishment from a separate account of punishment points. Indeed, evidence suggests that individuals tend to punish norm violations even when they are not affected by the norm violation themselves, but are watching social exchange situations from a third-party perspective (Fehr and Fischbacher 2004). Thus, we were interested in whether there were differences in brain activation during punishment acts where individuals might pursue some subjective benefit such as satisfaction through revenge (i.e., punishment for norm violations affecting one self, or *first-person punishment*) as compared to punishment acts where individuals seemingly do not pursue any subjective benefit (i.e., punishment for norm violations affecting other people, or *third-party punishment*). Hence, we compared two conditions where players B were either the recipients themselves or were "watching" interactions between the Dictators and some players C.

Our second question was whether caudate nucleus activation during punishment acts would be associated with effective rather than with altruistic punishment. Hence, we compared two conditions where the punishment—with equal costs—was either highly effective (strong punishment, resulting in a substantial reduction of the Dictator's payoff) or rather less effective (weak punishment, resulting in marginal reduction).

Analyses were performed for predefined regions of interest that were based on the literature reviewed above (see Fig. 12.2, left panel): Insula, ACC, and DLPFC as derived from the results by Sanfey et al. (2003) and Knoch et al. (2006); OFC as

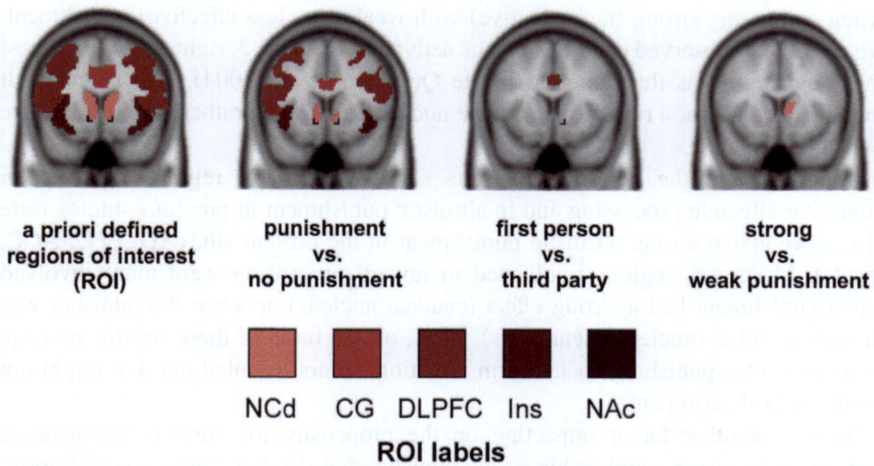

Fig. 12.2 Altruistic punishment in the dictator game: functional magnetic resonance imaging results from Strobel et al. (2011) at MNI coordinates (*xyz*) 10, 12, 0. The *first panel from the left* shows the regions of interest (ROIs) defined on the basis of the available literature (see text): NCd = caudate nucleus; CG = cingulate gyrus (displayed here is the anterior cingulate cortex, ACC); DLPFC = dorsolateral prefrontal cortex; Ins = insula; NAc = nucleus accumbens. The *next panels* show the clusters within these ROIs that were significant at a false discovery rate corrected level of significance of $P < 0.05$ for the contrasts "punishment > no punishment" (with all predefined ROIs being significantly more active in the punishment condition), "first person > third party", (where along with ACC, NAc was significantly more active in the first-person condition), and "strong > weak punishment" (with NCd being significantly more active in the strong punishment condition)

derived from the results of Koenigs and Tranel (2007; not displayed in Fig. 12.2 as unfortunately, due to technical reasons, OFC signals could not be captured properly in our study), and caudate nucleus as implicated in altruistic punishment by de Qvervain et al. (de Quervain et al. 2004). Moreover, as these authors argued for a role of reward processes in altruistic punishment, we additionally included another region of interest: the nucleus accumbens, which plays a prominent role in reward processing (see, e.g., Robbins and Everitt 1996).

To summarize the results, our behavioral data indicated that altruistic punishment showed medium correlations with self-reported altruistic tendencies as assessed with the Altruism facet of the revised NEO Personality Inventory (NEO-PI-R; Costa and McCrae 1992), thereby substantiating the notion that economic games such as the one employed here indeed capture individual behavioral tendencies that protrude over the laboratory. The imaging data revealed that, as expected, all predefined regions of interest were significantly more active in trials where our participants punished as compared to trials without punishment (see Fig. 12.2, second panel from the left). This substantiates the role of cognition-related areas (DLPFC, ACC) and emotion-related regions (Insula) in altruistic punishment. Furthermore, when comparing the first person with the third-party condition, nucleus accumbens (along with ACC) showed stronger activation in the first-person condition (Fig. 12.2, third panel from the left), and when comparing strong (i.e., effective) with weak (i.e., less effective) punishment, we selectively observed caudate nucleus activation (Fig. 12.2, right panel) in almost the same region as that reported by de Quervain et al. (2004). The latter result therefore argues for a role of the caudate nucleus in *effective* rather than in *altruistic* punishment per se.

Summarizing, the pattern of results shows that brain regions involved in cognitive-affective processing and in altruistic punishment in previous studies were also more active during altruistic punishment in the present study (DLPFC, ACC, insula). Moreover, regions implicated in reward processing were more involved when punishment had a strong effect (caudate nucleus) or when the punisher was directly affected (nucleus accumbens). Thus, on the basis of these results, revenge as a motive for punishing social norm violation cannot be ruled out, but might not be the only determinant.

Indeed, another factor impacting on the propensity for altruistic punishment appears to be group membership of the punisher and the defector of a social norm: Baumgartner et al. (2012) observed that individuals showed less third-party punishment of defecting members of their ingroup as compared to those of an outgroup. While punishing *outgroup* members correlated with the connectivity within a functional brain network implicated in sanction-related decision-making (right orbitofrontal an lateral prefrontal regions as well as right caudate nucleus), punishing *ingroup* members was correlated with the connectivity within a brain network involved in mentalizing processes. The authors interpreted the latter finding as presumably indicative of their subjects' efforts to understand the defective behavior of their ingroup members. Most interestingly, the authors could demonstrate that the activity in the mentalizing network was negatively correlated with the activity of network nodes of the punishment network, pointing to a suppressive influence of the former on the latter.

The study by Baumgartner et al. (2012), thus, broadens our perspective on altruistic punishment by shifting the focus from single brain regions and their presumed role on the modulation of behavior to brain *networks* and their correlated activity during altruistic punishment. This network perspective can be expected to considerably further our understanding of the processes underlying altruistic punishment. Yet, even this approach falls short of one crucial further aspect of the complex phenomenon under investigation: its process nature on a micro-timescale. Due to the low-temporal resolution of methods such as functional magnetic resonance imaging, fast computational processes underlying altruistic punishment—e.g., some calculation of a violation of a personal or social norm that needs to be performed ahead of cognitive-affective processes in order to trigger altruistic punishment—cannot readily be captured. Hence, methods that allow for a better temporal resolution are an important means to further our understanding of the neurocognitive processes during altruistic punishment. EEG is one such method (see the EEG chapter by Debener et al. in the methods chapter of this book for a detailed description of this method), as it provides a high-temporal resolution (albeit at the cost of a lower spatial resolution). Moreover, EEG is less costly, allowing for larger samples and, thus, for the detection of smaller effects as those typically to be detected in imaging studies.

12.9 The Feedback-Related Negativity as a Predictor of Altruistic Punishment

Besides EEG analyses in the frequency domain (see the above mentioned study by Knoch et al. 2010, who observed—presumably habitual—right frontal EEG alpha activity to be associated with the rejection of unfair offers in the Ultimatum Game), EEG analyses in the time domain are capable of capturing even very fast neural processes. By means of so-called event-related potentials (ERP), changes in neuronal activity following some critical event can be tracked at the millisecond level. The ERP approach has already been successfully applied to the study of the processes underling the propensity to reject unfair offers in the Ultimatum Game. One key study will be outlined in the following.

Hewig et al. (2011) investigated the role of the so-called feedback-related negativity (FRN) in Ultimatum Game bargaining. The FRN is a negativity in the ERP that is maximal over frontal recording sites. Based on source localization studies, it is assumed to be generated by the ACC when feedback in experimental conditions deviates from an individual's expectation and, thus, might be an indicator of processes underlying reinforcement learning (Holroyd and Coles 2002). In the case of the Ultimatum Game, unfair offers can be considered as deviations from some (individual or social) norm, which may prompt rejection of these offers, probably based on prediction error computations that are indexed by the FRN. Indeed, the authors could show that the rejection of unfair offers could be predicted by the magnitude of the FRN (to fair offers). Together with self-reported subjective emotion ratings and objectively measured skin conductance levels, 84 % of the rejection of unfair offers could be predicted. This supports the view that the detection of norm violations and the resulting (or at least accompanying) emotional processes as indexed here in emotional ratings and psychophysiological arousal measures trigger altruistic punishment.

So far unpublished own data based on a sample of 44 volunteers substantiate the results by Knoch et al. (2010) and Hewig et al. (2001): in an Ultimatum Game, the FRN significantly predicted the rejection rate (as in Hewig et al. 2011), but only for a first-person condition; in a third-party condition, it was resting frontal alpha (as in Knoch et al. 2010, but bilaterally) that predicted rejection rates. If replicable, these results would suggest that feedback-related and, thus, state-like (bottom–up generated) event-related perturbations in cortical (ACC) activity are more predictive of the rejection of unfair offers when the subjects are directly involved, while trait-like characteristics such as resting frontal EEG alpha activity are more important when subjects are acting as witnesses of unfair behavior in a perhaps more top–down controlled way.

12.10 Neuromodulatory Influences on Altruistic Behavior

While, as shown, there is ample evidence on the neural correlates of altruistic punishment, a so far less extensively investigated topic in this context is the role of neuromodulators in altruistic punishment. The already mentioned study by Crockett et al. (2010) is one of few studies that have implicated neuromodulators in altruistic punishment. Their result that serotonin depletion-induced increases in impulsive responding correlated with increased altruistic punishment can easily be reconciled with the reviewed evidence on the role of emotional responses to unfair behavior in economic games: Perceived norm violations trigger prepotent negative emotional responses to unfairness that—under conditions of heightened impulsivity and, thus, reduced inhibition of these prepotent responses via serotonin depletion—can outweigh economic considerations (as, e.g., in the context of the Ultimatum Game, it would economically reasonable to accept even the least nonzero offer).

However, evolutionarily, an enhanced impulsive responding that favored prepotent responses would not enhance cooperation, unless the prepotent response favored cooperation. Rather, in order to enhance cooperation, a violation of a social norm that hinders cooperation needs to be detected and correcting behaviors need to be at hand even at economic costs. Such behavior might be in addition otherwise rewarding, e.g., because a contingency between such behaviors and personal or social rewards had been learned. Here, another neuromodulator comes into play that has been associated with prediction error coding and reinforcement learning. FRN that was associated with altruistic punishment by Hewig et al. (2011) has been suggested as a reward prediction error signal and has also been associated with dopamine signaling (Holroyd and Coles 2002). Moreover, activity of brain regions activated during altruistic punishment, e.g., DLPFC, ACC, insula, and caudate nucleus—although likely to be serotonergically modulated as well—are prominent targets of dopaminergic projections from the midbrain that have been implicated in prediction error signaling (Schultz 1998) and reward processing (Robbins and Everitt 1996). Therefore, in view of the gene-culture coevolution assumption, genetic variation in dopamine function could be expected to impact on, first, neural responses to norm violations, and second, the expectation of possible future rewards derived from the punishment of unfair behavior, be it via the personal reward of satisfaction derived from revenge or via learned contingencies between punishment behavior and social rewards or both.

Therefore, in our study reviewed above (Strobel et al. 2011), we also examined the possible role of genetic variation of dopamine function in the modulation of neural responses during altruistic behavior. To this end, we included a widely studied genetic variation of dopamine function into our analyses: a G to A single nucleotide polymorphism in the gene encoding the dopamine-degrading enzyme catechol-O-methyltransferase (COMT) that results in the substitution of the amino acid valine (Val) by methionine (Met) at amino acid position 158 of the COMT enzyme. This so-called Val158Met polymorphism impacts on the COMT enzyme's thermostability: Met allele homozygotes exhibit only one-fourth of the COMT activity than Val/Val homozygotes, which likely results in higher levels of synaptic dopamine (Lachman et al. 1996; Chen et al. 2004). The *COMT* Val158Met polymorphism has been associated with prefrontally modulated cognitive and affective processing (see Mier et al. 2010, for a review and meta-analysis). Interestingly, it has also been implicated in reward processing, as in a study by Dreher et al. (2009), Met allele carriers exhibited higher activation in the ventral striatum and the DLPFC during reward anticipation.

While we could not identify a *COMT* genotype-related difference in DLPFC activation, we observed a punishment-related genotype effect at the level of nucleus accumbens activation: Carriers of the Met allele and, thus presumably higher synaptic dopamine activity, showed higher punishment-related nucleus accumbens activation. We interpreted this finding as indicative of a *COMT* Met allele-related bias of ventral striatal integration of input signals from DLPFC, ACC, insula, and other regions that results in higher reward anticipation for the decision to punish unfair behavior. Such an interpretation, however, would require that individuals have internalized contingencies between norm-enforcing behavior and social rewards, with *COMT* Met allele carriers being more susceptible to social signals of reward and, thus, more likely to adapt their behavior to attain such rewards (see Strobel et al. 2011).

While in our report, we regarded this finding as a first empirical support for the gene-culture coevolution assumption (which could be viewed as being further substantiated by the fact that the *COMT* Met allele appears to be a comparably recent variant unique to humans; Palmatier et al. 1999), several caveats need to be considered. First, our sample was comparably small (twenty-four subjects), and second, generalization of this effect on other forms of prosocial behavior is not readily possible given conflicting evidence from a study by Reuter et al. (2011) who observed *COMT* Met/Met genotype carriers to be *less* likely for charitable donation of money in an experimental setting. While this apparent discrepancy could be explained by different experimental settings, different internalized social norms for costly punishment of unfair behavior versus costly giving to a charity, and/or different expectations of personal or social rewards associated with the respective behaviors, further work is needed to resolve this discrepancy.

However, both these studies demonstrate a critical role of genetic variation in dopamine function in prosocial behavior, a role that is further substantiated by another study from our group: We observed carriers of a variant of the dopamine D4 receptor gene exon III polymorphism, which has been associated with impulsivity-related personality traits and behavioral tendencies (Ebstein et al. 1996; Kluger et al. 2002; Swanson et al. 2000), to report less altruistic behavior on the Altruism scale of the NEO-PI-R (Anacker et al. 2012).

Taken together, evidence on neuromodulatory influences on altruism/altruistic punishment is scarce and partly inconsistent, but nevertheless suggests that neuromodulators such as serotonin and dopamine play a prominent role in altruistic punishment. Yet, their exact roles in different contexts remain to be elucidated.

12.11 Conclusions and Further Directions

Taken together, the evidence reviewed in this chapter underscores the role of brain regions implicated in both cognitive processes, most likely such involving the inhibition of prepotent responses (Knoch et al. 2006, 2008, 2010) and affective processes in altruistic punishment (Koenigs and Tranel 2007). Negative emotional states such as anger (Sanfey et al. 2003; see also Seip et al. 2009), the expectation of satisfaction by revenge (de Quervain et al. 2004; Strobel et al. 2011), and, thus, spiteful motives (Jensen 2010) may drive altruistic punishment just as may group membership (Baumgartner et al. 2012). Other factors may influence altruistic punishment as well such as 'physical' factors like the size of societies (Marlowe and Berberesque 2008) or egalitarian motives even in the absence of cooperation to be reinforced (Dawes et al. 2007), but so far have not been examined neuroscientifically.

The mentioned cognitive-affective processes are likely preceded by a computation of a violation of a personal or societal norm, which to some extent can be captured by the magnitude of the FRN as a neural signature of a deviation from one's expectations (Hewig et al. 2011). A perceived norm violation then can trigger prepotent negative emotional responses to unfairness that under conditions of reduced inhibition of these prepotent responses can outweigh economic considerations. Such reduced inhibition may stem from increased impulsive responding, which may be due to altered serotonin activity as shown by Crockett et al. (2010), but could also be assumed to be linked to stress, as it was recently shown that acute psychosocial stress increased various prosocial behaviors (von Dawans et al. 2012), or to stable personality characteristics. They, in turn, might be partly genetically modulated, as might be the sensitivity to social signals of reward as speculated in the context of the study by Strobel et al. (2011) who observed an association of the *COMT* Val158Met polymorphism with nucleus accumbens activity during altruistic punishment. Thus, genetic variation in the function of neurotransmitters and neuromodulators might have a twofold impact on altruistic punishment: first, with regard to the acquisition and internalization of norms and the formation of prepotent responses when norms are violated, and second, with regard to the trait- or state-like propensity to inhibit or disinhibit such prepotent responses. Especially the latter issues are expected to attract attention in future studies.

Discerning readers will have noticed that in the present summary, terms like "may" or "might" are used quite often, as are phrases like "this points to" or "this suggests"—and, indeed, they were chosen intently: Neuroimaging and other neuroscientific techniques provide a valuable means for the study of the brain processes and neural mechanisms underlying altruistic behavior—but, itself, they cannot inform about the motives driving such behavior. The activity of some brain region or even a brain network during altruistic punishment can only suggest or point to such motives: If activity in a brain region X previously activated during reward expectation is also observed during altruistic punishment, one cannot readily deduce that altruistic punishment has to do with reward expectation (unless reward expectation is the only process that has been associated with the activity in X—which for a "multi-purpose device" such as our brain and for a complex cognitive-affective process such as reward processing is quite unlikely). One can only state that, because region X has also been implicated in reward processing, it cannot be ruled out that the activity in X during altruistic punishment might be suggestive of a role of reward expectation during altruistic punishment.

Researchers using neuroimaging and other neuroscientific approaches always need to be aware of the danger of such so-called "reverse inferences" (Poldrack 2006)—i.e., to infer a cognitive process underlying a given task from the activation of some brain region during that task, unless it is highly specific—but also of its potential use: to generate new hypotheses; and to adequately test these hypotheses, one cannot rely on elaborate neuroscientific tools alone, but need to employ carefully designed experimental paradigms, which rule out alternative explanations. Thus, the studies reviewed in this chapter have only begun to scratch at the surface of the complex phenomenon of altruistic punishment—yet, while its driving forces still remain to be elucidated, they have provided important clues and directions for future research.

References

Anacker K, Enge S, Reif A, Lesch K-P, Strobel A (2012) Dopamine D4 receptor gene variation impacts self-reported altruism. Mol Psychiatry (Epub ahead of print, 08.05.2012)

Aron AR, Robbins TW, Poldrack RA (2004) Inhibition and the right inferior frontal cortex. Trends Cogn Sci 8:170–177

Baumgartner T, Götte L, Gügler R, Fehr E (2012) The mentalizing network orchestrates the impact of parochial altruism on social norm enforcement. Hum Brain Mapp 33:1452–1469
Bechara A, Damasio H, Damasio AR (2000) Emotion, decision making, and the orbitofrontal cortex. Cereb Cortex 10:295–307
Bowles S, Gintis H (2004) The evolution of strong reciprocity: cooperation in heterogeneous populations. Theor Popul Biol 65:17–28
Boyd R, Gintis H, Bowles S, Richerson PJ (2003) The evolution of altruistic punishment. Proc Natl Acad Sci USA 100:3531–3535
Chen J, Lipska BK, Halim N, Ma QD, Matsumoto M, Melhem S, Kolachana BS, Hyde TM, Herman MM, Apud J, Egan MF, Kleinman JE, Weinberger DR (2004) Functional analysis of genetic variation in COMT: effects on mRNA, protein, and enzyme activity in postmortem human brain. Am J Hum Genet 75:807–821
Costa PT, McCrae RR (1992) Revised NEO personality inventory (NEO PI-R) and NEO five factor inventory. Psychol Assess Resour, Odessa
Crockett MJ, Clark L, Lieberman MD, Tabibnia G, Robbins TW (2010) Impulsive choice and altruistic punishment are correlated and increase in tandem with serotonin depletion. Emotion 10:855–862
Dawes CT, Fowler JH, Johnson T, McElreath R, Smirnov O (2007) Egalitarian motives in humans. Nature 446:794–796
de Quervain DJ, Fischbacher U, Treyer V, Schellhammer M, Schnyder U, Buck A, Fehr E (2004) The neural basis of altruistic punishment. Science 305:1254–1258
Delgado MR, Locke HM, Stenger VA, Fiez JA (2003) Dorsal striatum responses to reward and punishment: effects of valence and magnitude manipulations. Cogn Affect Behav Neurosci 3: 27–38
Dreher J-C, Kohn P, Kolachana B, Weinberger DR, Berman KF (2009) Variation in dopamine genes influences responsivity of the human reward system. Proc Natl Acad Sci USA 106: 617–622
Ebstein RP, Novick O, Umansky R, Priel B, Osher Y, Blaine D, Bennett ER, Nemanov L, Katz M, Belmaker RH (1996) Dopamin D4 receptor (D4DR) exon III polymorphism associated with the personality trait of Novelty Seeking. Nat Genet 12:78–80
Fehr E, Fischbacher U (2003) The nature of human altruism. Nature 425:785–791
Fehr E, Fischbacher U (2004) Third-party punishment and social norms. Evol Hum Behav 25: 63–87
Fehr E, Gächter S (2002) Altruistic punishment in humans. Nature 415:137–140
Gintis H (2003) The hitchhiker's guide to altruism: gene-culture coevolution, and the internalization of norms. J Theor Biol 220:407–418
Guala F (2012) Reciprocity: weak or strong? What punishment experiments do (and do not) demonstrate. Behav Brain Sci 35:1–15
Henrich J, McElreath R, Barr A, Ensminger J, Barrett C, Bolyanatz A, Cardenas JC, Gurven M, Gwako E, Henrich N, Lesorogol C, Marlowe F, Tracer D, Ziker J (2006) Costly punishment across human societies. Science 312:1767–1770
Herrmann B, Thöni C, Gächter S (2008) Antisocial punishment across societies. Science 319:1362–1367
Hewig J, Kretschmer N, Trippe RH, Hecht H, Coles MG, Holroyd CB, Miltner WH (2011) Why humans deviate from rational choice. Psychophysiology 48:507–514
Holroyd CB, Coles MGH (2002) The neural basis of human error processing: reinforcement learning, dopamine, and the error-related negativity. Psychol Rev 109:679–709
Jensen K (2010) Punishment and spite, the dark side of cooperation. Philos Trans R Soc B Biol Sci 365:2635–2650
Kluger AN, Siegfried Z, Ebstein RP (2002) A meta-analysis of the association between DRD4 polymorphism and novelty seeking. Mol Psychiatry 7:712–717
Knoch D, Pascual-Leone A, Meyer K, Treyer V, Fehr E (2006) Diminishing reciprocal fairness by disrupting the right prefrontal cortex. Science 314:829–832

Knoch D, Nitsche MA, Fischbacher U, Eisenegger C, Pascual-Leone A, Fehr E (2008) Studying the neurobiology of social interaction with transcranial direct current stimulation–the example of punishing unfairness. Cereb Cortex 18:1987–1990

Knoch D, Gianotti LR, Baumgartner T, Fehr E (2010) A neural marker of costly punishment behavior. Psychol Sci 21:337–342

Koenigs M, Tranel D (2007) Irrational economic decision-making after ventromedial prefrontal damage: evidence from the Ultimatum Game. J Neurosci 27:951–956

Lachman HM, Papolos DF, Saito T, Yu YM, Szumlanski CL, Weinshilboum RM (1996) Human catechol-O-methyltransferase pharmacogenetics: description of a functional polymorphism and its potential application to neuropsychiatric disorders. Pharmacogenetics 6:243–250

Marlowe FW, Berbesque JC (2008) More 'altruistic' punishment in larger societies. Proc R Soc B Biol Sci 275:587–590

Mier D, Kirsch P, Meyer-Lindenberg A (2010) Neural substrates of pleiotropic action of genetic variation in COMT: a meta-analysis. Mol Psychiatry 15:918–927

Montague PR, Lohrenz T (2007) To detect and correct: norm violations and their enforcement. Neuron 56:14–18

Ohtsuki H, Iwasa Y, Nowak MA (2009) Indirect reciprocity provides a narrow margin of efficiency for costly punishment. Nature 457:79–82

Palmatier MA, Kang AM, Kidd KK (1999) Global variation in the frequencies of functionally different catechol-O-methyltransferase alleles. Biol Psychiatry 46:557–567

Poldrack RA (2006) Can cognitive processes be inferred from neuroimaging data? Trends Cogn Sci 10:59–63

Reuter M, Frenzel C, Walter NT, Markett S, Montag C (2011) Investigating the genetic basis of altruism: the role of the COMT Val158Met polymorphism. Soc Cogn Affect Neurosci 6: 662–668

Robbins TW, Everitt BJ (1996) Neurobehavioural mechanisms of reward and motivation. Curr Opin Neurobiol 6:228–236

Sanfey AG, Rilling JK, Aronson JA, Nystrom LE, Cohen JD (2003) The neural basis of economic decision-making in the ultimatum game. Science 300:1755–1758

Schultz W (1998) Predictive reward signal of dopamine neurons. J Neurophysiol 80:1–27

Seip EC, van Dijk WW, Rotteveel M (2009) On hotheads and dirty harries: the primacy of anger in altruistic punishment. Ann N Y Acad Sci 1167:190–196

Singer T, Critchley HD, Preuschoff K (2009) A common role of insula in feelings, empathy and uncertainty. Trends Cogn Sci 13:334–340

Sober E, Wilson DS (1998) Unto others: the evolution and psychology of unselfish behavior. Harvard University Press, Cambridge

Strobel A, Zimmermann J, Schmitz A, Reuter M, Lis S, Windmann S, Kirsch P (2011) Beyond revenge: neural and genetic bases of altruistic punishment. NeuroImage 54:671–680

Swanson JM, Flodman P, Kennedy J, Spence MA, Moyzis R, Schuck S, Murias M, Moriarity J, Barr C, Smith M, Posner M (2000) Dopamine genes and ADHD. Neurosci Biobehav Rev 24:21–25

von Dawans B, Fischbacher U, Kirschbaum C, Fehr E, Heinrichs M (2012) The social dimension of stress reactivity: acute stress increases prosocial behavior in humans. Psychol Sci 23: 651–660

Part V
Translational and Developmental Approaches to Neuroeconomics

Chapter 13
Brain SEEKING Circuitry in Neuroeconomics: A Unifying Hypothesis for the Role of Dopamine-Energized Arousal of the Medial Forebrain Bundle in Enthusiasm-Guiding Decision-Making

Jaak Panksepp and Cristina G. Wilson

Abstract Affective shifts are critical factors in our decision-making with respect to all our survival concerns, including high cognitive ones such as those related to our economic investments and divestments. A critical emotional system, not commonly considered in neuroeconomics, is our primary process subcortical SEEKING system that regulates our exploratory-investigatory urges, including the eager anticipations of our higher mental processes. In economic decision-making, our SEEKING urge motivates us to consider the diverse opportunities and risks that are inherent in life-supportive decision-making. This state of mind, at normal levels of activity, energizes focus on cognitive details that can promote opportunities for success as well as avoid costly mistakes. However, excessive activity in this system may also promote faulty (addictive?) decision-making that is common in gambling, when hopes outweigh consideration of risks (as might be mediated by the cognitive representatives of FEAR and PANIC system). It is well known that all drug addictions are mediated by the feelings of euphoria that the SEEKING system can promote. Clearly, the ancient emotional systems of the brain need to be considered as motivators of neuroeconomic decisions, but they also need to be understood as primal motivations which need to be disciplined by higher decision-making capacities that emerge developmentally as a function of the losses and gains that have resulted from the vicissitudes of living in at times predictable but also unpredictable social (and physical) worlds. Without developmentally emergent cognitive discipline, the SEEKING system can promote delusional thinking.

J. Panksepp (✉)
Department of Integrative Physiology and Neuroscience, College of Veterinary Medicine, Washington State University, PO Box 647620, Pullman, WA 99164-762, USA
e-mail: jpanksepp@vetmed.wsu.edu

C.G. Wilson
Department of Psychology, Washington State University, Pullman, WA 99164-7300, USA

M. Reuter and C. Montag (eds.), *Neuroeconomics*, Studies in Neuroscience, Psychology and Behavioral Economics, DOI 10.1007/978-3-642-35923-1_13

As the Victorian Scottish philosopher and satirical writer Thomas Carlyle viewed it, economics was the "dismal science"—largely amounting to the flow of goods and money in highly competitive marketplaces, bringing both misery and riches, to those whose insights were deficient or insightful, respectively. During the past dozen years, the dismal social science of 'supply and demand' has been brightened considerably in the light of human, as well as cross-species neuroscience: we can now envision how economic decisions are crafted by both affective hopes and cognitive calculations of the brain's mind. However, there are many societal patterns that are still better fathomed by the flow of money as assessed by statistical (non-experienced) electronic computations rather than the software of the experienced mind, whether affective or cognitive. In the current chapter we argue that the perennial critical affective states of mind remain the intermediary brain decision-making processes, many shared by all mammals that we discuss here. We will largely focus on what has traditionally been called "the brain reward system," even though anyone paying attention should know that this system does not mediate sensory pleasures, but rather, the appetitive foraging, arising from a very precious brain system—the Medial Forebrain Bundle (that we call the SEEKING System) that allow us to pursue pleasures and satisfactions (including avoidance of negative events/states) with a joyful attitude that is full of *enthusiasm*.

Thereby, we will also focus heavily on the affective state-control contributions in decision-making, for that is where even the cross-species animal work provides considerable illumination. We will not consider the many cognitive "computations" that go into the making of life-supportive choices, but simply make the case that without the affective underpinnings, arising from lower (primary processes) of the brain, the cognitive "computations" would have no values to guide them. Our position will be evolutionary—namely, that individual economic decision-making could not operate without the enthusiasms, hopes, and wishes of the various evolutionary "layers" of the brain (Figs. 13.1, 13.2 and 13.3). We will designate the most fundamental substrates as being "primary" since they are constituted of various evolved emotional systems within our brains, that not only engender various affective states, but concurrently generate various states of action readiness. Thus, the mentally experienced aspects of primary process emotions mediate not only various *affective states*, but they are action oriented: Both features code for basic survival issues: all "good" feelings ("desirable" states-of-being at higher mental levels) predict that individuals are on survival trajectories, while all "bad" feelings (to be avoided) predict that one may be on the down-hill course of destruction (Panksepp 1982, 1998). Of course these simple minded, but all-important, feeling states need to guide learning and memory—namely, the deeply unconscious secondary processes (Solms and Panksepp 2012) that distribute feelings not only in external space and time (based on past experiences in the world), but also provide the raw materials for thinking. The tertiary processes of the brain, concentrated in thought-promoting neocortical reaches of the mind, is where higher, *cognitive* decision-making is finalized, after weighting the pros and cons for one's own interests and thriving.

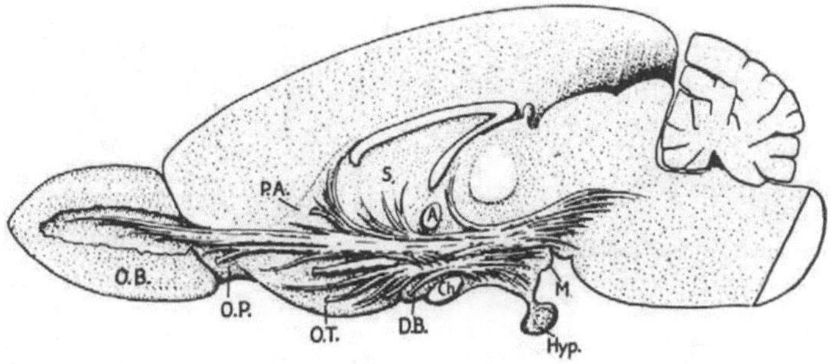

Fig. 13.1 A drawing of the Medial Forebrain Bundle (MFB), illustrating the main trajectory of the SEEKING system (classically referred to as "the brain reward system"). It runs up from the midbrain, through the lateral hypothalamus (LH), into more rostral neural regions. Other neural regions pictured: optic chiasm (Ch), olfactory bulbs (O.B.), olfactory peduncle (O.P.), paraolfactory area (P.A.), olfactory tract (O.T.), diagonal band of Broca (D.B.), anterior commissure (A), pituitary gland/the hypophysis (Hyp.), and mammillary bodies (M). (Figure from LeGros Clark et al. 1938)

While it is obvious that most human economic decision-making is heavily a neocortical thinking-related affair, we would be sorely mistaken to neglect the primary process affective (valuative) foundations upon which our higher decision-making processes were constructed in brain-evolution and individual development. Indeed, there are several conceptual kinds of intrinsic value systems. Some are homeostatically based, such as the basic consummatory mechanisms of HUNGER and THIRST, not to mention the many other intrinsic bodily needs such as thermoregulatory values (Cabanac 1992) and various other bodily survival states. Other affective value systems are sensorially based, such as the wonderful smells and tastes of the culinary world that still guide our post-modern "foraging" patterns in diverse food establishments, whether cafes, markets, or fancy restaurants. Yet, the category of affects that has the widest purview in everyday economic decision-making is the emotional values—the evolutionarily constructed, intrinsic needs of the brain itself that we will primarily focus on here.

If one had to select a single emotional system from the seven outlined by cross-species affective neuroscience research (Panksepp 1998)—namely SEEKING, RAGE, FEAR, LUST, CARE, PANIC and PLAY—that is more important for human decision-making than any other, it would surely be SEEKING. (Please note, capitalizations are used as a formal nomenclature convention for labeling "primary-process" psychobehavioral emotional/motivational systems of the brain). The SEEKING network is a general-purpose appetitive system that many of the other emotions and affects utilize, and it is the *only* one, if damaged severely, that compromises life itself for it depletes foraging/enthusiasm resources. This system courses through the Medial Forebrain Bundle (MFB, see Fig. 13.1 for original depiction) which was only recently visualized in humans with modern diffusion tensor imaging (Coenen, et al. 2012)—for overall summary of major connectivities see Fig. 13.2. When the MFB is damaged bilaterally it leads to an

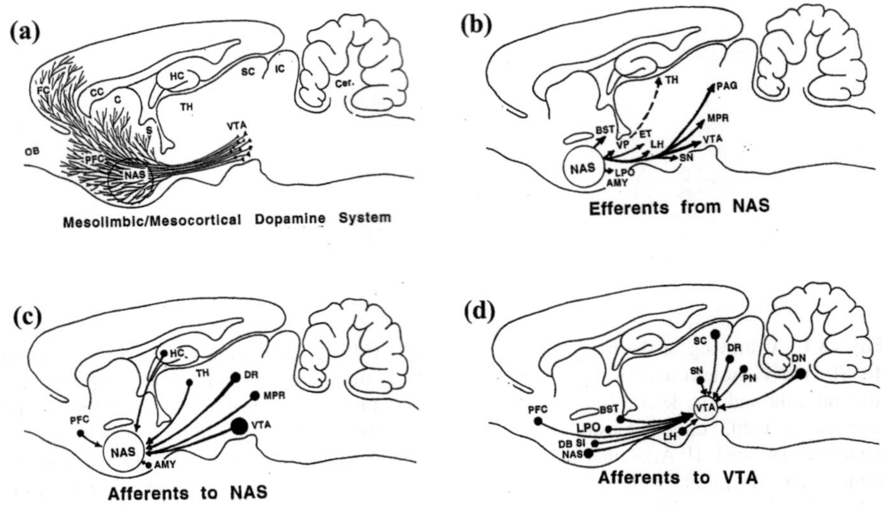

Fig. 13.2 Diagrams showing different aspects of the SEEKING system in the rat brain. **a** Shows ascending projections from the midbrain Ventral Tegmental Area (VTA) a10 dopamine neurons that innervate the nucleus accumebs (NAS) and prefrontal cortex (PFC) among other regions. **b** Efferents (descending) of the nucleus accumbens, mostly GABAergic. **c** The major afferent projections to the NAS. **d** Afferent projections to the VTA. Other abbreviations: Amygdala (AMY), bed nucleus of stria terminalis (BST), caudate-putamen (C), corpus callosum (CC), diagonal band of Broca (DB), dentate nucleus (DN), dorsal raphe (DR), entopeduncular nucleus (ET), frontal cortex (FC), hippocampus (HC), inferior colliculus (IC), lateral hypothalamus (LH), lateral preoptic area (LPO), mesopontine reticular nuclei (MPR), olfactory bulb (OB), periaqueductal gray (PAG), prefrontal cortex (PFC), parabrachial nucleus (PN), superior colliculus (SC), substantia innominate (SI), substantia nigra (SN), thalamus (TH), ventral pallidum (VP). (Figure from Ikemoto and Panksepp 1999)

amotivational organism that is no longer capable of fending for itself, and will die without nursing (Teitelbaum and Epstein 1962). It is small wonder that this system has been implicated in the burgeoning field of neuroeconomics (e.g., Knutson and Greer 2008; Glimcher 2011a, b), but not enough effort has been devoted to make clear that this general purpose emotional-motivational system, which encourages animals to explore and forage in their environments for all things needed for survival, has a much broader scope than neuroeconomics, and it certainly is poorly conceptualized as "the brain reward system" (the traditional historical label, that still lingers, despite its narrow-mindedness). In fact there are many rewarding systems in the brain, and the general-purpose SEEKING urge (for an extensive historical discussion, see Panksepp 1981) does not mediate sensory pleasure, but more likely feelings of "enthusiasm"—an essential psychological ingredient in getting what you want (which of course already reflects the transition from a primary aspect of mind to related higher mental processes).

Beside recognizing the above conceptual naming error, carried forward to this day (e.g., Haber and Knutson 2009), we must initially consider, in any disciplined discourse, that the brain (and the mind) is an evolutionarily layered organ, with at

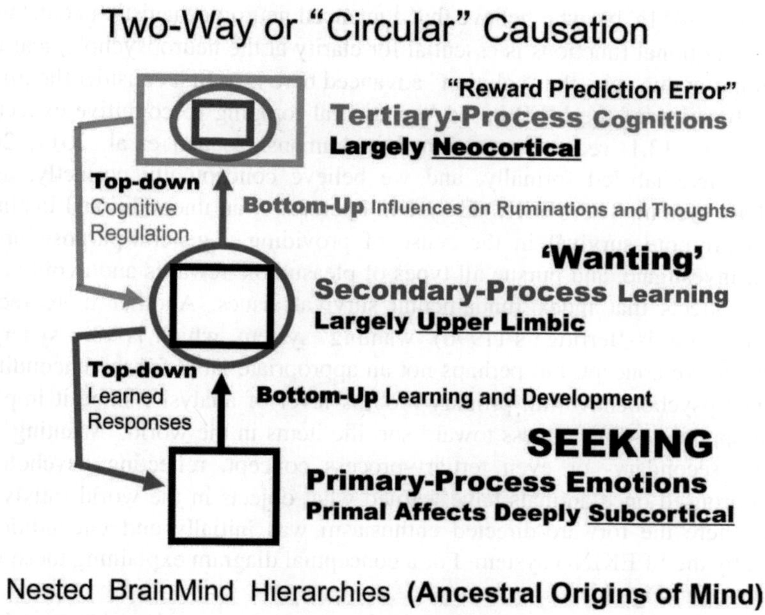

Fig. 13.3 Nested hierarchies of psychobehavioral control within the brain—a synopsis of the hierarchical bottom–up (*right*) and top–down (*left*) circular control that reflect evolutionary and developmental control in all primal brain emotional system. In order for higher Mind-Brain functions (such as neuroeconomic decision-making) to mature they initially have to be integrated with the lower affective Brain-Mind functions. Primary processes are depicted as squares (SEEKING level), secondary-process learning as circles (Berridge's "wanting" level of analysis), and tertiary processes (Schultz's "reward prediction level of analysis,") by rectangles. This schematic summarizes how bottom–up evolution of nested-hierarchies can integrate lower brain functions with higher brain functions, so as to eventually exert top–down regulatory control. Bottom–up controls prevail in early-infancy and early childhood development. Top–down control emerges as individuals approach adulthood

least three global progressions which, if not kept semantically clear, will lead to abundant communicative confusions. Indeed, as soon as one understands the conceptual problems with the classic hand-me-down concept of "the brain reward system" one is hard put not to consider alternative terminologies that can be deployed without psychological confusion, which respect not only behavioral brain functions but tightly related psychological phenomenologies as well. Although our goal is to discuss how this MFB-based brain network participates in human decision-making, especially in the economic realm, we must first deal with persistent conceptual issues that need to be addressed when we study of evolutionarily layered brain-mind processes.

For didactic convenience, we divide the brain and mind into three evolutionary layers of organization, as has been common in the history of neuroscience; for example Paul MacLean's (1990) "Triune Brain" view, commonly detested by behavioristic reductionists, but defended by those interested in the mind (Panksepp

1998, 2002, 2011a, b), who believe that a nuanced neuromechanistic understanding of brain emotional functions is essential for clarity at the neuropsychological level. Continuing in this vein, the "solution" advanced here is to first consider the primary process function of the MFB-based "behavioral-foraging to cognitive-expectancy system" (Fig. 13.1, recently envisioned in humans Coenen et al. 2011, 2012), which is here labeled formally, and we believe conceptually correctly, as the SEEKING system. The SEEKING system operates at an unconditional instinctual level to promote survival in the sense of providing a general-purpose urge to explore, investigate, and pursue all types of pleasurable rewards and avoid various negative affects that index fundamental survival issues. Another more recently suggested view is Berridge's (1996) 'wanting' system, which is very synergistic with the above concept, but perhaps not an appropriate label for the unconditional emotional-psychobehavioral, primary-process level of analysis, since it implies a learned appetitive-directedness toward specific items in the world. 'Wanting' is an excellent secondary- or even tertiary-process concept, reflecting psychological process aroused once animals have learned what objects in the world satisfy their needs, where the forward-directed enthusiasm was initially and unconditionally aroused by the SEEKING system. For a conceptual diagram explaining these levels of control, see Fig. 13.3.

Of course in neuroeconomics, where most consider human higher decision-making processes, investigators typically concern themselves with tertiary-process functions of the brain, a topic that has been superbly taken up by Rolls (2014) in his recent monograph on "Emotions and Decision-Making." The great interest in cognitive neuroscience, is to focus on such higher brain functions through human brain imaging, and clearly such higher order tertiary processes, cannot be easily studied in animals. Still, excellent neurophysiologists such as Wolfram Schulz, who pursued single-unit studies of ventral tegmental dopamine neurons—one of the headwaters of the SEEKING system (without acknowledging such a system)—found how robustly DA neurons responded to the anticipation of cue-predicted rewards, and how they diminished in activity when rewards were withheld, namely, what would traditionally be called a "frustration response." After being educated in psychologically relevant brain functions by an esteemed learning-oriented experimental psychologist, Tony Dickinson of Cambridge University, Schulz conceptualized this *frustration*-disappointment response of the ascending DA-pathways in computation terms, namely as a "reward prediction error" signal. That concept rapidly became a meme among computationally oriented scholars interested in higher order decision-making processes, who chose not to consider affective shifts in decision-making. In any event, those higher processes, highly dependent on what should be considered secondary processes (i.e., learning and memory) arise from diverse prior life experiences, which are constituted of various unconditional "rewarding" and "punishing" states of the nervous system (i.e., confrontation with various world events that generated primal internal "values", or affective states). The memory of those experiences, and related world outcomes prompt the SEEKING system to continue to pursue specific paths, because there are memories of past pleasurable or satisfying experiences that have already guided the appetitive learning of behavior patterns. Tangentially, from a practical human sociological point of view, this way of thinking supports our recommendation that teachers need to keep linking new learning to prior knowledge in ways that sustain *enthusiasm,* the key affect of the SEEKING system. Not only does the emerging cognitive sophistication reflect mental maturation, but it highlights how the brain comes to be an enthusiastic decision-making machine in the marketplace (aka, neuroeconomics). Thus, let us briefly consider the mechanisms of decision-making.

13.1 Economic and Other Decision-Making

Decision-making, though complex and varied, maintains a general input choice-feedback structure (Ernst and Paulus 2005). *Input* refers to the gathering and assessment of relevant decision information that assists in the formation of preferences. *Choice* refers to the behavior carried out and the anticipation of consequences; and *feedback* is the experience and appraisal of the consequences of the behavior. Cold (reflective) and hot (intuitive) processes differentially contribute to choice during each decision phase. Cold processes are typically slow, conscious, and rule governed, while hot processes are fast, automatic and associative. A distinction between hot and cold processes in judgment and decision making has been made by many scholars (e.g., Kahneman 2002; Slovic et al. 2005; Stanovich and West 2000), and there are different hypotheses as to how hot and cold influences interact (for a review see Evans and Stanovich 2013). For the purposes of this chapter we will simply assume that both hot and cold processes play important and neccessary roles in decision making, and avoid specific claims about how processes interact. During the input stage, emotionally salient information forms unreflective, but affectively driven choice preferences (hot), which with the emergence of higher cognitive episodic and prospective memories, can promote choice options that are consciously reasoned through (cold). Thus, during the choice stage there are bound to also be phenomenally conscious feelings of anticipation/motivation (hot), which can lead to cognitively driven choices (i.e., cold decisions). During the feedback stage, hot and cold processing are both used to integrate new information into existing knowledge/expectation structures. Historically, hot processes have been characterized as an impediment to rational choice behavior. The literature on judgement and decision making has primarily focused on hot processing as the source of decision bias (e.g., Kahneman and Frederick 2007), and there is substantial evidence that reliance on hot, intuitive influences can lead to sub-optimal choice. The view that emotion is an impediment to decision-making has also been perpetuated in popular culture: in Star Trek, for example, the Vulcan species is portrayed as logically superior to humans because of their ability to control and minimize emotional influences. However, modern neuroimaging studies provide strong evidence that affective processing in emotional integration (orbitofrontal cortex, amygdala), bodily arousal (anterior cingulate cortex), and anticipation/motivation substrates (ventral striatum—nucleus accumbens) can lead to more efficient and effective navigation of the decision environment (Levine 2009; Minati et al. 2012; Platt and Huettel 2008). Despite this evidence, few neural models of decision-making exist that incorporate affective substrates (Levine 2009), and little effort has been made to integrate them with existing animal models of basic emotional systems.

Our goal here is to frame our existing neuroscientific knowledge of decision-making within primary process emotion research, specifically the SEEKING system (Panksepp 1998; Panksepp and Biven 2012). To do so, we will first provide some more detailed background on the SEEKING system, focusing on the role of the nucleus accumbens since this substrate is important in the decision-making process. Then, building on this information, we will discuss the converging lines of evidence that support the role of the SEEKING system in each stage (input, choice, feedback) of decision-making: namely, enhanced activation of the nucleus accumbens to salient stimuli during input, delineation in activation between motivation and consumption during choice, and activation resulting in incentive learning during feedback. Finally, we will discuss the interaction of SEEKING and decision-making in the pathology of addiction.

13.2 The SEEKING System

The SEEKING system, also known as the brain reward system, was originally discovered by Olds and Milner (1954). In their seminal study, Olds and Milner (1954) found that rats would continuously press a lever that applied electrical stimulation to the septal area of the brain and the nearby nucleus accumbens. Since the stimulation was all that was needed to reinforce behavior, and reinforcement was equivalent to when natural rewards (e.g., food) were used, they concluded that these brain areas must process reward. An explosion of follow-up research has supported the role of these brain areas, especially the nucleus accumbens, in reward-related functions and dopamine appears to be the primary mediator of a distinct "rewarding" state in the brain. Specifically, researchers have found that animals will repeatedly self-administer dopamine agonists into the nucleus accumbens (Carlezon et al. 1995; Hoebel et al. 1983; Ikemoto et al. 1997; Phillips et al. 1994) and exhibit place preferences for environments where dopamine has been administered into the nucleus accumbens (Carr and White 1983, 1986; White et al. 1991). However, more recent research indicates that the dopamine-mediated activation of the nucleus accumbens is only instrumental in the anticipation/approach phase of reward; that is, it is necessary for the *seeking* of reward, but not the consumption (Ikemoto and Panksepp 1996, 1999). This conclusion is supported by lesion studies which have been unable to produce consistent deficits in reward consumption via nucleus accumbens damage (Liu et al. 1998; Sokolowski and Salamone 1998), and by participation of dopamine release within the nucleus accumbens during aversive (nonrewarding or punishing) experiences (Salamone 1994; Salamone et al. 1997). If the nucleus accumbens was involved in reward consumption we would expect consistent, large deficits in reward consumption with lesioning, and zero activation during punishment. Thus, the label "brain reward system" seems to be an inaccurate descriptor of the emotional process mediated by nucleus accumbens dopamine (Salamone et al. 2005).

Because of such problems and to better reflect the anticipatory nature of this process, Panksepp (1982, 1998) relabeled this foraging substrate the SEEKING system. Simply put, the SEEKING system motivates us to interact with our environment and to harvest the resources needed for survival (Panksepp and Biven 2012). Since the system is purely emotional/motivational, it does not change as a function of the behavioral goal; that is, activation of the system motivates organisms to engage with their surroundings, while not being tied to a specific reward/punishment or sensory modality, at least not until patterns of learning have emerged. In other words, the unconditional neural activation of this system has an enormous number of possible goal-directed manifestations after learning, including diverse economic decision-making processes in humans (Knutson and Greer 2008).

Anatomically, the hub of the SEEKING system is in the ventral tegmental area, and from there it projects to three major areas: the MFB and lateral hypothalamus, the nucleus accumbens, and the medial prefrontal cortex (Panksepp and Biven 2012). Within the nucleus accumbens, there are at least three distinct SEEKING manifestations: (i) flexible foraging responses which, with learning, lead to (ii) various unconscious habit responses which help program (iii) higher frontal cortical declarative-perceptual decision-making abilities. The declarative-perception system detects salient stimuli and can consciously contrast them with previously experienced stimulus-event relationships via projections from the nucleus accumbens to frontal cortical areas. The flexible SEEKING response systems (along with secondary and tertiary elaborations) is involved in unconditioned as well as eventually anticipatory/investigatory behaviors, which can be used as information for declarative-perceptual system, providing overall coherence for incentive learning. Thus, stimuli associated with previously life-sustaining or life-impairing events will result in higher nucleus accumbens activation leading to more energized approach or avoidance responses that are included in routine individual economic decision-making, but also establishing societal infectiousness within available marketplaces, including runs on stocks. Finally, the habit response system allows for the development and maintenance of procedural performance (i.e., stimulus-response learning); however, once habits/conditioning have been developed, nucleus accumbens activation may be no longer necessary for behavior expression, with evidence from animals indicating that behavioral control often shifts to dorsal-striatal (caudate-putamen) habitual responding. So, anticipation brought about by SEEKING guides our immediate approach/avoidance behaviors (flexible foraging responses), and creates/updates stimulus-event associations (declarative-perception) to facilitate learning until behavior often become automatic-procedural (habit response).

13.3 The Role of the SEEKING System in Economic and Related Decision-Making

Before detailing the evidence supporting the role of SEEKING in higher decision-making, it may be useful to briefly reiterate the difference between primary, secondary, and tertiary processes: Basic emotional systems, like SEEKING, are known as primary processes because they are evolutionarily older (subcortical) unconditional (i.e., "instinctual") structures that are homologous in animals (Panksepp 2010). Primary processes are expressed as pure feelings (entailing no necessary higher order analysis, but gradually, with learning, promoting that). For example, SEEKING activation in non-human animals is typified by apparently "purposeful" (i.e., coherent, intrinsically well organized) movements characterized by intense exploration of the environment, while deep-brain stimulation activation in humans can lead to feelings of eager, initially goalless enthusiasms and anticipation (Panksepp and Biven 2012) that can be sufficiently robust to lead to an antidepressive tonic elevation of mood, which is accompanied by hopes for the future (Panksepp et al. 2014; Schlaepfer et al. 2013; Wright and Panksepp 2012). Secondary processes are those that arise from simple (conditioned) learning and memory processes, which may be largely unconscious (Solms and Panksepp 2012). While some stimuli are inherently rewarding or punishing to animals, most stimulus-response relationships are learned via subcortical structures (Panksepp 2010). Finally, tertiary processes, like decision-making, are unique to animals possessing developed cortical structures (Panksepp and Biven 2012; Rolls 2014).

Of course, these levels of control are not independent, but highly interactive (Fig. 13.2). Primary processes can have a role within tertiary processes; for example, SEEKING system dopamine pathways only ascend to frontal cortical regions in "lower" mammals, but in "higher" mammals the pathways also ascend to sensory-perceptual cortices (Panksepp and Biven 2012). Thus, SEEKING has a role in higher order (tertiary) processes not so explicitly evident in "lower" mammals. Likewise, secondary processes usually have a role within tertiary processes since learning and memory are essential for most higher order cognitive functions. In decision-making (tertiary), learning (secondary) and SEEKING (primary) are necessary process components at each stage of the development of complex psychobehavioral processes.

Input stage. Compared to the other stages of decision-making, the role of SEEKING in the input stage of economic decision-making may seem to be relatively minor in mature organisms; the declarative-perception subsystem detects salient stimuli which shape cognitive preferences. Abundant evidence for such top–down processing comes from fMRI studies in humans (Hamman et al. 2004; Knutson et al. 2001a, b; Knutson et al. 2005; Zink et al. 2004). However, bottom–up regulation is still present, especially in the presence of unconditionally compelling stimuli. For instance, Hamman et al. (2004) found that nucleus accumbens activation was maximized in men and women when anticipating viewing sexual (salient) versus nonsexual interactions. Zink et al. (2004) extended this finding to monetary rewards showing that anticipation of salient (response-dependent) options elicited higher nucleus accumbens activation than anticipation of minimally salient (response-independent) options. The salience manipulation was confirmed via self-report and skin conductance response measurement: response-dependent options were reported to be more arousing and pleasurable, and elicited higher skin conductance responses than response-independent options. Similarly, Knutson et al. (2001a, b, 2005) found that activation of the nucleus accumbens during a monetary incentive delay task increased as a function of anticipation of reward magnitude. Activation was highest when anticipating large-magnitude gains and losses and was independent of option probability (Knutson et al. 2005). The declarative-perception SEEKING subsystem connects salient stimuli with previous experiences via neural projections to frontal cortices; thus, we would expect accompanying activation of these structures during stimuli detection. Of the aforementioned studies, only Zink et al. (2004) and Knutson et al. (2005) reported on the activation of cortical structures. Both found significant complementary activation of the orbitofrontal, dorsolateral prefrontal, and anterior cingulate cortices during detection of salient stimuli. These areas will be important to our later discussion of incentive learning during the feedback stage of decision-making.

Choice stage. What human brain imaging suggests is that the SEEKING system continues to play a substantial role even during the cognitively mediated choice stage of decision-making. Since the flexible SEEKING response subsystem initiates primary-process appetitive motivation/anticipation, it sustains its role even as higher order cognitive systems come into play. A large body of research supports differential neural activation of anticipation, and consumption during decision-making; with anticipation being primarily mediated by nucleus accumbens activity (Bjork et al. 2004; Knutson et al. 2001a, b; Schultz et al. 1993). In a series of human fMRI studies, Knutson et al. (2001a, b) demonstrated dissociations in neural activation during anticipation and consumption using a monetary incentive delay task; anticipation was correlated with nucleus accumbens activity while "consumption" was correlated with ventromedial frontal cortex activity. Using a similar procedure on both adolescents and adults, Bjork et al. (2004) also found activation of the nucleus accumbens during anticipation of monetary choice, regardless of age; although consumption was associated with medial prefrontal cortex activation. These results are consistent with electrophysiological findings in primates. Using monkeys trained on a delayed stimulus-response task, Schultz et al. (1993) found that dopamine projections from the ventral tegmental area to the nucleus accumbens were selectively activated during stimulus (juice) anticipation. Note that this is the exact anatomical projection of the SEEKING system described earlier, providing strong evidence for its role in decision-making.

Further evidence is provided by research detailing SEEKING during choice across different reward modalities. Since the SEEKING system is not modality specific, nucleus accumbens activation should be present in all reward types. Research using monetary rewards (Bjork et al. 2004; Knutson et al. 2001a, b) has already been discussed in detail, but utilization of other abstract rewards has produced analogous activation (Aharon et al. 2001; Azim et al. 2005). Nucleus accumbens activation was produced by Aharon et al. (2001) in anticipation of attractive faces, and by Azim et al. (2005) in anticipation of humorous stimuli. Studies using primary rewards in decision-making (as in Schultz et al. 1993) have also reported anticipatory nucleus accumbens activation (Berns et al. 2001; Gottfried et al. 2003; McClure et al. 2003; O'Doherty et al. 2002, 2004). Nucleus accumbens activation in humans was produced by Gottfried et al. (2003) in anticipation of pleasant olfaction, and by Berns et al. (2001), McCluree et al. (2003), and O'Doherty et al. (2002, 2004) in anticipation of juice. O'Doherty et al. (2004) also found that activation was produced regardless of task difficulty; that is, activation was the same during anticipation of passive and active choice. This result contrasts research by Salamone et al. (2005) and Salamone et al. (2007) which provided evidence for dopamine-mediated nucleus accumbens activation in rats during effortful (active) choice, but not passive choice. Conflict between these results may be a product of anatomical SEEKING differences between "lower" and "higher" mammals, as previously discussed; however, more research is needed to confirm the role of active versus passive choice on SEEKING anticipation.

There is also conflicting evidence for valence dissociation in anticipation of choice (Knutson and Cooper 2005). Since the SEEKING system is goalless, nucleus accumbens activation should be present in reward approach and punishment avoidance. Yet, while some research indicates no difference in neural activation between anticipation of reward and punishment (Rogers et al. 2004), other research suggests that the nucleus accumbens is only activated in anticipation of reward (Knutson et al. 2001a). Seymour et al. (2007) proposed a unifying interpretation of these conflicting results; namely, trial-by-trial fMRI analysis is not sophisticated enough to detect nucleus accumbens activation during anticipation of punishment. Using a temporal difference learning model, Seymour et al. (2007) found that anticipation of both reward and punishment are processed in the ventral striatum (housing the nucleus accumbens), but reward is processed in the anterior portion and punishment in the posterior portion. This anterior–posterior gradient interpretation supports recent findings in rat research. A series of studies by Reynolds and Berridge (2001, 2002, 2003) demonstrate that artificial excitation of anterior and posterior portions of the ventral striatum produces appetitive (i.e., feeding) and aversive (i.e., paw treading, burying) anticipatory behaviors, respectively. Presumably such differences should be evident also in human economic choices, reflecting wins and losses in the marketplace.

Finally, with regard to anticipatory choice, there is some debate over whether nucleus accumbens activation can be alternatively explained by motor demands. As previously mentioned, Zink et al. (2004) found greater activation of the nucleus accumbens during motor response-dependent (salient) input versus motor response-independent (passive) input; a finding that could be interpreted as the result of motor demands during anticipation of response-dependent choices (Knutson and Cooper 2005). However, other research shows that nucleus accumbens activation is still maintained in anticipatory choice during low-demand motor tasks (Ernst et al. 2004; Ramnani et al. 2004). Using a passive monetary task, Ramnari et al. (2004) found that anticipation of reward resulted in nucleus accumbens activation while passive experience of reward activated the medial prefrontal cortex. Ernst et al. (2004) found similar results using a risky choice wheel of fortune game; choice (initial wheel spin) activated the anterior cingulate cortex, parietal cortex, and supplementary motor area, and anticipation of reward (wheel spinning) activated the nucleus accumbens. Thus, there is fairly strong evidence that activation of the nucleus accumbens during the choice stage is due to SEEKING anticipation and not task-related motor demands.

Feedback stage. Information received during the feedback stage of decision-making activates the SEEKING system to modulate incentive learning. The flexible-response system is active in early learning, with salient feedback stimuli leading to more energized subsequent approach/avoidance responses; then, later in learning, once behavior becomes procedural, the habit response system is activated. This process is hypothesized to occur through a difference in dopamine neuron value coding between expected versus experienced outcomes; in other words, deviations from expectation act as learning signals, with greater deviations promoting greater learning (Montague et al. 1996). This hypothesis is supported by electrophysiological research in primates (Schultz et al. 1993), as well as neuroimaging studies in humans (Berns et al. 2001; McClure et al. 2003; Zink et al. 2004). Schultz et al. (1993) found that firing rates of dopamine neurons within the nucleus accumbens increase in response to unpredicted reward, are maintained in response to predicted reward, and decrease in response to expected reward that does not occur. Similarly, fMRI research shows greater activation of the nucleus accumbens during unpredicted outcome versus predicted outcome (Berns et al. 2001; McClure et al. 2003), with complimentary activation of frontal cortical structures (Zink et al. 2004).

Frontal cortical activation is both expected and necessary during this decision stage as information received via feedback is contrasted with experiences in memory, and learning schemas are updated. In particular, the orbitofrontal and dorsolateral prefrontal cortices appear to be important substrates in incentive learning (O'Doherty et al. 2003; Singer et al. 2004). Using a Prisoner's Dilemma game, Singer et al. (2004) found complimentary activation of the nucleus accumbens and the dorsolateral prefrontal cortex when subject's viewed faces of persons who had previously cooperated; activation further increased if cooperation was intentional versus nonintentional, but disappeared when viewing control faces. Alternatively, using a risky choice task, O'Doherty et al. (2003) found activation of the nucleus accumbens and the orbitofrontal cortex during feedback about monetary rewards; activation in the orbitofrontal cortex was magnified with reward salience. These results provide support for the role of the flexible-response SEEKING subsystem in incentive learning; and, in addition, they align with research indicating that the dorsolateral prefrontal cortex also operates in moral decisions (Greene et al. 2001), and the orbitofrontal cortex integrates reward information with future expectation (Siddiqui et al. 2008), both types of mentation being of obvious importance in economically significant decisions.

Since there is a tradeoff between the flexible and habit response SEEKING subsystems once learning becomes procedural, nucleus accumbens activation should be present early in learning, but decrease or disappear as experience grows.

Preliminary research in rats indicates this pattern (Cardinal et al. 2001; Schoenbaum and Setlow 2003). Specifically, Cardinal et al. (2001) found that lesions in the nucleus accumbens resulted in a preference for small immediate rewards over large future rewards, the potential result of impaired learning; and Schoenbaum and Setlow (2003) found that nucleus accumbens lesions impaired improvement of response latency in a learned discrimination task. At least one study has been conducted exploring this issue in humans: Using a monetary choice task, Tanaka et al. (2004) found that short-term reward prediction (early learning) was correlated with nucleus accumbens activation, while long-term reward prediction (procedural learning) was not. Thus, the proposed switch between flexible and habit response SEEKING subsystems is supported by the limited existing research.

In summary, the SEEKING system is integral in each stage of decision-making. During input, information relevant to choice is gathered, with salient stimuli leading to stronger SEEKING activation and formations of preference. In choice, motivation to engage with the decision environment and the anticipation of consequences, regardless of reward modality or valence, is driven by SEEKING. During feedback, consequences are assessed, with salient (unexpected) outcomes leading to greater SEEKING activation and incentive learning. Thus, SEEKING is part of normative, day-to-day choice experience; it operates continuously whether making financial, consumer, or health decisions. Yet, SEEKING also has significant clinical relevance; in particular, it appears to play an important role in the pathology of addiction.

13.4 Addiction, SEEKING, and Decision-Making

In a sense, individual economic decisions have relationship to addictive behaviors, which may help explain why cattle-type crowd behaviors may influence large scale boom and bust economic patterns. Addiction, either behavioral (e.g., gambling) or substance-linked (e.g., alcohol), is the process in which behavior becomes compulsive and takes precedence over other activities, leading to maladaptive decision-making (Robinson and Berridge 2000). Traditionally, addiction has been explained in terms of a positive/negative reinforcement model analogous to the brain reward hypothesis (Olds and Milner 1954), where the motivation to take drugs (craving) is brought about by either, (1) the positive rewarding effects of the drug experience, or (2) the desire to avoid negative effects of withdrawal (Robinson and Berridge 2000). Yet, as previously discussed, this hypothesis has limited explanatory value, and, with regards to addiction, cannot explain why addicts seek drugs when the amount will not be sufficient for pleasure, crave drugs before and after withdrawal, and continue to use drugs that do not produce strong, aversive withdrawal symptoms (e.g., psychostimulants). Conversely, some researchers posit that addiction occurs because the addictive stimuli are, in and of themselves, positive reinforcers (i.e., stimuli that increase the behaviors they immediately follow) (Jaffe 1992). However, this is hardly an explanation of addiction, just circular reasoning (e.g., addicts take drugs because drugs promote drug-taking).

Although these popular and established explanations of addiction are problematic, the mediating neural mechanism behind the addictive process is quite clear; namely, an increase in dopamine in the nucleus accumbens (Di Chiara 1999; Everitt and Robbins 2005). Given the role of the nucleus accumbens in SEEKING, and the compulsive "wanting" experienced by addicts ("a hallmark principle of addiction"; Everitt and Robbins 2005), it is reasonable to hypothesize that addiction is mediated by SEEKING. If this is the case, there should be dissociation between motivation to engage in the addictive behavior and consumption of the addictive stimuli, with the former being accompanied by nucleus accumbens activity. Evidence for such dissociation, specific to addiction, is provided by non-human animal research (Di Ciano and Everitt 2004; Hutcheson et al. 2001; Ito et al. 2004; Park et al. 2002). Hutcheson et al. (2001) found that lesioning the nucleus accumbens in rat's conditioned to self-administer heroin, impaired their heroin-SEEKING behavior. Ito et al. (2004) replicated these results using cocaine. Also using cocaine, Di Ciano and Everitt (2004) and Park et al. (2002) found that SEEKING was impaired by blocking dopamine receptors in the nucleus accumbens of rats conditioned to self-administer the drug. Collectively, these results indicate a role for nucleus accumbens-mediated SEEKING in the motivational aspects of addiction, and have implications for its treatment. If wanting of the addictive stimuli can be reduced via nucleus accumbens antagonism, then the cycle of addiction is broken. However, selectively impairing SEEKING for stimuli, and not creating a global deficit, may prove difficult and remains a challenge to be addressed.

The SEEKING system also appears to be directly involved in the pathology of addiction. Research has revealed that addiction produces reliable neuroadaptations to SEEKING structures such as the nucleus accumbens (Letchworth et al. 2001; Nader et al. 2002). Examining dopamine binding in monkeys conditioned to self-administer cocaine, Letchworth et al. (2001) found that extended exposure to cocaine produced increased binding in the nucleus accumbens. Using a similar procedure, Nader et al. (2002) found that dopamine 1 receptors increased in the nucleus accumbens after prolonged cocaine exposure, but decreased dopamine 2 receptor density. Thus, there appears to be a neural sensitization of the nucleus accumbens, mediated by dopamine 1 receptors, during addiction. In their incentive-sensitization theory of addiction, Berridge and Robinson (1998) propose that this sensitization is responsible for the compulsive wanting behavior in addiction. Sensitization pushes SEEKING into over-drive which is behaviorally displayed in addiction as focusing on the addictive stimulus-response relationship. For example, crack cocaine addicts who have run out of drugs exhibit a behavioral phenomenon known as "chasing ghosts" in which they compulsively search for and closely examine anything resembling a particle of the drug, despite reporting that they know it is a useless behavior (Berridge and Robinson 1998). This process is presumably reliant on continuing arousals from the SEEKING system, as it pulls information from previous associations from addictive behavior patterns to inform continued attempts at consumption. This can result in phenomena like place preference, where cravings for the addictive stimuli increase in relation to their previous contextual, spatial, or temporal associations (Robinson and Berridge 2000). In sum, the pathology of addiction may be dependent on the sensitization of the

SEEKING system, which produces compulsive wanting behavior and contributes to the disease's continuing vicious cycle.

The adaptation of the nucleus accumbens—a substrate essential to decision-making—during addiction results in some decision-making deficits in addicts. Most research on this topic has focused around performance on the Iowa Gambling Task (IGT), developed by Bechara et al. (1994). In this task, participants make choices between four decks over multiple trials; two decks are "good" leading to better outcomes over time, and two decks are "bad." The goal of the task is to learn deck properties via experience and maximize hypothetical winnings. Bechara and colleagues (2001, 2002a,b) found that addict's IGT performance was similar to patients with ventromedial prefrontal cortex (VMPFC) damage, both demonstrate "myopia" for the future by choosing mostly from the bad decks which yield high immediate gains but higher future losses. Interestingly, decision making impairments were not present in all addicts, between 32–37% had performance consistent with controls and this grouping was not predicted by differences in stimulus addiction (Bechara at al. 2001, 2002a). A follow-up study found that, of the addicts with poor IGT performance, a small sub-group (about 36% of sample) demonstrated insensitivity to future consequences consistent with VMPFC damage, but a much larger sub-group (about 64%) demonstrated a hypersensivity to reward (Bechara et al. 2002b).These results were interpreted in-line with classic explanations of addiction, where "myopia" for the future can arise from either primary or secondary processes. Primary processes generate a strong response to the delivery of the reward, which reinforces the selection of choices with high short term gains, but higher long term losses. Secondary processes produce the same pattern of choice by generating thoughts about gaining rewards. We propose that addiction-related decision making differences in the IGT may instead be the result of sensitisization of the SEEKING system. Compulsive wanting behavior produced by SEEKING could drive the selection of high short term reward choices, and since SEEKING is independent of the actual consumption of choice outcomes, addicts would continue to select high short term reward choices despite higher long term losses. However, more research on the impact of addiction on decision making is needed (with additional paradigms) to verify the role of SEEKING.

13.5 Conclusion

Our SEEKING system keeps us in a general state of engagement with the world; feelings of anticipation, regardless of the intended goal, drive us toward action. In decision-making, SEEKING motivates us to explore choice options and assists in learning, so choices can improve over time. Thus, without this SEEKING urge, we as decision-makers would be doomed to repeat our same mistakes over and over again. Addiction, which is characterized by a similar repetition of mistakes, may be the product of neural sensitization of the SEEKING system, leading to compulsive wanting behavior; however, more research on the exact interaction between the SEEKING urge and economic decision-making is needed.

From this perspective, it is important to remember that the modern era of neuroeconomics was strongly grounded in the study of animal emotional circuits. Perhaps the very first modern neuroeconomics study was done by Brian Knutson, soon after he departed from a year of postdoctoral work in the primary author's lab back in 1994, to work at a lab doing the first human-brain imaging at the American National Institutes of Health in Bethesda, Maryland. At our lab, he had learned much about the SEEKING system and its role in appetitive eagerness (Knutson et al. 1998, 1999, 2002), which provided lasting insight into the animal foundations of human nature. With remarkable enthusiasm, Brian translated those insights into the first brain imaging studies of neuroeconomic decision-making in humans (Knutson et al. 2000, 2001a, b), which has become a booming field of research, that has wisely sustained its linkages to the fundamental issues that have been best illuminated through animal research… albeit the market-share of discussions in the field still utilize the original hand-me-down and somewhat off-the-mark concept of "The Brain Reward System." All who have been paying attention realize that the brain has many rewarding systems, so neuroeconomics needs a more sophisticated discussion of what is actually happening psychologically in the brain (for abundant perspectives, see Glimcher 2011a, b; Glimcher and Fehr 2013), and the utility of "reward prediction error signals" although still the rage (e.g., Hart et al. 2014), does not well describe what the Mesolimbic Dopamine system is actually doing, psychologically, within the brain. It is essential to return to the actual affective processes that these subcortical neural systems actually engender within the brain and mind (e.g., Hayes et al. 2014; Panksepp 1998; Panksepp and Biven 2012).

So, returning to fundamentals, it is important to realize that the SEEKING system which lies at the very heart of organismic coherence (Teitelbaum and Epstein 1962), and which generates feelings of enthusiasm in both mice and men, guides all survival aspirations, including feelings that are often depleted in clinical depression (Panksepp et al. 2014, Panksepp 2015, 2016), as well as within in the larger societal purview of economic depressions. This does not mean that the boom and bust fluctuations of economic cycles have a critical subcortical basis, but simply that down deep within brain emotional systems, the economics of survival are still coded as affects. Those facets of mind are critical for individual feelings of well-being, and they do influence the flow of traffic in the diverse economic marketplaces invented by humans.

To the best of our knowledge there is nothing in the brain that evolved under the influence of economic pressures that emerged with the utilization of money as a shared standard of economic value. This is important to remember, for that means that the "dismal sciences" of neither macroeconomics nor microeconomics have strict rules within evolved regions of human brains. Thus, it may be worthwhile to return to a satirical remark from Thomas Carlyle: "There are good and bad times, but our mood changes more often than our fortune."

http://www.brainyquote.com/quotes/authors/t/thomas_carlyle.html#VXVSS6cLfzoe1oxl.99

References

Aharon I, Etcoff N, Ariely D et al (2001) Beautiful faces have variable reward value: fMRI and behavioral evidence. Neuron 32:537–551

Azim E, Mobbs D, Jo B et al (2005) Sex differences in brain activation elicited by humor. Proc Natl Acad Sci USA 102:16496–16501

Bechara A, Damásio H (2002a) Decision-making and addiction (Part I): Impaired activation of somatic states in substance-dependent individuals when pondering decision with negative future consequences. Neuropsychologia 40:1675–1689

Bechara A, Damásio AR, Damásio H et al (1994) Insensitivity to future consequences following damage to human prefrontal cortex. Cognition 50:7–15

Bechara A, Dolan S, Denburg N, Hindes A, Anderson SW, Nathan PE (2001) Decision-making deficits, linked to a dysfunctional ventromedial prefrontal cortex, revealed in alcohol and stimulant abusers. Neuropsychologia 39(4):376–389

Bechara A, Dolan S, Hindes, A (2002b) Decision-making and addiction (part II): myopia for the future or hypersensitivity to reward? Neuropsychologia 40:1690–1705

Berns GS, McClure SM, Pagnoni G et al (2001) Predictability modulates human brain response to reward. J Neurosci 21(8):2793–2798

Berridge KC (1996) Food reward: Brain substrates of wanting and liking. Neurosci Biobehav Rev 20:1–26

Berridge KC, Robinson TE (1998) What is the role of dopamine in reward: Hedonic impact, reward learning, or incentive salience? Brain Res Rev 28:309–369

Bjork JM, Knutson B, Fong GW et al (2004) Incentive-elicited brain activation in adolescents: Similarities and differences from young adults. J Neurosci 24(8):1793–1802

Cabanac M (1992) Pleasure: The common currency. J Theor Biol 155:173–200

Cardinal RN, Pennicott DR, Sugathapala CL et al (2001) Impulsive choice induced in rats by lesions of the nucleus accumbens core. Science 292:2499–2501

Carlezon WA, Devine DP, Wise RA (1995) Habit-forming actions of nomifensine in nucleus accumbens. Psychopharmacology 122(2):194–197

Carr GD, White NM (1983) Conditioned place preference from intra-accumbens but not intra-caudate amphetamine injections. Life Sci 33:2551–2557

Carr GD, White NM (1986) Anatomical disassociation of amphetamines rewarding and aversive effects: an intracranial micro-injection study. Psychopharmacology 89(3):340–346

Coenen VA, Schlaepfer TE, Maedler B et al (2011) Cross-species affective functions of the medial forebrain bundle-Implications for the treatment of affective pain and depression in humans. Neurosci Biobehav Rev 35(9):1971–1981

Coenen VA, Panksepp J, Hurwitz TA et al (2012) Human medial forebrain bundle (MFB) and anterior thalamic radiation (ATR): diffusion tensor imaging of two major subcortical pathways that may promote a dynamic balance of opposite affects relevant for understanding depression. J Neuropsychiatry Clin Neurosci 24:223–236

Di Chiara G (1999) Drug addiction as dopamine-dependent associative learning disorder. Eur J Pharmacol 375:13–30

Di Ciano P, Everitt BJ (2004) Direct interactions between the basolateral amygdala and nucleus accumbens core underlie cocaine-seeking behavior by rats. J Neurosci 24:7167–7173

Ernst M, Paulus MP (2005) Neurobiology of decision-making: a selective review from a neurocognitive and clinical perspective. Biol Psychiatry 58(8):597–604

Ernst M, Nelson EE, McClure EB et al (2004) Choice selection and reward anticipation: an fMRI study. Neuropsychologia 42:1585–1597

Evans JBT, Stanovich KE (2013) Dual-process theories of higher cognition: advancing the debate. APS 8(3):223–241

Everitt BJ, Robbins TW (2005) Neural systems of reinforcement for drug addiction: from actions to habits to compulsion. Nat Neurosci 8(11):1481–1489

Glimcher PW (2011a) Understanding dopamine and reinforcement learning: the dopamine reward prediction error hypothesis. Proc Natl Acad Sci USA 108:15647–15654
Glimcher PW (2011b) Foundations of neuroeconomic analysis. Oxford University Press, New York
Glimcher PW, Fehr E (eds) (2013) Neuroeconomics: decision making and the brain. Academic Press, London
Gottfried JA, O'Doherty J, Dolan RJ (2003) Appetitive and aversive olfactory learning in humans studied using event-related functional magnetic resonance imaging. J Neurosci 22(24):10829–10837
Grant S, Contoreggi C, London ED (1999) Drug abusers show impaired performance in a laboratory test of decision-making. Neuropsychologia 38:1180–1187
Greene JD, Sommerville RB, Nystrom LE et al (2001) An fMRI investigation of emotional engagement in moral judgment. Science 293(5537):2105–2108
Haber SN, Knutson B (2009) The reward circuit: linking primate anatomy and human imaging. Neuropsychopharmacology 35:4–26
Hamman SB, Herman RA, Nolan CL et al (2004) Men and women differ in amygdala response to visual sexual stimuli. Nat Neurosci 7(4):411–416
Hart AS, Rutledge RB, Glimcher PW et al (2014) Phasic dopamine release in the rat nucleus accumbens symmetrically encodes a reward prediction error term. J Neurosci 2014 34(3):698–704
Hayes DJ, Duncan NW, Xu J et al (2014) A comparison of neural responses to appetitive and aversive stimuli in humans and other mammals. Neurosci Biobehav Rev 45:350–368
Hoebel BG, Monaco AP, Hernandez EF et al (1983) Self-infusion of amphetamine directly into the brain. Psychopharmacology 81:158–163
Hutcheson DM, Parkinson JA, Robbins TW et al (2001) The effects of nucleus accumbens core and shell lesions on inravenous heroin self-administration and the acquisition of drug-seeking behavior under a second-order schedule of heroin reinforcement. Psychopharmacology 153:464–472
Ikemoto S, Panksepp J (1996) Dissociations between appetitive and consummatory responses by pharmacological manipulations of reward-relevant brain regions. Behav Neurosci 110:331–345
Ikemoto S, Panksepp J (1999) The role of nucleus accumbens dopamine in motivated behavior: a unifying interpretation with special reference to reward-seeking. Brain Res Brain Res Rev 31:6–41
Ikemoto S, Glazier BS, Murphy JM et al (1997) Role of D1 and D2 receptors in the nucleus accumbens in mediating reward. J Neurosci 17:8580–8587
Ito R, Robbins TW, Everitt BJ (2004) Differential control over cocaine-seeking behavior by nucleus accumbens core and shell. Nat Neurosci 7:389–397
Jaffe JH (1992) Current concepts of addiction. In: O'Brien CP, Jaffe JH (eds) Addictive states. Raven Press, New York
Kahneman D (2002) Maps of bounded rationality: a perspective on intuitive judgement and choice [Prize Lecture]. Retrieved from: http://www.nobelprize.org/nobel_prizes/economic-sciences/laureates/2002/kahnemann-lecture.pdf
Kahneman D, Frederick S (2007) Frames and brains: elicitation and control of response tendencies. Trends Cogn Sci 11(2):45–46
Knutson B, Cooper JC (2005) Functional magnetic resonance imaging of reward prediction. Curr Opin Neurol 18:1–7
Knutson B, Greer SM (2008) Anticipatory affect: neural correlates and con- sequences for choice. Philos Trans R Soc Lond B Biol Sci 363:3771–3786
Knutson B, Burgdorf J, Panksepp J (1998) Anticipation of play elicits vocalizations in juvenile rats. J Comp Psychol 112:65–73
Knutson B, Burgdorf J, Panksepp J (1999) High-frequency ultrasonic vocalizations index conditioned pharmacological reward in rats. Physiol Behav 66:639–643
Knutson B, Westdorp A, Kaiser E et al (2000) FMRI visualization of brain activity during a monetary incentive delay task. Neuroimage 12:20–27

Knutson B, Adams CM, Fong GW et al (2001a) Anticipation of increasing monetary reward selectively recruits nucleus accumbens. J Neurosci 21:159–164
Knutson B, Fong GW, Adams CM et al (2001b) Dissociation of reward anticipation and outcome with event-related fMRI. NeuroReport 12(17):3681–3687
Knutson B, Burgdorf J, Panksepp J (2002) Ultrasonic vocalizations as indices of affective states in rat. Psychol Bull 128:961–977
Knutson B, Taylor J, Kauman M et al (2005) Distributed neural representation of expected value. J Neurosci 25:4806–4812
LeGros Clark WE, Beattie J, Riddoch G et al (1938) The hypothalamus, morphological, functional, clinical and surgical aspects. Edinburgh: Oliver and Boyd
Letchworth SR, Nader MA, Smith HR et al (2001) Progression of changes in dopamine transporter binding site density as a result of cocaine self-administration in rhesus monkeys. J Neurosci 21:2799–2807
Levine DS (2009) Brain pathways for cognitive-emotional decision-making in the human animal. Neural Netw 22:286–293
Liu X, Strecker RE, Brener J (1998) Dopamine depletion in nucleus accumbens influences locomotion but not force and timing of operant responding. Pharmacol Biochem Behav 59:737–745
McClure SM, Berns GS, Montague PR (2003) Temporal prediction errors in a passive learning task activate human striatum. Neuron 38:339–346
Minati L, Grisoli M, Seth AK et al (2012) Decision-making under risk: a graph-based network analysis using functional MRI. NeuroImage 60:2191–2205
Montague PR, Dayan P, Sejnowski TJ (1996) A framework for mesencephalic dopamine systems based on predictive Hebbian learning. J Neurosci 16:1936–1947
Nader MA, Daunais JB, Moore T et al (2002) Effects of cocaine self-administration on striatal dopamine systems in rhesus monkeys: Initial and chronic exposure. Neuropsychopharmacology 27:35–46
O'Doherty JP, Deichmann R, Critchley HD et al (2002) Neural responses during anticipation of a primary taste reward. Neuron 33:815–826
O'Doherty J, Critchley H, Deichmann R, Dolan RJ (2003) Dissociating valence of outcome from behavioral control in human orbital and ventral prefrontal cortices. J Neurosci, 23(21):7931–7939
O'Doherty J, Dayan P, Schultz J et al (2004) Dissociable roles of the ventral and dorsal striatum in instrumental conditioning. Science 304:452–454
Olds J, Milner P (1954) Positive reinforcement produced by electrical stimulation of septal area and other regions of rat brain. J Comp Physiol Psychol 47(6):419–427
Panksepp J (1981) Hypothalamic integration of behavior: rewards, punishments, and related psychobiological process. In: Morgane PJ, Panksepp J (Eds.) Handbook of the hypothalamus, vol 3, Part A. Behavioral studies of the hypothalamus. New York: Marcel Dekker, pp 289–487
Panksepp J (1982) Toward a general psychobiological theory of emotions. Behav Brain Sci 5:407–467
Panksepp J (1998) Affective neuroscience: the foundation of human and animal emotions. Oxford University Press, New York
Panksepp J (2002) The MacLean legacy and some modern trends in emotion research. In: Cory GA, Gardner R Jr. (eds) The evolutionary neuroethology of Paul MacLean. Praeger, Westport, CT, pp ix–xxvii
Panksepp J (2010) Affective neuroscience of the emotional BrainMind: evolutionary perspectives and implications for understanding depression. Dialogues Clin Neurosci 12(4):533–545
Panksepp J (2011a) Cross-species affective neuroscience decoding of the primal affective experiences of humans and related animals. PLoS ONE 6(9):e21236
Panksepp J (2011b) The basic emotional circuits of mammalian brains: do animals have affective lives? Neurosci Biobehav Rev 35:1791–1804

Panksepp J (2015) Affective preclinical modeling of psychiatric disorders: taking imbalanced primal emotional feelings of animals seriously in our search for novel antidepressants. Dialogues in Clin Neurosci 17:363–379

Panksepp J (2016) The cross-mammalian neurophenomenology of primal emotional affects: From animal feelings to human therapeutics. J Comp Neurol 524:1624–1635

Panksepp J, Biven L (2012) The archaeology of mind: neuroevolutionary origins of human emotions. W. W. Norton & Company, New York

Panksepp J, Wright JS, Döbrössy MD et al (2014) Affective neuroscience strategies for understanding and treating depressions: from preclinical models to novel therapeutics. J Clinical Psych 2014(2):472–494

Park WK, Bari AA, Jey AR et al (2002) Cocaine administered into the medial prefrontal cortex reinstates cocaine-seeking behavior by increasing AMPA receptor-mediated glutamate transmission in the nucleus accumbens. J Neurosci 22(7):2916–2925

Phillips GD, Robbins TW, Everitt BJ (1994) Bilateral intra-accumbens self-administration of amphetamine: antagonism with intra-accumbens SCH-23390 and sulpiride. Psychopharmacology 114:477–485

Platt ML, Huettel SA (2008) Risky business: The neuroeconomics of decision-making under uncertainty. Nat Neurosci 11:398–403

Ramnani N, Elliot R, Athwal BS et al (2004) Prediction error for free monetary reward in the human prefrontal cortex. Neuroimage 23:777–786

Reynolds SM, Berridge KC (2001) Fear and feeding in the nucleus accumbens shell: rostrocaudal segregation of GABA-elicited defensive behavior versus eating behavior. J Neurosci 21:3261–3270

Reynolds SM, Berridge KC (2002) Positive and negative motivation in nucleus accumbens shell: bivalent rostrocaudal gradients for GABA-elicited eating, taste, and "liking"/"disliking" reactions, place preference/avoidance and fear. J Neurosci 22:7308–7320

Reynolds SM, Berridge KC (2003) Glutamate motivational ensembles in nucleus accumbens: rostrocaudal shell gradients of fear and feeding. Eur J Neurosci 17:2187–2200

Robinson TE, Berridge KC (2000) The psychology and neurobiology of addiction: an incentive-sensitization view. Addiction 95(2):91–117

Rogers RD, Ramnani N, MacKay C et al (2004) Distinct portions of anterior cingulate cortex and medial prefrontal cortex are activated by reward processing in separable phases of decision-making cognition. Biol Psychiatry 55:594–602

Rolls ET (2014) Emotion and decision-making explained. Oxford University Press, Oxford

Salamone JD (1994) The involvement of nucleus accumbens dopamine in appetitive and aversive motivation. Behav Brain Res 61:117–133

Salamone JD, Cousins MS, Snyder BJ (1997) Behavioral functions of nucleus accumbens dopamine: empirical and conceptual problems with the anhedonia hypothesis. Neurosci Biobehav Rev 21:341–359

Salamone JD, Correa M, Mingote SM et al (2005) Beyond the reward hypothesis: alternative functions of nucleus accumbens dopamine. Curr Opin Pharmacol 5:34–41

Salamone JD, Correa M, Farrar A et al (2007) Effort-related functions of nucleus accumbens dopamine and associated forebrain circuits. Psychopharmacology 191:461–482

Schlaepfer TE, Bewernick B, Kayser S et al (2013) Rapid effects of deep brain stimulation for treatment resistant depression. Biol Psychiat 73:1204–1212

Schoenbaum G, Setlow B (2003) Lesions of nucleus accumbens disrupt learning about aversive outcomes. J Neurosci 23(30):9833–9841

Schultz W, Apicella P, Ljungbergb T (1993) Responses of monkey dopamine neurons to reward and conditioned stimuli during successive steps of learning a delayed response task. J Neurosci 13(3):900–913

Seymour B, Daw N, Dayan P et al (2007) Differential encoding of losses and gains in the human striatum. J Neurosci 27:4826–4831

Siddiqui SV, Chatterjee U, Kumar D (2008) Neuropsychology of prefrontal cortex. Indian J Psychiatry 50(3):202–208

Singer T, Kiebel SJ, Winston JS et al (2004) Brain responses to the acquired moral status of faces. Neuron 41:653–662
Slovic P, Peters E, Finucane ML et al (2005) Affect, risk, and decision-making. Health Psychol 24 (4):35–40
Sokolowski JD, Salamone JD (1998) The role of accumbens dopamine in lever pressing and response allocation: Effects of 6-OHDA injected into core and dorsomedial shell. Pharmacol Biochem Behav 59:557–566
Solms M, Panksepp J (2012) The "Id" knows more than the "Ego" admits: Neuropsychoanalytic and primal consciousness perspectives on the interface between affective and cognitive neuroscience. Brain Sci 2:147–175
Stanovich KE, West RF (2000) Individual differences in reasoning: Implications for the rationality debate. Behav Brain Sci 23:645–726
Tanaka SC, Doya K, Okada G et al (2004) Prediction of immediate and future rewards differentially recruits cortico-basal loops. Nat Neurosci 7:887–893
Teitelbaum P, Epstein AN (1962) The lateral hypothalamic syndrome: Recovery of feeding and drinking after lateral hypothalamic lesions. Psychol Rev 69:74–90
White NM, Packard MG, Hiroi N (1991) Place conditioning with dopamine D1 and D2 agonists injected peripherally or into nucleus accumbens. Psychopharmacology 103:271–276
Wright JS, Panksepp J (2012) An evolutionary framework to understand foraging, wanting, and desire: the neuropsychology of the SEEKING System. Neuropsychoanalysis 14:5–39
Zink CF, Pagnoni G, Martin-Skurski ME et al (2004) Human striatal responses to monetary reward depend on saliency. Neuron 42:509–517

Chapter 14
The Psychology and Psychobiology of Simple Decisions: Speeded Choice and Its Neural Correlates

David K. Sewell and Philip L. Smith

Abstract In this chapter, we provide a tutorial review of the class of sequential sampling models of two-choice decision-making. These models, which have been developed in cognitive and mathematical psychology over the last 50 years, provide a detailed quantitative account of performance in simple, speeded choice tasks. The models explain the major findings from a wide variety of behavioral decision tasks, including the relationship between choice probabilities and response time (RT), the speed-accuracy tradeoff, the shapes of RT distributions, and the relative speed of correct and error responses. More recently, electrophysiological recordings from decision-related brain areas in awake behaving monkeys have revealed a correspondence between patterns of neural firing and the statistical processes of evidence accumulation assumed in the psychological models. We discuss the theoretical relationship between the cognitive process of evidence accumulation and neural firing rates and show how neural data can constrain behavioral models. Importantly, constraints from neurophysiological data can be used to test between models that are otherwise difficult to distinguish. The convergence of psychological theory and neurophysiological data suggests that a common theoretical and mathematical framework is sufficient to account for simple decision-making data at neural and behavioral levels of analysis.

14.1 The Psychology and Psychobiology of Simple Decisions: Speeded Choice and Its Neural Correlates

The simplest decision an organism can make involves choosing between two alternatives. The importance of such two-alternative forced choice (2AFC) decisions is reflected in the extent to which they have been studied in psychology and,

D.K. Sewell · P.L. Smith (✉)
The University of Melbourne, Melbourne, VIC, Australia
e-mail: philipls@unimelb.edu.au

D.K. Sewell
e-mail: d.sewell@uq.edu.au

M. Reuter and C. Montag (eds.), *Neuroeconomics*, Studies in Neuroscience, Psychology and Behavioral Economics, DOI 10.1007/978-3-642-35923-1_14

more recently, in behavioral economics. For both psychologists and economists, the fundamental theoretical questions involve *what* is chosen, as a function of the discriminability or desirability of the alternatives, *how long* it takes to make the choice, and what mechanisms determine the choice. A wealth of 2AFC data collected over the last 50 years has led to detailed analyses of the relationship between psychology's two ubiquitous dependent variables: choice probability and response time (RT). Probably the most familiar expression of this relationship is in the *speed-accuracy tradeoff*, which describes people's ability to trade speed against accuracy in decision-making tasks (e.g., Luce 1986; Wickelgren 1977). In psychology, the theoretical understanding of the choice-RT relationship has been guided by the development of the influential class of *sequential sampling models*. According to these models, decisions are made by summing, or accumulating, samples of noisy evidence over time. Statistical variability in the moment-to-moment quality of the evidence is assumed to reflect noise in the cognitive or neural processes that code for different response alternatives. A response is initiated only after sufficient evidence for a decision is obtained. Successful sequential sampling models, such as Ratcliff's (1978), Ratcliff and McKoon (2008) diffusion model, account for behavioral data at the level of choice probabilities and the shapes of correct and error RT distributions. The model's ability to provide very detailed accounts of data suggests that decision-making in simple decision tasks does indeed occur via a process of accumulating evidence to a criterion.

The characterization of decision-making as a process of accumulating evidence over time has attracted the attention of neuroscientists studying the neural correlates of decision-making. Using analogs of 2AFC decision tasks, in which monkeys respond to visually presented stimuli by making saccadic eye movements, neuroscientists have uncovered striking similarities between the firing rates of neurons in lateral intraparietal area (LIP), superior colliculus (SC), and the frontal eye fields (FEF) on the one hand, and the dynamics of evidence accumulation postulated by sequential sampling models on the other. Neurons in these brain areas are part of the oculomotor control circuit that is active when making saccadic eye movements. This invites the theoretical linking proposition (Schall 2004; Teller 1984) that activity in this circuit provides an online reflection of the accumulating evidence state prior to executing an eye movement. Indeed, patterns of neuronal firing rate data in regions involved in perceptual decision-making appear to reflect the temporal integration, or accumulation, of evidence over relatively long time scales (e.g., Shadlen and Newsome 1996; for a review see Gold and Shadlen 2007). The correspondence between psychological theory and neurophysiological data suggests that phenomena at both cognitive and neural levels of analysis may be understood in terms of a common theoretical and mathematical framework (Smith and Ratcliff 2004). The principal benefit of such correspondence is that insights at one level of analysis impose constraints on the other. The development of sequential sampling models from psychology has guided neurophysiological empirical investigations (e.g., Ratcliff et al. 2003, 2007), and neurophysiological data have proved diagnostic in testing between models that are otherwise difficult to distinguish on the basis of behavioral data alone (e.g., Purcell et al. 2010; Ratcliff et al. 2011).

In this chapter, we use models of simple decision-making as a link between psychological and neurophysiological levels of analysis. We focus particularly on the similarities between the time course of neural firing rates preceding an eye movement response and the time course of the process of evidence accumulation estimated from sequential sampling models. Although most of the studies we discuss involved neural data collected from monkeys performing simple perceptual decision tasks, sequential sampling models have also been successfully applied to behavioral data from other kinds of tasks involving higher order cognitive judgments including, but not limited to, recognition memory (e.g., Ratcliff 1978), lexical decision (e.g., Ratcliff et al. 2004), as well as tasks involving more complex value-based decisions (e.g., Busemeyer and Townsend 1993; Roe et al. 2001). For these kinds of "one-shot" tasks, a decision is usually made within a second or so after stimulus presentation. Because performance in these tasks is well described by sequential sampling models (e.g., Ratcliff and Smith 2004), we argue that the relationship between these sequential sampling models and neural processing dynamics likely extends beyond perceptual decision-making to other domains.

The chapter is organized into two parts. In the first part, we discuss the benchmark patterns of empirical results that have been used to evaluate competing sequential sampling models in psychology. We then provide a nontechnical tutorial overview of the main classes of sequential sampling models that have been developed in cognitive and mathematical psychology, and discuss how different models succeed or fail in addressing the benchmark results. Different models reflect differing assumptions about the architecture of the decision process: whether evidence is represented in continuous or discrete quantities, is sampled in continuous or discrete time, and whether there is competitive interaction among decision units. We discuss the empirical and theoretical implications of these distinctions in fairly general, qualitative, terms and omit the technical details. Readers interested in these details are referred to the sources cited herein.

As we discuss below, models that assume that evidence is accumulated by a diffusion process provide a better account of behavioral data than models that assume it is accumulated by other kinds of processes (Ratcliff and Smith 2004). This provides strong support for the class of diffusion models, as opposed to plausible alternatives like accumulator and counter models. However, more recent neurally inspired models have successfully combined elements of diffusion models with decision architectures traditionally used in accumulator and counter models (e.g., Ratcliff et al. 2007; Smith 2000; Usher and McClelland 2001). Throughout our discussion, we focus on Ratcliff's (1978), Ratcliff and McKoon (2008) diffusion model, as it has been tested extensively, and has been shown to provide a good account of decision-making in simple cognitive tasks (Ratcliff and Smith 2004). We then discuss recent theoretical developments showing that the simulated behavior of neural network circuits provides a biologically plausible basis for implementing a diffusion decision process in the brain (e.g., Smith 2010; Wang 2002). These studies provide a theoretical basis for understanding the neural computations that underpin decision processes, and provide an account of why diffusion models have proved successful in accounting for behavioral data.

In the second part of the chapter, we review findings that identify patterns of firing rates in LIP, SC, and FEF neurons as neural correlates of evidence accumulation processes. In a variety of perceptual decision tasks, shortly after stimulus presentation, firing rates in these brain areas exhibit systematic ramping to a threshold (e.g., Churchland et al. 2008; Hanes and Schall 1996; Purcell et al. 2010; Ratcliff et al. 2003, 2007, 2011; Shadlen and Newsome 1996, 2001), consistent with accumulation of evidence to a decision criterion. Much of the work we review here focuses on neural data collected from populations of macaque LIP neurons while the animals completed variants of a random dot motion direction discrimination task developed by Newsome, Shadlen, and colleagues (see Gold and Shadlen 2007, and Huk and Meister 2012, for reviews and discussion). In this task, a monkey is presented with a random dot motion stimulus, where some proportion of the dots move coherently in a single direction. To indicate the perceived direction of motion, the monkey makes a saccadic eye movement to a response target (see Fig. 14.1).

A striking aspect of the response of LIP neurons in tasks like these is that direction-specific firing rates remain elevated even after the stimulus is extinguished (e.g., Gnadt and Andersen 1988; Shadlen and Newsome 1996, 2001). When viewed in terms of the general theoretical distinction between categorical and precategorical stimulus representations, the asymptotic level of firing can be identified with formation of a response-related categorical representation of the stimulus, whereas the ramping of activity toward asymptote can be related to the evolving state of a precategorical representation. This is consistent with the idea that aspects of LIP

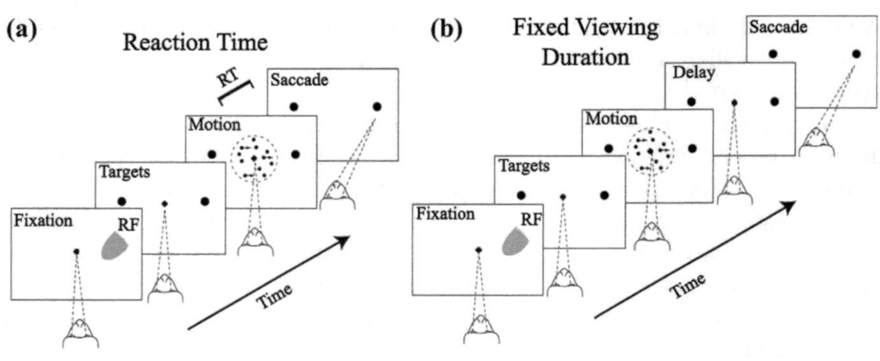

Fig. 14.1 Overview of two versions of the random dot motion direction discrimination task developed by Newsome, Shadlen, and colleagues. In the response time version of the task (*left panel*), a monkey fixates the cross in the center of the screen. Two saccade targets corresponding to different response alternatives are presented. One of the targets is positioned in the receptive field of an LIP neuron. The random dot stimulus is then presented. Once the monkey makes a decision about the direction of coherent motion in the stimulus, it makes an eye movement to one of the saccade targets. The procedure is similar in the fixed viewing duration version of the task (*right panel*) with the exception that there is an experimenter-enforced delay between offset of the random dot stimulus and when the monkey is allowed to make a response. Reproduced from Roitman and Shadlen (2002)

activity reflect memory for the outcome of a decision process. Importantly, this sustained neural activity can also be distinguished from activity related to planning a specific oculomotor response, reinforcing the idea that decision outcomes have explicit—and experimentally identifiable—neural correlates (Bennur and Gold 2011).

Beyond identifying neural correlates of the decision process, several studies have attempted to provide a joint account of behavioral and neural data through mathematical modeling (e.g., Purcell et al. 2010; Ratcliff et al. 2003, 2007, 2011). These studies bring to bear constraints from the neural data themselves, which have proved useful in distinguishing sequential sampling models that mimic each other at a purely behavioral level of analysis. We conclude the chapter with a brief discussion of outstanding questions, and future directions for research.

14.2 Sequential Sampling Models

In this section, we review the three major classes of sequential sampling models that have been studied in psychology. Following Ratcliff and Smith (2004), we summarize the key benchmark effects that have been used to assess the models, along with how well-different classes of models address such data. A complementary evaluation of sequential sampling models was conducted by Bogacz et al. (2006), who examined how closely different model classes approached optimal behavior—defined in terms of maximizing the rate of reward during an experiment—and the conditions under which the different models are formally equivalent. As we do not cover the broad scope of their review here, the interested reader is referred to their work for discussion.

All sequential sampling models of decision-making assume three common elements: (1) an encoded representation of the stimulus or evidence that drives the decision process, (2) a mechanism that integrates, or accumulates, evidence over time, and (3) a stopping rule that determines when to halt evidence accumulation and initiate an appropriate response. Figure 14.2 shows a taxonomy of sequential sampling models. Individual models outlined in the figure are obtained by varying assumptions about these common elements.

Historically, the major distinction among sequential sampling models was in terms of *decision architecture*, which refers to the number of simultaneous evidence totals assumed in the model, or equivalently, whether decisions are based on absolute or relative levels of evidence. Models based on random walk and diffusion processes—the left-hand branch in Fig. 14.2—assume only a single signed evidence total is accumulated over the course of a trial. The single evidence total implements a decision rule that is based on the *relative level of evidence*. For example, in Ratcliff's diffusion model, a decision is made only when the *difference* in the amount of evidence favoring one response over the other exceeds a criterion. By contrast, the decision architecture assumed by accumulator and counter models —the right-hand branch in Fig. 14.2—assumes that evidence for competing

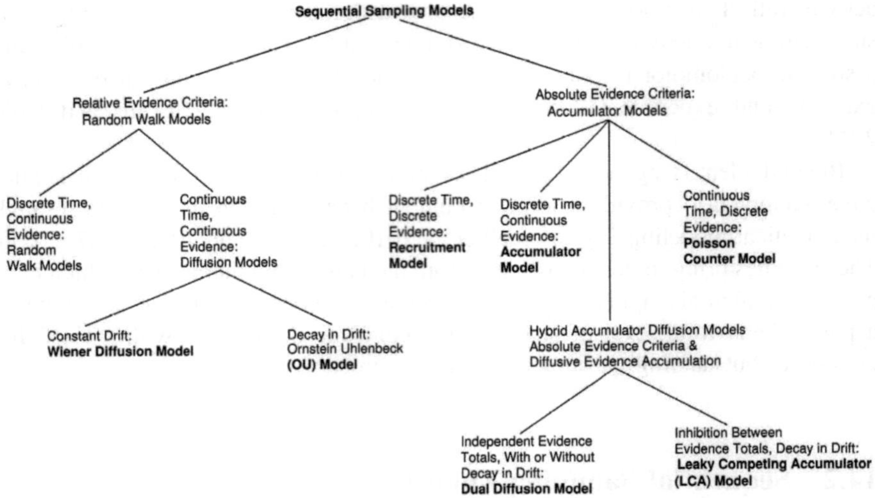

Fig. 14.2 Taxonomy of sequential sampling models. The branch on the left depicts random walk and diffusion models. The branch on the right depicts different accumulator and counter models. The class of models listed in the lower right of the figure are hybrid diffusion-accumulator models that combine aspects of diffusion models within a multi-accumulator architecture

responses is accumulated in separate evidence totals. According to these models, there is a separate accumulator for each response alternative. The value of each accumulator's evidence total reflects the level of evidence supporting that response. The first accumulator to accrue a criterion amount of evidence determines the response, implementing a decision rule based on *absolute levels of evidence*.

The second major distinction among sequential sampling models is whether time and evidence are represented as discrete or continuous quantities. These factors define the statistical properties of the evidence accumulation process. For example, random walk models (e.g., Edwards 1965; Laming 1968; Link and Heath 1975; Stone 1960) assume continuous-valued evidence that is sampled in discrete time steps. As we discuss below, different assumptions about decision architecture and the nature of the evidence accumulation process lead to differences in the abilities of models to predict key benchmark effects like the shapes of RT distributions and the relative ordering of correct and error response times. A third distinction, which applies to more recently developed hybrid diffusion-accumulator models (e.g., Usher and McClelland 2001; Ratcliff et al. 2007, 2011; Smith 2000), is whether accumulators for different response alternatives compete by mutually inhibiting one another. We defer discussion of this distinction to when we discuss neurophysiological data, as the relevant models are difficult to distinguish with just behavioral data (see Ratcliff and Smith 2004). Before discussing the different model classes, we review the key benchmark effects that must be addressed by any successful model of choice-RT.

14.2.1 Behavioral Benchmarks for Evaluating Sequential Sampling Models

In the cognitive and mathematical psychology literature, three benchmark effects have guided the development of models of decision-making. These are the speed-accuracy tradeoff, the right-skewed shape of RT distributions, and differences in the relative speeds of correct and error response times. Ratcliff and Smith (2004) examined how well different models accounted for these benchmark effects.

Speed-Accuracy Tradeoff. A common experimental manipulation in many kinds of tasks requires participants to differentially favor response speed and accuracy, and it is well known that participants are responsive to such instructions (Luce 1986; Wickelgren 1977). When speed is emphasized, responding is faster, but more error-prone. When accuracy is emphasized, accuracy increases, along with response time. Across conditions, the differences in performance can be quite striking. For example, accuracy varies by around 10 %, whereas mean RTs can differ dramatically across speed- and accuracy-emphasis conditions. In some cases, mean RTs in conditions emphasizing accuracy are double those in conditions emphasizing speed (Ratcliff and Rouder 1998). The pattern of changes in accuracy and mean RT across speed and accuracy conditions, when combined with the other empirical benchmark phenomena discussed below, impose powerful constraints on sequential sampling models.

Shape of Response Time Distributions. Accounting for the shapes of RT distributions has proved particularly diagnostic in evaluating sequential sampling models because the shapes of RT distributions impose more constraints than summary measures like mean RT. Typically, the mean and standard deviation of empirical RT distributions increase roughly in proportion to one another (Wagenmakers and Brown 2007). A characteristic property of RT distributions obtained from experiments with human participants is that they are right-skewed: Relative to median RT, the difference between the fastest RTs in different experimental conditions is much smaller than the difference between the slowest RTs. Moreover, as stimulus difficulty increases—for example, by increasing stimulus confusability—increases in mean RT are primarily associated with changes in the tails of the RT distribution. The fastest RTs—those that define the 'leading edge' of the RT distribution—change very little as difficulty increases, whereas the slowest RTs slow down dramatically.

Correct and Error Response Times. Early comparisons of sequential sampling models focused primarily on their ability to predict the empirical orderings of mean RTs for correct and error responses. In experiments where accuracy is stressed or the task is very difficult, error RTs tend to be slower than correct RTs. By contrast when speed is stressed, or task difficulty is low, error RTs tend to be faster than correct RTs (Luce 1986). Moreover, in some cases, there is a crossover pattern, in which error RTs are slower than correct RTs in some conditions, but faster in other conditions of the same experiment (e.g., Ratcliff and Rouder 1998; Ratcliff and Smith 2004; Ratcliff et al. 1999; Smith and Vickers 1988). Accounting for these different RT orderings has been one of the biggest theoretical challenges in developing sequential sampling models. We now introduce the major classes of sequential sampling models investigated by Ratcliff and Smith (2004), noting how well each model class addresses the benchmark phenomena.

14.2.2 Accumulator and Counter Models

According to *accumulator* and *counter models*, separate evidence totals for each response alternative are incremented until one of the totals exceeds a response criterion. In a 2AFC task, there are two separate evidence totals, one for each response alternative. In accumulator and counter models, the decision process can be conceptualized as a race between evidence accumulators to reach a decision threshold. The response is controlled by the accumulator that wins the race, implementing an *absolute decision rule.*

One of the earliest models of choice-RT was LaBerge's (1962) *recruitment model*, which assumed that evidence for different response alternatives is sampled in unit increments at discrete, equally spaced time steps. At each time step, one of the two counters is incremented by one unit. The probability that the counter receiving the increment is the one associated with a correct response is a function of the stimulus discriminability. When discriminability is high, one of the counters increments at a much faster rate than the other; when discriminability is low, the two counters increment at more similar rates. The class of accumulator (e.g., Audley and Pike 1965; Smith and Vickers 1988; Vickers 1970, 1979) and counter models (e.g., LaBerge 1994; Pike 1966; Smith and Van Zandt 2000; Townsend and Ashby 1983) can be viewed as extensions of the recruitment model. According to these models, evidence accumulation for different response alternatives proceeds in parallel.

The class of accumulator models developed by Vickers and colleagues (e.g., Smith and Vickers 1988; Vickers 1970, 1979) retained the recruitment model's assumption of a constant sampling rate in discrete time, but assumed continuous-valued evidence. Figure 14.3 illustrates the mechanics of the Vickers accumulator model. At each time step, evidence is sampled from a continuous distribution, which is usually assumed to be Gaussian, as shown in the figure—although Smith and Vickers (1988) also considered a model with a double-exponential distribution of evidence, and showed its predictions were similar to those of the Gaussian model. The strength of the evidence favoring different alternatives is controlled by the parameters of the evidence distribution. For a Gaussian distribution, the mean is set to μ, which is a function of the discriminability of the stimulus alternatives, and the standard deviation is set to 1. The standard deviation determines within-trial variability in the accumulation process, meaning that different samples will provide different levels of evidence. A sensory referent determines how evidence samples are classified. Samples falling above the referent increment one accumulator; samples falling below the referent increment the other accumulator. In either instance, the increment to the accumulator is the

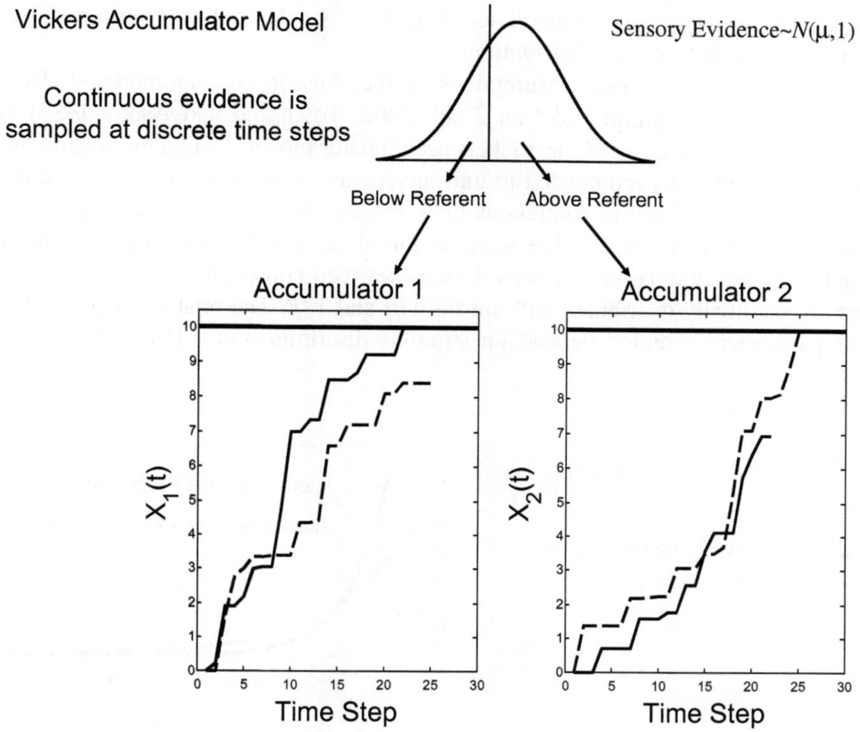

Fig. 14.3 Overview of the Vickers accumulator model. On each trial, continuous-valued evidence is sampled at a fixed rate, in discrete time. At each time step, a sample is drawn from a Gaussian distribution of sensory evidence with mean μ and standard deviation 1. The stimulus alternative the sample provides evidence for depends on the value of the sample relative to a sensory referent, and the corresponding accumulator is incremented by the appropriate amount. The response is controlled by the first accumulator to reach a criterion (set to 10 here). Response time is determined by the number of steps required to reach the criterion. Sample trajectories from two trials are shown. On one trial, the response is determined by Accumulator 1 (*solid line*), on the other trial, Accumulator 2 controls the response (*dashed line*)

absolute value of the difference between the sample and the referent. Once an accumulator has accrued a criterion amount of evidence, the corresponding response is initiated. To allow for trial-to-trial variability in the efficiency of stimulus processing (e.g., stimulus encoding), for the accumulator model and other models of different classes, some model parameters are allowed to vary across trials. Between-trial variability reflects the assumption of noise in the encoding process, which results in variability in the representations of nominally identical stimuli across different trials of an experiment. This assumption is a standard one in psychology and is fundamental to statistical decision frameworks such as signal detection theory (Green and Swets 1966). In the accumulator model, there is trial-to-trial variability in accumulation rates. Across trials, the accumulation rate μ is assumed to be Gaussian distributed with standard deviation σ_μ, making the model *doubly stochastic*. The accumulation rate for a single trial is determined by a random sample from this distribution.

The time and evidence assumptions of the Poisson counter model (LaBerge 1994; Pike 1966; Smith and Van Zandt 2000; Townsend and Ashby 1983) are complementary to those of the Vickers accumulator model. Unlike the accumulator model, evidence is accumulated in unit increments, as assumed in the recruitment model, but is sampled in continuous time. Figure 14.4 illustrates the properties of the Poisson counter model. The name of the model derives from the fact that, for each response alternative, the arrival times between consecutive evidence samples are exponentially distributed with means $1/\alpha$ and $1/\beta$. The relative magnitudes of the parameters α and β depend on stimulus discriminability. For each response

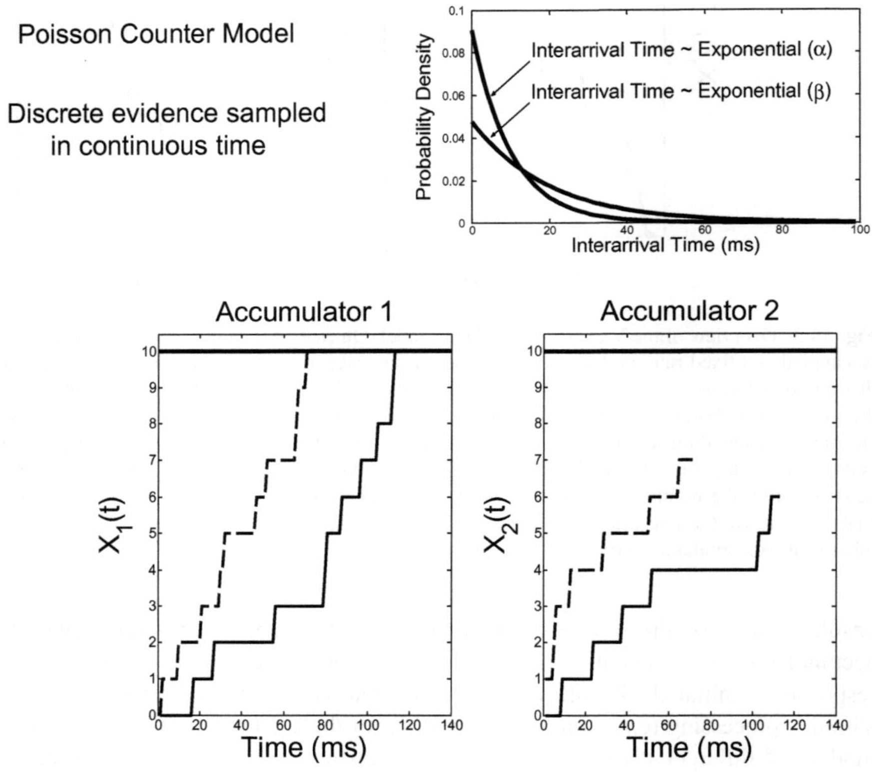

Fig. 14.4 Overview of the Poisson counter model. Discrete valued evidence is accrued in separate accumulators in continuous time. Waiting times between successive evidence samples are exponentially distributed with different mean interarrival times for the two alternatives. In this case, $\alpha > \beta$, resulting in shorter periods between arrival times for Accumulator 1. The response is determined by the first accumulator to accrue a criterion number of evidence samples, set here to 10. Response time is determined by the sum of the interarrival times for the fastest accumulator. Two simulated evidence accumulation trajectories are shown in the figure. In both cases, Accumulator 1 determines the response

alternative, the number of evidence samples accrued in some interval of time is therefore described by a Poisson process with respective intensities α and β. Like the Vickers accumulator model, when a criterion amount of evidence for one alternative has accrued, the corresponding response is initiated.

Speed-Accuracy Tradeoff. Accumulator and counter models, like all sequential sampling models, account for the speed-accuracy tradeoff in the same way, by varying the decision criterion across conditions. Under speed instructions, the criterion is relatively low, requiring only a small amount of evidence before a response is initiated. Because the accumulation process needs less evidence to reach the criterion, responses are faster. However, the low response criterion makes the process susceptible to decisions being initiated by random perturbations in the accumulation process, resulting in a greater proportion of errors. Under accuracy instructions, the decision criterion is relatively high. This means the accumulation process needs more evidence before a response can be initiated, resulting in longer RTs. However, responding becomes more accurate because the increased evidence sampling provides the accumulation process with more opportunity to average out the effects of momentary fluctuations due to noise.

Shapes of RT Distributions. Capturing the shapes of empirical RT distributions has proved especially difficult for counter and accumulator models. For example, the original recruitment model (LaBerge 1962), predicted negatively skewed RT distributions under some circumstances, and RT distributions that became more normally distributed with increasing response criterion. The latter prediction is inconsistent with data from studies emphasizing accuracy, which typically show that increases in RT with changing stimulus discriminability are primarily due to changes in the tails of the distributions rather than a change in central tendency (e.g., Ratcliff and Rouder 1998). In Ratcliff and Smith's (2004) evaluation of sequential sampling models, both the Poisson counter model and the Vickers accumulator model, were found to underestimate the extent of right skew in data, especially at low stimulus discriminabilities. Although the Vickers accumulator model could be made to better capture the shapes of RT distributions by assuming long-tailed (exponential) distributions of response criteria, the Poisson counter model could not. The reason for the difference is because the evidence totals in the Poisson counter model are independent of one another, whereas in the accumulator model there is an implied dependency between the totals because only one accumulator is incremented at any time step.

Correct and Error RTs. An appealing feature of the early accumulator and counter models (e.g., LaBerge 1962; Townsend and Ashby 1983; Vickers 1970) was that they predicted mean error RTs to be slower than correct RTs, as found in many studies with difficult to discriminate stimulus alternatives. However, as noted above, this is only one of the patterns that have been observed in data. Under speed-emphasis in tasks with highly discriminable stimuli, error RTs are typically faster than correct responses. Pike (1973) suggested that trial-to-trial variability in accumulation rate allowed counter and accumulator models to predict faster error RTs, but Ratcliff and Smith (2004) found that the error RTs predicted by both the Poisson counter model and the Vickers accumulator model were only marginally faster than correct RTs, and that both models underpredicted the magnitude of the effect. Ratcliff and Smith (2004) also found that the Poisson counter model was unable to account for crossover patterns in mean RT orderings. For the Vickers accumulator model, assuming between-trial variation in both the criterion and rate of evidence accrual can allow the model to account for crossover effects in some (Smith and Vickers 1988), but not all circumstances (Ratcliff and Smith 2004).

Summary of Accumulator and Counter Models. To summarize, accumulator and counter models only provide a partial account of the existing behavioral data. Although they address the speed-accuracy tradeoff and slow error RTs, they cannot provide an accurate account of the shapes of RT distributions without additional ad hoc assumptions about the distribution of response criteria, nor can they account for fast errors.

14.2.3 Diffusion and Random Walk Models

Unlike the counter and accumulator models discussed above, *random walk* and *diffusion models* assume a decision architecture involving a single signed evidence total, or equivalently, a *relative decision rule* (e.g., Edwards 1965; Laming 1968; Link and Heath 1975; Ratcliff 1978; Stone 1960). That is, evidence for one response alternative is simultaneously evidence against the other. The decision rule in these models can be viewed as initiating a response once the *difference* between two absolute evidence totals exceeds a criterion.

Early random walk models (Edwards 1965; Laming 1968; Stone 1960) were influenced by Wald's (1947) sequential probability ratio test (SPRT). The SPRT is an optimal way of choosing between two competing hypotheses: For any desired level of accuracy, the SPRT takes the minimum expected time to reach a decision. In the context of decision-making, the optimality property of the SPRT addresses costs associated with both time and accuracy, which is what initially recommended it as a psychological model of the decision process. The relationship between decision-making and optimality, as formalized in the SPRT, continues to influence researchers interested in the neuroscience of decision-making (Bogacz 2007; Bogacz et al. 2006; Gold and Shadlen 2001, 2002). The SPRT works by accumulating log-likelihood ratios in discrete time steps. To frame the SPRT in terms of perceptual decision-making, stimulus information is sampled at a constant rate by a mechanism that evaluates which response alternative is favored by each sample. In the SPRT, the evidence provided by each sample is computed by taking the log-likelihood of the ratio of probabilities of obtaining the sample, given the truth of each response alternative.

A limitation of early random walk models based on the SPRT was that they predicted mean RTs for correct and error responses would be equal, which is inconsistent with data. (A notable exception to this was a study by Green et al. (1983), who collected RTs from very highly practiced participants and found no difference between RTs for correct responses and errors, but their findings are atypical.) To account for the relative orderings of mean RTs for correct and error responses, (Link and Heath 1975), in their *relative judgment theory*, proposed a random walk model that omitted computation of likelihood ratios. In their model, noisy evidence samples are accumulated by the decision process directly, in much the same way as in the Vickers accumulator model. The relative judgment model is illustrated in Fig. 14.5. At the first time step, the initial state of evidence accumulation is set to z. The evidence total required to initiate a response is determined by a boundary separation parameter, a, which sets the decision criterion. At each time step, a new sample of evidence is accumulated, which moves the process toward either the upper or lower response boundary. When the process reaches one of the boundaries, the corresponding response is initiated.

Whereas random walk models sample evidence at a constant rate in discrete time, diffusion models sample evidence continuously in time (Ratcliff 1978; Ratcliff and McKoon 2008). In the *Wiener diffusion model* of Ratcliff, the accumulation of evidence can be described mathematically by the stochastic differential equation,

$$\mathrm{d}X_t = v\mathrm{d}t + \sigma \mathrm{d}W_t. \quad (14.1)$$

Equation 14.1 describes the change in accumulated evidence, $\mathrm{d}X_t$, in an infinitesimal time interval, $\mathrm{d}t$. Like the random walk model, evidence accumulation starts at some point, z, situated between response boundaries at a and 0. The mean rate at which evidence accumulates toward a response boundary is defined by the

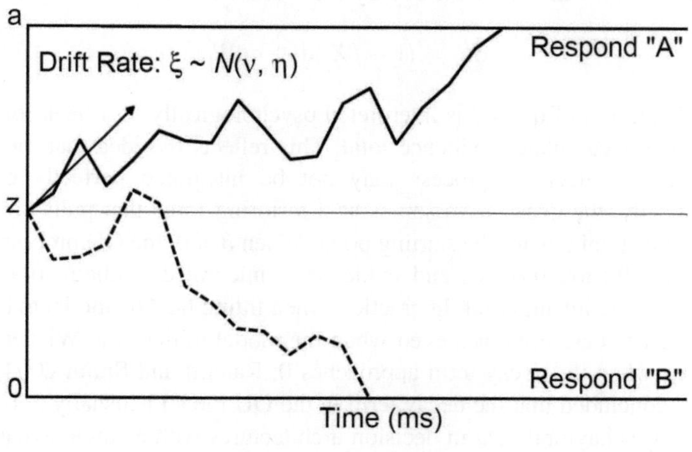

Fig. 14.5 Random walk and Wiener diffusion model. Relative evidence for two alternatives is accumulated from some starting point, z, toward one of two response boundaries at a and 0. The rate of evidence accumulation is defined by the drift rate of the diffusion process, which varies between trials according to a Gaussian distribution with mean ν and standard deviation η. The response is determined by the first boundary the process reaches. Two simulated evidence accumulation trajectories are shown. In one case, the upper boundary is reached, resulting in an '*A*' response (*solid line*). In the other, the lower boundary is reached, resulting in a '*B*' response (*dashed line*)

drift rate, which is normally distributed across trials with mean v and standard deviation η. This means the diffusion model, like the version of the accumulator model described by Smith and Vickers (1988) is doubly stochastic.

In Eq. 14.1, v controls deterministic changes in the quantity of accumulated evidence. In addition to this deterministic change, diffusion models incorporate moment-to-moment stochastic changes in the accumulated evidence. This is modeled as a white noise process, dW_t, which is the formal derivative of the Brownian motion, or Wiener diffusion, process. The variance of the stochastic part of Eq. 14.1 is determined by the *diffusion coefficient* of the process, σ^2. In Ratcliff's model, both the drift rate and diffusion coefficient are constant over the course of a trial, but other authors have proposed models with time dependencies in both drift rate and diffusion coefficient (e.g., Sewell and Smith 2012; Smith et al. 2010; Smith and Ratcliff 2009). The reason for introducing time dependencies is to try to capture the temporal properties of the perceptual and memory processes that encode the evidence in some cognitive tasks.

In principle, a limitation of the Wiener diffusion model is that, unless trial-to-trial variability in drift rate is assumed, the model allows accuracy to grow unboundedly as decision time increases (Smith 1995; Usher and McClelland 2001). This implies that decision-makers can achieve any desired level of accuracy, regardless of the discriminability of the stimulus alternatives, by setting sufficiently high-decision criteria. One way to impose a bound on accuracy without assuming between-trial parameter variability is to add a decay term to the drift. Augmenting the model in this way leads to an Ornstein-Uhlenbeck (OU) diffusion process (Busemeyer and Townsend 1993), which is described by the stochastic differential equation

$$\mathrm{d}X_t = (v - \beta X_t)\mathrm{d}t + \sigma \mathrm{d}W_t. \tag{14.2}$$

The $-\beta X_t$ term in Eq. 14.2 is interpreted psychologically as a 'leak' or temporal decay on the accumulated evidence total. This reflects the idea that the evidence available to the decision process may not be integrated perfectly over time. Mathematically, the decay term serves as a restoring force that pulls the accumulated evidence total toward the starting point. When $\beta = 0$, the OU process becomes the Wiener diffusion process, and evidence is integrated, without loss, over the course of an experimental trial. In practice, when fitting the OU model to behavioral data, the best fits are often achieved when the model mimics the Wiener diffusion model (i.e., when the decay term approaches 0; Ratcliff and Smith 2004). Ratcliff and Smith concluded that the decay term in the OU model is usually not necessary for modeling behavioral data in decision architectures with a single evidence total, but it can lead to improved fits in other decision architectures, like the dual diffusion model discussed subsequently (Smith and Ratcliff 2009).

Speed-Accuracy Tradeoff. As mentioned previously in relation to accumulator and counter models, diffusion and random walk models account for the speed-accuracy tradeoff by varying the position of the decision criterion. In these models, this is accomplished by allowing the boundary separation parameter to differ under speed and accuracy instructions. Under speed instructions, boundary separation is small, which means that, geometrically, the starting point of the decision process is much closer to the two response boundaries. The process does not need to accumulate as much discriminative evidence to reach a response boundary, resulting in reduced RT, but because the process is more susceptible to random moment-to-moment perturbations in the accumulation process, it also leads to a higher proportion of errors. Under accuracy instructions, boundary separation is large, meaning that the process must accrue a relatively large amount of discriminative evidence before reaching a response boundary. This results in longer RTs and higher accuracy, as the effects of moment-to-moment noise are small, relative to the distance between the starting point of the accumulation process and the response boundaries.

Shapes of RT Distributions. In contrast to counter and accumulator models, the shapes of RT distributions are naturally predicted by the geometry of diffusion processes. Both diffusion and random walk models predict right-skewed RT distributions without requiring any additional assumptions. The effects of between-trial noise can be illustrated by progressively decreasing drift rate and projecting accumulation trajectories onto one of the response boundaries (see Fig. 14.6a). When drift rate is incrementally decreased, the expected finishing times of the decision process becomes progressively longer. This geometric constraint on diffusion model predictions was discussed in detail by Ratcliff (2002), who showed that his diffusion model was unable to fit simulated data that were not right-skewed. This showed that the good performance of the model is not a function of its flexibility, but rather, because it predicts a restricted range of RT distribution shapes that happen to be precisely those usually found in empirical data.

More recently, Ditterich (2006) showed RT distribution data collected by Roitman and Shadlen (2002) from monkeys performing the motion direction discrimination task (see Fig. 14.1) were highly symmetrical, and could not be accommodated by Ratcliff's diffusion model. Ditterich's result is interesting as the monkey RT data from that task differ from data from human participants performing the same task, who produce right-skewed RT distributions that are well described by Ratcliff's model (Palmer et al. 2005; Ratcliff and McKoon 2008). However, highly symmetric RT distributions are sometimes found with human participants who are required to respond to an external deadline (e.g., Van Zandt et al. 2000). We discuss the implications of Ditterich's result in more detail below, but note that symmetrical RT distributions are not characteristic of *all* monkey data. For example, the RT distributions from monkeys performing brightness discrimination tasks and gap discrimination tasks are well described by diffusion models (Ratcliff et al. 2003, 2007, 2011).

Correct and Error RTs. Accounting for any differences between correct and error RTs proved difficult for early random walk models, which predicted equal mean RTs (Stone 1960). Indeed, an appealing feature of the early accumulator and counter models (e.g., LaBerge 1962; Townsend and Ashby 1983; Vickers 1970) was that they predicted mean error RTs that were slower than correct RTs. Ratcliff (1978) later showed that introducing trial-to-trial variability in drift rate allowed his diffusion model to predict error RTs that are slower than correct RTs. This prediction follows

(a)

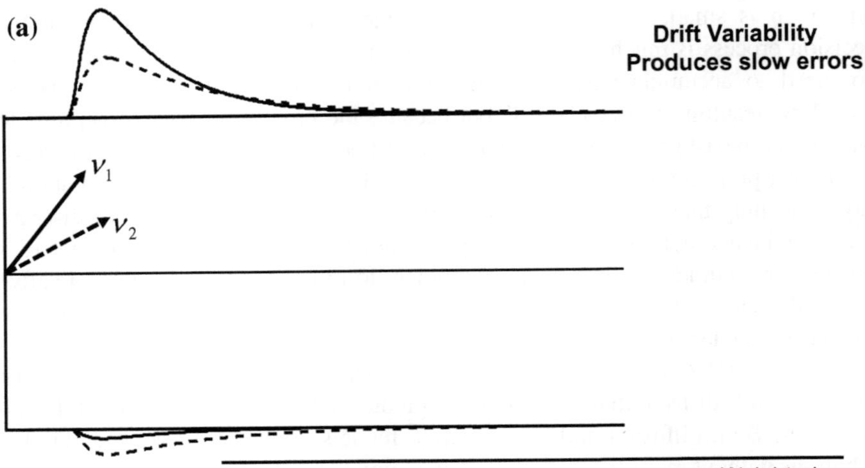

	v_1	v_2	Weighted Mean
Prop (Correct)	.92	.70	.81
MRT (Correct)	251 ms	341 ms	**290 ms**
Prop (Error)	.08	.30	.19
MRT (Error)	251 ms	341 ms	**321 ms**

(b)

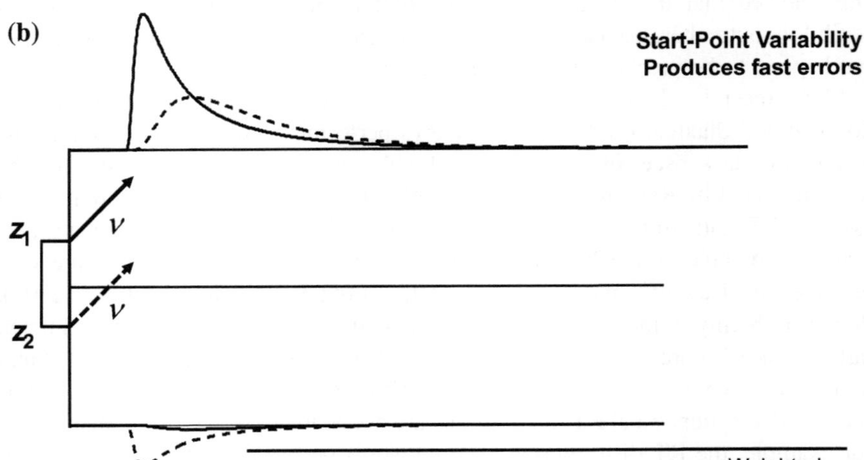

	z_1	z_2	Weighted Mean
Prop (Correct)	.93	.72	.83
MRT (Correct)	206 ms	349 ms	**268 ms**
Prop (Error)	.07	.28	.17
MRT (Error)	349 ms	206 ms	**233 ms**

◄ **Fig. 14.6** Illustration of how variability in drift rate (*top panel*) and start-point (*bottom panel*) determine the relative speed of correct and error responses in Ratcliff's diffusion model. In each panel, the top and bottom response boundaries are associated with correct and error responses, respectively. The *top panel* illustrates how trial-to-trial variability in drift rate allows the diffusion model to account for errors that are slower than correct responses. When drift rate varies across trials, some decision processes will be driven by a drift rate close to 0. For these trials, evidence accumulation will be slow, and more error prone. Because of the larger proportion of error trials driven by low drift rates, combining data from trials with different drift rates results in error responses that are, on average, slower than correct responses. To account for errors that are faster than correct responses, trial-to-trial variability in start-point is needed. When the decision process starts near the correct response boundary, errors will be very slow, but far less frequent. By contrast, when the process starts near the error boundary, the errors will be much faster, and occur with much higher probability. The result is an overall mean error RT that is less than that for correct responses

from the geometry of the diffusion process (see Fig. 14.6a). When there is between-trial variability in drift rate, accuracy, and RT for any individual trial is determined primarily by the drift rate sampled for that trial. When drift rate is low and takes on a value near 0, evidence accumulation is slow, and the probability of an error increases. When drift rate is high, evidence accumulation is fast, and the probability of an error is reduced. Combining data from individual trials with different drift rates results in slower mean error RTs because a greater proportion of the error responses are from trials with lower drift rates, and hence, slower RTs. This is illustrated in the table included in Fig. 14.6a. Diffusion models based on the OU process account for slow errors in the same way as Ratcliff's model.

Trial-to-trial variability in drift rate, however, is only a partial solution to the problem of orderings for correct and error RTs. To account for fast error responses, a different sort of trial-to-trial variability is required. For random walk models, Laming (1968) showed that trial-to-trial variability in starting point sufficed to predict mean error RTs that were faster than correct RTs. This was subsequently confirmed for Ratcliff's diffusion model (Ratcliff et al. 1999; Ratcliff and Rouder 1998; Ratcliff and Smith 2004). With variability in the starting point of the accumulation process, fast errors arise in the diffusion model for the same kinds of geometric reasons as slow errors are predicted with drift rate variability (see Fig. 14.6b). When the decision process starts near the incorrect response boundary, correct responses will be relatively slower and less frequent: The process must travel further to reach the correct boundary, and is more susceptible to random moment-by-moment perturbations leading it to terminate at the error boundary. By contrast, under the same conditions, error responses will be faster and more probable. The increase in the proportion of fast errors when the process starts near the incorrect boundary combined with the decrease in proportion of slow errors when the process starts near the correct boundary results in the overall mean error RT being faster than the mean correct RT (Ratcliff and Rouder 1998). To account for crossover patterns in mean RT orderings, the diffusion model requires both drift variability and start-point variability (Ratcliff et al. 1999; Ratcliff and Rouder 1998; Ratcliff and Smith 2004).

14.2.4 Hybrid Diffusion-Accumulator Models

More recently, models that combine elements of accumulator and diffusion models have been proposed. In these *hybrid models*, there are separate accumulators for different response alternatives, but unlike the Vickers (1970) accumulator model, the accumulation process is modeled as a diffusion process. Decision-making in these models is determined by the first of multiple racing diffusion processes to reach a criterion (Ratcliff et al. 2007, 2011; Ratcliff and Smith 2004; Smith 2000; Usher and McClelland 2001). Although the majority of hybrid models use an absolute decision rule, models with relative decision rules have been investigated by Ratcliff and Smith (2004), and make similar predictions. Hybrid models have been motivated, at least in part, by the pursuit of increased neural plausibility (e.g., Usher and McClelland 2001).

Perhaps the most well-known hybrid diffusion-accumulator model is the *leaky competing accumulator* (LCA) model of Usher and McClelland (2001; see Fig. 14.7). In this model, separate accumulators integrate evidence for different response alternatives. The accumulation process is modeled using a coupled pair of racing OU diffusion processes. In the LCA model, the two accumulators mutually inhibit each other with strength proportional to the amount of evidence in each accumulator. With two accumulators, i and j, the stochastic differential equation that describes the change of evidence in accumulator i in the LCA model is

$$\mathrm{d}X_i = (v_i - \beta X_i - kX_j)\mathrm{d}t + \sigma_i \mathrm{d}W_i, \quad i \neq j. \tag{14.3}$$

Ratcliff and Smith (2004) explored several alternatives to the LCA model that were all based on Eq. 14.3. For example, a racing accumulator model with leakage but no competition is obtained when $k = 0$, and the model reduces to a race between two independent OU diffusion processes. As these variations on the LCA model produced similar fits to behavioral data, Ratcliff and Smith (2004) concluded that the existing behavioral data were insufficient to distinguish different hybrid models.

Speed-Accuracy Tradeoff. Like other sequential sampling models, hybrid diffusion-accumulator models account for the speed-accuracy tradeoff by adjusting decision criteria in response to speed and accuracy instructions.

Shapes of RT Distributions. Because evidence accumulation is modeled as a diffusion process, hybrid models are able to account for the shapes of RT distributions in the same way as standard diffusion models. That is, the geometry of the accumulation process, which is characterized by within-trial noise, naturally produces right-skewed RT distributions without requiring additional assumptions.

Correct and Error RTs. Depending on specific assumptions, hybrid diffusion-accumulator models can account for the relative speeds of correct and error RTs in different ways. Mutual inhibition between accumulators accounts for slow error RTs in Usher and McClelland's (2001) LCA model, whereas variation in starting point—because it is accompanied by inhibition—allows it to predict fast error responses. Indeed, Ratcliff and Smith (2004) found the LCA model to be highly

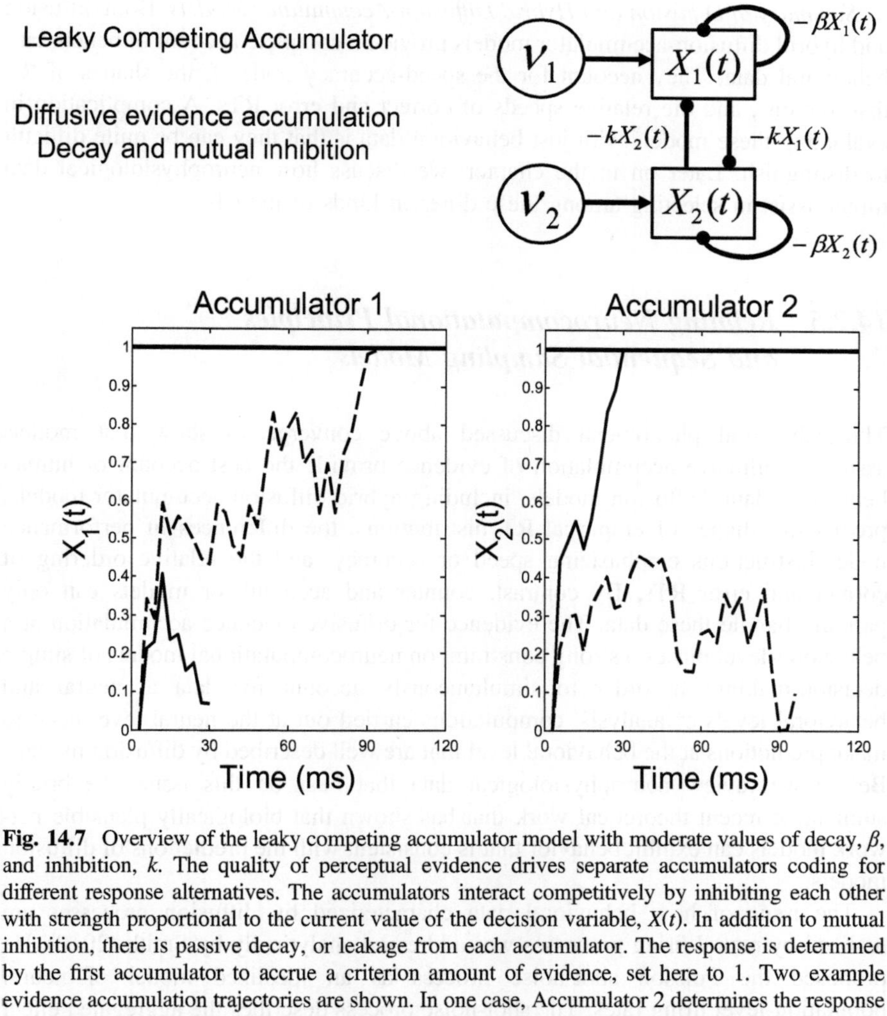

Fig. 14.7 Overview of the leaky competing accumulator model with moderate values of decay, β, and inhibition, k. The quality of perceptual evidence drives separate accumulators coding for different response alternatives. The accumulators interact competitively by inhibiting each other with strength proportional to the current value of the decision variable, $X(t)$. In addition to mutual inhibition, there is passive decay, or leakage from each accumulator. The response is determined by the first accumulator to accrue a criterion amount of evidence, set here to 1. Two example evidence accumulation trajectories are shown. In one case, Accumulator 2 determines the response (*solid lines*), and in the other, Accumulator 1 determines the response (*dashed lines*). The effect of inhibition can be seen in the accumulation trajectories. As the level of accumulated evidence for one response increases, the level of evidence for the other decreases

sensitive to initial start-point conditions. The accumulator that started with the larger amount of initial evidence almost always controlled the response for that trial. When there is no inhibition between accumulators, hybrid diffusion-accumulator models can predict slower error RTs if the accrual rates for each accumulator are constrained to sum to a constant, imposing the condition that the overall rate at which evidence becomes available to the decision process is constant. The mechanism for predicting slow errors is then identical to that in standard accumulator and counter models: The accrual rate for the error response is lower than that for the correct response. To predict fast errors, starting point variability is also required.

Summary of Diffusion and Hybrid Diffusion-Accumulator Models. Both diffusion and hybrid diffusion-accumulator models provide a complete account of benchmark behavioral data. They account for the speed-accuracy tradeoff, the shapes of RT distributions, and the relative speeds of correct and error RTs. A complication in evaluating these models with just behavioral data is that they can be quite difficult to distinguish. Later on in the chapter, we discuss how neurophysiological data might assist in selecting among these different kinds of models.

14.2.5 Relating Neurocomputational Principles and Sequential Sampling Models

The behavioral phenomena discussed above converge to show that models assuming diffusive accumulation of evidence provide the best account of human behavioral data. Diffusion models, including hybrid diffusion-accumulator models, predict the shapes of empirical RT distributions, the differences in performance under instructions emphasizing speed or accuracy, and the relative ordering of correct and error RTs. By contrast, counter and accumulator models can only partially handle these data. The evidence for diffusive evidence accumulation at a behavioral-level places a strong constraint on neurocomputational models of simple decision-making: In order to simultaneously account for data at neural and behavioral levels of analysis, computations carried out at the neural level need to make predictions at the behavioral level that are well described by diffusion models. Before we review neurophysiological data that bear on this issue, we briefly summarize recent theoretical work that has shown that biologically plausible network models can exhibit behavior that is consistent with the predictions of diffusion models.

The issue of how behavioral data characterized by diffusion processes are realized by neural firing processes was addressed analytically by Smith (2010). He proposed the Poisson shot-noise process as an idealized model of neural population-level firing rates. The shot-noise process describes the aggregated effect of a large number of small, independent, time-varying disturbances or perturbations, each of which occurs according to a Poisson process. In the shot-noise model, the Poisson process represents a sequence of action potentials in a bundle of neural fibers and the disturbances represent the flux in postsynaptic potentials in the population of cells on which the action potentials impinge. In Smith's model, pairs of excitatory-inhibitory shot-noise processes code evidence for different response alternatives. The difference between these pairs of processes is integrated until a criterion is reached, after which the corresponding response is initiated. Smith's analysis drew on the weak convergence properties of excitatory-inhibitory shot-noise pairs to an OU velocity process at high firing intensities. When the OU velocity process is integrated over time it becomes a model for behavioral-level evidence accumulation. At the long time scales of behavioral data, the integrated

OU process has similar statistics to the Wiener process in Ratcliff's (1978) diffusion model. Smith showed that simulated data generated by the shot-noise model were well fit by Ratcliff's model, providing evidence that aggregated neural processing dynamics give rise to evidence accumulation processes of the kind that can be characterized by diffusion models.

An issue that was not addressed by Smith's (2010) analysis is how the neural representation of accumulated evidence is maintained over time. Indeed, one of the difficulties in linking behavioral-level data with neural mechanisms arises from the discrepant time scales on which behavioral and neural processes operate. Models of behavioral decision-making data assume integration of evidence over time scales measured in hundreds of milliseconds, whereas intracellular integration processes, operate on time scales around an order of magnitude less—no more than 50 ms (Wang 2001, 2002; Wong and Wang 2006). The question then is how the long time scale integration needed to support decision-making is realized by short time scale neural processes. Smith's (2010) integrated OU model assumed long time scale integration as a theoretical primitive of the model, but gave no account of how such integration could be realized neurally.

Wang (2001, 2002), Wong and Wang (2006) has investigated models that attempt to address the time scale issue by incorporating slow reverberative feedback in the neural circuits involved in the decision process. In Wang's models, accumulated evidence is represented by patterns of firing rates in spiking neuron networks. These models, which incorporate recurrent activation and inhibitory feedback, produce attractor dynamics that result in patterns of firing activity that are sustained within the network. The sustained activity serves to maintain a short-term memory representation of the stimulus for the second or so required to make a decision, which is consistent with behavioral-level modeling of data from simple decision-making experiments (e.g., Ratcliff and Rouder 2000; Sewell and Smith 2012; Smith et al. 2004, 2010; Smith and Ratcliff 2009). Recent extensions of the spiking network model have incorporated a burst-cell mechanism for implementing detection of when the decision process reaches a response threshold, producing a more comprehensive account of how decision-making might be implemented in such circuits (Lo and Wang 2006).

The analysis of Wang and colleagues was complemented by a recent study by Smith and McKenzie (2011), who showed that very simple, stochastic, recurrent loop architectures exhibit the same dynamic properties as Wang's larger scale spiking network model. Smith and McKenzie extended the Poisson shot-noise model of Smith (2010) by adding a loop in which previously emitted spikes were maintained by recurrence, and new spikes were added by superposition. Whereas the dynamics of the Poisson shot-noise process served as a model of the short time scale integration of individual neurons, the recurrent loop dynamics served as a way of modeling long time scale dynamics exhibited by populations of neurons. Smith and McKenzie showed that the Poisson superposition model exhibited similar information accumulation properties to the integrated OU process derived from the shot-noise model. They also fit the recurrent loop model to group-averaged data from one of the experiments reported by Ratcliff and Smith (2004) and found the fit to be quantitatively similar to that of Ratcliff's diffusion model. These results support Smith's (2010) conjecture that behavioral data that are well described by a Wiener diffusion process may be generated neurally by an integrated OU process.

The analyses of Wang and colleagues (Lo and Wang 2006; Wang 2001, 2002; Wong and Wang 2006) and Smith and colleagues (Smith 2010; Smith and McKenzie 2011) provide insight into the relationship between sequential sampling models in psychology and computations performed in the underlying neural populations. Spiking network models and the Poisson shot-noise networks implement neurally plausible decision processes that characterize important aspects of behavioral data. In the next section of this chapter, we discuss how neurophysiological data and choice-RT modeling that directly incorporates neurophysiological data as input provide a complementary perspective on relating psychology and neuroscience.

14.3 Neurophysiological Correlates of Decision Processes

As noted in the introduction, a large number of studies have identified a link between patterns of neural firing rates in oculomotor brain circuits and the dynamics of evidence accumulation during decision-making. The key finding relating neural activity to evidence accumulation is the characteristic ramping of firing rates to a threshold level prior to initiating a saccade response. Using a variety of stimuli and tasks, it has been shown that neurons in FEF (Hanes and Schall 1996; Thompson et al. 1996, 1997), LIP (Shadlen and Newsome 1996, 2001), and SC (Ratcliff et al. 2003, 2007) all exhibit this ramping behavior. The consistency of the firing rate data in the face of considerable variation in stimulus and task properties suggests that this activity reflects the evolution of a visual decision-making process rather than simple sensory processing. In this section, we first discuss some of the research that has linked neural patterns of activity with cognitive representations of decision processes. We then focus on several recent studies that show how neural data can impose strong constraints on sequential sampling models, and can consequently be used to select among models that are very difficult to distinguish by behavioral data alone.

14.3.1 Identifying Decision-Related Neural Activity: The Case of LIP

A common task used to study the neural correlates of visual decision-making is the motion direction discrimination task developed by Newsome, Shadlen, and colleagues (see Fig. 14.1; for recent reviews, see Gold and Shadlen 2007; Huk and Meister 2012). Monkeys judge the global direction of motion of a random dot stimulus and make a saccade-to-target response to indicate their choice. Task difficulty is manipulated by varying the proportion of dots that move coherently in one direction. For these stimuli, it is well-established that motion coherence correlates closely with activity in macaque visual area MT (an extrastriate visual cortical area also known as V5), and that this activity forms the sensory basis for decision-making in this task. The response rate of MT increases roughly linearly with motion coherence, exhibits a time course that is tightly coupled with the onset of the stimulus, and quickly reaches a constant firing rate during stimulus viewing (e.g., Britten et al. 1993). However, MT firing rates are tightly linked to the presence of the stimulus on screen, and only correlate somewhat weakly with behavioral indices of judged direction of motion (Britten et al. 1996). For example, microstimulation of direction-selective MT neurons during stimulus viewing speeds responses favoring the preferred direction, and slows responses for the opposite direction (Ditterich et al. 2003), but the influence of microstimulation on behavioral choice becomes progressively weaker as the delay between stimulus offset and onset of microstimulation increases (Seidemann et al. 1998).

Perhaps the most striking distinction in activity patterns between MT and LIP neurons is that only the latter exhibit gradual ramping of firing rates to a threshold following presentation of a stimulus (e.g., Shadlen and Newsome 1996, 2001), making LIP activity, but not MT activity, consistent with the kinds of evidence accumulation processes assumed by sequential sampling models. A simple hypothesis linking LIP activity with evidence accumulation is that the two are related linearly (e.g., LIP activity reflects discriminative perceptual evidence, integrated over time). The implication is that the rate of increase in LIP activity—or more generally, activity closely related to the decision process—would be reflected in the rate of evidence accumulation, corresponding to the *drift rate* in a diffusion decision model (e.g., Ratcliff 1978; Ratcliff and McKoon 2008), which indexes the average quality of stimulus information driving the decision process, and sets the average rate at which the evidence total changes over time. Although drift rate is commonly estimated as a free parameter from data in human behavioral studies, for the motion direction discrimination task, it is possible to derive estimates of drift rate directly from the motion coherence of the stimulus because the neural representation in MT is well approximated by a linear transformation of the motion signal (Britten et al. 1993). Deriving drift rates from the motion signal in this way has proved successful in modeling data from the motion direction discrimination task (Ditterich 2006; Mazurek et al. 2003). We now discuss several strands of research that provide convergent support for directly linking neural firing rates with the process of evidence accumulation.

Figure 14.8 shows neural firing rate data averaged across 104 LIP neurons of macaques performing the random dots task from a study by Shadlen and Newsome (2001). Monkeys viewed the stimulus for a variable amount of time before it was extinguished. After a variable retention interval, during which fixation had to be maintained, the monkeys indicated their response by making a saccadic eye movement. The data are presented in two ways: In the left panel, firing rate for different levels of motion coherence is presented relative to the onset of the

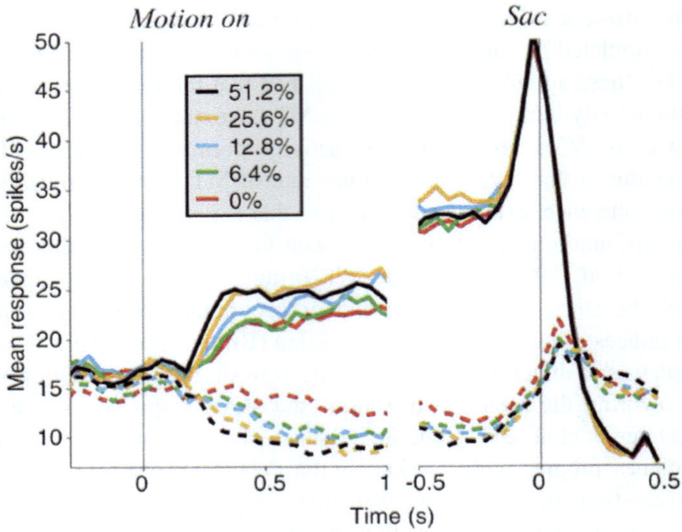

Fig. 14.8 Population response of 104 LIP neurons recorded from macaques performing the motion discrimination task. *Solid lines* are for responses to saccade targets within the neuron's receptive field, *dashed lines* are for responses to targets outside the receptive field. The *left panel* shows firing rates for stimuli at different levels of motion coherence, aligned on stimulus onset. The *right panel* shows firing rates aligned on the saccade response. Reproduced from Shadlen and Newsome (2001)

stimulus. In the right panel, the neural response is presented relative to initiation of the saccade. The key findings are that (1) increasing the viewing time increases the level of spiking activity, (2) increases in spike rate occur faster with higher levels of motion coherence, and (3) prior to initiating a saccade, spike rates reach a common asymptote regardless of motion coherence. Results such as these led Gold and Shadlen (2001) to suggest LIP neurons might be performing a 'read out' function from sensory neurons in MT, integrating relative levels of evidence for different response alternatives. Although these results are intriguing, a limitation of the original version of the random dots task was that monkeys made their decisions after an experimenter-enforced delay, which meant the task did not yield the kind of RT data that have been used to test sequential sampling models in psychology.

Roitman and Shadlen (2002) subsequently investigated an RT version of the motion direction discrimination task that allowed the monkey to respond immediately upon making a decision. The findings from the RT task replicated the main findings from the original task: LIP firing rates increased as a function of stimulus viewing time at a rate determined by motion coherence, reaching a common threshold firing rate prior to making a saccade. Because monkeys initiated a saccade

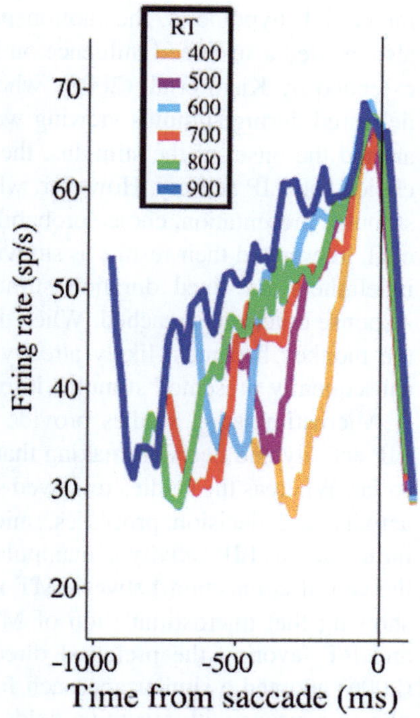

Fig. 14.9 Population response of LIP neurons recorded from macaques performing the response time version of the motion discrimination task. Responses are grouped according to behavioral response time, and are aligned on initiation of a saccade response. Note that the rate at which activity approaches the threshold firing rate varies according to response time. Reproduced from Roitman and Shadlen (2002)

upon making a decision, additional RT-related aspects of the data could be examined. Consistent with the notion of LIP activity reflecting evidence accumulation to a criterion, both behavioral RTs and LIP ramping activity were systematically dependent on motion coherence. At higher levels of coherence, LIP firing rates more rapidly approached asymptote, and behavioral RTs were shorter. The relationship between behavioral RT and ramping of LIP activity is shown in Fig. 14.9, which plots response-aligned LIP activity associated with different RT bands. Whereas the slowest RTs exhibit gradual ramping of LIP activity approximately 700–800 ms prior to a saccade, the fastest RTs are associated with LIP ramping commencing only 300–400 ms prior to a response.

The issue of whether LIP represents the time integral of perceptual evidence was tested experimentally by Huk and Shadlen (2005). They presented monkeys with a random motion texture overlaid with the familiar random dots stimulus. During stimulus viewing, motion coherence was briefly perturbed by adding a 100 ms pulse of coherent motion to the texture. The direction of the motion pulse either matched or opposed that of the random dot stimulus. Consistent with the time integration hypothesis, the motion pulse not only biased decision outcomes, but also exerted a sustained influence on LIP activity for up to 800 ms. This result was extended by Kiani et al. (2008), who showed that the timing of the motion pulse delivered during stimulus viewing was critical. When the motion pulse was added around the onset of the stimulus, the direction of motion affected both behavioral choice and LIP activity. However, when the pulse was delivered toward the end of stimulus presentation, choice probability and LIP firing rates were unaffected. Kiani et al. interpreted their results as showing that evidence accumulation does not occur indefinitely for fixed duration stimuli, but that integration only occurs until a response criterion is reached. When the motion pulse was presented late in the trial, the monkey had most likely already made a decision, and apparently discounted subsequently presented stimulus information.

Microstimulation studies provide further insight into the relationship between LIP activity and decision-making that go beyond the correlational results discussed so far. Whereas the studies reviewed so far have observed a correlation between LIP activity and decision processes, microstimulation allows for more direct causal influences, as LIP activity is manipulated directly. Ditterich et al. (2003) established the causal connection between MT neurons and direction of motion decisions by showing that microstimulation of MT neurons induced a bias in choice behavior and RT favoring the preferred direction of the stimulated neuron. Hanks et al. (2006) adopted a similar approach for neurons in LIP. By applying microstimulation to neurons with receptive fields overlapping the response target, Hanks et al. showed that decision-making could be biased in a way similar to that demonstrated by Ditterich et al. Crucially, however, microstimulation to LIP produced smaller effects on choice and RT than MT stimulation. This is consistent with the idea that MT codes for the sensory data that provides the input to the decision process, whereas LIP codes for the value of the decision variable itself (i.e., the current level of accumulated evidence). In diffusion model terms, MT stimulation affects the drift rate of the diffusion process, which characterizes the quality of the perceptual evidence and determines the rate of evidence accumulation at the decision stage. By contrast, LIP stimulation does not influence drift rate, but instead adds a constant to the decision variable, moving the process closer to one response boundary. This is tantamount to biasing the starting point of the accumulation process without influencing the quality of the perceptual evidence or the rate of evidence accumulation.

An alternative way to use microstimulation to investigate decision-making is to use it to evoke a saccade that is spatially orthogonal to the locations of the response targets and to assess the extent to which the evoked saccade deviates toward one response or another (Gold and Shadlen 2000, 2003). If the motor program expressing the monkey's decision is incrementally updated based on the level of accumulated evidence, then the deviations of the evoked saccades should systematically relate to the monkey's final, voluntary, saccade response. The logic of this approach resembles that of the response-signal procedure used with human participants (e.g., Ratcliff 1988, 2006; Wickelgren 1977). In the response-signal task, decision-making is interrupted by a response signal, which prompts the observer to immediately make a response. In the case of evoked saccades, microstimulation serves as a response signal, which elicits a response directly. Gold and Shadlen (2000) applied microstimulation to FEF neurons while monkeys performed the random dots task. By varying the timing of the microstimulation, they were able to map the time course of evoked saccade trajectories, and indirectly trace the growth of decision-related information. Motion coherence and viewing duration had strong and systematic effects on the deviation of the evoked saccade toward the response target ultimately selected by the monkey: Stimuli with higher motion coherence that were viewed for longer produced larger deviations in the evoked saccades.

Results from experiments using the motion direction discrimination task provide convergent support for the idea that LIP firing rates reflect the time integral of perceptual evidence. Manipulations of the input to the decision process, via the stimulus itself (e.g., Huk and Shadlen 2005; Kiani et al. 2008) or brain areas responsible for coding stimulus properties (Ditterich et al. 2003) influence both the speed and accuracy of responses. Importantly, these behavioral effects are reflected in the firing rates of LIP neurons, and are detectable even when the nature of the response mapping is not known in advance of stimulus onset (Bennur and Gold 2011). That firing rates preceding a saccadic response increase to a common threshold level (Roitman and Shadlen 2002; Shadlen and Newsome 1996, 2001) lends strong support to the idea that LIP activity reflects bounded evidence accumulation, as postulated by diffusion models.

14.3.2 Diffusion Modeling of Decision-Related Neural Activity

The striking similarity between decision-related neural activity and the accumulation of noisy evidence to a criterion invites the obvious question of whether a single model can simultaneously account for both neural and behavioral data. In the context of the motion direction discrimination task, a number of studies have shown that key aspects of the behavioral data can be accounted for within a diffusion or hybrid diffusion-accumulator modeling framework. Mazurek et al. (2003) proposed a hybrid diffusion-accumulator model driven by two populations of simulated MT neurons coding for different response alternatives. The time integral of the difference in activation between MT populations was modeled with two evidence accumulators. The levels of activation in the accumulators were assumed to correspond to the spike rates of neural populations in LIP. The first accumulator to reach a criterion determined the response. The model is closely related to Ratcliff's (1978) diffusion model in that the inputs to the two accumulators are anticorrelated —evidence for one alternative is evidence against the other. Behaviorally, the Mazurek et al. model successfully accounted for accuracy data and mean RTs for correct responses. The model was also able to capture presaccadic ramping of neural activity to a threshold. However, it was not able to predict error RTs, nor was it fit to RT distributions. Ditterich (2006) subsequently showed that the model made incorrect predictions for the RT distributions from Roitman and Shadlen's (2002) study. He modified the model by transforming the difference in MT activations using a sigmoidal time-dependent gain function prior to integration in LIP. By modifying the LIP input in this way, Ditterich showed that the model could account for choice probabilities, correct and error RT distributions, and the pattern of LIP activations simultaneously.

Ditterich's (2006) success with the modified Mazurek et al. (2003) model is interesting because Ratcliff's (1978) diffusion model has been shown to account for behavioral data from the motion direction discrimination task with human participants (Palmer et al. 2005; Ratcliff and McKoon 2008). More generally, Ratcliff's diffusion model (and the related dual diffusion model), has been shown to account for data from monkeys performing other simple decision tasks, such as gap discrimination (Ratcliff et al. 2003) and brightness discrimination (Ratcliff et al. 2007, 2011). The important distinction between the Ratcliff diffusion model and that of Ditterich is that the drift rate in the former is constant over the course of a trial, whereas it is time-varying in the latter. By scaling the drift rate and diffusion coefficient by a common time-dependent gain—affecting v and σ in Eq. 14.1—the model was able to account for the error RT data and produce RT distributions that were more symmetrical than they otherwise would be. The intuition for why the shapes of the distributions change in this way is that the moment-to-moment changes in evidence accumulation become larger in size (due to the increase in drift rate), but more irregular (due to the increase in noise in the accumulation process brought on by the change in the diffusion coefficient). This increases the likelihood that momentary noise will abruptly terminate the process by pushing it toward one of the decision boundaries. Because gain increased as a sigmoid function of time, slower RTs would be affected more than faster RTs, terminating faster than they would under stationary drift. Changing the drift and diffusion coefficients in this way serves to truncate the tails of the RT distributions, resulting in more symmetrical predictions. Psychologically, this account is of interest because it suggests that humans and macaques may be performing the task somewhat differently. One possibility is that monkeys performing the motion direction discrimination task may reduce their decision criterion over the course of a trial. Smith (1995, 2000) pointed out that many diffusion processes with time-varying drift rate (such as the OU process) can be modeled as processes with constant drift, but time-varying decision criteria—a form of model mimicry. Although this was not Ditterich's preferred account for his result, time-varying response criteria have been proposed in other tasks like visual search, where exhaustive search of distractor arrays can be very time-consuming (e.g., Thornton and Gilden 2007). Ditterich's explanation for time-dependent drift, which was couched in terms of maximizing reward rate, is consistent with the idea of observers seeking a balance between time spent on a single trial and reward rate within an experiment by manipulating their response criterion (cf. Bogacz 2007; Bogacz et al. 2006). Whether dynamically adjusting response criterion within a single trial provides a theoretically complete account of maximizing reward rate remains to be seen.

Whereas Ditterich (2006) successfully modeled behavioral and spike rate data from LIP, Ratcliff, and colleagues have examined neural data from SC alongside behavioral data from a variety of perceptual tasks (Ratcliff et al. 2003, 2007, 2011). SC is part of the same oculomotor decision circuit as LIP, and some SC neurons display many similar ramping to threshold characteristics as their LIP counterparts, as we discuss below. Ratcliff et al. (2003) investigated the relationship between the information processing dynamics of the diffusion model to SC firing rates in monkeys performing a gap discrimination task. In Ratcliff's diffusion model, a single accumulator integrates *relative* evidence for different response alternatives. By simultaneously recording from SC neurons associated with each response alternative, Ratcliff et al. (2003) were able to operationalize relative evidence in neural terms as the difference in firing rates. By fitting the standard two-barrier diffusion model to the choice-RT data, and then simulating sample evidence accumulation trajectories from the fitted model, they were able to compare the accumulation of relative evidence in the diffusion model with the difference in firing rates between competing SC neurons. An interesting feature of their data was that the time course of the growth of the difference in SC spike rates—a neural representation of accumulated discriminative evidence—was strongly predictive of behavioral RT (see Fig. 14.10; cf. Roitman and Shadlen 2002). Fast and slow responses were related to fast and slow growth of the difference in firing rates, respectively. Figure 14.10 shows simulated evidence accumulation pathways for fast, slow, and intermediate responses against (relative) SC firing rates. The model very closely reproduces the key features of the data, lending support to the idea that SC neurons are accumulating evidence in a manner similar to how the diffusion model characterizes the decision process.

A limitation of the Ratcliff et al. (2003) study is that the standard diffusion model is silent on the firing rates of the *individual* SC neurons that code for the different response alternatives. By showing a correspondence between neural data and evidence accumulation trajectories in the two-barrier diffusion model, Ratcliff et al. (2003) could only relate its information processing dynamics to the *difference* in firing rates. To address the neural data in more detail, and to generalize their findings to a different perceptual task, Ratcliff et al. (2007) examined firing rates of individual SC neurons in macaques performing a brightness discrimination task. To account for these data, they used a hybrid diffusion-accumulator model, suggested by Smith (2000) and developed by Ratcliff and Smith (2004), which they termed the *dual diffusion model*. This models the decision process as a race between two independent diffusion processes to reach a criterion. The resulting model can be thought of as a diffusion process implementation of an independent race model, like the Poisson counter model discussed previously. After fitting the dual diffusion model to the behavioral choice-RT data, Ratcliff et al. (2007) simulated decision trajectories from the fitted model. Not only was the model able to account for the individual neural firing rates, it was also able to account for the difference in firing rates between SC neurons. This result provided a more direct interpretation of the

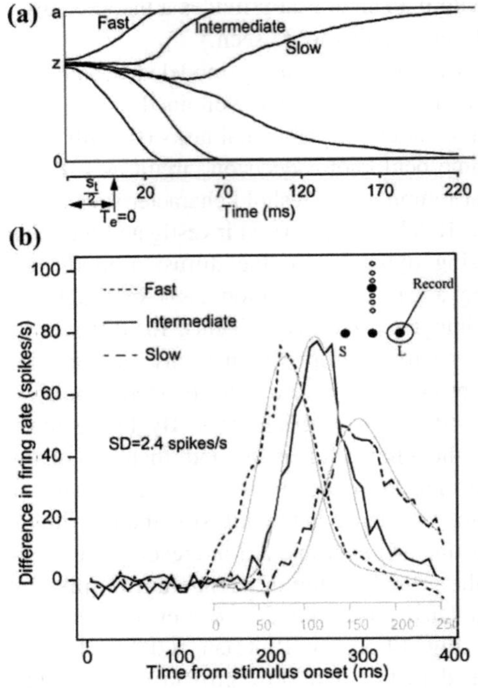

Fig. 14.10 Comparison of simulated evidence accumulation trajectories from Ratcliff's diffusion model with population response of SC neurons recorded from macaques completing a gap discrimination task. The *top panel* shows simulated accumulation trajectories for fast, intermediate, and slow responses based on model parameters derived from fitting the diffusion model to the behavioral data. The *bottom panel* shows the difference in accumulation trajectories for the two response alternatives (*gray lines*) alongside the difference in firing rates for neurons coding for different response alternatives (*black lines*). Reproduced from Ratcliff et al. (2003)

original Ratcliff et al. (2003) result, supporting the idea that the accumulation of discriminative evidence is reflected in the relative firing rates of different neural populations.

14.3.3 Using Neural Data to Constrain Model Selection

The sequential sampling models developed in cognitive psychology have now reached a point at which they are able to quantitatively account for all the key features of choice-RT data observed behaviorally (Ratcliff and Smith 2004). As noted by Ratcliff and McKoon (2008), this is a remarkable accomplishment for psychology, where developing quantitatively precise theories is intrinsically difficult. Part of the success of sequential sampling models has come from adding between-trial parameter variability like variability in start-point (Laming 1968; Ratcliff et al. 1999), and drift rate (Ratcliff 1978). This necessarily increases the complexity (and flexibility) of the models. A consequence of this increased complexity is that it has become quite difficult to distinguish models purely on the basis of fit to behavioral data. As Ratcliff and Smith reported, a variety of hybrid diffusion-accumulator models could produce fits to behavioral data that were quantitatively very similar to Ratcliff's diffusion model. Importantly, these hybrid models emphasized different psychological mechanisms. For example, Usher and McClelland's (2001) LCA model was shown to account for behavioral data by assuming some combination of competition between accumulators and leakage (or decay) of evidence during the decision process. Although fits of these models typically show that decay is not needed to account for behavioral data (i.e., best-fitting estimates of decay parameters are typically quite small, exerting minimal influence on model behavior; Ratcliff and Smith 2004), it is possible that decay may operate at a neurophysiological level, but remain undetectable in behavioral data. Indeed, the effects of short time scale decay that characterizes cellular membrane potentials may be masked by long time scale integration that serves to sustain spiking activity in a cortical network. The long time scale integration assumed by cognitive models explains why Ratcliff's diffusion model can fit data generated by a Poisson shot-noise model, as shown by Smith (2010), even though the former does not involve decay, whereas the latter does. Given the biological reality of short time scale decay at the cellular level, the key question concerning the role of decay in choice-RT models is whether there is an additional level of decay that operates over the longer time scales that characterize the time course of the decision process.

By collecting neural firing rate data, it becomes possible to evaluate claims that long time scale decay and mutual inhibition are needed to account for neural data. Ratcliff et al. (2011) investigated the role of inhibition, recording from pairs of SC neurons that coded for different response alternatives in a brightness discrimination task and found no evidence of inhibition among SC neuron pairs. When conditioned on the firing rate of the neuron coding for the target response, there was little difference in the firing rate for the neuron coding the nontarget response, whereas an inhibition account would predict a decrease. Ratcliff et al. also found that the dual diffusion and LCA models provided similarly good fits to the behavioral choice-RT data, but that the amount of inhibition predicted by the LCA model was minimal. As these conclusions were based on fits to behavioral data, this result might have been expected on the basis of similar fits reported by Ratcliff and Smith (2004). To investigate whether inhibition was needed to account for the neural data, simulated evidence accumulation trajectories were generated from both models. As with the behavioral data, both models closely captured the underlying neural dynamics. Because the dual diffusion model assumes no inhibition, and the simulated LCA trajectories also incorporated minimal inhibition, the result suggests that inhibition was not needed to account for the neural data. Indeed, by increasing levels of inhibition in the LCA model and generating simulated evidence accumulation trajectories, Ratcliff et al. showed that the model produced poorer fits to behavioral data and, contrary to the data, predicted negatively correlated firing rates in target nontarget SC neuron pairs. They concluded that if inhibition were involved in decision-making in SC neurons, the strength of the inhibition was so low as to be undetectable.

The studies by Ratcliff et al. (2003, 2007, 2011) illustrate one of the advantages of using both behavioral and neural data in model testing. By fitting models to behavioral data and examining whether the model dynamics resemble the underlying neural dynamics, it is possible to draw strong conclusions about the relationship between the underlying cognitive and neural processes. In the case of simple decision-making, combining analysis of neural data with mathematical modeling of behavioral data provides support for the linking hypothesis that evidence accumulation processes described by cognitive sequential sampling models are functionally related to firing rates in the underlying neural populations. By examining data at the neural level, it is also possible to resolve model mimicry problems that appear at the behavioral level. Whereas hybrid diffusion-accumulator models are able to provide competitive fits to behavioral data, they do so by invoking different psychological mechanisms (e.g., mutual inhibition). While, it is possible to appeal to neural plausibility to argue for the inclusion of certain processing mechanisms, the requirement to account for neurophysiological and behavioral data simultaneously allows these assumptions to be tested directly.

Neurally Constrained Modeling. A new development in integrating neural and cognitive-behavioral levels of analysis is that of *neurally constrained modeling* of choice-RT data (e.g., Purcell et al. 2010). Like the approach of Ratcliff et al. (2003, 2007, 2011), who showed that models fitted to behavioral data also predict neural firing rates, this approach investigates the relationship between firing rates and sequential sampling models. However, instead of treating accumulation rates as free parameters, this approach uses empirical spike trains recorded from early sensory areas as inputs to the decision process. By using neural data directly to constrain the model parameters, this approach provides a direct way to test assumptions about how neurons coding for different perceptual representations influence subsequent decision process.

Purcell et al. (2010) compared a range of sequential sampling models on their ability to account for combined behavioral and neurophysiological data from macaques performing a visual search task. In this task, the monkeys were required to make a saccade response to a singleton target stimulus embedded in an array of distractors (e.g., a red target among green distractors). Behavioral accuracy and RT data were collected. While the monkeys performed the task, recordings from two kinds of FEF neurons were made. Single-unit recordings from visual neurons in FEF, which were assumed to code for the perceptual representations of the stimuli, were collected. The spike trains recorded in these neurons were used as the input to the decision process in the models. Recordings were also made from movement neurons in FEF, which are responsible for executing the saccade response. Purcell et al. assumed that spike rates for these neurons coded for the integration of perceptual evidence over time.

Purcell et al. (2010) considered a number of different model architectures including correlated, independent, and competitively interacting diffusion processes, and examined the effect of different assumptions on the predicted evidence accumulation process. Evidence accumulation was either perfect (as in Ratcliff's diffusion model; Eq. 14.1), leaky (as in the OU diffusion process; Eq. 14.2), or required a threshold level of visual FEF activity prior to the onset of evidence accumulation by movement neurons, which they termed a *gated accumulation model*. Purcell et al. found that both leaky and gated accumulation models provided the best account of the behavioral data. Because visual FEF neurons were assumed to provide a continuous stream of perceptual evidence to movement neurons—including the period before the neuron successfully discriminated the target—the models that assumed perfect integration failed to account for behavioral data, as they integrated input noise for the bulk of the trial. By contrast, the leaky and gated accumulation models were able to account for the behavioral data because leakage and the gating threshold counteracted the effects of input noise early in the trial. Using model parameters that optimized fit to behavioral data, Purcell et al. simulated evidence accumulation trajectories and compared them with the firing rates of movement neurons during the task. They considered several aspects of the movement neuron dynamics: the time at which spike activity increased from baseline levels, the speed with which spike rate increased, the baseline firing rate, and the threshold firing rate corresponding with initiation of a saccade. Whereas the leaky and gated accumulation models both provided good fits to the behavioral data, only the gated accumulation model was able to account for the full suite of neural data.

The work of Purcell et al. (2010) shows the value of neural constraints in evaluating sequential sampling models. However, the properties of their proposed gating mechanism have yet to be investigated in other settings. Their gating mechanism represents one possible way to limit the accumulation of noise prior to stimulus presentation, first discussed by Laming (1968) and recently investigated by Ratcliff and Smith (2010). Ratcliff and Smith investigated decision-making in a task in which letter stimuli were presented in dynamic noise and found that the effect of noise was to shift the leading edge of the RT distribution later in time. The shift was interpreted as a delay in the onset of evidence accumulation by the decision process. Their results showed that the onset of evidence accumulation by the decision process must be sensitive to the temporal stability of the stimulus representation. When stimuli are embedded in dynamic noise, the process of forming a stable representation of the stimulus is slowed, resulting in delayed onset of decision-making. It is not yet clear whether Purcell et al.'s gating mechanism can also provide an explanation of the delay in the onset of evidence accumulation produced by noise in Ratcliff and Smith's (2010) task.

To summarize, by making very simple assumptions about how neural activity relates to coding of perceptual evidence, it is possible to simultaneously model data at both behavioral and neurophysiological levels of analysis. From a theoretical perspective, addressing both kinds of data has become increasingly important, as different kinds of model architectures have proved equally capable of accounting for behavioral data (Ratcliff and Smith 2004). Although there are a number of open theoretical issues that remain to be settled, this approach to analyzing combined neural and behavioral data constitutes a significant methodological and theoretical development.

14.4 Summary and Conclusions

In this chapter, we have attempted to provide an overview of the development of different classes of sequential sampling models in cognitive and mathematical psychology, and how they relate to data collected from single-unit recordings in neural oculomotor circuits. Sequential sampling models are among the most rigorously tested models in psychology, and after many decades of refinement, the best models can account for all of the major behavioral phenomena of interest (Ratcliff and McKoon 2008; Ratcliff and Smith 2004). One of the side effects of this theoretical progress is that several classes of models (e.g., diffusion and hybrid diffusion-accumulator models) have been shown to provide comparably good fits to behavioral data, and are consequently very difficult to distinguish. This is a salient issue because the assumptions of some models—such as those involving decay and mutual inhibition—are motivated by appeals to neural plausibility. While we appreciate such considerations, we note that such assumptions, like any theoretical assumptions, must be subject to empirical test. It is clear from recent work that there is much to be gained from the theoretical exchange between psychologists and neuroscientists interested in decision-making. The rich history of choice-RT research in cognitive and mathematical psychology provides a theoretical framework for studying the neural basis of decision-making. Reciprocally, constraints from neural data provide an opportunity to test and refine decision models in ways that would not have been possible using behavioral data alone.

14.4.1 Outstanding Questions and Future Directions

Despite the substantial progress that has been made over the last decade or so in identifying neural correlates of decision processes and relating these two components of successful sequential sampling models, there are a number of issues that remain open. We briefly discuss three of these issues here.

One Decision Architecture or Many? Although the studies reviewed above have consistently shown links between neural firing rates in oculomotor areas and evidence accumulation rates in decision models, there are subtle differences in some of the specific details of models associated with different brain areas. For example, for SC neurons in monkeys performing a brightness discrimination task, Ratcliff et al. (2011) found no functional role for mutual inhibition among decision neurons, suggesting that models like the LCA, which ascribe a central role to inhibition, may not be appropriate for modeling decision processes executed in SC. An open question is whether other brain areas show evidence for inhibition at the level of decision processes, and whether an LCA-like model might be suitable for neurons outside of SC.

Recent work by Purcell et al. (2012), extended their previous study of FEF movement neurons in monkeys performing a visual search task. Whereas, their initial study only allowed modeling of correct RT distributions (Purcell et al. 2010) their follow-up examined RT distributions for both correct and error responses using the same neurally constrained modeling approach. In contrast to what Ratcliff et al. (2011) found for SC neurons, Purcell et al. (2012) found support for a gated accumulation model that involved mutual inhibition among accumulators. The implication of these studies is that decision-making architectures and dynamics, though fundamentally very similar, as they draw on common principles of evidence accumulation, may vary in significant ways across different brain areas. Additional work is needed to better understand how and why neural decision architectures might vary.

Modeling Evidence Integration. The results from the neurally constrained modeling approach of Purcell et al. (2010, 2012) provide support for a simple linking proposition about the neural representation of perceptual evidence. Using empirical spike train data to drive a decision mechanism is an important first step in developing a comprehensive neurocomputational account of decision-making. However, this is only a partial solution. As the work of Wang and colleagues (e.g., Lo and Wang 2006; Wang 2001, 2002; Wong and Wang 2006) and Smith and McKenzie (2011) highlighted, the question remains as to how temporal *integration* of evidence is realized neurocomputationally. Owing to the large differences in time scales between individual neurons involved in decision-making and the overall decision-making process itself, the neural mechanism that integrates evidence over behaviorally relevant time scales needs to be developed further. As with other theoretical approaches (e.g., Ditterich 2006; Mazurek et al. 2003; Purcell et al. 2010, 2012; Ratcliff et al. 2003, 2007, 2011; Smith 2010), temporal integration of perceptual evidence is assumed to be a theoretical primitive of the models. The spiking neuron models developed by Wang and colleagues and recurrent loop models, like Smith and McKenzie's, suggest some theoretical possibilities for how evidence might be integrated over long time scales. Combining such a theoretical decision architecture with the neurally constrained approach provides an interesting avenue for further development, as it could provide a strong test of the principles of recurrence and superposition used by Smith and McKenzie.

Neural Correlates of Bias and Reward. In addition to the work reviewed above that has investigated neural correlates of evidence accumulation, recent work has begun to use choice-RT modeling to investigate the neural correlates of prior knowledge and bias effects in simple decision tasks in humans (e.g., Bode et al. 2012; Forstmann et al. 2010; Mulder et al. 2012) and monkeys (e.g., Rao et al. 2012; Rorie et al. 2010).

For example, Rao et al. (2012) studied LIP responses of monkeys performing the motion direction discrimination task. Using cues indicating the most likely direction of motion for an upcoming stimulus, they observed systematic shifts in LIP firing rates following cue presentation, but no effect on the rate at which activity approached a threshold firing rate. A similar effect on LIP response was observed by Rorie et al. (2010) who used a reward-based manipulation. For both studies, choice-RT modeling revealed an effect on the start-point of the evidence accumulation process.

In human studies, Forstmann et al. (2010) had people complete a cued version of the motion discrimination task similar to that used by Rao et al. (2012). Using fMRI, Forstmann et al. found that cues produced changes in cortico-striatal brain circuits, but only after accounting for the effects of response bias—as measured by fitting an accumulator model to behavioral data (Brown and Heathcote 2008). This work has recently been extended by Mulder et al. (2012), who manipulated prior probability with cues and reward using payoffs. Their fMRI analysis revealed common effects of bias and reward on cortico-parietal decision circuits, which produced changes in response bias—indexed by the estimated starting point of evidence accumulation—in Ratcliff's diffusion model.

In a similar vein, Bode et al. (2012) investigated sequence effects in a perceptual decision-making task, involving stimuli presented briefly in noise. They combined a model-based behavioral analysis with a pattern-classification analysis of EEG data. Using Ratcliff's diffusion model, Bode et al. showed that sequence effects in the behavioral data were consistent with changes in the starting point of the evidence accumulation process, reflecting a bias favoring the response made on the previous trial. Consistent with this response bias account, it was also found that prestimulus activity in the EEG trace was predictive of people's responses on catch-trials where only noise was presented.

Taken together, these investigations into the neural correlates of bias and prior knowledge effects add to the growing body of evidence that both neural and behavioral choice-RT data can be accommodated by a common sequential sampling framework.

14.4.2 Conclusion

With the recent refinement of techniques for recording spike rates from monkeys performing simple perceptual tasks, it has become possible to directly evaluate modeling assumptions made on the basis of neural plausibility. A number of research groups, using a variety of tasks and methods, have provided converging evidence that the spiking dynamics of neurons in FEF, LIP, and SC bear a striking resemblance to the evidence accumulation dynamics postulated by sequential sampling models of decision-making. By recording from oculomotor areas known to be involved in the processing of perceptual stimuli and preparation of saccadic responses, one can evaluate the degree of convergence between cognitive theory and neural data. Assumptions made in cognitive models have provided a basis for testing hypotheses about how the brain integrates perceptual evidence over time, and, importantly, how neural data can be used as a tool for model selection.

Acknowledgments This work was supported by ARC Discovery Grant DP110103406 awarded to P.L. Smith and R. Ratcliff. We thank Simon Lilburn for helpful comments on a previous version of the manuscript. David Sewell is now at the University of Queensland.

References

Audley RJ, Pike AR (1965) Some alternative stochastic models of choice. Br J Math Stat Psychol 18:207–225

Bennur S, Gold JI (2011) Distinct representations of a perceptual decision and the associated oculomotor plan in the monkey lateral intraparietal area. J Neurosci 31:913–921

Bode S, Sewell DK, Lilburn SD, Forte JD, Smith PL, Stahl J (2012) Predicting perceptual decision biases from early brain activity. J Neurosci 32:12488–12498

Bogacz R (2007) Optimal decision-making theories: linking neurobiology with behaviour. Trends Cogn Sci 11:118–125

Bogacz R, Brown E, Moehlis J, Holmes P, Cohen JD (2006) The physics of optimal decision making: a formal analysis of models of performance in two-alternative forced-choice tasks. Psychol Rev 113:700–765

Britten KH, Shadlen MN, Newsome WT, Movshon JA (1993) Responses of neurons in macaque MT to stochastic motion signals. Vis Neurosci 10:1157–1169

Britten KH, Newsome WT, Shadlen MN, Celebrini S, Movshon JA (1996) A relationship between behavioral choice and the visual response of neurons in macaque MT. Vis Neurosci 13:87–100

Brown SD, Heathcote A (2008) The simplest complete model of choice response time: Linear ballistic accumulation. Cogn Psychol 57:153–178

Busemeyer JR, Townsend JT (1993) Decision field theory: a dynamic-cognitive approach to decision making in an uncertain environment. Psychol Rev 100:432–459

Churchland AK, Kiani R, Shadlen MN (2008) Decision-making with multiple alternatives. Nat Neurosci 11:693–702

Ditterich J (2006) Evidence for time-variant decision making. Eur J Neurosci 24:3628–3641

Ditterich J, Mazurek ME, Shadlen MN (2003) Microstimulation of visual cortex affects the speed of perceptual decisions. Nat Neurosci 6:891–898

Edwards W (1965) Optimal strategies for seeking information: models for statistics, choice reaction times, and human information processing. J Math Psychol 2:312–329

Forstmann BU, Brown S, Dutilh G, Neumann J, Wagenmakers EJ (2010) The neural substrate of prior information in perceptual decision making: a model-based analysis. Frontiers in Human Neuroscience 4:1–12

Gnadt JW, Andersen RA (1988) Memory related motor planning activity in posterior parietal cortex of macaque. Exp Brain Res 70:216–220

Gold JI, Shadlen MN (2000) Representation of a perceptual decision in developing oculomotor commands. Nature 404:390–394

Gold JI, Shadlen MN (2001) Neural computations that underlie decisions about sensory stimuli. Trends in Cognitive Sciences 5:10–16

Gold JI, Shadlen MN (2002) Banburismus and the brain: decoding the relationship between sensory stimuli, decisions, and reward. Neuron 36:299–308

Gold JI, Shadlen MN (2003) The influence of behavioral context on the representation of a perceptual decision in developing oculomotor commands. J Neurosci 23:632–651

Gold JI, Shadlen MN (2007) The neural basis of decision making. Annu Rev Neurosci 30: 535–574

Green DM, Swets JA (1966) Signal detection theory and psychophysics. Wiley, New York

Green DM, Smith AF, von Gierke SM (1983) Choice reaction time with a random foreperiod. Percept Psychophys 34:195–208

Hanes DP, Schall JD (1996) Neural control of voluntary movement initiation. Science 274: 427–430

Hanks TD, Ditterich J, Shadlen MN (2006) Microstimulation of macaque area LIP affects decision-making in a motion discrimination task. Nat Neurosci 9:682–689

Huk AC, Meister MLR (2012) Neural correlates and neural computations in posterior parietal cortex during perceptual decision-making. Front Integr Neurosci 6:1–13

Huk AC, Shadlen MN (2005) Neural activity in macaque parietal cortex reflects temporal integration of visual motion signals during perceptual decision making. J Neurosci 25: 10420–10436

Kiani R, Hanks TD, Shadlen MN (2008) Bounded integration in parietal cortex underlies decisions even when viewing duration is dictated by the environment. J Neurosci 28:3017–3029

LaBerge D (1962) A recruitment theory of simple behavior. Psychometrika 27:375–396

LaBerge D (1994) Quantitative models of attention and response processing in shape identification tasks. J Math Psychol 38:198–243

Laming DRJ (1968) Information theory of choice reaction time. Wiley, New York

Link SW, Heath RA (1975) A sequential theory of psychological discrimination. Psychometrika 40:77–105

Lo C-C, Wang X-J (2006) Cortico-basal ganglia circuit mechanism for a decision threshold in reaction time tasks. Nat Neurosci 9:956–963

Luce RD (1986) Response times: Their role in inferring elementary mental organization. Oxford University Press, Oxford

Mazurek ME, Roitman JD, Ditterich J, Shadlen MN (2003) A role for neural integrators in perceptual decision making. Cereb Cortex 13:1257–1269

Mulder MJ, Wagenmakers EJ, Ratcliff R, Boekel W, Forstmann BU (2012) Bias in the brain: a diffusion model analysis of prior probability and potential payoff. J Neurosci 32:2335–2343

Palmer J, Huk AC, Shadlen MN (2005) The effect of stimulus strength on the speed and accuracy of a perceptual decision. Journal of Vision 5:376–404

Pike AR (1966) Stochastic models of choice behaviour: response probabilities and latencies of finite Markov chain systems. Br J Math Stat Psychol 21:161–182

Pike AR (1973) Response latency models for signal detection. Psychol Rev 80:53–68

Purcell BA, Heitz RP, Cohen JY, Schall JD, Logan GD, Palmeri TJ (2010) Neurally constrained modeling of perceptual decision making. Psychol Rev 117:1113–1143

Purcell BA, Schall JD, Logan GD, Palmeri TJ (2012) From salience to saccades: multiple-alternative gated stochastic accumulator model of visual search. J Neurosci 32: 3433–3446

Rao V, DeAngelis GC, Snyder LH (2012) Neural correlates of prior expectations of motion in the lateral intraparietal and middle temporal areas. J Neurosci 32:10063–10074

Ratcliff R (1978) A theory of memory retrieval. Psychol Rev 85:59–108

Ratcliff R (1988) Continuous versus discrete information processing: modeling accumulation of partial information. Psychol Rev 95:238–255

Ratcliff R (2002) A diffusion model account of response time and accuracy in a brightness discrimination task: fitting real data and failing to fit fake but plausible data. Psychon Bull Rev 9:278–291

Ratcliff R (2006) Modeling response signal and response time data. Cogn Psychol 53:195–237

Ratcliff R, McKoon G (2008) The diffusion decision model: theory and data for two-choice decision tasks. Neural Comput 20:873–922

Ratcliff R, Rouder JN (1998) Modeling response times for two-choice decisions. Psychol Sci 9:347–356

Ratcliff R, Rouder JN (2000) A diffusion model account of masking in two-choice letter identification. J Exp Psychol Hum Percept Perform 26:127–140

Ratcliff R, Smith PL (2004) A comparison of sequential sampling models for two-choice reaction time. Psychol Rev 111:333–367

Ratcliff R, Smith PL (2010) Perceptual discrimination in static and dynamic noise: the temporal relation between perceptual encoding and decision making. J Exp Psychol Gen 139:70–94

Ratcliff R, Van Zandt T, McKoon G (1999) Connectionist and diffusion models of reaction time. Psychol Rev 106:261–300

Ratcliff R, Cherian A, Segraves MA (2003) A comparison of macaque behavior and superior colliculus neuronal activity to predictions from models of two-choice decisions. J Neurophysiol 90:1392–1407

Ratcliff R, Gomez P, McKoon G (2004) A diffusion model account of the lexical decision task. Psychol Rev 111:159–182

Ratcliff R, Hasegawa YT, Hasegawa RP, Smith PL, Segraves MA (2007) Dual diffusion model for single-cell recording data from the superior colliculus in a brightness-discrimination task. J Neurophysiol 97:1756–1774

Ratcliff R, Hasegawa YT, Hasegawa RP, Childers R, Smith PL, Segraves MA (2011) Inhibition in superior colliculus neurons in a brightness discrimination task? Neural Comput 23:1790–1820

Roe RM, Busemeyer JR, Townsend JT (2001) Multialternative decision field theory: a dynamic connectionist model of decision making. Psychol Rev 108:370–392

Roitman JD, Shadlen MN (2002) Response of neurons in the later intraparietal area during a combined visual discrimination reaction time task. J Neurosci 22:9475–9489

Rorie AE, Gao J, McClelland JL, Newsome WT (2010) Integration of sensory and reward information during perceptual decision-making in lateral intraparietal cortex (LIP) of the macaque monkey. PLoS ONE 5:1–21

Schall JD (2004) On building a bridge between brain and behavior. Annu Rev Psychol 55:23–50

Seidemann E, Zohary E, Newsome WT (1998) Temporal gating of neural signals during performance of a visual discrimination task. Nature 394:72–75

Sewell DK, Smith PL (2012) Attentional control in visual signal detection: effects of abrupt-onset and no-onset stimuli. J Exp Psychol Hum Percept Perform 38:1043–1068

Shadlen MN, Newsome WT (1996) Motion perception: seeing and deciding. Proc Nat Acad Sci 93:628–633

Shadlen MN, Newsome WT (2001) Neural basis of a perceptual decision in the parietal cortex (area LIP) of the rhesus monkey. J Neurophysiol 86:1916–1936

Smith PL (1995) Psychophysically principled models of visual simple reaction time. Psychol Rev 102:567–593

Smith PL (2000) Stochastic dynamic models of response time and accuracy: a foundational primer. J Math Psychol 44:408–463

Smith PL (2010) From poisson shot noise to the integrated Ornstein-Uhlenbeck process: neurally principled models of information accumulation in decision-making and response time. J Math Psychol 54:266–283

Smith PL, McKenzie CRL (2011) Diffusive information accumulation by minimal recurrent neural models of decision-making. Neural Comput 23:2000–2031

Smith PL, Ratcliff R (2004) Psychology and neurobiology of simple decisions. Trends Neurosci 27:161–168

Smith PL, Ratcliff R (2009) An integrated theory of attention and decision making in visual signal detection. Psychol Rev 116:283–317

Smith PL, Van Zandt T (2000) Time-dependent poisson counter models of response latency in simple judgment. Br J Math Stat Psychol 53:293–315

Smith PL, Vickers D (1988) The accumulator model of two-choice discrimination. J Math Psychol 32:135–168

Smith PL, Ratcliff R, Wolfgang BJ (2004) Attention orienting and the timecourse of perceptual decisions: response time distributions with masked and unmasked displays. Vision Res 44:1297–1320

Smith PL, Ellis R, Sewell DK, Wolfgang BJ (2010) Cued detection with compound integration-interruption masks reveals multiple attentional mechanisms. J Vis 10:1–28

Stone M (1960) Models for choice-reaction time. Psychometrika 25:251–260

Teller DY (1984) Linking propositions. Vision Res 10:1233–1246

Thompson KG, Hanes DP, Bichot NP, Schall JD (1996) Perceptual and motor processing stages identified in the activity of macaque frontal eye field neurons during visual search. J Neurophysiol 76:4040–4055

Thompson KG, Bichot NP, Schall JD (1997) Dissociation of visual discrimination from saccade programming in macaque frontal eye field. J Neurophysiol 77:1046–1050

Thornton TL, Gilden DL (2007) Parallel and serial processes in visual search. Psychol Rev 114:71–103

Townsend JT, Ashby FG (1983) Stochastic modeling of elementary psychological processes. Cambridge University Press, Cambridge

Usher M, McClelland JL (2001) The time course of perceptual choice: the leaky, competing accumulator model. Psychol Rev 108:550–592

Van Zandt T, Colonius H, Proctor RW (2000) A comparison of two response time models applied to perceptual matching. Psychon Bull Rev 7:208–256

Vickers D (1970) Evidence for an accumulator model of psychophysical discrimination. Ergonomics 13:37–58

Vickers D (1979) Decision processes in visual perception. Academic Press, New York

Wagenmakers EJ, Brown S (2007) On the linear relation between the mean and the standard deviation of a response time distribution. Psychol Rev 114:830–841

Wald A (1947) Sequential analysis. Wiley, New York

Wang X-J (2001) Synaptic reverberation underlying mnemonic persistent activity. Trends Neurosci 24:455–463

Wang X-J (2002) Probabilistic decision making by slow reverberation in cortical circuits. Neuron 36:955–968

Wickelgren WA (1977) Speed-accuracy tradeoff and information processing dynamics. Acta Psychol 41:67–85

Wong K-F, Wang X-J (2006) A recurrent network mechanism of time integration in perceptual decisions. J Neurosci 26:1314–1328

Chapter 15
A Neurocognitive Perspective on the Development of Social Decision-Making

Geert-Jan Will and Berna Güroğlu

Abstract In this chapter, we review evidence for the hypothesis that developmental changes in cognitive control and perspective taking are crucial in understanding age-related changes in social behavior. Studies that have examined the developmental roots of prosocial behavior using experimental economic games show that other-oriented concern and a preference for fairness emerge early in development. Continued development of intentionality understanding and strategic behavior in bargaining situations suggest that perspective taking and cognitive control undergo extended development and continue to contribute to changes in social behavior well into adolescence. Functional neuroimaging studies have shown that these behavioral changes are accompanied by an increased recruitment of brain regions implicated in cognitive control (e.g., dorsolateral prefrontal cortex) and perspective taking (e.g., temporoparietal junction). Together these studies show that developmental changes in cognitive control and perspective taking and their underlying neural circuitry are associated with progressively more strategic thinking and an increased incorporation of other's perspectives into social decision-making across development.

Beginning early in ontogeny humans show levels of sociality that surpass those of other species (Tomasello and Vaish 2013). For example, 1-year-old toddlers help others to achieve a goal by picking up objects that are needed to successfully complete an action without any explicit request or reward (Warneken and Tomasello 2006). Despite the early emergence of key social tendencies, social

G.-J. Will (✉) · B. Güroğlu
Institute of Psychology, Leiden University,
Wassenaarseweg 52, 2333 AK Leiden, The Netherlands
e-mail: g.j.will@fsw.leidenuniv.nl

B. Güroğlu
e-mail: BGuroglu@fsw.leidenuniv.nl

G.-J. Will · B. Güroğlu
Leiden Institute for Brain and Cognition (LIBC), Leiden University
Medical Center, P.O. Box 9600, Leiden 2300 RC, The Netherlands

M. Reuter and C. Montag (eds.), *Neuroeconomics*, Studies in Neuroscience, Psychology and Behavioral Economics, DOI 10.1007/978-3-642-35923-1_15

behavior continues to develop and grow in complexity across childhood and adolescence. Children expand their behavioral repertoire with social tactics, such as teasing and deception, and they increasingly learn to take other people's feelings into account in their social responses (Burnett and Blakemore 2009b). For instance, whereas preschoolers mainly use deception to their own benefit, school-aged children increasingly start to use deception to protect other's feelings (e.g., telling 'white lies') (Talwar et al. 2007).

Developmental changes in social behavior are thought to be supported by developmental changes in general cognitive functions like impulse control and specific social cognitive functions such as the ability to adopt another person's perspective (e.g., perspective taking). Cognitively controlling impulses is of crucial importance for the regulation of social behavior and continues to develop across childhood and adolescence (Davidson et al. 2006; Rueda et al. 2005). Children acquire a core component of perspective taking when they develop an understanding that other people's mental states might differ from their own (Wellman et al. 2001). Even though this 'theory of mind' emerges before adolescence, more advanced forms of perspective taking needed to act on the understanding of other people's mental states continue to develop during adolescence (Dumontheil et al. 2010; Selman 1980). This rather protracted development of both impulse control and perspective taking is likely to contribute to developmental changes in social behavior across adolescence and into adulthood. This implication is central in the neurobiological models of social development which posit that continued structural development of the brain is associated with functional changes in brain networks implicated in cognitive control and social cognition, which in turn contribute to developmental changes in social behavior (Blakemore 2008; Crone and Dahl 2012; Nelson et al. 2005).

In this chapter, we review evidence for the hypothesis that the gradual development of impulse control and perspective-taking skills are associated with progressively more strategic thinking and an increased incorporation of other's intentions in social decision-making. In the following sections, we first describe why paradigms from behavioral economics provide valuable tools to study developmental changes in social behavior and its underlying mechanisms (Sect. 15.1). Subsequently, we describe the age-related behavioral changes in these games (Sect. 15.2), followed by evidence linking these behavioral changes to children's developing abilities to control selfish impulses and to take other people's perspective (Sect. 15.3). Next, we focus on functional neuroimaging studies showing that social decision-making in adults relies on separable, but interacting, networks in the brain (Sect. 15.4). Finally, we review recent neuroimaging studies demonstrating differential development of the brain networks involved in social decision-making (Sect. 15.5), supporting the proposition that increased intentionality understanding and strategic motivations in social decision-making are associated with developmental changes in these networks.

15.1 Why Use Economic Games to Study Social Development?

To investigate the psychological and neural mechanisms underlying social decision-making, psychologists and neuroscientists have turned to game theoretical paradigms derived from behavioral economics (Rilling and Sanfey 2011). These paradigms offer a context of social interactions where the decisions people make have actual consequences for their own and their interaction partner's outcomes. Two of these games, namely the Ultimatum Game and the Dictator Game, have proven to be valuable tools to study concerns about fairness (for a detailed description of these games please see Chap. 20 in this book). In these two-player exchange games one player (i.e., the proposer) is given a set of valuable rewards, such as money, candy or stickers and is given the opportunity to propose a split of the rewards between themselves and a second player. In the Ultimatum Game, the second player (i.e., the responder) can either accept or reject the proposal. If the proposal is accepted, both players receive their part of the stake as proposed. In case of rejection, neither of the players receives anything (Güth et al. 1982). The Dictator Game is different in the sense that the responder (i.e., the recipient in this case) does not have the power to reject the proposal and thus passively receives the amount of rewards that the first player transfers (Forsythe et al. 1994).

Game theoretical models assume that humans are rational decision-makers who act to maximize personal outcomes (Camerer 2003). Accordingly, game theory predicts that Ultimatum Game proposers would make the smallest offer possible and that responders would accept any offer greater than zero. However, findings show that (adult) proposers and responders do not follow the game theoretical predictions: proposers offer most often an equal split and responders usually reject offers smaller than 20 % of the stake (Camerer 2003). In the Dictator Game, there is no possibility for reciprocation or retribution for the recipient, so game theory would predict that proposers would keep the entire set of rewards to themselves. Interestingly, proposers in the Dictator Game rarely act in accordance with these predictions. Adult humans transfer on average 20–30 % of the stake to anonymous others with 50 % of the stake typically being one of the most frequently occurring offers (Forsythe et al. 1994; Hoffman et al. 1994). These deviations from the game theoretical predictions suggest that people not only have an interest in maximizing their own payoffs but also have a concern for the other person's outcomes. Importantly, whereas the positive offers in the Dictator Game reflect other-regarding concern, the comparatively larger offers in the Ultimatum Game suggest that strategic considerations aimed at reducing the possibility of rejection also play a role in decisions about fairness. In addition, the consistently found rejections of unfair Ultimatum Game offers suggest an aversive response to receiving less than the proposer (known as 'disadvantageous inequity aversion') and rejection of the offer possibly provides the responder with a way of correcting such inequity (Fehr and Schmidt 1999).

Using these games for developmental research offers several advantages (Gummerum et al. 2008a, b). First, an important advantage is that the same paradigm can be used across a wide age range (from children as young as 3 years old to adults), enabling meaningful comparisons between different age groups. Second, the structured nature of the games makes it possible to quantify complex social behavior, which makes them useful for neuroimaging research. Third, these games allow for experimental manipulations where subcomponents of social decision-making, such as understanding another person's intentions and controlling selfish impulses, can be disentangled. Such subcomponents of decision-making might be differentially sensitive to developmental change. For example, emotional reactions to unfairness might mature earlier than an understanding of an interaction partner's intentions, which might depend on slowly developing cognitive functions. Psychological and neural mechanisms underlying such subprocesses can further be investigated by relating age-related and age-independent individual differences in behavior and neural activation to external measures of cognitive control (e.g. inhibition tasks) or perspective taking (e.g. 'theory of mind' tasks). By doing so, one can examine how different cognitive functions and their underlying neural substrates are involved in developmental changes in social behavior.

15.2 Development of a Preference for Fairness

Concern for another person's wellbeing has strong developmental roots and emerges at very young ages. Twelve- to 18-month-old infants willingly engage in instrumental helping of an adult who has dropped (Warneken and Tomasello 2006) or misplaced (Liszkowski et al. 2008) an object that is needed to complete an action and during the second year of life toddlers start to comfort others in distress (Zahn-Waxler et al. 1992). Nonetheless, infants and toddlers are much more reluctant to show prosocial behavior when it is costly, i.e., when they have to give up some of their own possessions to benefit another person (Svetlova et al. 2010). Developmental studies employing the Dictator Game have shown that although children tend to keep most of the resources to themselves, the size of their donations increase with age between the ages of 3 and 8 (Benenson et al. 2007; Blake and Rand 2010; Smith et al. 2013) and by age 9 no longer differs from donations made by adults (Gummerum et al. 2008a, b; Güroğlu et al. 2009; Steinbeis et al. 2012).

Interestingly, this developmental increase in costly sharing is not due to developmental differences in explicit knowledge about what constitutes a fair (i.e., in most cases equal) distribution of resources. Infants as young as 15 months already expect resources to be distributed equally as indicated by prolonged eye gazes in situations when resources are distributed unequally between two recipients compared to situations where both recipients receive an equal amount of resources (Schmidt and Sommerville 2011; Sloane et al. 2012). A recent study showed that, although 3-year-olds do not differ from 8-year-olds in their judgments about what constitutes an equal division of rewards, they still tend to keep more than half of the rewards to themselves in a Dictator Game and the willingness to give away half of the rewards increases between the ages of 3–8 (Smith et al. 2013). Furthermore, converging evidence from developmental investigations of rejections of unequal distributions confirms that the willingness to incur costs to avoid unequal outcomes ("I'd rather receive nothing than less than the other") increases between age 3 and 8 (Blake and McAuliffe 2011). Also when distributing resources, 8-year-olds appear not to choose a distribution that favors a peer; even when this has no consequences for their own outcomes (Fehr et al. 2008; Shaw and Olson 2012; Shaw et al. 2013). Taken together, these findings suggest that a developing sense of fairness makes children increasingly enforce equality when this is costly, but that it does not make them necessarily more generous or tolerant of higher outcomes for a peer.

Investigations of proposer behavior in the Ultimatum Game show that not only a prosocial concern about the other person's profits or equal outcomes plays a role in fairness considerations, but also that strategic considerations aimed at reducing the possibility of rejection come into play. That is, adults tend to offer higher shares of the stake (closer to an equal split of the rewards) when the second player can punish unfair offers (e.g., by rejecting them in an Ultimatum Game) (Fehr and Fischbacher 2004; Spitzer et al. 2007). As such, the difference in Ultimatum Game and Dictator Game offers provides a measure of strategic social behavior. During late childhood (age 7–10), children start making higher Ultimatum Game proposals compared to Dictator Game allocations, but their Ultimatum Game proposals are still smaller than those proposed by adults (Harbaugh et al. 2003). During adolescence, the difference between Ultimatum Game and Dictator Game offers becomes progressively greater, suggesting a developmental increase in strategic behavior across adolescence (Güroğlu et al. 2009; Leman et al. 2009). The results from these studies also demonstrate that the increasing discrepancy between Ultimatum Game and Dictator Game offers is driven by increasingly higher Ultimatum Game offers and that Dictator Game offers made by children in late childhood do not differ from adult Dictator Game offers (Güroğlu et al. 2009; Steinbeis et al. 2012). Taken together these studies show that a prosocial tendency to share resources with another person emerges early in development, but also that social behavior becomes increasingly strategic across childhood and adolescence.

15.3 Cognitive Mechanisms Underlying Developmental Change in Strategic Social Behavior: Impulse Control and Perspective Taking

Strategic bargaining depends on the notion that unfair Ultimatum Game proposals can be punished, while Dictator Game proposals cannot. Crucially, strategic bargaining—assessed as the difference between Ultimatum Game and Dictator Game offers—develops across childhood and is associated with the developing capacity to control impulses (Steinbeis et al. 2012). That is, the difference between the number of rewards transferred in the Ultimatum Game and in the Dictator Game increases between the ages of 6 and 14 (see Fig. 15.1a) and irrespective of age, children, and adults who were better at controlling a prepotent motor response in a stop-signal reaction time (SSRT) task, also showed more strategic bargaining (see Fig. 15.1b). These findings suggest that strategic social behavior relies on the capacity to implement behavioral control over a selfish impulse of keeping all resources to oneself in situations where selfish behavior can be punished.

In addition, it has been argued that proposers have to take the responder's perspective in order to infer what kind of offers are likely to be rejected (Singer 2006; Singer and Fehr 2005). Indeed, 4–5 year old children who passed a false-belief task (a task to probe the acquisition of a 'theory of mind'), more often proposed a fair offer in the Ultimatum Game than same-aged peers who failed to pass this task (Takagishi et al. 2010). Furthermore, children with deficits in perspective taking such as children with autism spectrum disorders tend to propose self-serving unfair offers in the Ultimatum Game (Sally and Hill 2006). Interestingly, a prosocial tendency to share at least some part of one's resources with a peer in a Dictator Game is no different in children with autism spectrum disorders, suggesting that perspective-taking abilities are especially important when social interactions have a strategic component.

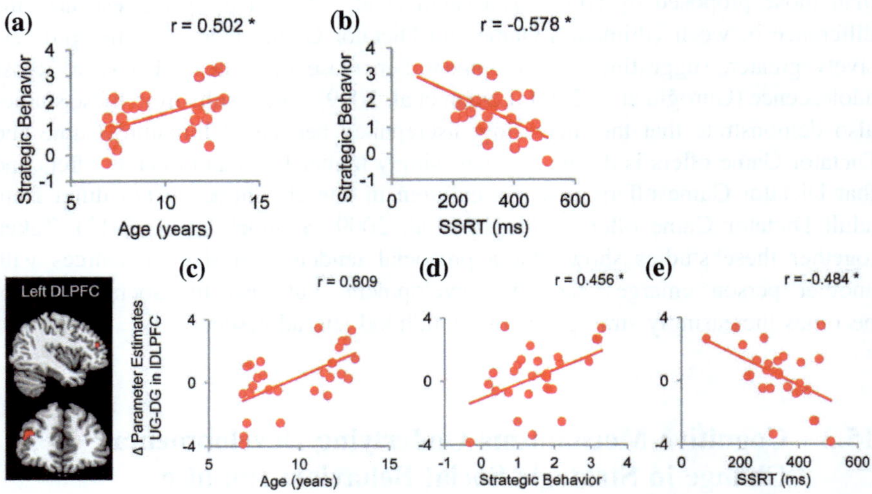

Fig. 15.1 Age-related changes in strategic behavior and recruitment of the left dorsolateral prefrontal cortex (dlPFC): **a** Strategic behavior (Ultimatum Game (UG) proposals—Dictator Game (DG) proposals) increased with age; **b** More strategic behavior was associated with better performance on a measure of impulse control (lower stop-signal reaction times [SSRTs] represent enhanced impulse control); **c** Older children recruited the left dlPFC to a larger extent when making offers in the UG compared to the DG; **d** More strategic behavior was associated with higher activation in left dlPFC when making offers in the UG compared to the DG; **e** Higher levels of impulse control were associated with higher activation levels in left dlPFC when making offers in the UG compared to the DG. Adapted from Steinbeis et al. (2012) reprinted with permission

Further evidence for a role of perspective taking in decisions about fairness comes from studies that have shown that identical unfair Ultimatum Game offers (in terms of monetary outcomes) are rejected at different rates, depending on the alternative offer that was available to the proposer (Falk et al. 2003). Specifically, an unfair offer is less often rejected when the proposer had no better alternative (e.g., a less unfair distribution of the stake) compared to cases where the proposer had a fair alternative to share the stake equally. This suggests that responders not only judge the fairness of an offer by its absolute value or the relative profits in comparison with the proposer's profits, but also in terms of the proposer's intentions behind an unfair offer. Several studies examining developmental differences in responses to unfair Ultimatum Game offers with varying alternative options indeed showed interesting age differences in such intentionality understanding in fairness considerations (Güroğlu et al. 2009; Sutter 2007). A comparison of four age groups in distinct phases of development (9-year-old preadolescents, 12-year-old early adolescents, 15-year-old mid-adolescents and 18-year-old late adolescents/young adults) showed that rejection rates of an unfair offer where the proposer had no other alternative decreased between the ages of 9 and 18. Furthermore, 9-year-olds rejected monetarily identical unfair offers regardless of whether the proposer had a fair alternative, no alternative or an even more unfair alternative. With increasing age, adolescent proposers and responders flexibly adapted their bargaining behavior in accordance with the alternative that is available to an unfair distribution, suggesting an age-related increase in the incorporation of the proposer's intentionality behind an unfair offer ("it is unfair, but there was no better alternative") into the decision-making process (Güroğlu et al. 2009).

The role of perspective taking in social decision-making has also been investigated using another economic game called the Trust Game (for a detailed description of Trust Game please see Chap. 20 in this book) (Berg et al. 1995). In the Trust Game, the first player (the trustor) is given the choice of either splitting the stake with a second player (the trustee) or transferring the entire endowment to the trustee and let the trustee split the stake. When the trustor decides to trust the trustee by transferring everything, the stake is multiplied (usually by 3 or 4). The trustee can reciprocate trust by sharing this higher stake equally, or defect trust and keep all the money. Developmental studies have shown that the frequency of trusting the second player continues to increase during adolescence (Sutter and Kocher 2007; van den Bos et al. 2010). Furthermore, young adults and older adolescents show higher levels of reciprocity than early adolescents and children (van den Bos et al. 2010). Moreover, these age differences are most pronounced in situations where the trustor takes a larger risk of losing money by trusting the second player (van den Bos et al. 2010). Trust-decisions become riskier when the amount of money that can be lost in case of defection increases. In adults, riskier trust-decisions are met by higher levels of reciprocity, which possibly reflects a recognition of the trustor's positive intentions and an appreciation of the risk the trustor took by investing in the trustee (Malhotra 2004; Pillutla et al. 2003). Van den Bos et al. (2010) showed that 9-year-olds did not reciprocate more when the trustor took a larger risk than when he/she took a relatively lower risk. This 'risk-dependent' reciprocity gradually increased across adolescence, again suggesting a continuing increase in the sensitivity to other people's intentions well into adulthood.

Taken together, behavioral studies employing different economic exchange paradigms consistently show that cognitive development related to both impulse control and perspective taking play a crucial role in understanding age-related changes in social behavior. In the following sections, we will review results from neuroimaging studies that show that developmental changes in neural circuitry implicated in impulse control and perspective taking contribute to developmental changes in social decision-making.

15.4 Neural Networks Involved in Social Decision-Making

Neuroimaging studies have elucidated a role for three distinct, but interacting, brain networks in social decision-making: a basic affective network, a cognitive regulatory network, and a 'mentalizing' network (see Fig. 15.2) (Rilling and Sanfey 2011). We will first briefly summarize findings that provide support for the notion that these three networks contribute to social decision-making. Subsequently, we will review the evidence from developmental functional magnetic resonance imaging (fMRI) studies that show that these networks are differentially sensitive to developmental change. Findings from these studies support the hypothesis that asynchronous development of these systems is associated with age-related increases in strategic social behavior and intentionality understanding in social interactions.

First, neural structures implicated in the processing of basic positive and negative affect, such as the anterior insula (Sanfey et al. 2003), ventral striatum (Tabibnia

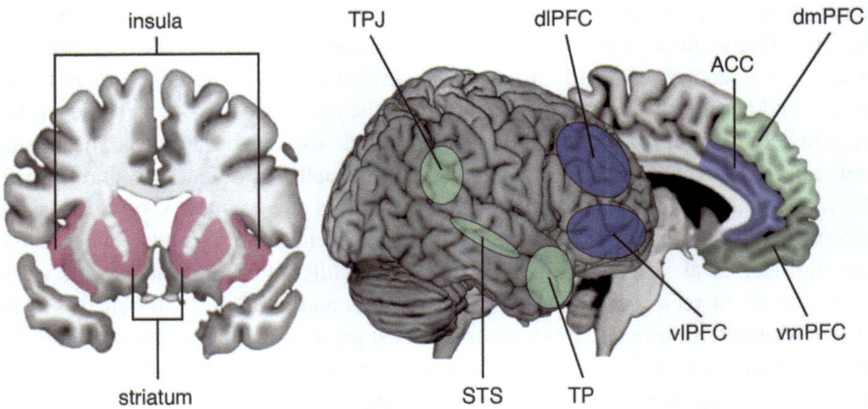

Fig. 15.2 Schematic representation of brain networks involved in social decision-making: basic affective network (*pink*), cognitive-regulatory network (*blue*) and mentalizing network (*green*). TPJ = Temporoparietal junction; STS = Superior Temporal Sulcus; TP = Temporal pole; dlPFC = dorsolateral Prefrontal Cortex; vlPFC = ventrolateral Prefrontal Cortex; ACC = Anterior Cingulate Cortex; dmPFC = dorsomedial Prefrontal Cortex; vmPFC = ventromedial Prefrontal Cortex

et al. 2008), and the amygdala (Haruno and Frith 2010) are involved in biasing social decisions, i.e., whether certain social stimuli should be approached (associated with a positive emotional signal) or avoided (associated with a negative emotional signal). For example, increased activation of the anterior insula, a brain region that is involved in encoding representations of the physiological state of the body and negative affect, such as disgust, anger, and sadness, has been associated with unreciprocated trust (Rilling et al. 2008) and receiving unfair offers in an Ultimatum Game (Sanfey et al. 2003). Interestingly, the anterior insula is activated not only when people receive unfair offers, but also when people observe someone else receiving an unfair offer (Corradi-Dell'Acqua et al. 2013) and when people have to divide resources unequally themselves (Hsu et al. 2008). In contrast, activation of the ventral striatum, a region important for processing rewards, has been associated with mutual cooperation in a prisoner's dilemma (Rilling et al. 2002, 2004b) and receiving an equal split of the stake in the Ultimatum Game (Tabibnia et al. 2008). Based on such findings, it has been argued that brain structures involved in basic emotion processing might play a role in signaling pleasantness (ventral striatum) and unpleasantness (anterior insula) of social interactions and consequently might give rise to the maintenance or elimination of such interactions.

Second, brain regions that are involved in the processing of basic positive and negative affect interact with a cognitive regulatory network including the dorsal anterior cingulate cortex (dACC) and regions in the prefrontal cortex (PFC), such as the ventrolateral prefrontal cortex (vlPFC) and the dorsolateral prefrontal cortex (dlPFC) (Rilling and Sanfey 2011). Activation in this cognitive regulatory network has been associated with cognitive control over selfish impulses and allows individuals to act in a goal-directed manner when there is a conflict between self-interest and social norms (Knoch et al. 2006; Sanfey et al. 2003). For example, activation in lateral regions of the PFC has been associated with strategic bargaining (Spitzer et al. 2007) and temporarily disrupting activity in the dlPFC using repetitive transcranial magnetic stimulation decreases rejection rates of unfair offers in an Ultimatum Game, while leaving explicit fairness judgments unaffected (Knoch et al. 2006). These findings suggest that control-related brain regions are of crucial importance for the regulation of (strategic) social behavior.

Third, when making social decisions, affective, and cognitive regulatory regions interact with a third system, namely the 'mentalizing' network. The mentalizing network includes the left and right temporoparietal junction (TPJ), superior temporal sulci, ventral, and dorsal regions of the medial PFC and the temporal poles (Frith and Frith 2010; Saxe et al. 2004). Regions in this network are consistently identified in tasks that probe reasoning about other people's mental states (i.e., mental state reasoning or *mentalizing*), for instance when people have to infer other people's thoughts, beliefs or desires (Blakemore et al. 2007; Saxe and Kanwisher 2003). Moreover, taking other people's perspective in economic exchange has repeatedly been associated with activation in regions of the mentalizing network, such as the TPJ (Güroğlu et al. 2010) and the dorsomedial PFC (Rilling et al. 2004a; van den Bos et al. 2009).

15.5 Understanding Changing Social Behavior from a Developmental Neuroscience Perspective

Longitudinal research examining changes in brain structure over time has shown that different brain regions reach maturity at different ages (Giedd et al. 1999; Gogtay et al. 2004; Shaw et al. 2008; Sowell et al. 2003). That is, sensorimotor regions in the occipital and parietal lobes reach maturity first, followed by other parts of the cortex in a posterior to anterior direction. In particular, the dlPFC and the TPJ are among the brain regions latest to fully mature, developing well into early adulthood, which in turn might (partially) explain a similar protracted developmental pattern in their associated functions, such as cognitive control (dlFPC) and perspective taking (TPJ). Indeed, models of functional brain development have posited that structural brain development might underlie emerging contributions of later maturing brain networks to social behavior (Blakemore 2008; Crone and Dahl 2012). Importantly, whereas affective networks including subcortical brain structures might reach maturity during childhood or puberty, regions of the cognitive regulatory network and the mentalizing network show continued structural changes well into the second and third decades of life (Goddings et al. 2013; Mills et al. 2012). This interplay between structural and functional brain development could underlie a developmental asynchrony between earlier maturing affective reactions to unfairness (associated with activity in basic affective network) and continued development of strategic considerations and intentionality understanding in social decision-making (associated with later maturing cognitive regulatory and mentalizing networks).

To investigate developmental changes in the neural networks involved in fairness-related decision-making Güroğlu et al. (2011) examined the neural correlates of intentionality understanding in reactions to unfairness in four phases of development: 10-year-olds, 13-year-olds, 16-year-olds, and young adults aged 20. Their results showed that age was positively associated with TPJ and dlPFC activity when participants were confronted with an unfair offer where the proposer had no alternative to making an unfair offer (see Fig. 15.3a). Rejection rates of such unfair offers decreased across adolescence, which again suggests that with age, adolescents become increasingly sensitive to the proposer's intentions behind an unfair proposal. Furthermore, mediation analyses showed that age-related decreases in rejection rates in this 'no alternative' condition were fully mediated by activation in the dlPFC (Fig. 15.3b) and the TPJ (Fig. 15.3c). Moreover, no developmental differences were observed in dACC and bilateral insula activation during reactions to unfair proposals. Together these findings suggest that the detection of violations of fairness norms and underlying neural responses in the insula and dACC mature prior to entering adolescence and that the continued development of intentionality understanding in fairness decisions across adolescence is accompanied by age-related increases in neural activity in brain regions important for perspective taking (i.e., TPJ) and impulse control (i.e., dlPFC).

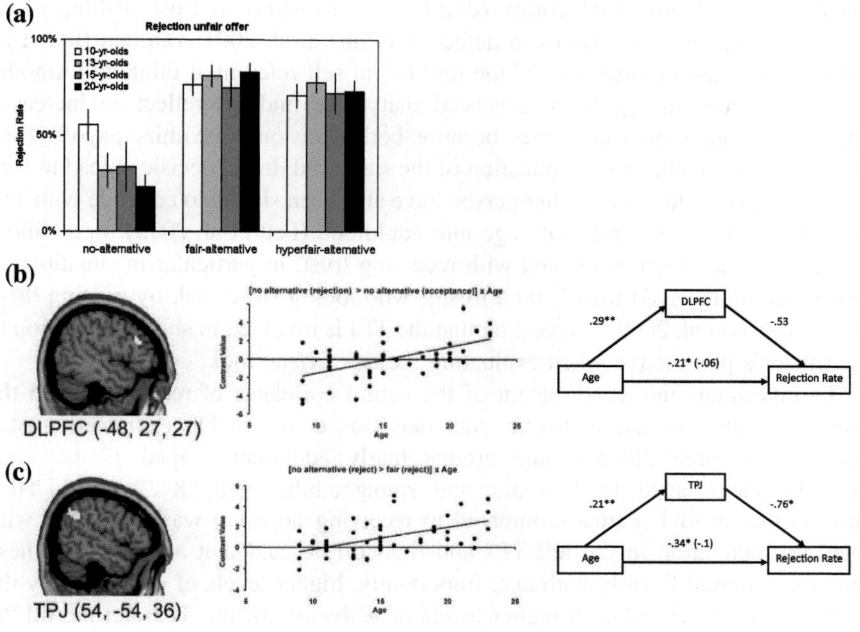

Fig. 15.3 Age-related changes in intentionality understanding in fairness are mediated by age-related increases in recruitment of the dorsolateral prefrontal cortex (dlPFC) and the temporoparietal junction (TPJ): **a** Rejection rates of unfair offers when the proposer could not make a fair offer decrease with age; **b** Rejection of unfair offers when the proposer could not make a fair offer is associated with increased recruitment of the dlPFC (**b**) and TPJ (**c**); the age-related changes in behavior are mediated by neural activation in these regions. Adapted from Güroğlu et al. (2011) reprinted with permission

The importance of the emerging contribution of dlPFC to the development of strategic social behavior was elegantly demonstrated by Steinbeis et al. (2012) in a study where they asked children (aged 6–13) to be a proposer in both the Ultimatum Game (where unfair offers can be punished) and a Dictator Game (where there is no sanction to unfair offers). They showed that activity in both left and right dlPFC when making Ultimatum Game proposals compared to Dictator Game proposals correlated positively with two measures of strategic behavior: (1) the difference between Ultimatum Game and Dictator Game offers (Fig. 15.1d) and (2) the difference between Ultimatum Game offers and the proposers' beliefs about the smallest acceptable offer to the responder. Moreover, they also showed that activity in the left dlPFC when making Ultimatum Game proposals compared to Dictator Game proposals increases between the ages of 6 and 13 (Fig. 15.1c).

The involvement of brain regions in both the cognitive regulatory network (e.g., dlPFC) and the mentalizing network (e.g., the dorsomedial prefrontal cortex and the TPJ) in social interactions has also been studied using the Trust Game. Neuroimaging studies with adult participants have demonstrated the involvement of the dorsomedial prefrontal cortex (dmPFC) in decisions to trust (Rilling et al. 2004a), as well as in decisions to defect (McCabe et al. 2001; van den Bos et al. 2009). Given the involvement of the dmPFC in self-referential thinking (Amodio and Frith 2006), it has been suggested that these findings reflect an increased attention to one's own outcomes because both decisions maximize payoffs (i.e., trust-decisions lead to a multiplication of the stake and defect decisions lead to sure gains). Decisions to trust another person have also been shown to coincide with TPJ activation, which increases with age into adulthood (Fett et al. 2013). In addition, TPJ activity has been associated with receiving trust, in particular in situations in which people received trust from a trustor who took a larger risk by trusting them (van den Bos et al. 2009), suggesting that the TPJ is involved in shifting attention to the trustor's perspective when evaluating the risk he/she took.

To investigate the development of the neural correlates of reciprocity and the role of perspective taking herein, van den Bos et al. (2011) examined trustee behavior in three different age groups (early adolescents aged 12–14 years; mid-adolescents aged 15–17 years; and young adults aged 18–22 years). They showed that receiving trust (compared to receiving no trust) was associated with increased activation in the left TPJ and right dlPFC and that activation in these regions increased linearly with age. Importantly, higher levels of risk taken by the trustor were associated with higher levels of activation of the TPJ and the dlFPC during reciprocity choices. Moreover, they showed that participants of all ages activated the dmPFC during defection, but that early adolescents also activated the dmPFC when they reciprocated trust. This latter result corroborates findings from developmental neuroimaging studies that show an age-related decrease in dmPFC activity across adolescence during mentalizing in 'theory of mind' tasks (Blakemore et al. 2007; Moriguchi et al. 2007) and an age-related increase in functional specificity of the TPJ to processing information about people's mental states compared to other forms of social information (Gweon et al. 2012; Saxe et al. 2009). It has been suggested that the age-related shift in the relative contributions of the dmPFC and the TPJ to social reasoning might tip early adolescents toward more self-oriented choices (associated with higher mPFC activation) and late adolescents toward more other-oriented choices (associated with higher TPJ activation) (Crone 2013).

15.6 Conclusions and Future Directions

In this chapter, we reviewed evidence for the notion that the abilities to control impulses and to take others' perspectives when making social decisions undergo extended development, and that these behavioral changes can be traced to brain networks involved in social decision-making developing at different rates. Much of this evidence comes from studies employing experimental paradigms with economic games, which have proven to be valuable tools for studying the development of social behavior and in particular for successfully dissecting subprocesses involved in social decision-making. Behavioral studies show that other-oriented concern and a preference for fairness have strong developmental roots. Greater sensitivity to others' intentions and more strategic behavior in bargaining situations provide evidence that continued development in perspective taking and impulse control contribute to changes in social behavior that occur across adolescence. Finally, these behavioral changes are accompanied by an increased recruitment of regions involved in impulse control (e.g., dlPFC) and perspective taking (e.g., TPJ) in decisions where perspectives of interaction partners have to be weighed against self-interest and social norms.

While elucidating developmental differences in recruitment of the dlPFC and TPJ and their involvement in social decision-making is a crucial first step, many fundamental questions remain unanswered. First, it is important not only to understand how the different brain regions (such as the TPJ or the dlPFC) are differentially recruited across development, but also how these regions interact and communicate with one another. For example, increased functional connectivity between regions of the mentalizing network (e.g., pSTS/TPJ) and brain structures implicated in the computation of value (e.g., ventral MPFC) has been associated with higher levels of prosocial behavior in adults (Hare et al. 2010). Functional connectivity in the cognitive regulatory network (Fair et al. 2008) and the mentalizing network (Burnett and Blakemore 2009a; Klapwijk et al. 2013) changes across adolescence, suggesting that developmental changes in functional connectivity may further contribute to changes in social behavior.

Second, the majority of developmental functional neuroimaging studies are based on cross-sectional data sets with participants of different ages. Although cross-sectional studies are an excellent first step to demonstrate developmental *differences*, there is a great need for longitudinal studies of social brain development. Longitudinal designs rule out the role of possible cohort differences and can give us insight on actual developmental *changes* within participants. One of the major questions in the field of developmental neuroimaging centers around the specific contributions of maturational processes relative to environmental or societal influences on the development of (social) behavior. For example, children who are accepted by their peers during childhood express higher levels of prosocial behavior and show advanced development of empathy and 'theory of mind' compared to children who are rejected by their peers (Slaughter et al. 2002). Longitudinal studies can provide insights on how developmental trajectories of individual characteristics (such as long-term peer acceptance or rejection by peers) are related to social cognitive development and how they relate to developmental trajectories of both brain structure and function. An increased understanding of this intricate interplay between a dynamic social context and a maturing brain will be crucial for developing interventions that can help children and adolescents in navigating their social worlds.

Acknowledgments The authors thank Eduard Klapwijk and Amy Skerry for helpful comments on an earlier version of this chapter and Anouk Haegens for help with the figures.

References

Amodio DM, Frith CD (2006) Meeting of minds: the medial frontal cortex and social cognition. Nat Rev Neurosci 7:268–277

Benenson J, Pascoe J, Radmore N (2007) Children's altruistic behavior in the dictator game. Evol Hum Behav 28:168–175

Berg J, Dickhaut J, McCabe K (1995) Trust, reciprocity, and social history. Game Econ Behav 10:122–142

Blake PR, McAuliffe K (2011) "I had so much it didn't seem fair": Eight-year-olds reject two forms of inequity. Cognition 120:215–224

Blake PR, Rand DG (2010) Currency value moderates equity preference among young children. Evol Hum Behav 31:210–218

Blakemore SJ (2008) The social brain in adolescence. Nat Rev Neurosci 9:267–277

Blakemore SJ, den Ouden H, Choudhury S, Frith C (2007) Adolescent development of the neural circuitry for thinking about intentions. Soc Cogn Affect Neurosci 2:130–139

Burnett S, Blakemore SJ (2009a) Functional connectivity during a social emotion task in adolescents and in adults. Eur J Neurosci 29:1294–1301

Burnett S, Blakemore SJ (2009b) The development of adolescent social cognition. Ann N Y Acad Sci 1167:51–56

Camerer C (2003) Behavioral game theory: experiments in strategic interaction. Princeton University Press, Princeton, NJ

Corradi-Dell'Acqua C, Civai C, Rumiati RI, Fink GR (2013) Disentangling self- and fairness-related neural mechanisms involved in the ultimatum game: an fMRI study. Soc Cogn Affect Neurosci 8:424–431

Crone EA (2013) Considerations of fairness in the adolescent brain. Child Dev Perspect 7:97–103

Crone EA, Dahl RE (2012) Understanding adolescence as a period of social-affective engagement and goal flexibility. Nat Rev Neurosci 13:636–650

Davidson MC, Amso D, Anderson LC, Diamond A (2006) Development of cognitive control and executive functions from 4 to 13 years: Evidence from manipulations of memory, inhibition, and task switching. Neuropsychologia 44:2037–2078

Dumontheil I, Apperly I, Blakemore S (2010) Online usage of theory of mind continues to develop in late adolescence. Developmental Science 13:331–338

Fair DA, Cohen AL, Dosenbach NU, Church JA, Miezin FM, Barch DM, Raichle ME, Petersen SE, Schlaggar BL (2008) The maturing architecture of the brain's default network. Proc Natl Acad Sci USA 105:4028–4032

Falk A, Fehr E, Fischbacher U (2003) On the nature of fair behavior. Econ Inq 41:20–26

Fehr E, Fischbacher U (2004) Third-party punishment and social norms. Evol Hum Behav 25:63–87

Fehr E, Schmidt KM (1999) A theory of fairness, competition, and cooperation. Quart J Econ 114:817–868

Fehr E, Bernhard H, Rockenbach B (2008) Egalitarianism in young children. Nature 454:1079–1083

Fett AK, Gromann PM, Giampietro V, Shergill SS, Krabbendam L (2013) Default distrust? An fMRI investigation of the neural development of trust and cooperation. Soc Cogn Affect Neurosci

Forsythe R, Horowitz JL, Savin NE, Sefton M (1994) Fairness in simple bargaining experiments. Game Econ Behav 6:347–369

Frith U, Frith C (2010) The social brain: allowing humans to boldly go where no other species has been. Philos Trans R Soc B Biol Sci 365:165–176

Giedd JN, Blumenthal J, Jeffries NO, Castellanos FX, Liu H, Zijdenbos A, Paus T, Evans AC, Rapoport JL (1999) Brain development during childhood and adolescence: a longitudinal MRI study. Nat Neurosci 2:861–863

Goddings AL, Mills KL, Clasen LS, Giedd JN, Viner RM, Blakemore SJ (2013) The influence of puberty on subcortical brain development. Neuroimage
Gogtay N, Giedd JN, Lusk L, Hayashi KM, Greenstein D, Vaituzis AC, Nugent TF 3rd, Herman DH, Clasen LS, Toga AW, Rapoport JL, Thompson PM (2004) Dynamic mapping of human cortical development during childhood through early adulthood. Proc Natl Acad Sci USA 101:8174–8179
Gummerum M, Hanoch Y, Keller M (2008a) When child development meets economic game theory: an interdisciplinary approach to investigating social development. Hum Dev 51:235–261
Gummerum M, Keller M, Takezawa M, Mata J (2008b) To give or not to give: Children's and adolescents' sharing and moral negotiations in economic decision situations. Child Dev 79:562–576
Güroğlu B, van den Bos W, Crone EA (2009) Fairness considerations: increasing understanding of intentionality during adolescence. J Exp Child Psychol 104:398–409
Güroğlu B, van den Bos W, Rombouts SA, Crone EA (2010) Unfair? It depends: neural correlates of fairness in social context. Soc Cogn Affect Neurosci 5:414–423
Güroğlu B, van den Bos W, van Dijk E, Rombouts SA, Crone EA (2011) Dissociable brain networks involved in development of fairness considerations: understanding intentionality behind unfairness. Neuroimage 57:634–641
Güth W, Schmittberger R, Schwarze B (1982) An experimental analysis of ultimatum bargaining. J Econ Behav Organ 3:367–388
Gweon H, Dodell-Feder D, Bedny M, Saxe R (2012) Theory of mind performance in children correlates with functional specialization of a brain region for thinking about thoughts. Child Dev 83:1853–1868
Harbaugh WT, Krause K, Liday SG (2003) Bargaining by children. Working paper
Hare TA, Camerer CF, Knoepfle DT, Rangel A (2010) Value computations in ventral medial prefrontal cortex during charitable decision making incorporate input from regions involved in social cognition. J Neurosci 30:583–590
Haruno M, Frith CD (2010) Activity in the amygdala elicited by unfair divisions predicts social value orientation. Nat Neurosci 13:160–161
Hoffman E, Mccabe K, Shachat K, Smith V (1994) Preferences, property-rights, and anonymity in bargaining games. Game Econ Behav 7:346–380
Hsu M, Anen C, Quartz SR (2008) The right and the good: distributive justice and neural encoding of equity and efficiency. Science 320:1092–1095
Klapwijk ET, Goddings A-L, Burnett Heyes S, Bird G, Viner RM, Blakemore S-J (2013) Increased functional connectivity with puberty in the mentalising network involved in social emotion processing. Horm Behav 64:314–322
Knoch D, Pascual-Leone A, Meyer K, Treyer V, Fehr E (2006) Diminishing reciprocal fairness by disrupting the right prefrontal cortex. Science 314:829–832
Leman PJ, Keller M, Takezawa M, Gummerum M (2009) Children's and adolescents' decisions about sharing money with others. Soc Dev 18:711–727
Liszkowski U, Carpenter M, Tomasello M (2008) Twelve-month-olds communicate helpfully and appropriately for knowledgeable and ignorant partners. Cognition 108:732–739
Malhotra D (2004) Trust and reciprocity decisions: the differing perspectives of trustors and trusted parties. Organ Behav Hum Decis Process 94:61–73
McCabe K, Houser D, Ryan L, Smith V, Trouard T (2001) A functional imaging study of cooperation in two-person reciprocal exchange. Proc Natl Acad Sci USA 98:11832–11835
Mills KL, Lalonde F, Clasen LS, Giedd JN, Blakemore SJ (2012) Developmental changes in the structure of the social brain in late childhood and adolescence. Soc Cogn Affect Neurosci
Moriguchi Y, Ohnishi T, Mori T, Matsuda H, Komaki G (2007) Changes of brain activity in the neural substrates for theory of mind during childhood and adolescence. Psychiatry Clin Neurosci 61:355–363

Nelson E, Leibenluft E, McClure E, Pine D (2005) The social re-orientation of adolescence: a neuroscience perspective on the process and its relation to psychopathology. Psychol Med 35:163–174

Pillutla MM, Malhotra D, Murnighan JK (2003) Attributions of trust and the calculus of reciprocity. J Exp Soc Psychol 39:448–455

Rilling J, Sanfey A (2011) The neuroscience of decision-making. Annu Rev Psychol 62

Rilling J, Gutman D, Zeh T, Pagnoni G, Berns G, Kilts C (2002) A neural basis for social cooperation. Neuron 35:395–405

Rilling JK, Sanfey AG, Aronson JA, Nystrom LE, Cohen JD (2004a) The neural correlates of theory of mind within interpersonal interactions. Neuroimage 22:1694–1703

Rilling JK, Sanfey AG, Aronson JA, Nystrom LE, Cohen JD (2004b) Opposing BOLD responses to reciprocated and unreciprocated altruism in putative reward pathways. NeuroReport 15:2539–2543

Rilling JK, Goldsmith DR, Glenn AL, Jairam MR, Elfenbein HA, Dagenais JE, Murdock CD, Pagnoni G (2008) The neural correlates of the affective response to unreciprocated cooperation. Neuropsychologia 46:1256–1266

Rueda MR, Rothbart MK, McCandliss BD, Saccomanno L, Posner MI (2005) Training, maturation, and genetic influences on the development of executive attention. Proc Natl Acad Sci USA 102:14931–14936

Sally D, Hill E (2006) The development of interpersonal strategy: autism, theory-of-mind, cooperation and fairness. J Econ Psychol 27:73–97

Sanfey AG, Rilling JK, Aronson JA, Nystrom LE, Cohen JD (2003) The neural basis of economic decision-making in the ultimatum game. Science 300:1755–1758

Saxe R, Kanwisher N (2003) People thinking about thinking people. The role of the temporo-parietal junction in "theory of mind". Neuroimage 19:1835–1842

Saxe R, Carey S, Kanwisher N (2004) Understanding other minds: linking developmental psychology and functional neuroimaging. Annu Rev Psychol 55:87–124

Saxe RR, Whitfield-Gabrieli S, Scholz J, Pelphrey KA (2009) Brain regions for perceiving and reasoning about other people in school-aged children. Child Dev 80:1197–1209

Schmidt MF, Sommerville JA (2011) Fairness expectations and altruistic sharing in 15-month-old human infants. PLoS ONE 6:e23223

Selman R (1980) The growth of interpersonal understanding. Academic Press, New York

Shaw A, Olson KR (2012) Children discard a resource to avoid inequity. J Exp Psychol Gen 141:382–395

Shaw P, Kabani NJ, Lerch JP, Eckstrand K, Lenroot R, Gogtay N, Greenstein D, Clasen L, Evans A, Rapoport JL, Giedd JN, Wise SP (2008) Neurodevelopmental trajectories of the human cerebral cortex. J Neurosci 28:3586–3594

Shaw A, Montinari N, Piovesan M, Olson KR, Gino F, Norton MI (2013) Children develop a veil of fairness. J Exp Psychol Gen

Singer T (2006) The neuronal basis and ontogeny of empathy and mind reading: review of literature and implications for future research. Neurosci Biobehav Rev 30:855–863

Singer T, Fehr E (2005) The neuroeconomics of mind reading and empathy. Am Econ Rev 95:340–345

Slaughter V, Dennis MJ, Pritchard M (2002) Theory of mind and peer acceptance in preschool children. Br J Dev Psychol 20:545–564

Sloane S, Baillargeon R, Premack D (2012) Do infants have a sense of fairness? Psychol Sci 23:196–204

Smith CE, Blake PR, Harris PL (2013) I should but I won't: why young children endorse norms of fair sharing but do not follow them. PLoS ONE 8:e59510

Sowell ER, Peterson BS, Thompson PM, Welcome SE, Henkenius AL, Toga AW (2003) Mapping cortical change across the human life span. Nat Neurosci 6:309–315

Spitzer M, Fischbacher U, Herrnberger B, Grön G, Fehr E (2007) The neural signature of social norm compliance. Neuron 56:185–196

Steinbeis N, Bernhardt BC, Singer T (2012) Impulse control and underlying functions of the left DLPFC mediate age-related and age-independent individual differences in strategic social behavior. Neuron 73:1040–1051

Sutter M (2007) Outcomes versus intentions: on the nature of fair behavior and its development with age. J Econ Psychol 28:69–78

Sutter M, Kocher MG (2007) Trust and trustworthiness across different age groups. Game Econ Behav 59:364–382

Svetlova M, Nichols SR, Brownell CA (2010) Toddlers' prosocial behavior: from instrumental to empathic to altruistic helping. Child Dev 81:1814–1827

Tabibnia G, Satpute AB, Lieberman MD (2008) The sunny side of fairness: preference for fairness activates reward circuitry (and disregarding unfairness activates self-control circuitry). Psychol Sci 19:339–347

Takagishi H, Kameshima S, Schug J, Koizumi M, Yamagishi T (2010) Theory of mind enhances preference for fairness. J Exp Child Psychol 105:130–137

Talwar V, Murphy SM, Lee K (2007) White lie-telling in children for politeness purposes. Int J Behav Dev 31:1–11

Tomasello M, Vaish A (2013) Origins of human cooperation and morality. Annu Rev Psychol 64:231–255

van den Bos W, van Dijk E, Westenberg M, Rombouts SARB, Crone EA (2009) What motivates repayment? Neural correlates of reciprocity in the Trust Game. Soc Cogn Affect Neurosci 4:294–304

van den Bos W, Westenberg M, Van Dijk E, Crone E (2010) Development of trust and reciprocity in adolescence. Cogn Dev 25:90–102

van den Bos W, van Dijk E, Westenberg M, Rombouts SARB, Crone EA (2011) Changing brains, changing perspectives: the neurocognitive development of reciprocity. Psychol Sci 22:60–70

Warneken F, Tomasello M (2006) Altruistic helping in human infants and young chimpanzees. Science 311:1301–1303

Wellman HM, Cross D, Watson J (2001) Meta-analysis of theory-of-mind development: the truth about false belief. Child Dev 72:655–684

Zahn-Waxler C, Radke-Yarrow M, Wagner E, Chapman M (1992) Development of concern for others. Dev Psychol 28:126–136

Chapter 16
Neuroeconomic Approaches in Mental Disorders

S. Lis and P. Kirsch

Abstract Approaches from neuroeconomy have recently received increased attention in the investigation of mental disorders. In this chapter, we will give an overview of concepts and paradigms from neuroeconomics that have been applied in different mental disorders and summarize first results in this emerging field. We focus, thereby, on 'social decision-making' which constitutes one of the main concepts of neuroeconomy. First findings suggest that these approaches may prove to be promising research tools in the investigation of social functioning, which is a prominent symptom domain in many mental disorders. In contrast to self- and observer-based questionnaires that have provided information on how interaction behaviour is subjectively perceived by the patients themselves or their social environment, the variety of exchange games from behavioural economy allow for a direct and, thus, unbiased assessment of interaction behaviour. So far, findings suggest that neuroeconomic tools are suited to uncover alterations in social interaction behaviour in mental disorders, such as anxiety disorders, depression, borderline personality disorder, attention-deficit/hyperactivity disorder, but also schizophrenia and autism. However, the investigation of social interaction behaviour in mental disorders poses particular challenges. Deficits in basal or complex cognitive functions, such as working memory, deficits in basal social cognitive processes, such as the recognition of emotional facial expressions and a lower socioeconomic status due to long periods of illness and unemployment can be assumed to affect interaction behaviour. These have to be disentangled for the purpose of characterizing social decision-making in mental disorders and understanding the causes underlying its alterations. The combination of neuroeconomic

S. Lis (✉)
Institute of Psychiatric and Psychosomatic Psychotherapy,
Central Institute of Mental Health, Medical Faculty Mannheim,
University of Heidelberg, Mannheim, Germany
e-mail: stefanie.lis@zi-mannheim.de

P. Kirsch
Department of Clinical Psychology, Central Institute of Mental Health,
Medical Faculty Mannheim, University of Heidelberg, Mannheim, Germany
e-mail: Peter.Kirsch@zi-mannheim.de

M. Reuter and C. Montag (eds.), *Neuroeconomics*, Studies in Neuroscience, Psychology and Behavioral Economics, DOI 10.1007/978-3-642-35923-1_16

approaches and their elaborated quantitative models with methods from experimental psychology und cognitive neurosciences seems a promising avenue to achieve this goal.

16.1 Introduction

Neuroeconomics has been defined as a discipline originated by merging economics, psychology and neuroscience (Glimcher and Rustichini 2004), combining their theoretical and methodological approaches. From behavioural economics, this new discipline inherited the demand to define a unified theory of human decision-making or even human behaviour. However, when taking mental disorders and the related altered behaviours into account, it becomes obvious that this demand is almost unapproachable within this context. Clinical sciences are far away from defining single theories which account for all behavioural alterations, even when only focusing on one particular mental disorder. Therefore, it is obvious that neuroeconomic approaches to mental disorders cannot provide theories explaining their whole etiology. However, with the emergence of neuroeconomics, a number of central concepts, mainly adopted from economics, came into the focus of neuroscience. These may provide a theoretical and methodological framework and, therefore, be of great value when describing impaired behaviours in patients suffering from mental disorders and searching for their neurobiological underpinnings.

A variety of avenues by which neuroeconomic approaches can foster the investigation of alterations in decision-making and its cerebral correlates in mental disorders are discussed in a recent series of commentaries addressing disorders, such as depression (Ernst 2012), anxiety disorders (Hartley and Phelps 2012), addiction (Monterosso et al. 2012) and attention-deficit/hyperactivity disorder (Sonuga-Barke and Fairchild 2012). A comprehensive review on the usability of neuroeconomics for psychiatric research (Hasler 2012) characterized these central concepts, such as utility, expectation, uncertainty and gain-loss-asymmetry in relation to mental disorders. Hasler (2012) outlined the link between specific neuroeconomic parameters and behavioural correlates which bear a special importance within the context of different mental disorders. These parameters, such as specific coefficients of value function, time discounting function or utility function, rely on a broad database of studies in healthy subjects. They constitute not only specific quantitative measurements of behaviour, but also allow for linking them to certain neurobiological systems, such as specific cerebral structures or neurotransmitter systems.

Three different main domains can be distinguished in neuroeconomics: 'social decision-making', 'decision-making under risk and uncertainty' and 'intertemporal choice' (Loewenstein et al. 2008). From these domains, social decision-making has attracted most attention in the context of psychiatric research. However, the methodology to investigate this domain of functioning still needs to be developed. It can be assumed that neuroeconomic approaches will be able to fill the gap by providing ecologically valid paradigms that allow the probe of those functions regarded as crucial for most mental disorders (Sharp et al. 2012): Alterations in social interaction constitute a significant symptom domain, e.g. in social phobia or borderline personality disorder (BPD), they modulate the extent of symptoms within other domains of psychopathology, or affect recovery and long-term outcomes (see King-Casas and Chiu 2012). This chapter will, therefore, focus on the application of neuroeconomic approaches to social functioning in mental disorders.

The importance of alterations in social cognition for the psychopathology, and especially for the often pronounced impairments of everyday functioning in the context of mental disorders, has been noticed with growing interest during the last years. Recently, it has been emphasized that the study of social cognition should be extended to a second person perspective (Schilbach et al. 2013): it allows to take into account that particular social cognitive deficits in mental disorders might only be observable in real interpersonal interactions as they are used in neuroeconomic games. However, the possible usefulness of neuroeconomic approaches, particularly using the approach of social exchange games, to study social decision-making received attention only very recently. Due to this, the number of available studies that apply such games in the investigation of mental disorders remains sparse. This particularly holds true when compared to the boom of these approaches in the social neurosciences in general. However, the interest in these approaches is momentarily drastically increasing which becomes evident in the growing number of both empirical studies and theoretical papers that emphasize the usefulness of these approaches for the understanding of mental disorders in the context of cognitive, affective and social neuroscience (Kishida et al. 2010; Hasler 2012; Kishida and Montague 2012; Sharp et al. 2012).

Beyond the availability of well-founded computational models of social exchanges, these approaches may constitute ecologically valid and well-defined test settings to study single social cognitive processes and their importance for social interactions. Social exchanges require a multitude of different cognitive, affective and social processes. Thus, Carter (2012) emphasized that social 'choice behaviour relies on the integration in the brain of a diverse array of specialized processes that vary according to changing context' (see also Loewenstein et al. 2008). With that, neuroeconomic approaches allow the study of single specific social cognitive functions, such as emotion recognition processes or mentalizing that are assumed to be altered in mental disorders. However, alterations of these processes are here investigated in a complex interplay with other processes during social interactions. This is in contrast to analyzing them in isolation in a highly artificial manner as it is done, e.g. in traditional multiple choice emotion recognition tasks. Beyond the consideration of specific social cognitive processes being part of a complex functional network, this approach extends the study of social cognition by taking the goal, i.e. the 'why', of processing into account in that it investigates overt or covert behaviour within a specific social interaction context (Willems 2011).

Interestingly, while there are a number of recent behavioural studies using social exchange games in clinical research, only a small number of them have applied neuroscientific methodology. In the next section, we will summarize the findings separately for a number of mental disorders. Finally, we will discuss their applicability for clinical social neuroscience.

16.2 Exchange Games in Specific Mental Disorders

From the variety of paradigms available, the majority of studies have applied approaches that focus on the assessment of trust and fairness and the perception of social norms.

The most often applied exchange games are the ultimatum game (UG), and the trust game (TG) (for a detailed description of economic games please see Chap. 2 of this book). These approaches were applied in the investigation of alterations of social interaction behaviour in schizophrenia, depressive disorders, borderline personality disorder, anxiety disorders, social phobias, psychopathy, attention-deficit-hyperactivity disorder and autism spectrum disorder. The majority of studies focus on virtual interaction situations within dyads, in which the behaviour of an alleged social interaction partner is simulated by computer algorithms. Studies that actually measure the behaviour of real partners constitute a minority (e.g. Agay et al. 2008; Chiu et al. 2008; King-Casas et al. 2008; Zhang et al. 2012).

Studies differ in regard to whether exchanges are investigated as a single exchange with several partners or as multi-round games, i.e. as repeated exchanges with the same partner, shifting the emphasis from certain cognitive processes (e.g. trust) to others (e.g. the maintaining of interpersonal relations involving the building of reputation). In most exchange games, the roles of the partners differ in regard to the cognitive processes involved, e.g. taking the role of the investor or the trustee during the TG, or of the proposer or the responder in the UG. Studies differ in regard to whether the patient's behaviour is studied in fixed or varying roles within the single exchange games, e.g. BPD patients always represent the trustees in a multi-round trust game (e.g. King-Casas et al. 2008), or schizophrenic patients act as proposers as well as responders in an UG (e.g. Agay et al. 2008). Several studies suggest that the occurrence of alterations in social behaviour depends on the specific role during the exchange game, as e.g. patients show behavioural alterations as responders but not as proposers (e.g. Agay et al. 2008). Thus, it is important to take the details of the realization of neuroeconomic approaches into account when assessing heterogeneity and homogeneity in the still limited number of neuroeconomic studies in mental disorders.

16.2.1 Schizophrenia

Agay et al. (2008) conducted the first study with schizophrenic patients applying the UG as a neuroeconomic exchange game. They found a comparable exchange behaviour in patients and healthy subjects in the role of the responder, but alterations of behaviour in the patients' group in the role of the proposer. After rejection of an offer, both groups increased their offer in the second round. However, in case of an accepted offer, healthy subjects reduced their offer in the subsequent trial, while the schizophrenic patients did not adjust their behaviour. Approximately 20 % of the patients exhibited nonstrategic behaviour, in that they further reduced their offer after a rejection or increased their offer after acceptance in the previous trial. Such behaviour was observed in none of the healthy subjects. A study by Billeke et al. (2015) supported this finding of altered strategic behaviour in schizophrenic patients when taking the role of the proposer during an UG. In general, the patients made more often hyperfair offers and showed a higher variation of behaviour towards the alleged human co-players, but also when playing with a computer. During the course of the interaction, the pattern of adaptation to risk was opposite in healthy subjects and patients when taking the social and nonsocial condition into account. Moreover, the authors were able to link these differences to alterations in neural processing during the anticipation of the responder's choice, i.e. to brain oscillatory activity in frontal and temporoparietal brain regions.

Other studies suggest that schizophrenic patients differ also as responder during UG exchanges: patients accepted more unfair offers than healthy subjects (Csukly et al. 2011; Wischniewski and Brune 2011). Such a behaviour was also observed in subjects with a high level of interpersonal schizotypal symptoms (van't Wout and Sanfey 2011). These results have been interpreted as an indication of a reduced sensitivity towards unfairness to their own disadvantage in schizophrenic patients (Wischniewski and Brune 2011), and a blunted emotional response (van't Wout and Sanfey 2011). However, it seems worth pointing out that the socioeconomic status of patients that suffer from chronic mental disorders is quite often reduced compared to healthy controls. Thereby, the value of the gain at risk may differ between groups and thus, differentially influence the exchange behaviour of these groups. Until now, the majority of studies do not take this presumably important factor into account.

Wischniewski and Brune (2011) suggest that in general, the exchange behaviour of schizophrenic patients seems to be determined by the same rules as that of healthy subjects (see also de la Asuncion et al. 2015). In both groups, rejection rates increased with the unfairness of offers. Similarly, both groups equalled in regard to the likelihood of punishing unfairness of a proposer when they watched two individuals during the dictator game (DG). Based on their finding, they conclude that moral value appreciation is not affected in schizophrenia.

In contrast, Csukly et al. (2011) found higher acceptance rates of unfair proposals and lower acceptance rates with fair proposals in the schizophrenic patients during an UG. This contradicts an overall increased tendency to accept offers independent of their height due to a higher need caused by a lower socioeconomic status. In contrast, these data agree with a general lack in differential reactions during social interactions. A reduced influence of the partner's fairness has been confirmed by subsequent studies: investment were not influenced by the trustee's fairness during an UG (de la Asuncion et al. 2015) or during a trust game (Lis et al. 2011; see also Fett et al. (2012) for similar findings in a mixed group of patients with different psychotic disorders). Beyond the partners' fairness, emotional facial expressions may provide socially relevant context information and may serve as cues towards the partners' future behaviour. Studies that investigated the influence of emotional expressions on the adaptation of behaviour, consistently suggest impairments in schizophrenia (Csukly et al. 2011; Lis et al. 2011; de la Asuncion et al. 2015). All studies revealed that the interaction behaviour is modulated by the facial expression of the social partner in healthy subjects, but not in schizophrenic patients. This holds true for the responder behaviour in the UG and the investor behaviour in the TG. The behavioural alteration towards facial expressions seems to be particularly relevant to the case in which they serve to guide behaviour during social interactions, since no deficits were observed in a simple emotion evaluation task (Lis et al. 2011). However, caution is necessary when interpreting these findings as deficits specific for using emotional cues to adapt social behaviour. None of the studies applied a control condition to exclude, as an underlying mechanism, more general cognitive impairments in the adaptation of behaviour to complex cues, i.e. to context information without a social and emotional content.

16.2.2 Depressive Disorders

During the last few years, the interest in neuroeconomic tasks has tremendously increased in the investigation of social dysfunctioning in depressive disorders (see for a recent review Wang et al. 2015). In contrast to studies in schizophrenia, studies focused more often on the affective responses to fair and unfair reactions.

As it was reported for patients suffering from schizophrenia, alterations in the behaviour of patients with major depressive disorder depend on the role taken during an UG (Destoop et al. 2012). While no alterations were observed when the patients acted as the responder, they offered a higher percentage of their account when in the role of the proposer compared to healthy subjects (see similar findings Scheele et al. 2013). Destoop et al. (2012) interpreted this behaviour as a strategy to avoid rejection which agrees with an enhanced rejection sensitivity in individuals with depressive disorders (Rosenbach and Renneberg 2011).

However, findings are not consistent regarding the behaviour when in the role of the responder during UG. In line with findings from Radke et al. (2013b), Wang et al. (2014) reported, in general, a lower rate of acceptance in patients with major depressive disorders (MDD). Moreover, in the case of unfair offers, there was no differential response towards an alleged human co-player in contrast to a computer control condition in MDD, suggesting that this behavioural alteration is not characteristic for social interactions: while healthy participants particularly rejected more often the very unfair offers from a human partner, MDD patients rejected very unfair offers from both the human partner and the computer (Wang et al. 2014). In contrast, Scheele et al. (2013) found more rejections only in the case of slightly unfair offers. When studying student populations with depressive symptoms, no behavioural alterations were found (Gradin et al. 2015; see also Pulcu et al. 2015) or even increased acceptance rates (Harle et al. 2010). This suggests that altered interaction behaviour in clinical samples is not solely attributable to the effects of depressive symptoms.

Alterations of exchange behaviour seem to not be caused by impaired judgements of fairness in general. Data from Scheele et al. (2013) found no evidence towards altered perception of fairness, although data of Wang et al. (2014) suggest that patients may tend to be more sensitive in the case of very unfair offers when fairness is evaluated independently of an interaction.

Several studies consistently revealed altered emotional responses during interactions in depressive disorders. Unfair offers during an UG elicited stronger negative emotions (Harle et al. 2010) and higher guilt (Pulcu et al. 2015) and defection during a PD evoked stronger feelings of betrayal (Gradin et al. 2016). Moreover, depressive participants reported lower satisfaction in the case of cooperative behaviour of both the participant and the partner during a PD (Gradin et al. 2016) and reduced levels of happiness to fair offers during an UG (Gradin et al. 2015). These alterations were accompanied by alterations in brain activation during interactions suggesting altered processing of fairness and reward as well as linked emotional responses. Depressive participants showed a reduced activation in the left dorsolateral prefrontal cortex, which was correlated with the experience of increased guilt when the participant defected and the partner cooperated during PD (Gradin et al. 2016). Increasing engagement of the nucleus accumbens and the dorsal caudate with increasing fairness of an offer during UG was attenuated in depressive participants resulting in a lower activation in these brain area compared to healthy subjects (Gradin et al. 2015). In contrast, increasing unfairness of offers was linked to increasing activation of the dorsal anterior cingulate and the insula in both groups (Gradin et al. 2015).

Zhang et al. (2012) used a modified trust game to investigate the effects of altruism and risk-taking on social interaction behaviour in depressive subjects. All subjects played as trustees. They were informed that during each exchange they would play with a different real co-player, while actually, the investments were determined by the experimental setting. In this modification of the trust game, the investor asked the trustee to repay a specific ratio of the amount at his/her disposal ("beneficial" with 20 %, "equal" with 50 % and "unfair" with 80 %). Participants could choose to repay exactly the amount requested, more than requested ("altruistic" behaviour) or less than requested ("deceptive" behaviour). Additionally, the probability (25 or 75 %) with which the investor would detect whether this investment differed from the requested amount was given. If deception was detected, the monetary units earned in the single trial were confiscated as punishment. Subjects with a depressive disorder differed qualitatively in their preference of choice options, in that they chose altruistic repayments in the case of beneficial and equal requests less often, and also deceptive repayments in the case of unfair requests less often. Thus, effects of previous exchanges seem to be weaker in depressive patients who were less generous in the case of benevolent partners, as well as less punishing in the case of malevolent partners. Beyond that, risk of deception affected the patients' behaviour as well: they chose deceptive responses less frequently than healthy subjects when the risk for detection was low. This finding was interpreted as indicating enhanced risk avoidance in depressive disorders, but also suggests that these patients might have problems with integrating information from different domains, such as risk-taking, inferring the intentions of others and decision-making. When looking on deceptive behaviours of the participants themselves, Shao et al. (2015) identified a reduced efficiency in the integration of lateral prefrontal-striatal/limbic networks in MDD patients, although during this modified trust game behavioural alterations in MDD could not unequivocally be confirmed.

Besides studies that point to alterations in specific aspects of interaction behaviour during economic tasks in depressive disorders, others suggest that decision-making in social interactions is determined by mechanism similar to those in healthy subjects. Examples are the effects of emotional expressions and the influence of offer alternatives (Radke et al. 2013b; see also for an attenuated effect of smiling faces as cues for fair behaviour of a trustee in case of a comorbid depressive disorder in BPD, Franzen et al. 2011).

Data suggest that alterations in social interaction behaviour in depressive disorders can be quantified by exchange games. However, future studies have to address the underlying causes of these alterations, such as an increased rejection sensitivity as suggested by Destoop et al. (2012), or an inability to convert anger into a social action like a 'fight back' towards a partner linked to neuroanatomical alterations in nucleus caudatus or the anterior insula activations as suggested by Zhang et al. (2012). Moreover, interaction behaviour may be differentially affected by the participants' gender, possibly modulated by specific symptoms, such as suicidal ideations (Caceda et al. 2014). It seems obvious that exchange games have to be combined with approaches from experimental psychology that stress different cognitive and social cognitive processes during an exchange situation to understand the mechanisms responsible for altered interaction behaviour and altered emotional responses.

16.2.3 Anxiety Disorders

Applying an UG, Grecucci et al. (2012) found altered behaviour of subjects with anxiety disorder in the responder's, but not the proposer's, role. Patients accepted more unfair offers than healthy subjects. This effect depended on the clinical subtype, being observable in those patients with generalized anxiety disorder, but not with panic disorder. Additionally, the pharmacological treatment modulated interaction behaviour: treatment with serotonin reuptake inhibitors normalized these behavioural alterations. The authors discuss the relevance of interpersonal factors and assertiveness in decision-making of patients with anxiety disorders.

As an alternative approach, a TG was used to investigate social interaction behaviour of patients with social anxiety disorders in the role of the investor (Sripada et al. 2009). Interaction during a multi-round trust game played with a human partner was associated to an attenuated activation in medial prefrontal cortex in patients with an anxiety disorder compared to healthy subjects (Sripada et al. 2009). The authors linked this result to deficits in mentalizing, since no comparable pattern of altered cerebral activation was observed when playing with a computer. However, during the computer game, subjects knew they had a 50 % chance to receive a fair repay, while there were three virtual 'human' partners that differed in regard to their probability of splitting the investment (with a repayment of 75, 50 or 25 %). Thus, the alleged human interaction was linked to a higher uncertainty of exchange behaviour, a factor well known to influence decision-making during exchange games.

A study by McClure et al. (2007) addressed social interaction behaviour in adolescents with anxiety and mood disorders (A/D) who were free of psychotropic medication during the time of testing. The authors applied the Prisoners' Dilemma (PD) to directly measure pro-social, submissive and hostile or competitive behaviours together with self-reports on anger and the experience of cooperativeness. Alterations in play behaviour suggest a priority on maintaining positive interpersonal interactions by choosing pro-social behaviour, but not conflict avoidance: A/D patients more often cooperated compared to their healthy controls when the co-player had cooperated in the preceding round. However, there were no differences between groups in regard to reactions to noncooperative behaviour of the co-players. In contrast, a PD study by Rodebaugh et al. (2013) reported that patients with generalized social anxiety give less during a PD interaction and tended to show less reaction to defection. Interestingly, female patients reported a higher extent of anger after defection by the partner (Sripada et al. 2009). Yet, this self-report was not accompanied by an increased probability for choosing a defecting strategy in the following round. As the authors note, at first glance, such an altered interaction pattern does not appear to be the reason for interpersonal dysfunction. However, social interactions are complex feedback-oriented systems. Within these systems, comparable behavioural patterns might be perceived by social partners in different manners depending on modulating factors, such as expectations and expectation violations intertwined with specific personality traits. Until now, there has been a lack of studies that address the mechanisms of the interpretation of social actions, such as the interpretation of self-protective behaviour as limited self-disclosure or an enhanced vulnerability by reduced self-assertiveness, especially in the context of a mental disorder in one of the interaction partners.

16.2.4 Personality Disorders

16.2.4.1 Borderline Personality Disorder

Deficits in social interactions constitute one of the central psychopathological features of BPD. The diagnostic criteria of the DSM-IV-TR describe the impairments in social relationships as 'frantic efforts to avoid real or imagined abandonment', 'pattern of unstable and intense interpersonal relationships characterized by alternating between extremes of idealization and devaluation' and 'inappropriate, intense anger or difficulty controlling anger'. A multitude of studies confirm the existence of alterations in social relationships in BPD (for review see Lis and Bohus 2013).

The most prominent neuroeconomic study in BPD was done by King-Casas et al. (2008). They applied a 10-round trust game with real interactions, while measuring the BOLD response. While the investor was always a healthy subject, the trustees' role was taken by a BPD patient or a healthy subject. Dyads including a BPD trustee showed less cooperation over the course of the game. This was due to those exchange rounds in which the investor transferred a very low amount of monetary units. In these rounds, healthy trustees often repaid a high sum. Such behaviour might aim at encouraging the investor to be more cooperative and transfer greater investments in subsequent trials. In contrast, BPD trustees reacted with a low repayment, i.e. they showed a less generous behaviour and, thereby, an inability to "repair" the cooperation. These findings are in line with those applying a PD: BPD patients showed a lower proportion of cooperative responses than healthy controls and patients with a bipolar disorder when playing with an alleged partner (Saunders et al. 2015). Since the partner's behaviour was preprogrammed to follow a tit-for-tat strategy, a break of cooperation by the BPD patient was followed by a defection by the partner. This choice may result in a permanent break of cooperation as described by King-Casas et al. (2008) if the patients did not initiate a repair of the relation themselves by switching to a cooperative behaviour after defection of the partner. This implies that particularly reconciliation after problems in social relations may represent a core deficit in this mental disorder. However, further studies must test whether this interpretation for the emergence of uncooperative behaviour in specific social constellations holds true.

The study by King-Casas et al. (2008) linked alteration in overt interaction behaviour in BPD to alterations in cerebral activation, especially a lack in the modulation of the activation in the anterior insula by the height of the received investment in the BPD group. The authors interpreted this finding as a lack in the perception of the violation of social norms in the behaviour of social partners, and speculated that this might be linked to negative expectations during social interactions. However, a subsequent study by Franzen et al. (2011) suggests that the perception of social norms does not differ between BPD patients and controls. When asked to assess the fairness of a trustee's behaviour during a trust game, the BPD patients' ratings differentiated the degrees of fair and unfair behaviour in the same manner as healthy subjects. This suggests that the ability of recognizing unfairness is uncompromised in BPD. Further studies are necessary to clarify whether the findings of King-Casas et al. (2008) can indeed be explained by an altered expectancy on the fairness of social partners in BPD.

It can be assumed that context information is essential for the formation and adaptation of expectancies in social encounters. Findings of Franzen et al. (2011) suggest that BPD patients use context information to guide social interaction behaviour differently from healthy subjects. Social context might be given by a potential break-down of a cooperation (see Meyer-Lindenberg 2008), but equally by social cues, such as the emotional facial expressions of interaction partners. In healthy subjects, the presence of emotional facial cues strongly affects interaction behaviour: they ignore the overall fairness of an interaction partner and transfer equal investments to fair and unfair partners with varying emotional expressions (Franzen et al. 2011; Lis et al. 2011, 2013). In contrast, BPD patients adapt their investment behaviour not only to the emotional facial cues signalling the expected repayment in the single round, but also simultaneously integrate the preceding experiences with the fairness of a trustee, i.e. they transferred less money to unfair than to fair partners, while simultaneously taking the specific provided social cues in a single round into account. An altered influence of emotional facial expression on exchange behaviour was also found by Polgar et al. (2014): while healthy subjects accepted more offers during an UG when the partner displayed a happy facial expression, this held true for the BPD patients only in the case of fair offers. Thereby, BPD patients accepted more offers overall than healthy controls.

The only study with a nonsocial control task applied a trust game (Unoka et al. 2009). Over the course of five rounds, healthy investors increased the amount of transferred monetary units, while BPD patients did not change their investment behaviour. This finding suggests that BPD patients failed to develop trust. However, this study did not provide any feedback on the interaction partner's behaviour during the game. Thus, one may argue that it does not measure interaction behaviour, defined as a two-way encounter during which the behaviour of one individual influences that of the other and that uncertainty about the behaviour of the partner may contribute to this finding. Nevertheless, Unoka et al. (2009) confirmed that the behavioural alteration seen in BPD is specific for a social encounter, since no comparable behaviour was observed in a control task that asked subjects to play with a computer.

The findings of the various studies using the trust game in BPD all suggest altered interaction behaviour. Thus, further studies seem promising to gain further knowledge on the underlying mechanisms and modulating variables. A recent study by Lonnqvist et al. (2012) points to the importance of personality traits for the behavioural pattern seen by King-Casas et al. (2008): the inability to repair ruptured cooperation seems to be linked to a combination of high neuroticism and low agreeableness, i.e. the interplay of two personality dimensions identified by the five-factor model (see also Thielmann et al. 2014).

Although these findings suggest a high importance of personality traits for the exchange behaviour in BPD, other findings suggest a different link between personality traits and exchange behaviour in healthy individuals and BPD subjects. Wischniewski and Brune (2012) applied the dictator game and found a diverging effect of personality traits in BPD compared to healthy subjects. In the healthy group, selfish behaviour was linked to high Machiavellianism and low agreeableness. An opposite pattern was observed in the BPD group. Additionally, high extraversion and high openness went along with less costly punishment and higher neuroticism was linked with higher punishment in the BPD group, but not in healthy subjects. The authors interpreted these results in BPD as—as they state

—'angry retaliation' which is motivated by identification with the victim's perspective, suggesting different motivations for the observed behaviour in BPD patients and healthy subjects.

In an alternative approach, Bartz et al. (2011) investigated the effects of the 'pro-social' peptide oxytocin on interaction behaviour in BPD. They applied a three-round "assurance game" to investigate cooperation and trust. In this social dilemma paradigm, the virtual partner was programmed to behave cooperatively. Comparably to the PD, the subject can decide whether to cooperate or defect, and the gain is determined by the interplay of cooperation and defection of both players. BPD patients cooperated more often than healthy individuals. This is in accordance with the findings of Franzen et al. (2011) that point to overall higher investments in BPD. Surprisingly, oxytocin did not modulate the exchange behaviour. However, it affected whether the participant expected the co-player to be trustworthy. Again, surprisingly, oxytocin did not increase trustfulness, but led to less trusting expectations in BPD patients. Although this finding suggests that the oxytocin system might be affected in BPD, the data of Bartz et al. (2011) do not suggest a simple lack in trust explained by a lower oxytocin level. Instead, the authors propose that oxytocin may affect the salience of social cues, and, thereby, trigger positive or negative social emotions based on the social repertoire of the individual and/or the social context. It has to be emphasized that oxytocin modulates the self-reported expectations towards a social partner, but not the actual exchange behaviour. This discrepancy in the self-report of emotional experiences and motivations and actual interaction behaviour underlines the importance of applying interaction paradigms to extend our understanding of social dysfunction in mental disorders. It agrees with other studies, e.g. McClure et al. (2007), that similarly depicts a gap between self-perceived behaviour and the actually chosen social actions. Considering the often discussed bias in perceiving oneself and others that is assumed to exist in BPD, but also in other mental disorders, such as depression, this discrepancy seems to be expectable and should be taken into account in investigating alteration in social functioning, as well as in the development of psychotherapeutic interventions and cognitive training programs (see also Lis and Bohus 2013).

16.2.4.2 Psychopathy

Although psychopathic behaviour is characterized by the violation of moral standards and many studies have aimed at identifying deficits in moral reasoning, the number of studies that try to capture alterations of moral behaviours with neuroeconomic tasks is sparse. Koenigs et al. (2010) investigated quite a small sample of six primary and six secondary psychopaths compared to 22 non-psychopaths with a DG and an UG. Only primary psychopaths offered lower amounts during the DG than during the UG. Beyond that, they accepted fewer unfair offers in the role of the responder in the UG. Since the authors found a comparable behavioural pattern in subjects with lesions in the vmPFC, they argued in favour of an involvement of this cerebral structure in primary psychopathy and hypothesized that deficits in regulating anger and frustration may underlie these alterations. However, a subsequent study of the behaviour in the role of the responder during an UG did not replicate altered behaviour in criminal offenders with psychopathy (Radke et al. 2013a). Further studies are necessary that disentangle cognitive and affective processing during exchange behaviour together with its psychophysiological and neuronal correlates.

16.2.5 Attention-Deficit/Hyperactivity Disorder

Although high levels in interpersonal dysfunction are well known to characterize adults with Attention-Deficit/Hyperactivity Disorder (ADHD), up to date, only a few studies have addressed alterations in social cognitive processes and used exchange games to study interpersonal functioning in this clinical group. Lis et al. (2013) used a trust-game to explore behavioural alteration during social interactions in adults with ADHD. They manipulated the overall fairness together with the availability of social cues that signalled the likely fairness of the repayment during a single exchange in four virtual trustees. ADHD patients were equally generous toward all social interaction partners independent of their partners' fairness. This behaviour was beneficial in respect to maximizing a gain when playing with a fair partner, but resulted in a disadvantage when playing with unfair partners. Alterations were not caused by deficits in more basal cognitive or social cognitive processes, since these abilities were controlled for in the study. Thus, a general inattentiveness as a cause of interaction problems in ADHD, as it has been discussed in the past, has been excluded in this study. One line of interpretation of these findings is that of altered reward sensitivity in ADHD, which results in overemphasizing value, but ignoring the probability of reward. Thereby, it agrees with other approaches applying paradigms from behavioural economics that do not focus on social exchanges, but on decision-making under risk and uncertainty or intertemporal choice (Sonuga-Barke and Fairchild 2012).

A recent study identified the relevance of a comorbid conduct disorder for alterations in adolescents with ADHD (Northover et al. 2015). This subgroup rejected moderately unfair offers in the role of the responder during an UG more often than ADHD patients without this comorbid disorder or healthy controls. Thereby, the rejection rate was closely linked to the severity of the comorbid conduct disorder. The authors propose deficits in emotion regulation as underlying mechanism in this subgroup of ADHD patients.

16.2.6 Autism Spectrum Disorder

As in psychopathy or ADHD, the number of studies applying an exchange game in autism spectrum disorder (ASD) is sparse. In contrast to the large number of studies in schizophrenia, depression or BPD, this seems surprising since alterations in social cognition and social interaction behaviour form a core symptom domain of

ASD. Thus, the DSM-IV defines qualitative impairments in social interaction as one of the diagnostic criteria which may manifest itself in a lack of social or emotional reciprocity.

The study by Chiu et al. (2008) is one of the few actual neuroeconomic studies investigating social exchange behaviour in mental disorders. The authors measured BOLD responses during a 10-round iterative TG. Adolescents or young adults with high-functioning ASD and healthy subjects took the role of the trustee while playing with a healthy subject as investor. The ASD patients' behavioural pattern during the exchange equalled that of the healthy controls, but differed from healthy subjects regarding hemodynamic response patterns along the anterior-posterior axis of the cingulate cortex. The authors postulate that a TG trial can be divided into two phases. They separate a decision of the partner, in this case the decision of the investor, from the decision of the player his/herself, in this case the decision to repay a specific ratio of the monetary units as trustee. Neural activity revealed robust "self" and "other" response patterns along cingulate cortex for these distinct phases in healthy subjects. In contrast, the ASD patients' "self" response pattern, i.e. a relatively greater activation in middle cingulate domains and less activation in the anterior and posterior ends of the cingulate cortex, was absent. The observed activation during the "self" decision phase in ASD equalled that of healthy subjects during a non-social control condition, i.e. playing with a computer. Thereby, the attenuation of the neural "self"-pattern was stronger with the extent of symptom severity in the ASD patients. The alterations of brain activation were confined to the "self"-related decisions since the brain patterns during the investor's decision were comparable in both groups.

In the past, autism has been primarily linked to impairments in Theory-of-Mind tasks and the inability to infer other's emotions and intentions. However, the authors proposed that their data might be interpreted as an impairment in the ability to represent their own intentions, but an intact ability to represent the intentions of social partners. This shifts the locus of dysfunction to introspection and self-referential processes in ASD. Nevertheless, it seems worth considering that an alternative interpretation might be that the self-phase is also determined by inferring the intentions of others, in that the own repay might influence the future investment behaviour of the partner. Comparably, the processing during the investors' decision phase might be affected more strongly by an evaluation of the actual observable investment behaviour and less by the inference of future intentions of the partner.

Altered social interaction behaviour in high-functioning autism (HFA) seems to not be caused by deficits in moral reasoning (Li et al. 2014): HFA children differentiated between hypothetical characters described in stories and even assessed individuals harming others as worse than healthy control children. Nevertheless, these judgements did not affect the patients' cooperation in a subsequent PD. In contrast, healthy children were more cooperative towards partners to whom they had previously attributed more positive moral features. These findings suggest that autism might be linked to a lack in taking context information into account. Whether this may also hold true for the use of other cues, such as emotional facial expressions, has to be investigated in future studies.

The relevance of studies using neuroeconomic tasks for the explanation of everyday functioning in autism has been shown by Edmiston et al. (2015). They linked altered engagement of neural structures during a PD to playground behaviour of autistic children. Both ASD and healthy subjects did not differ in the ratios of cooperative and defective responses when they played the PD with a child they had previously met during a naturalistic peer interaction playground paradigm. However, the engagement of a social salience network including brain regions such as the insula, the temporoparietal junction and the caudate was attenuated in ASD children, particularly during defection of a human partner. Beyond these alterations specific for social interaction, altered neural activations, particularly in the insula, were observed during defection in the PD when subjects interacted with both a human partner and a computer during a nonsocial control condition. Across groups, both the engagement of the insula as well as the temporoparietal junction during defection by a human partner was linked to behaviour and stress responsivity during the playground paradigm. These findings suggest that the identification of altered neural activations during standardized economic tasks actually contributes to a deeper understanding of dysfunctional behaviour in naturalistic social interactions.

16.3 Summary

As Carter (2012) stated, ‘it is unlikely that there will be neuroeconomic disorders’. Nevertheless, the limited number of studies that applied economic exchange games in clinical research to date paints a promising picture of these approaches as a toolbox for the investigation of alterations in social functioning in mental disorders. So far, findings can be summarized in that exchange games are suited to uncover alterations in social interaction behaviour together with its neuronal correlates. Beyond that, these approaches enable the investigation of specific social cognitive functions as emotion recognition or Theory of Mind embedded in an ecologically valid context: single processes are analyzed as components within a broader functional network of diverse cognitive and social cognitive processes that influence each other with the final goal to successfully cope with social relationships. These processes have to be continuously adjusted during the interaction depending on the partner’s action and reaction. This becomes particularly important if economic tasks are not realized as one-shot games, but require a repeated interaction with the same partner over multiple rounds. Further, studies have to address the involved cognitive and social cognitive processes in more detail. Thereby, exchange games with virtual partners and a controlled variation of single modulating factors within a carefully chosen experimental design might provide a useful tool to get further insight into the underlying mechanisms and the identification of the impairment linked to a specific function and its relevance for a specific mental disorder. However, due to the still very small number of empirical studies, it is too early to assess whether the theoretical frameworks and experimental approaches will contribute to the understanding of the causes underlying mental disorders. This holds true especially since—from a methodological point of view—the existing studies are rather heterogeneous in regard to, e.g. sample sizes, concomitant psychopharmacological treatments, different comorbidities and the control of potentially modulating factors, such as socioeconomic status and cognitive impairments. All these factors may be of particular relevance when investigating impairments of social interactions in mental disorders (see Fig. 16.1) and add to heterogeneity of findings that goes beyond that of the specific realization of exchange games with its consequences for the related cognitive functions.

It seems worth summarizing some aspects that might improve future research with economic exchange games in mental disorders. First of all, one of the main advantages of this scientific approach is that it provides testable quantitative models. None of the studies in clinical research so far have made the attempt to use this strength to prove theoretical assumptions about dysfunctions in specific domains of interaction behaviour. Closely connected to this is the lack of studies that use quantitative parameters of these models to actually link them to altered behavioural correlates and neurobiological systems. In general, the majority of studies strongly relied on the measurement of overt behaviour and did not apply neuroscientific methods. Thus, studies that aim to characterize the alterations in

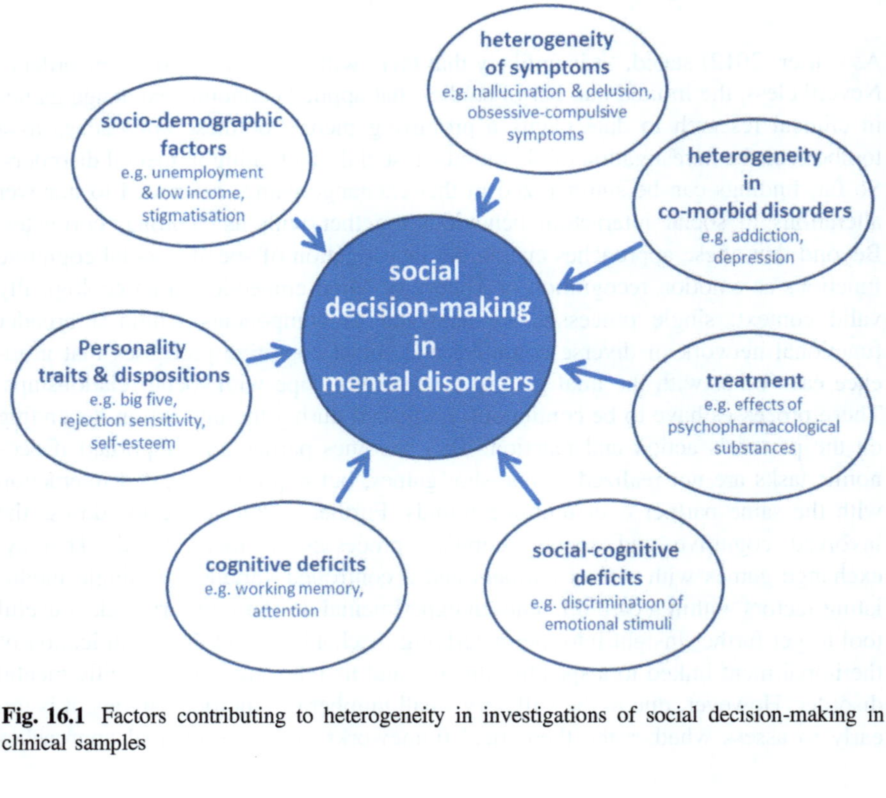

Fig. 16.1 Factors contributing to heterogeneity in investigations of social decision-making in clinical samples

mental disorders by using these strengths of neuroeconomic approaches are strongly needed.

Beyond that, exchange games aim towards measuring interaction behaviour. However, social interaction is defined as a dynamic sequence of social actions between individuals or groups. These actions are continuously adjusted according to the perceived past and anticipated future behaviour of the interaction partner. This implies that an analysis of interaction behaviour requires the analysis of interdependencies of social actions between partners. Yet, not all of the studies that claim to apply exchange games actually take these interdependencies into account, in that they do not provide feedback about the partner's behaviour or focus more on general performance than on the adaptation to the partner's behaviour. This might affect findings of alterations in mental disorders since taking the second person's perspective into account seems to be essential for the detection of alterations in social cognitive processes. Furthermore, only few studies measured interactions between real partners. One might argue that virtual partners allow for an experimental control of the partner's behaviour and are, thereby, the basis to gain insight into the relevance of specific social cues or behavioural pattern for the development of an interaction. We fully agree with this argumentation, since—particularly in clinical samples—the patients' behaviour might lead to specific interactional patterns that prevent the analysis of factors that modulate interactions and might constitute promising target behaviours for therapeutic interventions. Nevertheless, the programming of virtual partners requires a precise knowledge in regard to realistic interaction behaviours to allow for plausible modelling and experimental manipulation of behaviour of virtual partners. To achieve this goal, we need a close orchestration between the different sub-disciplines combined in the field of neuroeconomy to improve approaches in clinical research based upon the knowledge of healthy interaction behaviour.

Finally, it has to be mentioned that clinical samples are often characterized by dysfunctions in quite basal cognitive processes that might affect basal processes of perception and action, but also more complex functions such as working memory or attention. The separation of alterations genuinely linked to social cognition from those in other domains of cognitive functioning constitutes a particular challenge in the use of approaches from neuroeconomy in clinical research. So far, studies that apply appropriate nonsocial control condition constitute the minority.

Although we have to deal with a variety of problems in the use of neuroeconomic approaches in clinical research, a growing number of studies add to our understanding of the important domain of social functioning in mental disorders. In the past, the main sources of information about social interaction behaviour in clinical samples have been self- and observer-based questionnaires that provide information on how interaction behaviours are subjectively perceived. Neuroeconomic approaches add objective behavioural and neurobiological descriptions of the interaction behaviour itself and thus allow a new perspective in the investigation of interaction problems in mental disorders.

References

Agay N, Kron S et al (2008) Ultimatum bargaining behavior of people affected by schizophrenia. Psychiatry Res 157(1–3):39–46
Bartz J, Simeon D et al (2011) Oxytocin can hinder trust and cooperation in borderline personality disorder. Soc Cogn Affect Neurosci 6(5):556–563
Billeke P, Armijo A et al (2015) Paradoxical expectation: oscillatory brain activity reveals social interaction impairment in schizophrenia. Biol Psychiatry 78(6):421–431
Caceda R, Moskovciak T et al (2014) Gender-specific effects of depression and suicidal ideation in prosocial behaviors. PLoS One 9(9):e108733
Carter CS (2012) Neuroeconomics: sharpened tools of value for clinical cognitive and affective neuroscience. Biol Psychiatry 72(2):82–83
Chiu PH, Kayali MA et al (2008) Self responses along cingulate cortex reveal quantitative neural phenotype for high-functioning autism. Neuron 57(3):463–473
Csukly G, Polgar P et al (2011) Are patients with schizophrenia rational maximizers? Evidence from an ultimatum game study. Psychiatry Res 187(1–2):11–17
de la Asuncion J, Docx L et al (2015) Abnormal emotion processing, but intact fairness and intentionality considerations during social decision-making in schizophrenia. Front Psychol 6:1058
Destoop M, Schrijvers D et al (2012) Better to give than to take? Interactive social decision-making in severe major depressive disorder. J Affect Disord 137(1–3):98–105
Edmiston EK, Merkle K et al (2015) Neural and cortisol responses during play with human and computer partners in children with autism. Soc Cogn Affect Neurosci 10(8):1074–1083
Ernst M (2012) The usefulness of neuroeconomics for the study of depression across adolescence into adulthood. Biol Psychiatry 72(2):84–86
Fett AK, Shergill SS et al (2012) To trust or not to trust: the dynamics of social interaction in psychosis. Brain 135(Pt 3):976–984
Franzen N, Hagenhoff M et al (2011) Superior 'theory of mind' in borderline personality disorder: an analysis of interaction behavior in a virtual trust game. Psychiatry Res 187(1–2):224–233
Glimcher PW, Rustichini A (2004) Neuroeconomics: the consilience of brain and decision. Science 306(5695):447–452
Gradin VB, Perez A et al (2015) Abnormal brain responses to social fairness in depression: an fMRI study using the ultimatum game. Psychol Med 45(6):1241–1251
Gradin VB, Perez A et al (2016) Neural correlates of social exchanges during the Prisoner's Dilemma game in depression. Psychol Med 46(6):1289–1300
Grecucci A, Giorgetta C et al (2012) Anxious ultimatums: how anxiety disorders affect socioeconomic behaviour. Cogn Emot 27:230–244
Harle KM, Allen JJ et al (2010) The impact of depression on social economic decision making. J Abnorm Psychol 119(2):440–446
Hartley CA, Phelps EA (2012) Anxiety and decision-making. Biol Psychiatry 72(2):113–118
Hasler G (2012) Can the neuroeconomics revolution revolutionize psychiatry? Neurosci Biobehav Rev 36(1):64–78
King-Casas B, Chiu PH (2012) Understanding interpersonal function in psychiatric illness through multiplayer economic games. Biol Psychiatry 72(2):119–125
King-Casas B, Sharp C et al (2008) The rupture and repair of cooperation in borderline personality disorder. Science 321(5890):806–810
Kishida KT, Montague PR (2012) Imaging models of valuation during social interaction in humans. Biol Psychiatry 72(2):93–100
Kishida KT, King-Casas B et al (2010) Neuroeconomic approaches to mental disorders. Neuron 67 (4):543–554
Koenigs M, Kruepke M et al (2010) Economic decision-making in psychopathy: a comparison with ventromedial prefrontal lesion patients. Neuropsychologia 48(7):2198–2204

Li J, Zhu L et al (2014) The relationship between moral judgment and cooperation in children with high-functioning autism. Sci Rep 4:1–6

Lis S, Bohus M (2013) Social interaction in borderline personality disorder. Curr Psychiatry Rep 15(2):338

Lis S, Schönwetter T et al (2011) Impaired integration of social-emotional information in schizophrenia during a simulated social interaction. Nervenheilkunde 30:385–393

Lis S, Baer N et al (2013) Social interaction behavior in ADHD in adults in a virtual trust game. J Atten Disord 20:335–345

Loewenstein G, Rick S et al (2008) Neuroeconomics. Annu Rev Psychol 59:647–672

Lonnqvist JE, Verkasalo M et al (2012) Personality disorder categories as combinations of dimensions: Translating cooperative behavior in borderline personality disorder into the five-factor framework. J Pers Disord 26(2):298–304

McClure EB, Parrish JM et al (2007) Responses to conflict and cooperation in adolescents with anxiety and mood disorders. J Abnorm Child Psychol 35(4):567–577

Meyer-Lindenberg A (2008) Trust me on this. Science 321(5890):778–780

Monterosso J, Piray P et al (2012) Neuroeconomics and the study of addiction. Biol Psychiatry 72 (2):107–112

Northover C, Thapar A et al (2015) Emotion regulation in adolescent males with attention-deficit hyperactivity disorder: testing the effects of comorbid conduct disorder. Brain Sci 5(3):369–386

Polgar P, Fogd D et al (2014) Altered social decision making in borderline personality disorder: an ultimatum game study. J Pers Disord 28(6):841–852

Pulcu E, Thomas EJ et al (2015) Social-economical decision making in current and remitted major depression. Psychol Med 45(6):1301–1313

Radke S, Brazil IA et al (2013a) Unfair offers, unfair offenders? Fairness considerations in incarcerated individuals with and without psychopathy. Front Hum Neurosci 7:406

Radke S, Schafer IC et al (2013b) Do different fairness contexts and facial emotions motivate 'irrational' social decision-making in major depression? An exploratory patient study. Psychiatry Res 210(2):438–443

Rodebaugh TL, Shumaker EA et al (2013) Interpersonal constraint conferred by generalized social anxiety disorder is evident on a behavioral economics task. J Abnorm Psychol 122(1):39–44

Rosenbach C, Renneberg B (2011) Rejected, excluded, ignored: the perception of social rejection and mental disorders—a review. Verhaltenstherapie 21:87–98

Saunders KE, Goodwin GM et al (2015) Borderline personality disorder, but not euthymic bipolar disorder, is associated with a failure to sustain reciprocal cooperative behaviour: implications for spectrum models of mood disorders. Psychol Med 45(8):1591–1600

Scheele D, Mihov Y et al (2013) A negative emotional and economic judgment bias in major depression. Eur Arch Psychiatry Clin Neurosci 263(8):675–683

Schilbach L, Timmermans B et al (2013) Toward a second-person neuroscience. Behav Brain Sci 36(4):393–414

Shao R, Zhang HJ et al (2015) The neural basis of social risky decision making in females with major depressive disorder. Neuropsychologia 67:100–110

Sharp C, Monterosso J et al (2012) Neuroeconomics: a bridge for translational research. Biol Psychiatry 72(2):87–92

Sonuga-Barke EJ, Fairchild G (2012) Neuroeconomics of attention-deficit/hyperactivity disorder: differential influences of medial, dorsal, and ventral prefrontal brain networks on suboptimal decision making? Biol Psychiatry 72(2):126–133

Sripada CS, Angstadt M et al (2009) Functional neuroimaging of mentalizing during the trust game in social anxiety disorder. NeuroReport 20(11):984–989

Thielmann I, Hilbig BE et al (2014) Willing to give but not to forgive: Borderline personality features and cooperative behavior. J Pers Disord 28(6):778–795

Unoka Z, Seres I et al (2009) Trust game reveals restricted interpersonal transactions in patients with borderline personality disorder. J Pers Disord 23(4):399–409

van't Wout M, Sanfey AG (2011) Interactive decision-making in people with schizotypal traits: a game theory approach. Psychiatry Res 185(1–2):92–96
Wang Y, Zhou Y et al (2014) Impaired social decision making in patients with major depressive disorder. BMC Psychiatry 14:18
Wang Y, Yang LQ et al (2015) Game theory paradigm: a new tool for Investigating social dysfunction in major depressive disorders. Front Psychiatry 6:128
Willems RM (2011) Re-appreciating the why of cognition: 35 years after Marr and Poggio. Front Psychol 2:1–5
Wischniewski J, Brune M (2011) Moral reasoning in schizophrenia: An explorative study into economic decision making. Cogn Neuropsychiatry 25:1–16
Wischniewski J, Brune M (2012) How do people with borderline personality disorder respond to norm violations? Impact of personality factors on economic decision-making. J Pers Disord 27:531–546
Zhang HJ, Sun D et al (2012) Impaired social decision making in patients with major depressive disorder. Brain Behav 2(4):415–423

Part VI
Applied Neuroeconomics

Chapter 17
Consumer Neuroscience and Neuromarketing

Bernd Weber

Abstract Consumer choices represent one of the most abundant decisions people make. Within the field of neuroeconomics a focus has developed investigating the behaviour of consumers at the interface of psychology, marketing and neuroscience. While in the earlier years mostly the underlying computational neurobiological processes of simple, value-based consumer choices have been increasingly understood, more recently applied questions have received increased attention. Is it possible to derive additional information on consumers' preferences by integrating neurophysiological data in addition to classical market research to increase its predictive value for real market behaviour? What are the most promising neurophysiological tools enabling new insights in consumer behaviour and choices? This chapter tries to illustrate this broad spectrum of consumer neuroscience, from understanding to prediction of behaviour.

17.1 Introduction

The investigation of human decision processes as a multidisciplinary approach between fields like economics, psychology and neuroscience has over recent years extended its agenda to the study of consumer decisions. The focus on this important domain of individual decisions has gained a lot of attention by media and commercial companies, being it justified or certainly in many ways also hyped. Mainly driven by the scientific community, the field has somewhat parted into a commercial focus with specific market research questions, which is usually termed

B. Weber (✉)
Department of Epileptology, University Hospital Bonn,
Sigmund-Freud Str. 25, 53127 Bonn, Germany
e-mail: bernd.weber@ukb.uni-bonn.de

B. Weber
Center for Economics and Neuroscience (CENs), University of Bonn,
Nachtigallenweg 86, 53127 Bonn, Germany

M. Reuter and C. Montag (eds.), *Neuroeconomics*, Studies in Neuroscience, Psychology and Behavioral Economics, DOI 10.1007/978-3-642-35923-1_17

neuromarketing, and a more research-driven agenda, *consumer neuroscience* (Hubert 2010; Plassmann et al. 2010).

Many different motives are involved when investigating consumer decisions. The motivation of marketers is mainly driven by the hopes to have a new tool at their hands which will provide new—and hopefully more cost-effective—insights into the needs and preferences of consumers. More classical tools of market research, e.g. focus groups, questionnaires or simulated choices and market tests all have their disadvantages (and advantages), which explains the need of marketers to look for new methods and insights (Ariely and Berns 2010).

On the other hand also policy makers show a growing interest into the mechanisms driving human (consumer) decisions, incorporating models and theories from different disciplines. Initiatives like *Mindspace* (http://www.instituteforgovernment.org.uk) try to inform policy makers based on the understanding of individual decision processes about possible effects and effectiveness of interventions and regulations.

One of the strongest motivations of academic consumer researchers to get involved in neuroscience is the understanding that models based on *biology-free assumptions* may not be as useful as previously thought. Most models in social sciences, i.e. also in economics and marketing, are ignoring insights about the *black-box* brain or biology per se. While it may seem intuitive to assume that knowledge about biological processes is useful to inform behavioural researchers, there are strong notions against this assumption (Gul and Pesendorfer 2008), but also strong support with even special issues in renowned marketing journals (Shiv and Yoon 2012).

Within this chapter I want to highlight three major topics for which the application of neuroscientific tools and insights may be useful for investigating consumer behaviour. The first topic is the growing understanding of valuation processes in the brain. How does the brain assign values to specific choice options and how are they integrated into actual choices. Second, it is of utmost importance to understand individual differences, i.e. heterogeneity in choices between (and within) consumers to be able to better model and predict their behaviour and reactions to interventions. This leads to the third issue I want to tackle in this chapter, i.e. the use of neuroscientific tools for the prediction of choices and behavioural changes.

I want to conclude with an outlook of how the two fields, i.e. neuroscience and behavioural consumer research may establish an even stronger cooperations in the future.

17.2 Neuroscience of Valuation and Decision Biases

To perform a consumption decision, values to each choice option of the menu have to be assigned and compared. Over recent years our knowledge about how the brain performs these processes has grown tremendously (Rangel et al. 2008; Rangel and Hare 2010; Bartra et al. 2013).

In one of the earliest studies in the field of consumer neuroscience, Brian Knutson and colleagues had subjects perform simple purchase decisions of items in a functional MRI environment (Knutson et al. 2007). They observed that during the time of viewing the product, activity in the nucleus accumbens (NAcc) was positively correlated with the product preference. Activation in the medial prefrontal cortex (mPFC) was positively correlated with the price-differential, i.e. the net value of the product during price presentation. During the actual choice, bilateral anterior insula activity was negatively correlated with choice, while ventromedial prefrontal activity (vmPFC) was positively correlated with the decision to buy the product.

A variety of follow-up studies have used this framework to better understand valuation processes in the brain. It is a very consistent finding that the choice value of an option is coded in the vMPFC (Plassmann et al. 2007; Chib et al. 2009; Rangel and Hare 2010; Fehr and Rangel 2011).

Very interesting studies building on valuation processes have shown how willpower or self-control can modulate the choice value presented in the vmPFC. Todd Hare and colleagues had dieters perform choices between healthy and unhealthy food items in an MRI scanner. Subjects with high as well as low self-control abilities were investigated. The authors again observed a goal-value signal of the respective products in the vmPFC which was down regulated if subjects were able to successfully exert self-control by activity in the dorsolateral prefrontal cortex (dlPFC). Similar results have been observed in a more recent study by Hutcherson et al. (2012) showing the effect of cognitive regulation strategies on value signals in the vmPFC and dlPFC during food choice.

A fascinating phenomenon in consumer behaviour is the effect of external cues on consumption experience and behaviour, nicely termed *marketing placebo effects* by Shiv and colleagues (Shiv et al. 2005). The beliefs of consumers about aspects of a product like quality, prices, brands or packaging can influence the perception of the product itself, even above and beyond the pure physical consumption (Ariely and Norton 2009). Expectancies in the consumers mind about the consumption of products may even influence the efficacy of medical products (Waber et al. 2008). In this study, Waber and colleagues presented subjects with differently priced placebo analgesic pills and could show that the placebo which was given at a higher price had a stronger effect on pain reduction than lower priced placebos. The application of neuroscientific methods now allows not only to describe this effect but to actually show the neural mechanisms underlying these placebo effects and to discern which processes in the brain are actually influenced by these different placebos (Atlas and Wager 2012; Geuter et al. 2012; Wager and Atlas 2013).

In a study performing wine-tasting in an fMRI environment, Plassmann and colleagues showed that price information can influence the hedonic experience of wines (Plassmann et al. 2008). The authors presented the same wines at different prices via a pumping system to the subjects in the scanner. The identical wine led to higher activation in the orbitofrontal cortex and increased taste-pleasantness ratings when presented at a higher price. The high price, probably perceived as a quality signal, actually changed the perception of the same physical properties during taste.

Other kinds of external cues have also been shown to influence neural processes underlying valuations of products. Linder and colleagues performed a study on organic food labelling and could show that organic labelling, in comparison to conventional food, led to an increase in willingness to pay for food items accompanied by increased activity in valuation regions, i.e. the NAcc and the dlPFC (Linder et al. 2010).

All of these studies show that the understanding of human valuation processes leads to an increased knowledge about consumer behaviour. Beyond the effects on choices and ratings, neuroscience methods allow to observe the processes which are actually influenced during marketing actions. This may lead to different models and predictions about consumer behaviour.

One nice example is a study by Wadhwa et al. (2008). They investigated the effect of food sampling on subsequent consumer behaviour. They hypothesized that increasing reward-related activity (and therefore motivation-related dopamine activity) by having subjects sample high-valued food items should increase the consumption of other products. This was actually the case. Subjects who received samples with high incentive value (like a flavoured beverage) not only consumed more of other drinks (like Pepsi) but even unrelated but rewarding activities, like a massage or chocolate.

17.3 Understanding Heterogeneity of Subjects

Consumers differ in their needs and preferences. They do not only differ between individuals but also between different time points within themselves. Understanding the sources of these inter- as well as intraindividual differences is highly relevant for consumer researchers. Marketers have traditionally tried to address these issues by population segmentation based on classic demographic data, such as income, education, geographic location, sex, etc. (Keller 1993). More recent approaches include attitudes of consumers towards brands or products (Churchill Jr and Iacobucci 2009). What these models are missing are the individual heterogeneity in the decision process itself or the underlying biology, variables that today are in the focus of neuroscience (Hariri 2009) and which might help to improve consumer segmentation (Yoon et al. 2012).

People differ with respect to their underlying biology which influences their personality and behaviour (Benjamin et al. 2008). In the domain of decision-making it has for example been shown that variations in the dopamine-related gene COMT influence the ability of subjects to adapt to changing environments in a reward-learning task or their susceptibility to confirmation biases (Krugel et al. 2009; Doll et al. 2011). It is especially important but beyond the scope of this chapter to note that these genetic variations do not deterministically lead to differences in behaviour, but that the environment has a strong impact on the effects of the genetic variations (Caspi and Moffitt 2006).

Recent developments in statistical analyses of functional MRI data have also shown that the way in which people differ when performing cognitive tasks can be visualized and used to classify people into distinct groups (Poldrack et al. 2009).

Venkatraman et al. (2012) have applied this idea to the domain of brand management. They propose a framework for a neural market segmentation which goes beyond classical market research approaches and which accounts for additional heterogeneity of consumers, invisible to classical measures. The authors propose to increase the differentiation of consumers not only based on their behaviour or questionnaire data but to integrate neural data. They assume that consumers showing identical observable behaviour might show differences in underlying neural processes. This heterogeneity could be used to further sub segment consumers.

Intraindividual differences, e.g. related to stress, hormonal or vigilance levels are also of interest to consumer research. In a classical study, Ariely and Loewenstein (2006) investigated the effect of sexual arousal on different decision-making and judgment tasks. When being sexually aroused the male college students exhibited higher willingness to engage in risky or morally questionable behaviour as compared to a non-aroused state. Stress effects on decision-making have also gained a lot of focus in recent years (Starcke and Brand 2012). Porcelli and Delgado (2009) showed for example that people under stress showed a stronger reflection effect, i.e. an increase of risk taking for financial decisions in the loss domain and a decrease in the gain domain.

Hunger also strongly influences our decision-making. Mehta and colleagues examined neural responses to food stimuli in a satiated and hungry state. They observed that reactions in valuation regions of the brain to food stimuli in a satiated state were significantly reduced and that the amount of reduction predicted food choice in a buffet after the experiment (Mehta et al. 2012). The effect of hunger does not only hold for acute choices. Read and van Leeuwen showed that hungry people preferred unhealthy over healthy food products significantly even when making advance choices, i.e. choices that would be implemented in a week from the decision itself (Read and Van Leeuwen 1998).

Taken together, inter- as well as intraindividual heterogeneity and its biological foundations are very important to understand consumer decision-making. The combination of different approaches and methods, i.e. hormonal measures, genetics and neuroimaging, will help to further clarify our understanding of this heterogeneity.

17.4 Prediction of Choice and Behavioural Change

Beyond the improvement of understanding consumer behaviour, the prediction of choices and changes in behaviour are important goals in consumer research. In the above mentioned study of Knutson and colleagues, the authors tried to examine the additional value of neural measures to predict actual choices of the subjects

(Knutson et al. 2007). In this first attempt they showed that adding neural measures to a logistic regression including self-reported preferences and net values of products did actually increase the predictive power of the model beyond the self-report measures. Although the improvement of the model by adding functional MRI data was only marginal it did show that there may be an application of fMRI in predicting choices of consumers [and later reanalysis of the data did actually improve its predictive value (Grosenick et al. 2008)]. Tusche and colleagues used a multivariate classifier approach on fMRI data to show that preferences even for unattended stimuli could be detected in brain activation significantly above chance (Tusche et al. 2010). Functional MRI classifiers are algorithms which are trained to distinguish different patterns of neural activity related to different cognitive states. These classifiers can then be used to analyse unknown fMRI data and "read-out" the cognitive states the participants are in. They presented pictures of cars to subjects in the scanner and tried to predict later (hypothetical) choices on preferences for the cars based, on neural activity during the picture presentation. The regions which provided information for the classifier were again the mPFC and the insular cortex, consistent with previous findings. Levy and colleagues used functional MRI activity in the NAcc and mPFC during the presentation of products without the need to perform any choice to predict later binary choices between the products (Levy et al. 2011).

One study by Berns and Moore took this approach even one step further and tried to predict real market data based on activation levels in a sample of subjects. In a first study Berns and colleagues showed that activity in the NAcc and OFC while subjects listened to songs in an MRI scanner, correlated with individual preferences for the songs (Berns et al. 2010). In a follow-up study, the authors presented data showing that activity in these regions actually correlated to the number of albums sold in the US market through 2010 (Berns and Moore 2012). These results very strongly suggest the possible use of fMRI data even in smaller samples for prediction of real market data. But certainly replications also in other domains are necessary.

Besides the ability to predict consumer choices, the effects of market campaigns on the behaviour of consumers are another very important topic. A line of research using health campaigns investigated the use of functional MRI data to predict the individual effectiveness of the advertisements on the actual behavioural change of subjects. In a first study using ads from a health campaign on sunscreen use, Falk and colleagues presented ads for sunblocker use to subjects (Falk et al. 2010). The authors observed that activity in the mPFC while watching the ads explained 23 % of the variance in actual sunscreen use over the next week beyond the variance explained by self-reported attitudes and intentions. In a similar study, using smoking ads, the authors showed that again activity in the mPFC while viewing smoking-cessation ads doubled the explained variance in levels of carbon monoxide as a measure for recent smoking compared to self-stated attitudes and intentions (Falk et al. 2011).

17.5 Outlook

In the present chapter I wanted to show that there is growing evidence suggesting the additional value of functional MRI data (or other neuroscience methods) in understanding and predicting consumer behaviour. While certainly more studies are needed, especially relating to real market data, there is little doubt that the combination of neuroscience, psychology and economics in consumer research has already proven its usefulness and will become more established in the field of consumer science. One has to be aware that the knowledge created in this research should not only be used by market researchers and companies but to inform policy makers to build an environment that helps consumers to perform the choices that best fit to their needs and preferences.

References

Ariely D, Berns GS (2010) Neuromarketing: the hope and hype of neuroimaging in business. Nat Rev Neurosci 11:284–292

Ariely D, Loewenstein G (2006) The heat of the moment: the effect of sexual arousal on sexual decision making. J Behav Decis Making 19:87–98

Ariely D, Norton MI (2009) Conceptual consumption. Annu Rev Psychol 60:25p

Atlas LY, Wager TD (2012) How expectations shape pain. Neurosci Lett 520:140–148

Bartra O, McGuire JT, Kable JW (2013) The valuation system: a coordinate-based meta-analysis of BOLD fMRI experiments examining neural correlates of subjective value. Neuroimage 76:412–427

Benjamin J, Ebstein RP, Belmaker RH (2008) Molecular genetics and the human personality. American Psychiatric Pub

Berns GS, Moore SE (2012) A neural predictor of cultural popularity. J Consum Psychol 22: 154–160

Berns GS, Capra CM, Moore S, Noussair C (2010) Neural mechanisms of the influence of popularity on adolescent ratings of music. Neuroimage 49:2687–2696

Caspi A, Moffitt TE (2006) Gene–environment interactions in psychiatry: joining forces with neuroscience. Nat Rev Neurosci 7:583–590

Chib VS, Rangel A, Shimojo S, O'Doherty JP (2009) Evidence for a common representation of decision values for dissimilar goods in human ventromedial prefrontal cortex. J Neurosci 29:12315–12320

Churchill GA Jr, Iacobucci D (2009) Marketing research: methodological foundations. Cengage Learning

Doll BB, Hutchison KE, Frank MJ (2011) Dopaminergic genes predict individual differences in susceptibility to confirmation bias. J Neurosci 31:6188–6198

Falk EB, Berkman ET, Mann T, Harrison B, Lieberman MD (2010) Predicting persuasion-induced behavior change from the brain. J Neurosci 30:8421–8424

Falk EB, Berkman ET, Whalen D, Lieberman MD (2011) Neural activity during health messaging predicts reductions in smoking above and beyond self-report. Health Psychol Off J Div Health Psychol Am Psycholog Assoc 30:177–185

Fehr E, Rangel A (2011) Neuroeconomic foundations of economic choice recent advances. The Journal of Economic Perspectives 25:3–30

Geuter S, Eippert F, Attar CH, Büchel C (2012) Cortical and subcortical responses to high and low effective placebo treatments. NeuroImage

Grosenick L, Greer S, Knutson B (2008) Interpretable classifiers for FMRI improve prediction of purchases. IEEE Trans Neural Syst Rehabil Eng 16:539–548
Gul F, Pesendorfer W (2008) The case for mindless economics. In: The foundations of positive and normative economics, pp 3–39
Hariri AR (2009) The neurobiology of individual differences in complex behavioral traits. Annu Rev Neurosci 32:225
Hubert M (2010) Does neuroeconomics give new impetus to economic and consumer research? J Econ Psychol 31:812–817
Hutcherson CA, Plassmann H, Gross JJ, Rangel A (2012) Cognitive regulation during decision making shifts behavioral control between ventromedial and dorsolateral prefrontal value systems. The Journal of Neuroscience 32:13543–13554
Keller KL (1993) Conceptualizing, measuring, and managing customer-based brand equity. J Mark 1–22
Knutson B, Rick S, Wimmer GE, Prelec D, Loewenstein G (2007) Neural predictors of purchases. Neuron 53:147–156
Krugel LK, Biele G, Mohr PN, Li SC, Heekeren HR (2009) Genetic variation in dopaminergic neuromodulation influences the ability to rapidly and flexibly adapt decisions. Proc Natl Acad Sci USA 106:17951–17956
Levy I, Lazzaro SC, Rutledge RB, Glimcher PW (2011) Choice from non-choice: predicting consumer preferences from blood oxygenation level-dependent signals obtained during passive viewing. J Neurosci 31:118–125
Linder NS, Uhl G, Fliessbach K, Trautner P, Elger CE, Weber B (2010) Organic labeling influences food valuation and choice. Neuroimage 53:215–220
Mehta S, Melhorn SJ, Smeraglio A, Tyagi V, Grabowski T, Schwartz MW, Schur EA (2012) Regional brain response to visual food cues is a marker of satiety that predicts food choice. Am J Clin Nutr 96:989–999
Plassmann H, O'Doherty J, Rangel A (2007) Orbitofrontal cortex encodes willingness to pay in everyday economic transactions. J Neurosci 27:9984–9988
Plassmann H, O'Doherty J, Shiv B, Rangel A (2008) Marketing actions can modulate neural representations of experienced pleasantness. Proc Natl Acad Sci USA 105:1050–1054
Plassmann H, Yoon C, Feinberg FM, Shiv B (2010) Consumer neuroscience. In: Wiley international encyclopedia of marketing. Wiley
Poldrack RA, Halchenko YO, Hanson SJ (2009) Decoding the large-scale structure of brain function by classifying mental states across individuals. Psychol Sci 20:1364–1372
Porcelli AJ, Delgado MR (2009) Acute stress modulates risk taking in financial decision making. Psychol Sci 20:278–283
Rangel A, Hare T (2010) Neural computations associated with goal-directed choice. Curr Opin Neurobiol
Rangel A, Camerer C, Montague PR (2008) A framework for studying the neurobiology of value-based decision making. Nat Rev Neurosci 9:545–556
Read D, Van Leeuwen B (1998) Time and desire: The effects of anticipated and experienced hunger and delay to consumption on the choice between healthy and unhealthy snack food. Organ Behav Hum Decis Process 76:189–205
Shiv B, Yoon C (2012) Integrating neurophysiological and psychological approaches: towards an advancement of brand insights. Journal of Consumer Psychology 22:3–6
Shiv B, Carmon Z, Ariely D (2005) Placebo effects of marketing actions: consumers may get what they pay for. J Market Res (JMR) 42:11p
Starcke K, Brand M (2012) Decision making under stress: a selective review. Neurosci Biobehav Rev 36:1228–1248
Tusche A, Bode S, Haynes JD (2010) Neural responses to unattended products predict later consumer choices. J Neurosci 30:8024–8031
Venkatraman V, Clithero JA, Fitzsimons GJ, Huettel SA (2012) New scanner data for brand marketers: How neuroscience can help better understand differences in brand preferences. J Consum Psychol 22:143–153

Waber RL, Shiv B, Carmon Z, Ariely D (2008) Commercial features of placebo and therapeutic efficacy. JAMA J Am Med Assoc 299:1016–1017
Wadhwa M, Shiv B, Nowlis SM (2008) A bite to whet the reward appetite: the influence of sampling on reward-seeking behaviors. J Marketing Res 45:403–413
Wager TD, Atlas LY (2013) How is pain influenced by cognition? Neuroimaging weighs in. Perspect Psychol Sci 8:91–97
Yoon C, Gonzalez R, Bechara A, Berns G, Dagher A, Dubé L, Huettel S, Kable J, Liberzon I, Plassmann H, Smidts A, Spence C (2012) Decision neuroscience and consumer decision making. Mark Lett 23:473–485

Part VII
Neuroscience Methods in Neuroeconomics

Chapter 18
Skin Conductance Measures in Neuroeconomic Research

Dominik R. Bach

Abstract Skin conductance responses (SCR) are an established component of the psychological methods toolkit, and increasingly popular in neuroeconomics. This chapter discusses how SCR are generated by the sympathetic nervous system, the underlying central processes, and provides practical guidelines for SCR research. These guidelines are based on the existing methodological literature and recommendations by the Society for Psychophysiological Research. Analysis strategies for SCR are presented in the light of contemporary, model-based approaches that yield optimal statistical power to make inference on central states. The chapter then gives an overview over applications of SCR in neuroeconomics and outlines current research directions. Because emotional, cognitive, and motor processes can all elicit SCR, interpretation in economic experiments is sometimes challenging. It is therefore recommended to experimentally control possible cognitive and motor confounds. Finally, it would be useful to complement SCR with other peripheral measures of sympathetic/parasympathetic activity, in particular heart period and pupil size.

18.1 Background

Sweat glands primarily serve thermoregulation. However, as many people have experienced during exams, sweating in some parts of the body increases under particular states of affective or cognitive arousal. This has been termed 'emotional sweating' and occurs especially in palmar, plantar, facial, genital and axillary regions (Boucsein 2012). Palms and soles have received most scientific interest, as they are easy to access. Across the body, sweat glands are innervated by sympathetic—and to a very limited extent, parasympathetic—nerve fibres. From their end

D.R. Bach (✉)
University of Zurich, Zurich, Switzerland
e-mail: dominik.bach@bli.uzh.ch

D.R. Bach
University College London, London, UK

M. Reuter and C. Montag (eds.), *Neuroeconomics*, Studies in Neuroscience, Psychology and Behavioral Economics, DOI 10.1007/978-3-642-35923-1_18

terminals, acetylcholine is released as transmitter and diffuses locally to reach the sweat glands. These nerves are termed sudomotor as they drive the sweat glands, and are orchestrated by the hypothalamus, which thereby controls thermoregulation, in line with a general function in maintaining homeostasis. For control of emotional sweating in particular, the hypothalamus is an output relay under control of higher subcortical and cortical areas (Boucsein 2012; Critchley 2002).

Changes in sweat expulsion can be easily measured by recording the conductance of the skin. In a nutshell, more sweat means more electrolytes which results in lower resistance. Commonly, conductance is reported, i.e. the reciprocal of resistance. Different indicators of emotional sweating have been developed; two are widely used. One is the amplitude of short, phasic conductance increases with 1-2 s latency to an event, and 30–60 s duration; these are usually termed event-related or evoked skin conductance responses (SCR). SCR are regarded as indicators of event-related arousal. When SCR occur spontaneously in the absence of external events, they are called spontaneous fluctuations (SF) or non-specific SCR (NS. SCR), and their number per time unit (nSF) is a common measure of tonic arousal (Boucsein 2012). SCR and SF have a roughly similar shape (see Fig. 18.1).[1]

Phasic SCR are elicited by a wide range of stimuli, including pain, emotionally arousing (positive and negative) stimuli, events requiring cognitive resources, or motor actions. Equally, SF is observed in various situations, including anticatory anxiety and mental load. The reader is referred to Boucsein (2012) for an in-depth review. Thus, there is no specific cognitive process causing SCR.

One may relate the amplitude of event-related SCR, or the number of SF, to a state of the sudomotor (sympathetic) nervous system, which we have termed (phasic or tonic) 'sympathetic arousal' (Bach and Friston 2013). It is currently not precisely known whether sympathetic arousal is globally regulated, or whether arousal is specific to sympathetic subsystems such as the sudomotor system.[2] Sympathetic arousal is a descriptive but useful construct: it relates more directly to psychological processes that elicit it than the observed SCR themselves. This is because the latter are also influenced by a number of peripheral factors that induce measurement noise. Contemporary model-based analysis methods seek to estimate sympathetic arousal from observed data, and relate these estimates to experimental manipulations, as reviewed in Bach and Friston (2013). In the field of behavioural economics, research on what elicits sympathetic arousal is ongoing, and is reviewed in the final part of this chapter.

---

[1]Further measures of tonic arousal include the tonic skin conductance level (SCL) and area under the curve (AUC) which I will not discuss in this chapter—interested readers are referred to (Bach et al. 2010c; Boucsein 2012).

[2]The fact that SCR indices can dissociate from heart rate might indicate that there is no global sympathetic arousal (Boucsein 2012); but the heart rate is under control of the parasympathetic system as well.

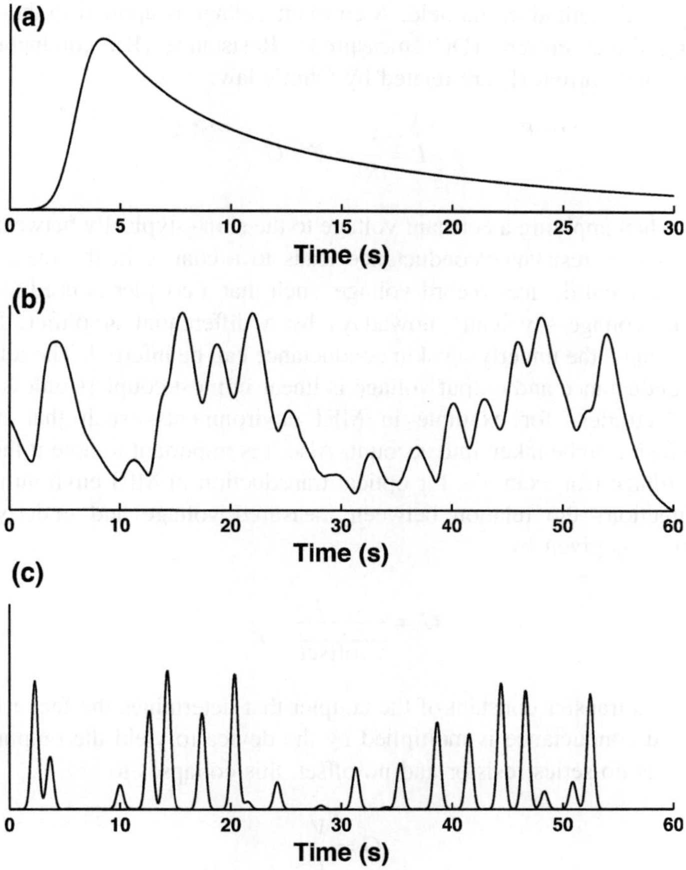

Fig. 18.1 **a** Shape of a typical skin conductance response after band-pass filtering. This canonical shape was derived from 1278 phasic SCR in 64 individuals (Bach et al. 2010b). **b** Example for a skin conductance time series. This is a 60 s data segment from an experiment involving public speaking anticipation (Bach and Erdmann 2007). **c** From the observed skin conductance time series, the (unobservable) sudomotor nerve activity can be estimated. The example here used a model for spontaneous fluctuations (SF), which specifies that sudomotor activity occurs in a finite number of compact bursts. One can see that the ensuing sudomotor nerve time series has a higher temporal resolution: for example, the first bump in the skin conductance time series is likely to be caused by two subsequent bursts of sudomotor firing. Tonic sympathetic arousal, the underlying unobservable construct, can now be quantified as the number of sudomotor bursts per time unit

18.2 Methodology

The methodology of acquiring SCR is fairly standard nowadays; interested readers are referred to Boucsein (2012) for technical details and recent developments, and to Boucsein et al. (2012) for current recommendations by the Society for Psychophysiological Research (SPR). Exosomatic DC recording is the most commonly used method in the field. A constant voltage is applied to the skin, and the ensuing direct current (DC) measured.[3] Resistance (R), conductance (G), voltage (V) and current (I) are related by Ohm's law:

$$I = \frac{V}{R} = V * G$$

Hence, when applying a constant voltage to the skin—typically between 0.5 and 3 V, a change in resistance/conductance leads to a change in the measured DC. Most measurement devices record voltage, such that a coupler is used to translate current into voltage—typically nowadays by a differential amplifier. From the recorded voltage, the underlying skin conductance can be inferred. The relationship between conductance and output voltage is linear in most couplers unless one uses resistive electrodes—for example in MRI environments—such that the series resistance R_S has to be taken into account. Also it is important to note if there exists an offset voltage (for example, for optical transduction in MRI environment). The transfer function—the relation between measured voltage and underlying skin conductance—is given by

$$G = \frac{1}{\frac{c}{V\text{-offset}} - R}$$

Here, c is a transfer constant of the coupler that determines the factor by which the measured conductance is multiplied by the device to yield the output voltage. When there is no series resistor and no offset, this collapses to

$$G = \frac{V}{c}$$

If the transfer function is linear (i.e. in the absence of series resistors), it is possible to analyse the output data of the coupler without knowledge of the transfer function, and report results in arbitrary units. However, it is good practice to report results in conductance units and this is mandatory when absolute cutoff values are used (for example for counting nSF). Sometimes, the coupler sensitivity is adjusted to individual SCL, such that each individual might have a different transfer function.

Palm and sole possess the highest density of sweat glands responding to non-thermoregulatory central influences, with no difference in SCR shape (Bach et al. 2010b). Typical measurement sites are thenar/hypothenar, or medial/distal phalanges of two different fingers—distal phalanges show higher SCR (Boucsein

[3]This is different from the infrequently used methods of endosomatic recording, which measures the skin potential response (SPR) without applying an external voltage, and from exosomatic measurements with alternating voltage, measuring alternating current (AC). A further method, not recommended by the SPR, is to use constant current, measuring voltage.

2012). If both hands are needed to make responses, or in a magnetic resonance imaging (MRI) scanner, it might be useful to record from the inner arch of the foot (Boucsein 2012). According to consensus guidelines, there is no need to pre-treat the recording sites; in particular, no soap should be used. For theoretical reasons and out of practical considerations, it is recommended to use Ag/AgCl cup electrodes filled with skin-isotonic gel (usually 0.3–0.5 % NaCl) (Hygge and Hugdahl 1985). There is no recommendation on electrode size; standard electrodes available from lab equipment providers are 6–10 mm in diameter. In between-subject designs, it is important to keep electrode type, size, gel and ambient temperature/humidity constant, as these can influence measurements (Boucsein 2012).

18.3 Data Acquisition and Preprocessing

Traditionally, it was standard to filter analogue data before conversion to a digital signal, in order to allow low sampling rates and to reduce the dynamic range of responses. Contemporary acquisition systems with large dynamic range and cheap memory make this strategy dispensable, and allow for direct A/D-conversion and storage of raw data. When unfiltered data are A/D-converted, it is important to choose a sampling frequency generously higher than the signal of interest—to reduce aliasing of high noise frequencies. The signal of interest is in a frequency range below 5 Hz, and 100 Hz would be a typical sampling rate. Filters can then be applied digitally (see below).

After data acquisition, it is useful to visually inspect the time series for artefacts. In practice, subject motion, loose electrodes, and equipment malfunction often cause artefacts. In fMRI environments, two additional sources of artefacts are gradient noise and currents induced by subject motion in the static field. The first usually has a high frequency and is therefore easier to filter out. Subject motion in the static field is a more important concern and should be minimized by restricting subject motion and fixing wires wherever possible. It is current practice to band-pass filter the data before analysis. A typical low-pass cutoff would be 5 Hz, to reduce high-frequency noise and allow downsampling to 10 Hz. A typical high-pass cutoff is 0.015 Hz, in order to reduce slow signal drifts. Model-based analysis prefers slightly different filters (Bach et al. 2013; Staib et al. 2015).

18.4 Analysis

Traditionally, researchers used paper-and-pencil approaches to measure the amplitude of SCR peaks in a post-stimulus time-window, or to count SF over a period of time (Boucsein 2012). This approach makes the assumption that SCR occur in a particular post-stimulus window, and that their peak amplitude or number is at least monotonically related to a psychological process of interest. Early computerized analysis packages softwares emulated this strategy. Over the last two decades, more sophisticated methods emerged. A practical motivation was that developments in cognitive neuroscience often mandated rapid event-related experimental designs. SCR have a long tail (about 30–60 s) such that they overlap in such circumstances (Barry et al. 1993). This has lead to the development of model-based methods (Alexander et al. 2005; Bach et al. 2009, 2010a, b, c, 2011; Benedek and Kaernbach 2010a, b; Lim et al. 1997). These methods estimate, by different algorithms, the underlying sudomotor nerve activity, which offers a higher temporal resolution and makes peak separation easier. It turned out that causal models of this sort have a number of more fundamental advantages over traditional data analysis (Bach and Friston 2013).

In particular, any 'operational' definition of dependent measures makes assumptions. To neuroeconomists, SCR are not interesting by themselves, but because they relate to central processes. The precise relation between SCR and central processes is not precisely specified in operational approaches, and assumptions on this relation may vary from publication to publication. Specifying these implicit assumptions mathematically in a causal model has two important consequences: first, it is then possible to empirically test the assumptions. Second, we can now infer the central process quantitatively by model inversion. This allows the experimenter to test quantitative relationships of this central process with other variables—asking, for example, whether outcome variance or entropy better explains sympathetic arousal.

A major faultline runs between two particular model-based approaches (Bach and Friston 2013). The first one contains only a mapping from sudomotor nerve activity to skin conductance (peripheral model). Algorithms using this approach deterministically compute a time series of sudomotor activity directly from observed SCR. This sudomotor time series does not directly relate to the experiment and needs to be further analysed using peak-scoring methods (Alexander et al. 2005; Benedek and Kaernbach 2010a, b). A freely available software representing this approach is Ledalab. The second approach contains a peripheral model *and* a neural model—the latter constraining what sudomotor activity to expect in a given experiment. Probabilistic algorithms then estimate the most likely parameters of the neural model, given the data and a sudomotor model (Bach et al. 2009, 2010b, 2011; Bach and Staib 2015). This is similar to standard methods for fMRI analysis (see Chap. 20). The software PsPM (Psychophysiological Modelling) represents this approach.

Probabilistic model-based methods offer a further advantage over traditional peak-scoring methods: higher precision. For example, specifying a neural model of when to expect SCR, renders estimation of SCR amplitude more precise. Sympathetic arousal estimates from probabilistic approaches have been shown to separate different experimental conditions better than peak-scoring (Bach 2014; Bach et al. 2013; Staib et al. 2015), and also better than deterministic methods, such as Ledalab (Bach 2014; Green et al. 2014). In the following, I will refer only to methods using probabilistic model inversion.

Crucially, probabilistic model inversion is the process of finding the most likely parameters of a model, given some data and some prior knowledge. The goal of this procedure is not to precisely explain the observed SCR data using many parameters. Instead, the method establishes plausible estimates of sympathetic arousal and ignores variance that is presumably not caused by experimental manipulation. Therefore, probabilistic model inversion uses, to varying degree, information about the experimental timing, and plausible prior assumptions about the sudomotor system. It is therefore important to think about what model to use and what information to give to the model inversion process. The more information the model has about known causes of sudomotor nerve responses, the more precise the estimated sympathetic arousal estimates will be.

Because probabilistic models aim to incorporate as much knowledge as possible into the model without being biased with respect to research hypotheses, there are different models for different situations. Phasic sympathetic arousal causes short firing bursts of the sudomotor nerve (SN). On a practical level, one needs to distinguish two cases. One is when SN firing occurs with constant latency after a short (<1 s) stimulus—we have termed this 'evoked SN responses', and the ensuing peripheral responses 'evoked SCR'. Phasic arousal corresponds to the amplitude of the SN response and can be estimated in a general linear convolution framework under the assumption of a constant latency (Bach et al. 2009, 2010b, 2013), much like in fMRI analysis (see Chap. 20). Another situation is when SN responses are elicited by experimental events, but their onset is not precisely known and needs to be estimated, too. Such event-related SN responses can be modelled using nonlinear methods; we have implemented this in the inversion framework of dynamic causal modelling (often used to estimate effective connectivity, an altogether different application) (Bach et al. 2010a; Staib et al. 2015). Tonic sympathetic arousal, on the other hand, can be quantified as the number of SN responses per time unit. Estimation of this number has been implemented in the inversion framework of dynamic causal modelling (Bach et al. 2011), or as a computationally faster alternative, with a matching pursuit algorithm (Bach and Staib 2015).

18.5 Neuroeconomic Research: The 'Somatic Marker Hypothesis' and Beyond

Traditionally, SCR and SF have been used in the study of emotion and reinforcement learning, and in applied and clinical psychology. In the context of neuroeconomics, one application is therefore to infer emotional influences on value-based decisions. An important proposal for such research has been the 'somatic marker hypothesis' (SMH). It is discussed here because it has generated a wave of SCR research in economic decision-making.

The SMH states that decisions are guided by preconscious signals elicited by each contemplated choice option (Damasio 1994). These signals are conveyed to the decision-making system by a mechanism that orchestrates particular bodily states, so called 'somatic markers'. These somatic markers are perceived by the decision-making system and preconsciously bias a decision into the right direction before conscious knowledge on the best strategy is formed. This means, instead of directly wiring the preconscious signal to the decision-making system, this signal takes a loop over a somatic state. In rehearsed situations, the anticipation of somatic markers suffices to bias a decision, such that the somatic markers are not actually generated. Of the three stages of this model (preconscious appraisal, elicitation of somatic states, perception of these states influences a decision), the first one is more or less established. Preconscious appraisal mechanisms are an integral aspect of a wide range of appraisal theories of emotions (Lazarus 1982, 1984; Scherer et al. 2001); these however usually assume that the preconscious appraisal directly influences decisions or actions. Also, the second stage—elicitation of somatic states —is an established finding: emotions have physiological correlates (Coan and Allen 2007), and some have proposed that the subjective experience of emotions is based on the perception of these body states (James 1884; Reisenzein 1983; Schachter and Singer 1962). The interesting aspect of the SMH was that the perception of body states guides value-based decisions, and that this is required for optimal decision-making.

This proposal was tested in a series of studies using the Iowa Gambling task (IGT) in which participants repeatedly bet on one of four lotteries (realised as card decks) and learn, via feedback, the statistics of each card deck, see, e.g. (Bechara et al. 1996, 1997). Two card decks have a negative expected value, the other two a positive expected value. As soon as participants can gauge the expected loss and gain, it would be optimal to mostly bet on the decks with positive outcome, although some exploration might still be necessary in order to learn more about the negative decks (Dunn et al. 2006). In studies in which SCR were recorded (Bechara et al. 1996, 1997), anticipatory SCR during decision and outcome anticipation were higher on trials on which a deck with negative expected outcome was chosen, compared to a positive deck. During the course of learning, these increased SCR were observed when participants started preferring to bet on the positive decks but before participants verbally reported knowledge of the advantageous strategy (Bechara et al. 1997). That is, they mostly made advantageous choices and had increased SCR when making disadvantageous choices, but when asked what they knew about the task, they did not state different value of the decks, or a coherent strategy. Patients with ventromedial prefrontal lesions lesions who do not learn to pick the card decks with positive expected outcome did not show this SCR signal (Bechara et al. 1996, 1997).

The SMH has received much attention and also criticism (Dunn et al. 2006). Here I will only discuss whether SCR constitute a somatic marker that is used for decision-making. Two hypotheses appear relevant: (a) Advantageous decisions in the IGT are taken before explicit knowledge is reported; hence there must be preconscious decision biases. (b) These preconscious biases are conveyed by somatic states; i.e. somatic states have a causal influence on decisions.

The first point is somewhat unclear in the original studies because people were not explicitly questioned about the statistics of the four card decks, in order to assess their conscious knowledge. They were only asked to report anything they had gathered about the task (Bechara et al. 1997). In further experiments with more explicit questions, people were actually aware of the statistics in the IGT at the moment when they started making advantageous decisions (Bowman et al. 2005; Maia and McClelland 2004). This may mean that preconscious signals are not required to make optimal decisions in this task. The SMH authors have addressed these findings by arguing that somatic markers would even guide decision-making when there *was* conscious knowledge (Bechara et al. 2005). To substantiate this assertion, one could investigate whether somatic markers are more closely related to decisions than explicit knowledge. Indeed, explicit knowledge *and* elevated SCR during choice of bad decks in the IGT have been reported (Guillaume et al. 2009). However, it is not known whether SCR explain variance in the decisions that is unexplained by declarative knowledge. In summary, as yet there is no clear evidence that preconscious biases—conveyed via somatic states or otherwise—are used to guide behaviour in the IGT, over and above an influence of declarative knowledge.

The second question is whether potential preconscious influences are conveyed via somatic states during deliberation (the 'body loop') (Damasio 1994). Of course it would be highly inefficient if the brain computed values preconsciously, sent a message to the body, and used sensors to pick up those body states to come to a deliberate decision; the brain could wire this signal directly to the decision-making system (Rolls 1999). Indeed, according to the SMH, preconscious signal may also be conveyed by anticipation of somatic states (the 'as-if'-loop) (Damasio 1994). For the body loop, somatic states would have to be generated *before* a decision is being made. It turns out that the IGT studies are not ideally suited to determine whether SCR were generated *before* or *after* decisions. The reason is that participants were instructed to ponder their decisions for at least 15 s and were then instructed by the experimenter to rapidly indicate their selection (Bechara et al. 1996, 1997) such that it is not known *when* participants made a decision in their mind. Hence, SCR could have been elicited before or after a decision. For example, in a simple Go/NoGo task, incorrect choice are followed by larger SCRs before feedback is given (Whitney et al. 2007). This appears to reflect post-decision error monitoring and is not a pre-decision response.

One conclusion that might be drawn from the original IGT study is that the anticipation of negative monetary outcomes is reflected in larger SCR. However, there is a twist to the IGT: in the original version, the two lotteries with negative mean outcomes are associated with a larger outcome variance than the two lotteries with positive expected outcome. In another replication study, this confound was reversed: here, the two positive lotteries were associated with the larger outcome variance. In this study, larger SCR were observed in responses to choosing the positive lotteries (Tomb et al. 2002). Taking this and the initial study together, this suggests that sympathetic arousal might be associated with variance of the predictive distribution rather than with its mean—variance is a measure of the predictive uncertainty (see Chap. 6 in this book). However, a study that independently varied mean and variance of outcome distributions has reported somewhat inconclusive results on this question (Yen et al. 2012).

18.6 Current Use of SCR in Neuroeconomic Research, and Framework for Future Studies

Phasic SCR are currently used in a plethora of neuroeconomic investigations; very few studies report tonic SF. There is still research on the implications of the somatic marker hypothesis (Botvinick and Rosen 2009; Crone and van der Molen 2007; Hinson et al. 2006; Miu et al. 2008; Visagan et al. 2012; Wagar and Dixon 2006). Many studies examine affective influences on decision-making, for example influences of empathy (Hein et al. 2011), negative emotional arousal (Sarlo et al. 2012; Sokol-Hessner et al. 2009; Van't Wout et al. 2006), anxiety (Engelmann et al. 2015), and stress (Heereman and Walla 2011). Other researchers were interested in the opposite relation: the effect of decision outcomes on affective state (Bediou et al. 2011). Some investigators have used SCR to index the success of aversive learning (Delgado et al. 2011), to assess the integration of positive and negative outcomes (Talmi et al. 2009), or to measure cognitive effort (Salvia et al. 2012).

Other research has elucidated economic variables that impact on SCR. SCR amplitude covaries with expected outcome (Studer and Clark 2011; Studer et al. 2016; Wu et al. 2016), with outcome undertainty in various forms of gambles (de Berker et al. 2016; Tomb et al. 2002), and with counterfactual reasoning after decision outcomes (Wu et al. 2016).

The richness of these different approaches reflects the relatively unspecific nature of phasic sympathetic arousal and points to a methodological difficulty: sympathetic arousal measures can only be interpreted if there is experimental control over its antecedents, and competing antecedents can be kept constant. For example, if phasic sympathetic arousal is taken to index emotional processes, then cognitive effort, vigour of motor responses, and other sources of sympathetic arousal need to be controlled. Importantly, economic decisions can involve different levels of cognitive effort as a possible confound—depending, for example on the number of choice possibilities, or the difficulty of computations to determine statistics of outcome distributions.

An interesting issue is the specificity of body states. One corollary of the SMH is that bodily states can convey highly specific economic signals. No such specificity has yet been found. However, this idea relates to the early proposal of emotion specificity in bodily signals, discussed at least since James (1884). The combination of contemporary multivariate techniques has brought some fresh air into this research field (Stephens et al. 2010). In order to better identify pattern of bodily states in economic decision-making paradigms, it would be useful to acquire other indicators of autonomic arousal together with SCR. Recently, model-based analysis methods have been developed for event-related changes in heart period (Castegnetti et al. 2016; Paulus et al. 2016), respiration (Bach et al., 2016), and in pupil size (Korn and Bach 2016). These approaches are simple to implement, and this may facilitate the use of such measures in neuroeconomic research

To summarise, it is desirable to experimentally constrain possible sources of sympathetic arousal such as cognitive effort and motor responses, in neuroeconomic research. As measure of sympathetic arousal, SCR may be complemented by other peripheral measures of autonomic nervous activity such as heart period or pupil size. This could also lead to a better understanding of the pattern of bodily states induced by economic variables. Nevertheless, SCR, and the underlying sympathetic arousal, can be a very useful measure in decision-making experiments, and it is not surprising that it is currently as popular as ever.

Acknowledgments The author thanks Deborah Talmi, Joel Winston, Michael Gaebler, Lyudmyla Kovalenko, and Matthias Staib, for helpful comments on a first draft of this manuscript.

References

Alexander DM, Trengove C, Johnston P et al (2005) Separating individual skin conductance responses in a short interstimulus-interval paradigm. J Neurosci Methods 146(1):116–123

Bach DR (2014) A head-to-head comparison of SCRalyze and ledalab, two model-based methods for skin conductance analysis. Biol Psychol 103C:63–68

Bach DR, Erdmann G (2007) Influences of habitual and situational bodily symptom focusing on stress responses. Cogn Emot 21(5):1091–1101

Bach DR, Friston KJ (2013) Model-based analysis of skin conductance responses: towards causal models in psychophysiology. Psychophysiology 50(1):15–22

Bach DR, Staib M (2015) A matching pursuit algorithm for inferring tonic sympathetic arousal from spontaneous skin conductance fluctuations. Psychophysiology 52(8):1106–1112

Bach DR, Flandin G, Friston K et al (2009) Time-series analysis for rapid event-related skin conductance responses. J Neurosci Methods 184(2):224–234

Bach DR, Daunizeau J, Friston KJ et al (2010a) Dynamic causal modelling of anticipatory skin conductance responses. Biol Psychol 85(1):163–170

Bach DR, Flandin G, Friston KJ et al (2010b) Modelling event-related skin conductance responses. Int J Psychophysiol 75(3):349–356

Bach DR, Friston KJ, Dolan RJ (2010c) Analytic measures for quantification of arousal from spontaneous skin conductance fluctuations. Int J Psychophysiol 76(1):52–55

Bach DR, Daunizeau J, Kuelzow N et al (2011) Dynamic causal modeling of spontaneous fluctuations in skin conductance. Psychophysiology 48(2):252–257

Bach DR, Friston KJ, Dolan RJ (2013) An improved algorithm for model-based analysis of evoked skin conductance responses. Biol Psychol 94(3):490–497

Bach DR, Gerster S, Tzovara A, Castegnetti G (2016) A linear model for event-related respiration responses. Journal of Neuroscience Methods 270:174–155

Barry RJ, Feldmann S, Gordon E et al (1993) Elicitation and habituation of the electrodermal orienting response in a short interstimulus interval paradigm. Int J Psychophysiol 15(3): 247–253

Bechara A, Tranel D, Damasio H et al (1996) Failure to respond autonomically to anticipated future outcomes following damage to prefrontal cortex. Cereb Cortex 6(2):215–225

Bechara A, Damasio H, Tranel D et al (1997) Deciding advantageously before knowing the advantageous strategy. Science 275(5304):1293–1295
Bechara A, Damasio H, Tranel D et al (2005) The Iowa gambling task and the somatic marker hypothesis: some questions and answers. Trends Cogn Sci 9(4):159–162; discussion 162–164
Bediou B, Mohri C, Lack J et al (2011) Effects of outcomes and random arbitration on emotions in a competitive gambling task. Front Psychol 2:213
Benedek M, Kaernbach C (2010a) A continuous measure of phasic electrodermal activity. J Neurosci Methods 190(1):80–91
Benedek M, Kaernbach C (2010b) Decomposition of skin conductance data by means of nonnegative deconvolution. Psychophysiology 47(4):647–658
Botvinick MM, Rosen ZB (2009) Anticipation of cognitive demand during decision-making. Psychol Res 73(6):835–842
Boucsein W (2012) Electrodermal activity. Springer, New York
Boucsein W, Fowles DC, Grimnes S et al (2012) Publication recommendations for electrodermal measurements. Psychophysiology 49(8):1017–1034
Bowman CH, Evans CE, Turnbull OH (2005) Artificial time constraints on the Iowa gambling task: the effects on behavioural performance and subjective experience. Brain Cogn 57(1):21–25
Castegnetti G, Tzovara A, Staib M et al (2016) Modeling fear-conditioned bradycardia in humans. Psychophysiology 53(6):930–939
Coan JA, Allen JJB (2007) Handbook of emotion elicitation and assessment. Oxford University Press, Oxford
Critchley HD (2002) Electrodermal responses: what happens in the brain. Neuroscientist 8(2):132–142
Crone EA, van der Molen MW (2007) Development of decision making in school-aged children and adolescents: evidence from heart rate and skin conductance analysis. Child Dev 78 (4):1288–1301
Damasio AR (1994) Descartes error: emotion, reason and the human brain. Avon, New York
de Berker AO, Rutledge RB, Mathys C et al (2016) Computations of uncertainty mediate acute stress responses in humans. Nat Commun 7:10996
Delgado MR, Jou RL, Phelps EA (2011) Neural systems underlying aversive conditioning in humans with primary and secondary reinforcers. Front Neurosci 5:71
Dunn BD, Dalgleish T, Lawrence AD (2006) The somatic marker hypothesis: a critical evaluation. Neurosci Biobehav Rev 30(2):239–271
Engelmann JB, Meyer F, Fehr E et al (2015) Anticipatory anxiety disrupts neural valuation during risky choice. J Neurosci 35(7):3085–3099
Green SR, Kragel PA, Fecteau ME et al (2014) Development and validation of an unsupervised scoring system (autonomate) for skin conductance response analysis. Int J Psychophysiol 91 (3):186–193
Guillaume S, Jollant F, Jaussent I et al (2009) Somatic markers and explicit knowledge are both involved in decision-making. Neuropsychologia 47(10):2120–2124
Heereman J, Walla P (2011) Stress, uncertainty and decision confidence. Appl Psychophysiol Biofeedback 36(4):273–279
Hein G, Lamm C, Brodbeck C et al (2011) Skin conductance response to the pain of others predicts later costly helping. PLoS One 6(8):e22759
Hinson JM, Whitney P, Holben H et al (2006) Affective biasing of choices in gambling task decision making. Cogn Affect Behav Neurosci 6(3):190–200
Hygge S, Hugdahl K (1985) Skin conductance recordings and the NaCl concentration of the electrolyte. Psychophysiology 22(3):365–367
James W (1884) What is an emotion? Mind 9(34):188–205
Korn CW, Bach DR (2016) A solid frame for the window on cognition: Modeling event-related pupil responses. J Vis 16(3):28

Lazarus RS (1982) Thoughs on the relations between emotion and cognition. Am Psychol 37 (9):1019–1024
Lim CL, Rennie C, Barry RJ et al (1997) Decomposing skin conductance into tonic and phasic components. Int J Psychophysiol 25(2):97–109
Maia TV, McClelland JL (2004) A reexamination of the evidence for the somatic marker hypothesis: what participants really know in the Iowa gambling task. Proc Nat Acad Sci USA 101(45):16075–16080
Miu AC, Heilman RM, Houser D (2008) Anxiety impairs decision-making: psychophysiological evidence from an Iowa gambling task. Biol Psychol 77(3):353–358
Paulus PC, Castegnetti G, Bach DR (2016) Modelling event-related heart period responses. Psychophysiology 53(6):837–846
Reisenzein R (1983) The Schachter theory of emotion: two decades later. Psychol Bull 94(2):239–264
Rolls ET (1999) The brain and emotion. Oxford University Press, Oxford
Salvia E, Guillot A, Collet C (2012) Autonomic nervous system correlates to readiness state and negative outcome during visual discrimination tasks. Int J Psychophysiol 84(2):211–218
Sarlo M, Lotto L, Palomba D et al (2012) Framing the ultimatum game: gender differences and autonomic responses. Int J Psychol 48(3):263–271
Schachter S, Singer JE (1962) Cognitive, social, and physiological determinants of emotional state. Psychol Rev 69:379–399
Scherer KR, Schorr A, Johnstone T (2001) Appraisal processes in emotion: theory, methods, research. Oxford University Press, New York
Sokol-Hessner P, Hsu M, Curley NG et al (2009) Thinking like a trader selectively reduces individuals' loss aversion. Proc Nat Acad Sci USA 106(13):5035–5040
Staib M, Castegnetti G, Bach DR (2015) Optimising a model-based approach to inferring fear learning from skin conductance responses. J Neurosci Methods 255:131–138
Stephens CL, Christie IC, Friedman BH (2010) Autonomic specificity of basic emotions: evidence from pattern classification and cluster analysis. Biol Psychol 84(3):463–473
Studer B, Clark L (2011) Place your bets: psychophysiological correlates of decision-making under risk. Cogn Affect Behav Neurosci 11(2):144–158
Studer B, Scheibehenne B, Clark L (2016) Psychophysiological arousal and inter- and intraindividual differences in risk-sensitive decision making. Psychophysiology 53(6):940–950
Talmi D, Dayan P, Kiebel SJ et al (2009) How humans integrate the prospects of pain and reward during choice. J Neurosci 29(46):14617–14626
Tomb I, Hauser M, Deldin P et al (2002) Do somatic markers mediate decisions on the gambling task? Nat Neurosci 5(11):1103–1104; author reply 1104
Van't Wout M, Kahn RS, Sanfey AG et al (2006) Affective state and decision-making in the ultimatum game. Exp Brain Res 169(4):564–568
Visagan R, Xiang A, Lamar M (2012) Comparison of deck- and trial-based approaches to advantageous decision making on the Iowa Gambling Task. Psychol Assess 24(2):455–463
Wagar BM, Dixon M (2006) Affective guidance in the Iowa gambling task. Cogn Affect Behav Neurosci 6(4):277–290
Whitney P, Hinson JM, Wirick A et al (2007) Somatic responses in behavioral inhibition. Cogn Affect Behav Neurosci 7(1):37–43
Wu Y, Van Dijk E, Aitken M et al (2016) Missed losses loom larger than missed gains: electrodermal reactivity to decision choices and outcomes in a gambling task. Cogn Affect Behav Neurosci 16(2):353–361
Yen NS, Chou IC, Chung HK et al (2012) The interaction between expected values and risk levels in a modified Iowa gambling task. Biol Psychol 91(2):232–237

Chapter 19
Electroencephalography: Current Trends and Future Directions

Stefan Debener, Cornelia Kranczioch and Maarten De Vos

Abstract Since Hans Berger, a German psychiatrist, discovered in 1929 at Jena that brain activity can be recorded from the human scalp noninvasively, the electroencephalogram (EEG) has fascinated generations of neuroscientists. How complex mental functions such as attention, memory, language, or decision-making are implemented in the brain are fundamental questions in cognitive neuroscience, and the EEG technology provides information, which may be crucial in answering those questions. Indeed, understanding the neural underpinnings of cognitive functions would help developing effective treatments for patients suffering from various brain dysfunctions. We argue here that, among maybe a dozen or so technologies, the scalp-recorded EEG among the most informative ones when it comes to understanding the mind–brain relationship. Moreover, next generation EEG technology is unobtrusive, smartphone operated, and useable outside of the laboratory environment. This helps to advance applied fields such as brain–computer interfaces, neuroergonomics, and neurorehabilitation.

S. Debener (✉) · C. Kranczioch · M. De Vos
Neuropsychology Lab, Department of Psychology, Faculty of Medicine and Health Sciences, University of Oldenburg, 26111 Oldenburg, Germany
e-mail: Stefan.debener@uni-oldenburg.de

S. Debener · C. Kranczioch · M. De Vos
Center for Neurosensory Science and Systems, University of Oldenburg, 26111 Oldenburg, Germany

S. Debener · M. De Vos
Cluster of Excellence Hearing4all, University of Oldenburg, 26111 Oldenburg, Germany

M. De Vos
Institute of Biomedical Engineering, University of Oxford, Oxford OX3 7DQ, UK

M. Reuter and C. Montag (eds.), *Neuroeconomics*, Studies in Neuroscience, Psychology and Behavioral Economics, DOI 10.1007/978-3-642-35923-1_19

19.1 Introduction

Modern EEG technology has unique advantages and, for many applications, these advantages put EEG into the driver's seat when compared to all other brain activity monitoring technologies. EEG signals measure brain activity with millisecond resolution, and EEG technology is potentially mobile. We speculate that, in the near future, these advantages will make it possible to develop truly intelligent assistive technologies. Future EEG technology will make it possible to make use of implicit information derived from brain activity monitoring in various work and clinical settings. Moreover, based on EEG it will also be possible to modify ones' brain activity patterns and aim for better cognitive performance and emotion regulation, either to compensate for existing dysfunctions, or to further enhance normal cognitive skills.

This chapter provides an introductory overview of acquiring EEG signals. This will be followed by an introduction to event-related potentials (ERP), which has been the gold standard for the study of cognitive functions via EEG. However, scientists also discovered that event-related brain oscillations provide information that is neglected by the ERP approach. Access to this type of information requires a different analysis approaches. How brain oscillations may serve human cognition is currently an important field of neuroscience research. We will illustrate the relevance of considering both types of information, ERPs and oscillations, by discussing the attentional blink (AB) phenomenon, a short moment of visual inattention. A person is usually not aware of those brief moments of inattention, but lapses of attention can have dramatic, life-changing consequences, for instance while driving a car. Following this example, future developments and applications of the EEG technology will be presented. Specifically we argue that further miniaturization, combined with a real-time EEG analysis approach, may enable us to monitor brain activity, and thus the neural correlates of mental processes, outside of the lab and in real-life situations. This development will lead to new assistive technologies, for instance for achieving vigilance and cognitive performance monitoring. Our own first steps toward those mobile EEG applications will be summarized, and the requirements toward achieving those goals will be defined.

19.2 EEG Acquisition and Recording Environment

EEG signals are typically recorded with electrodes attached to the scalp. These electrodes are connected to a shoebox-sized biopotential amplifier controlled by a computer, which also stores the data for subsequent analysis. The recorded signals reflect voltage fluctuations over time, with amplitudes in the microvolt range, and frequencies in the range of near DC to approximately 80 Hz. It is important to understand that the recorded signal consists of a mixture of an unknown number of brain and non-brain sources. The non-brain source contributions are called artifact and could be of technical origin (e.g., line noise interference, electrode motion, etc.) or reflect non-brain electrical activity (e.g., electrical heart beat activity; eye blinks, etc.). A typical recording consists of the (linear) summation of all instantaneously active artifact and non-artifact sources. Figure 19.1 shows a typical raw EEG recording. In addition to artifacts, oscillatory brain activity is evident. Approximately 10/s posterior alpha oscillations have been related to visual perception, vigilance, and attention (Klimesch 2012) but are difficult to quantify during the occurrence of artifacts. The simultaneous recording from various scalp sites (typically ranging between 32 and 128) enables a careful spatial examination of the data and helps to tease apart the signal of interest (e.g., brain oscillations) from other contributions. EEG alpha activity is prominent in eyes closed, awake resting conditions. Alpha activity has a direct relationship to several cognitive processes (Klimesch 2012), and shows an abnormal spatial pattern in for instance depression (Debener et al. 2000), to name only one condition that has been linked to this prominent brain oscillation.

Figure 19.1 inset also shows a typical recording cap, which is needed to hold electrodes at predefined locations on the scalp. In most lab conditions, a separate amplifier and several computers are needed for data acquisition and the control of experimental events (e.g., presentation of sounds or visual events) and the collection of subject behavior (e.g., button presses to target events). As illustrated, usually two or more experimental stimulations are presented and the brain activity in response to those events is analyzed (color-coded time windows). Most research-EEG recordings are conducted with this type of setup and in a highly controlled laboratory environment, to allow for constant light, temperature, and noise conditions. Importantly, subjects are usually required to minimize movement during recording, as movement of electrodes and/or electrode cables causes massive

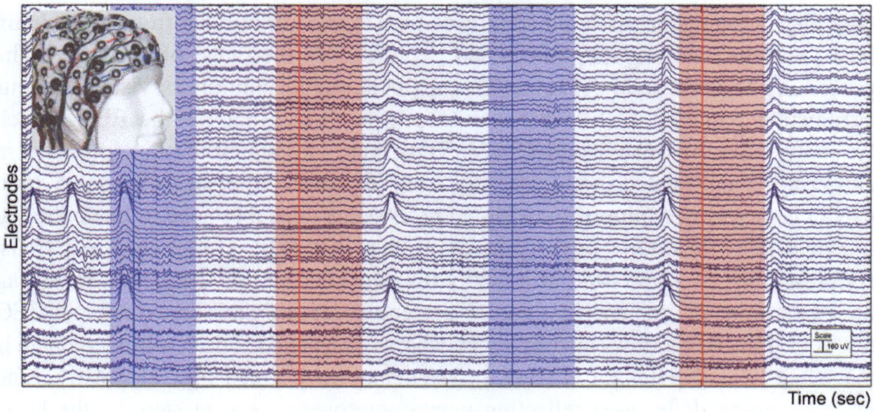

Fig. 19.1 Ongoing EEG recording (64 electrodes, 10 s). Large-amplitude deflections refer to eye blink artifact. Approximately 10/s oscillatory activity is spatially and temporally focused and reflects brain activity in the EEG alpha (8–13 Hz) frequency range. Colored shaded areas indicate epochs of interest for event-related analysis, such as ERPs

interference artifacts and usually renders analysis of the underlying brain activity next to impossible. All body movements underlying overt behavior are controlled by muscle activity. The corresponding electrical muscle activity can be measured as electromyogram (EMG) and propagates to the EEG electrodes, thus posing another strong source of artifact. For these reasons, most behavioral observations that have been linked to EEG-recorded brain activity are based on heavily constrained behaviors such as button presses, not on real-life behavior. As we will show later, more recent advances in EEG amplifier technology have made it possible to overcome this important drawback. This is a fundamental development, since all other currently available noninvasive brain activity and neuroimaging technologies suffer from the very same limitation—they do not tolerate movement, and thus it is poorly understood how natural mental processes and behavior relate to brain activity.

Another practical drawback of the EEG technology is that the application of electrodes takes time, as gel is applied and skin needs to be prepared to ensure a stable contact between electrodes surface and skin. Since the upper layer of the skin is poorly conductive, EEG signal quality can be improved by careful abrasion of the upper skin layer (by using a cotton web and abrasive paste with a peeling effect). This lowers electrode impedance and thereby attenuates the influence of electromagnetic environmental noise on the resulting EEG recordings. Depending on the number of electrodes used, this process can easily take between 20 and 90 min. However, recent developments in amplifier and electrodes design aim at overcoming the limitations of gel use and skin preparation.

19.3 Event-Related Potentials

Ongoing, or spontaneous brain activity is based on contributions from many brain areas, and the resulting amplitudes can easily exceed the range of $\pm 100\ \mu V$. The processing of a particular event on the other hand, say an auditory or visual stimulus as illustrated in Fig. 19.1, contributes much smaller signals to the EEG, which typically range between 1 and 15 μV. Thus there is a signal-to-noise problem in analyzing event-related brain activity, and improving the signal-to-noise ratio (SNR) is the primary aim of most signal processing steps that need to be applied to recorded EEG data. One way of improving the SNR is by averaging across trials. By presenting the same event repeatedly (as illustrated in Fig. 19.1), and recording the exact point in time each event was presented, it is possible to average all EEG traces or trials (sometimes also called epochs, or segments). The resulting average is the event-related potential (ERP), a systematic wave consisting of positive and negative going deflections reflecting neurocognitive processing steps in the brain time-locked to the stimulus presentation. Figure 19.2 illustrates the averaging process and the underlying concept, the additive ERP model. As illustrated, the averaging process on the one hand reveals time-locked brain responses, which are assumed to take place of every single-trial level. On the other hand, the averaging

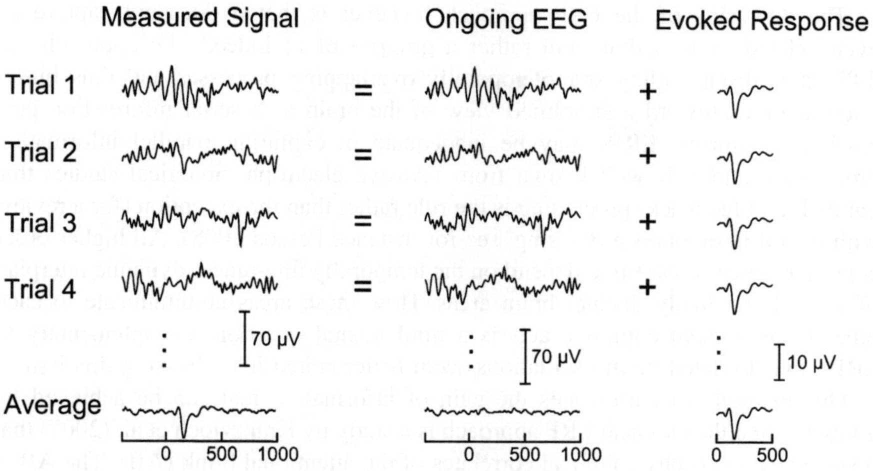

Fig. 19.2 The additive ERP model assumes that the event-related response is independent from ongoing brain activity, and that the evoked response is invariant across trials, that is, across repeated presentation of the same event. Averaging across trials therefore improves the signal-to-noise ratio, as ongoing activity sums up to near zero (i.e., is "averaged out"), whereas the evoked response (i.e., the ERP) is preserved, hence across trials it "comes out of noise"

process suppresses all non time-locked activities like the ongoing EEG, and thereby increases the SNR. In effect, this averaging is necessary to make the ERP visible, or make it "come out" of the noise.

Trial-averaged ERPs are easy to calculate and very reliable. They provide a rich set of information about mental processes related to perception, cognition, and emotion. Indeed, experimental task manipulations often come with specific and reliable changes in ERP deflections, and the underlying ERP components are very informative with regard to the processing (and processing stage) differences between the experimental conditions applied. With this approach it is often possible to tell whether the difference in performance between two task conditions is due to an early perceptual processing, subsequent object processing, later decision-making or a motor output preparation level of processing. Drawing similar inferences from behavioral observations alone, or from other noninvasive measures of human brain functions (such as functional magnetic resonance imaging) is typically not possible. As an example, in Hewig et al. (2011) it was investigated why humans often deviate from rational choices when making decisions. In this experiment the Ultimatum and Dictator games were used to examine the role of affect on decision-making. It was found that the feedback-related negativity, an ERP component related to reinforcement learning, predicts the decision to reject unfair offers in the ultimatum game. These and further findings led to objective evidence that decision-making is guided by subjective affective markers (Hewig et al. 2011). It is easy to see the practical potential of this result: by feeding back brain signals such as the feedback-related negativity, individuals could be trained to follow rational rather than affective signals and learn to gain control over implicit decision-making processes.

The downside of the ERP approach however is that it does not capture all event-related brain activity but rather a glimpse of it. Indeed, the approach has difficulties disentangling spatiotemporally overlapping processes and thus biases interpretations toward a simplified view of the brain as a serial information processing machinery. ERPs may be inadequate in capturing parallel information processing, and it is well known from invasive electrophysiological studies that parallel (and feedback) processing is the rule rather than the exception (for a review with regard to emotion processing, see for instance Pessoa 2008). All higher order cognitive processes seem to depend on the temporally fine-tuned, dynamic interplay of several regionally distinct brain areas. How these areas communicate to each other to implement cognitive acts is a fundamental question. Complementary to ERPs, event-related brain oscillations seem better suited in addressing this issue.

One example that illustrates the gain of information that can be achieved by going beyond the classical ERP approach is a study by Kranczioch et al. (2007) that investigated neurophysiological correlates of the attentional blink (AB). The AB is a temporary deficit in visual awareness that can arise when chunks of important information have to be picked out of a continuous stream of irrelevant information. In the lab such a situation is created by presenting a stream of pictures at a rate of about 10 per second. A volunteer is asked to detect or identify two of the pictures. For example, the stream may consist of black capital letters, and the participant is asked to identify a single green capital letter presented within the stream and to detect whether a black X follows the green letter. Typically volunteers easily identify the green letter, denoted as the first target or T1, while they frequently miss the subsequently presented black X, denoted as T2. This deficit can be observed in particular if T2 follows T1 by about 200–500 ms, but it is often absent or greatly reduced if T2 follows T1 by about 100 ms (cf. Fig. 19.3a).

One interesting aspect of the AB is that it does not occur invariantly. That is, even if in two runs of the laboratory task the stimulus stream and the two targets are virtually identical, in the first run a volunteer might detect T2 while in the second she might miss it. Kranczioch et al. (2007) where interested in whether this difference is reflected in the P3 ERP component, which is related to attentional resource allocation. They observed that when T2 was detected it was associated with a larger P3 then when it was missed. Importantly, the P3 evoked by T1 showed the opposite pattern in that it was smaller when T2 was missed as compared to when it was detected. This pattern of results is illustrated in Fig. 19.3b. This finding indicates that the distribution of attentional resources between T1 and T2 relates to whether T2 falls victim to the AB or not. But what might cause the different distribution of resources?

In their study Kranczioch et al. also investigated studied event-related brain oscillations, both before and during the appearance and processing of T1 and T2, asking why in some trials participants detect T2 while in others they miss it. The most striking difference between trials in which T2 was detected and in which T2 was missed was found in the dynamics of connectivity patterns between electrodes. Already before the appearance of T1, EEG activity before T2 would be missed was characterized by a connectivity pattern suggesting a strong focus on processing the stream of nontargets. In contrast, runs in which T2 would be detected where

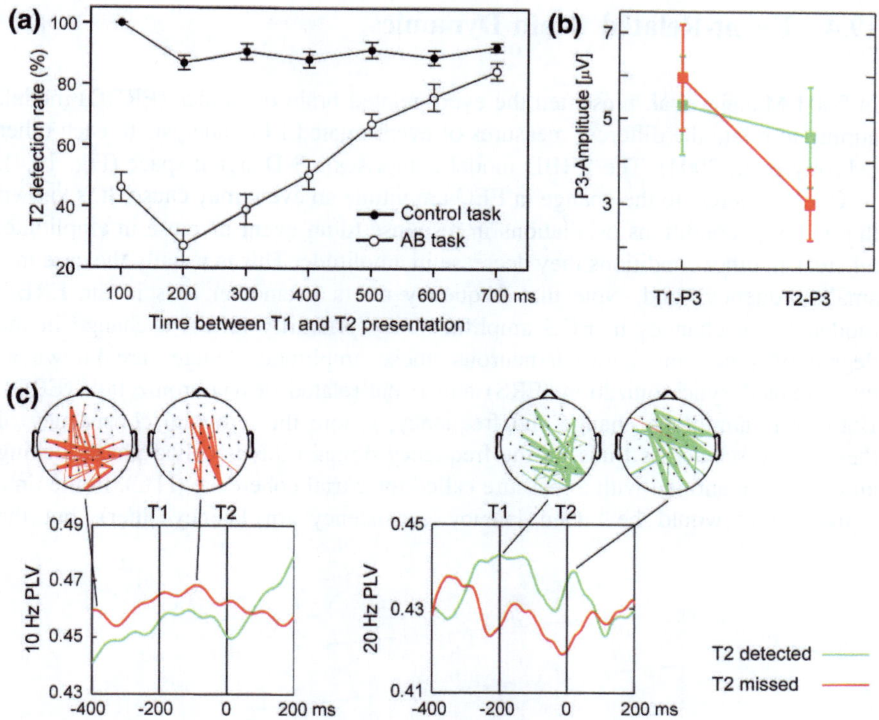

Fig. 19.3 **a** Behavioral pattern in a typical attentional blink (AB) task as compared to a control task. In the AB task T1 and T2 are task relevant. T2 detection rate is reduced as compared to a control task in which only T2 is task relevant while T1 can be ignored. When in the AB task T2 is presented at around 100 ms after T1, T2 detection is frequently found to be less impaired then at longer intervals. Figure adapted with permission from Kranczioch et al. (2007). **b** Amplitudes of the P3 ERPs evoked by T1 and T2 in trials in which T2 was detected (*green*) and in trials in which T2 was missed (*red*). Figure adapted with permission from Kranczioch et al. (2007). **c** Spatiotemporal connectivity pattern observed for trials in which T2 was detected (*green*) and in which T2 was missed (*red*). The measure of connectivity is the phase-locking value (PLV). Trials in which T2 was missed where characterized by a relatively higher PLV at a frequency of around 10 Hz that started well before the presentation of T1. Trials in which T2 were detected were characterized by an increase in PLV at a frequency of around 20 Hz, in particular at and around target presentation. The head plots illustrate the corresponding spatial connectivity patterns at exemplary time points. Figure adapted with permission from Kranczioch et al. (2007)

associated with a connectivity pattern compatible with a stronger focus on the task, that is, inhibiting the nontargets and searching for and processing of the targets (Fig. 19.3b). These and related findings (e.g., Gross et al. 2004) provide clear experimental evidence for the idea that whether an AB occurs is related to the state of the brain an incoming target meets, which in turn depends on the dynamic interplay of distinct processes and brain areas. This insight was made possible solely by complementing the classical ERP approach by the idea of event-related brain dynamics, a concept that will be in the focus of the next section.

19.4 Event-Related Brain Dynamics

In 2004 Makeig et al. presented the event-related brain dynamics (ERBD) model, aiming to relate the different measures of event-related EEG analysis to each other (Makeig et al. 2004). The ERBD model comprises a 3-D signal space (Fig. 19.4).

One axis refers to the change in EEG amplitude an event may cause. It is known that in some conditions oscillations in response to an event increase in amplitude, whereas in other conditions they decrease in amplitude. This is usually the case in a small frequency band. Note that frequency refers to another axis in the ERBD model. Since changes in EEG amplitude may primarily reflect a change in the degree of synchrony between neurons, those amplitude changes are known as event-related synchronization (ERS) and event-related desynchronization (ERD). Both axes, amplitude change and frequency, ignore the temporal consistency of these processes across trials. In the frequency domain this is called phase-locking and can be quantified with a measure called inter-trial coherence (ITC). In the time domain, this would be called latency consistency (or latency jitter), but the

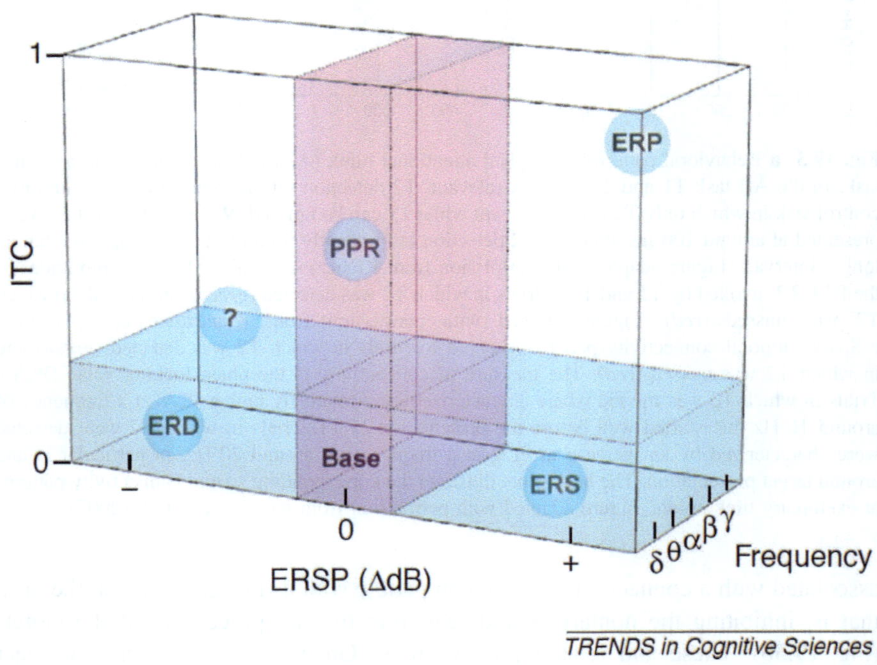

Fig. 19.4 Event-related brain dynamics model. Event-related brain responses can take any position within this 3-D volume. The additive ERP, as illustrated in Fig. 19.2, captures the upper right-hand corner, as it adds power (*x*-axis) in a very small frequency range (*z*-axis), and this is consistent across trials (*y*-axis). ITC = inter-trial coherence; ERSP = event-related spectral perturbation; ERS/ERD = event-related synchronization/desynchronization; PPR = partial phase resetting. Figure adapted from Makeig et al. (2004) with permission from the publisher

frequency-domain procedure offers the advantage of looking at this measure for different frequencies independently. ITC spans the third dimension in the ERBD. To the extend that the additive ERP model as illustrated in Fig. 19.2 applies, the ERP takes the upper right-hand corner of the ERBD: it adds power to the ongoing brain activity, is usually limited to the low frequency range, and happens with very high consistency across trials. It is however also known that the reorganization of phase in the absence of power changes can contribute to the trial-averaged ERP (Makeig et al. 2004), which is called partial phase resetting (PPR). While difficult to prove, PPR has been implicated as a crucial cortical operation for several cognitive functions. For instance, recent neurocognitive theories of speech processing assume that PPR of ongoing oscillations in auditory cortex is required to process connected speech signals.

Another important aspect of the ERBD model is that, in order to detect and study regionally circumscribed or local brain dynamics with scalp-recorded EEG, it is necessary to first unmix, or linearly decompose the measured data into its underlying source signals. This effort can only by a good guess, or a modeling effort, as the true source configuration of the measured data is generally unknown. The ERBD suggested the use of independent component analysis (ICA) to unmix brain signals from artifacts. ICA also promises to tease apart different brain signals from each other (Makeig et al. 2004). While all unmixing approaches including ICA come with theoretical pitfalls and practical difficulties and are far from perfect, the ICA approach has been found to be particularly helpful in SNR improvement in the absence of trial averaging (De Vos et al. 2012). That is, the ICA decomposition is achieved on the continuous, or single-trial data, and thereby enables to study single-trial brain responses. Single-trial analysis of EEG data is fascinating for at least two reasons. First, it allows us to relate physiology to mental processes on a trial-to-trial basis. Since physiological and behavioral measures fluctuate across trials, as illustrated in the AB above, the trial-by-trial approach provides a means to study how closely they are coupled to each other. Second, the single-trial analysis approach enables a (near) real-time access to the brain response. This is a fundamental requirement for brain–computer interface and neuro-feedback procedures, where users get access to their online brain signals to operate a technical device or learn to modulate their brain signals by means of studying its online feedback.

19.5 Brain–Computer Interfaces

The brain–computer interface (BCI) technology offers a direct communication pathway between brain activity and a technical device (for an overview, see REF BOOK Springer). BCIs could be used to restore the loss of voluntary control over muscles, as in locked-in patients, which can learn to communicate with a BCI. Probably the most popular BCI application is the EEG P3 matrix speller. The P3 or P300 signal is a positive going centroparietal deflection typically lasting several hundred milliseconds and peaking at approximately 300 ms post stimulus onset.

The P3 can be detected on a single-trial level and is partly under user control. Task-relevant events (i.e., those events an individual considers important, or those that have some importance in a particular context) drive the P3 amplitude and different attentional processes give rise to different subcomponents of the P3 (Debener et al. 2005). A common view is that the P3 reflects the voluntary deployment of attentional resources (see Polich 2007, for review). Accordingly, in the BCI framework, this EEG signal can be used to control a device. In the P3-speller rows and columns of a 6 × 6 matrix of letters are randomly flashed, and users selectively focus their attention on a particular letter that is to be selected, for instance by silently counting the number of occurrences this letter is flashed. Using a machine learning approach it can be decoded from the brain signals which letter a subject attended to, albeit with less than 100 % accuracy on a single-trial level.

Bringing BCI paradigms to practical application is challenging, however, for a number of reasons. Among others, further challenges are caused by the practical limitations inherent in the EEG technology. For instance, dedicated hardware is required, electrode caps have to be positioned, gel has to be applied and skin prepared. Another, maybe even more serious challenge is that the SNR is usually not good enough to operate the P3 speller without averaging over a few trials. That is, it takes in practice several seconds to spell a single letter with the P3-speller, making it rather cumbersome to write for instance a full sentence by thought alone. A further challenge is the paradigm itself and the mental process required to generate a brain signal that can be detected. In the P3-speller paradigm it is not the thought of the user, but rather the selective attention to a single letter, that evokes a detectable brain response. BCI paradigms relying on brain signals evoked by some events are called reactive BCI (Zander and Kothe 2011). In contrast, active BCIs do not require a particular external stimulus. Motor imagery BCIs are an example of an active BCI. Here brain signals related to the imagery of a particular motor act can be detected and the signal generated used to operate a BCI or a neuro-feedback system. The corresponding sensorimotor ERD signal can be implemented in the absence of an external stimulus, hence the name active BCI. Both active and reactive BCIs however rely on the user to voluntarily generate a brain signal, and this is a demanding, effortful task. A more practical, highly innovative complementary approach would be to use brain signals that are generated without explicit user control to operate a BCI. To name a few, those signals could be evoked by an automatic orienting response (thereby revealing information about automatic attention processes), an emotional response (reflecting the emotional involvement of a user), or an error-related signal (informing about whether a user became aware of an incorrect behavior). Note that these signals are generated without explicit user awareness or voluntary contribution, hence the name passive BCI.

It is beyond the scope of this chapter to cover the achievements made in the field of BCI research and outline the practical limitations of this approach. However, one important limitation refers to virtually all EEG applications, including BCI. The recordings have to be done in the absence of gross movements. In other words, EEG recordings are confined to the laboratory, or rather artificial setups outside of a laboratory, and participants cannot behave naturally during recording. In the next chapter, we summarize our first efforts toward a fully mobile EEG system aiming to overcome this limitation. With miniaturized hardware a mobile EEG can be build that is largely tolerant to participant movement. At present only prototypes of such systems are available, but they already demonstrate that it is possible to develop a fully mobile BCI system. For the near future this promises cognitive enhancement assistive technology, achieved by means of a mobile, passive BCI approach.

19.6 Toward a Fully Mobile EEG System

Several years ago, we discovered that the modification and combination of consumer EEG hardware with standard laboratory EEG electrodes and caps enable creating a very small, lightweight EEG system (Debener et al. 2012). The resulting amplifier is shown in Fig. 19.5. It has less than half of the size and weight of a modern smartphone and can be mounted onto an electrode cap. Moreover, because of the amplifier technology used, electrode preparation is minimal and it thus takes only very few minutes to set up a 14 channel recording session with our system.

Signal transmission to a nearby computer or smartphone is possible by means of a wireless protocol. Because of the very low weight and wireless transmission users

Fig. 19.5 A new wireless, fully head-mounted EEG system developed at Oldenburg University, Germany (Debener et al. 2012). The system allows for truly mobile EEG recordings, data can be recorded with a smartphone. Note that the cable is used for sound stimulus delivery, not for signal acquisition

can freely move around nearly unconstrained during recording. Importantly, the distance between recording electrodes and amplifier is very short, and all cables are tightly fixed onto the cap, avoiding movement of cables relative to each other. In our experience these two measures taken together dramatically reduce the amount of interference artifact that otherwise heavily corrupts conventional, wired EEG recordings.

In a first validation study (Debener et al. 2012), we recorded EEG data from 16 participants performing a selective auditory attention task. In one condition data were recorded indoors, in a seated office scenario. In the other condition, subject went for a walk outdoors on Oldenburg university campus. Each recording session lasted approximately 11 min and order was balanced out across subjects. In this auditory oddball task subjects had to silently count the occurrence of a rare target tone, which generates a P3 ERP response (e.g., Debener et al. 2005). We asked two questions: First, is it possible to measure the trial-averaged P3 response while subjects walk outdoors. Second, is it possible to reliably identify the P3 even on a single-trial level? Our results revealed clear and reliable P3 ERPs for both indoor seated and outdoor walking conditions. Interestingly, the outdoor walking P3 ERP was approximately 30 % smaller in amplitude. Further analyses led us to conclude that this was mainly due to a larger degree of cognitive distraction in the outdoor condition. This strongly suggests that we can monitor brain activity meaningfully, as the degree of cognitive distraction on the primary task (auditory selective attention) could be quantified. Moreover, using a single-trial analysis approach with ICA as preprocessing and linear discriminant analysis for the statistical evaluation, we also found that the P3 response to auditory target events can be reliably identified above chance level, both in indoor and outdoor conditions. To the best of our knowledge, this is the first study demonstrating a single-trial EEG analysis of cognitive potentials during a freely moving outdoor recording condition. In our most recent protocols, we made use of more advanced wireless amplifier technology (www.braintrain.com) and developed screen-printed electrodes, which can be worn around the ear (Fig. 19.6). We found that unobtrusive, printed electrodes can give rise to good EEG signal quality (Debener et al. 2015).

The P3 response to single trials is at present probably the most popular brain signal for steering a BCI. Thus we believe that our mobile EEG studies and the new sensor technology represent a step forward toward truly mobile BCI systems. Current technology features smartphone operates signal acquisition (Stopczynski et al. 2013) and stimulus presentation already (Mathôt et al. 2012). The next steps will be to close the BCI loop, which requires adding real-time signal processing on smartphone and the delivery of meaningful feedback signals to the user (imagine an alertness monitoring device analyzing brain signals for patterns of sleepiness and sending warning signals when necessary). In any case, the combination of a miniaturized, wireless head-mounted EEG amplifier with printed electrodes and smartphone-based signal recording may be a substantial step forward toward a fully mobile BCI system. Future generations of EEG amplifiers could be made even smaller and fit behind the ear, like current hearing aids. The remaining challenge

Fig. 19.6 The cEEGrid, a flex-printed array of miniature EEG electrodes placed with an adhesive around the ear. This approach enables, unobtrusive, near invisible EEG acquisition (Debener et al. 2015)

then is to close the BCI loop by showing that (passive) BCI applications are feasible in daily-life situations.

19.7 Conclusion

Cognitive neuroscience research relies heavily on noninvasive procedures for the study of the neural correlates of human cognition and behavior. Unfortunately all currently established procedures require that human volunteers remain still during recording and minimize (head and body) movement. Accordingly, the risk of a cognitive neuroscience relying on "finger movements" (Baumeister et al. 2007) as representing natural behavior is substantial. How well laboratory results can be generalized to real-life scenarios should therefore be questioned. We propose here that recent advances in technology, such as the advent of powerful smartphones, will allow us to overcome this limitation and explicitly test for the ecological validity of neurocognitive and neuroeconomic theories. We argue that laboratory cognitive neuroscience needs to be complemented with a new era, field cognitive neuroscience. We envision that this exciting new possibility will also help bridging the gap between fundamental and applied cognitive neuroscience approaches.

The most promising candidate for achieving this goal is the EEG technology. It is beyond doubt that EEG signals provide rich information about cognitive functions, from temporal fluctuations of attention to implicit influences on decision-making processes and prediction of future behavior. Well-established procedures such as cognitive ERPs can now be detected on a single-trial level thanks to advanced signal processing (Blankertz et al. 2011), enabling the near real-time access to brain correlates of cognition. EEG technology is well established, noninvasive and relatively low priced. As our work shows that next generation EEG systems offer the unique potential of being near invisible (Debener et al. 2015) and tolerating daily-life actions (Debener et al. 2012). Future EEG technology could be made very small and user friendly. We envision transparent EEG systems. Note that the concept of transparency is bidirectional. It refers to a user forgetting about wearing the technology (like many forgetting they wear glasses) and individuals in the environment not noticing someone to use EEG technology. In addition to these practical requirements, a key to success will be advanced signal processing. Spatial filter procedures such as ICA help de-noising the data to some extent, but more powerful, robust and adaptive procedures are needed to reliably decode brain signals, achieve high information transfer, and quickly adapt to different recording situations. We believe that the small, wireless and smartphone-based BCI technology holds great potential for basic research in cognitive neuroscience, clinical applications and may constitute the basis for future applied assistive technologies.

References

Baumeister RF, Vohs KD, Funder DC (2007) Psychology as the science of self-reports and finger movements. Whatever happened to actual behavior? Perspect Psychol Sci 2(4):397–403

Blankertz B, Lemm S, Treder M, Haufe S, Müller KR (2011) Single-trial analysis and classification of ERP components–a tutorial. Neuroimage 56(2):814–825

De Vos M, Thorne JD, Yovel G, Debener S (2012) Let's face it, from trial to trial: comparing procedures for N170 single-trial estimation. Neuroimage 63(3):1196–1202

Debener S, Beauducel A, Nessler D, Brocke B, Heilemann H, Kayser J (2000) Is resting anterior EEG alpha asymmetry a trait marker for depression? Findings for healthy adults and clinically depressed patients. Neuropsychobiology 41(1):31–37

Debener S, Makeig S, Delorme A, Engel AK (2005) What is novel in the novelty oddball paradigm? Functional significance of the novelty P3 event-related potential as revealed by independent component analysis. Brain Res Cogn Brain Res 22(3):309–321

Debener S, Minow F, Emkes R, Gandras K, de Vos M (2012) How about taking a low-cost, small, and wireless EEG for a walk? Psychophysiology 49(11):1449–1453

Debener S, Emkes R, De Vos M, Bleichner M (2015) Unobtrusive ambulatory EEG using a smartphone and flexible printed electrodes around the ear. Sci Rep 5:16743. doi:10.1038/srep16743

Gross J, Schmitz F, Schnitzler I, Kessler K, Shapiro K, Hommel B, Schnitzler A (2004) Modulation of long-range neural synchrony reflects temporal limitations of visual attention in humans. Proc Natl Acad Sci USA 101(35):13050–13055

Hewig J, Kretschmer N, Trippe RH, Hecht H, Coles MG, Holroyd CB, Miltner WH (2011) Why humans deviate from rational choice. Psychophysiology 48(4):507–514

Klimesch W (2012) Alpha-band oscillations, attention, and controlled access to stored information. Trends Cogn Sci 16(12):606–617

Kranczioch C, Debener S, Maye A, Engel AK (2007) Temporal dynamics of access to consciousness in the attentional blink. Neuroimage 37(3):947–955

Makeig S, Debener S, Onton J, Delorme A (2004) Mining event-related brain dynamics. Trends Cogn Sci 8(5):204–210

Mathôt S, Schreij D, Theeuwes J (2012) OpenSesame: an open-source, graphical experiment builder for the social sciences. Behav Res Methods 44(2):314–324. doi:10.3758/s13428-011-0168-7

Pessoa L (2008) On the relationship between emotion and cognition. Nat Rev Neurosci 9:148–158

Polich J (2007) Updating P300: an integrative theory of P3a and P3b. Clin Neurophysiol 118 (10):2128–2148

Stopczynski A, Stahlhut C, Larsen JE, Petersen MK, Hansen LK (2013) The Smartphone Brain Scanner: A Mobile Real-time Neuroimaging System arXiv e-print 2013: 2013arXiv1304.0357S

Zander T, Kothe C (2011) Towards passive brain–computer interfaces: applying brain–computer interface technology to human–machine systems in general. J Neural Eng 8 025005 (5 pp)

Chapter 20
Functional Magnetic Resonance Imaging (fMRI)

Sebastian Markett

Abstract Functional magnetic resonance imaging (fMRI) is undoubtedly one of the most common techniques used in the cognitive neurosciences and neuroeconomics. The methods section of fMRI papers are oftentimes filled with jargon. We hope to clarify this jargon by defining and explaining the most fundamental concepts. The present chapter has been written to target a broad audience of scholars and students and explains the principles of fMRI: The reader will learn what signals are measured in fMRI, how this measure relates to neural activity, and how fMRI data are most commonly analyzed. This includes a brief summary of physical, physiological, and statistical ideas. We further present a comprehensive step by step guide through a typical fMRI data analysis to provide scholars and students with the appropriate knowledge to understand basic fMRI methodology in research papers and to judge whether the presented analysis is meaningful and appropriately protected against the most common pitfalls in the field of neuroimaging.

20.1 Introduction

Functional magnetic resonance imaging (fMRI) has been the major backbone of the cognitive neurosciences since their very early days. Therefore, it is of little wonder that this method has become extremely popular in the field of neuroeconomics as well. A search with the keywords "functional magnetic resonance imaging" and "neuroeconomics" carried out in Google scholar in early 2016 returned 3120 hits, approximately a quarter of the hits of a search for "neuroeconomics" on its own. What has made this research approach so popular? There are four certain reasons: (a) fMRI has an excellent spatial resolution that allows for the precise anatomical location of neural activation within the brain (b) fMRI comes with sufficient temporal resolution to detect neural correlates of behavior on the basis of experimental trials (c) fMRI is very sensitive and can therefore measure subtle differences in

S. Markett (✉)
Department of Psychology, University of Bonn, Bonn, Germany
e-mail: sebastian.markett@uni-bonn-diff.de

M. Reuter and C. Montag (eds.), *Neuroeconomics*, Studies in Neuroscience, Psychology and Behavioral Economics, DOI 10.1007/978-3-642-35923-1_20

neural activation between experimental conditions which is a prerequisite to test theories on human behavior, and (d) fMRI is noninvasive and safe to use in human research participants because it does not require any pharmacological contrast agents or the lowering of signal detecting devices into the cranium.

There is probably a fifth reason less persuasive to the critical scholar but with great impact for the presentation of research results: fMRI outputs beautiful and intuitively comprehendible images. Even though there are many different ways to present fMRI data, the most common approach to visualize the results is to mark activated regions with red- and yellow-colored blobs on an otherwise greyscale brain. It is these images that has led to the popular notion that fMRI enables the researcher to observe the living brain in action. This might be true to some extent but carries one misconception: The colored blobs themselves are no physiological signals returned by the MRI scanner.[1] They stand at the end of many time-consuming processing and statistical analysis steps and are nothing more than statistical parameters that reflect differences in signal strength between experimental conditions. It is only after the analysis that these statistical parameters are color-coded and then spatially overlaid on a three-dimensional image of the brain. This explains why this approach to neurophysiological data has been labeled statistical parametric mapping and the resulting images statistical parametric maps (Friston et al. 1995).

This chapter is organized in two parts. The first part will focus on the fundamentals of fMRI to answer the question what signal fMRI scanners actually measure and how this signal relates to psychological processes. Understanding the fMRI signal requires some basic knowledge of physics and cell physiology which we hope to cover up in a comprehensive way for readers who have a background in behavioral economics and psychology. The second part will focus on data analysis and will deal with the processing pathway from the raw fMRI data that come out of the MRI scanner to the well-known statistical parametric maps mentioned earlier. Understanding the analysis of fMRI data requires some basic knowledge in psychological experimental design and statistics which we hope to cover up in a comprehensive way for readers without a background in behavioral economics or psychology.

20.2 Fundamentals of fMRI

fMRI is a four-letter acronym. In the introduction, we have already dealt with the fourth letter, the i for imaging, and established that fMRI outputs images of the brain. As a fact, fMRI outputs functional images of the brain. This is what the f stands for and it means that the images acquired in an fMRI scanner allow for inferences on brain function, in this case on neural activity. The opposite (or better

[1]fMRI is a special case of magnetic resonance imaging. Functional (fMRI) and structral (MRI) images are acquired on the same scanning device.

the compliment) to fMRI would be structural MRI, an approach not sensitive for brain function but for brain anatomy (see the chapter by Christian Gaser in this edition for the role of structural MRI in the context of neuroeconomics). The remaining two letters in fMRI, the m and the r, stand for magnetic and resonance respectively. They refer to the means by which an fMRI scanner acquires the images: fMRI scanners measure magnetic properties of atomic nuclei in the brain which they accomplish by applying magnetic fields oscillating at the resonant frequency of these nuclei. We will come back to this later in more detail.

The main question we seek to answer in this part of the chapter is how fMRI scanners measure neural activity. As a matter of fact, we can answer this question quite easily on the spot: They do not. This information might come surprising because we usually speak of neural activity revealed by fMRI but it is true: fMRI scanners do not measure neural activity directly. What they do measure, however, are magnetic properties of brain tissues that depend on physiological processes that are most strongly correlated with the neural activity underlying psychological processes.

20.2.1 The Magnet

All fMR imaging starts with a magnet. We have established earlier that fMRI relies on the measurement of magnetic properties of atomic nuclei in the brain. This may sound odd at first glance: If you ever tried to attach a magnet (like the ones that people use to stick notes to their fridge) to your head it will probably come off instantly. This is because the head and the brain have no magnetic properties by themselves. What has magnetic properties, however, are the nuclei of atoms in the brain. MR image acquisition is based on the fact that some atomic nuclei spin around themselves. Hydrogen—the most abundant atom in the brain—has such a spinning nucleus and can therefore be measured by MRI. The nucleus of a hydrogen atom consists of only one positively charged proton. Because of its positive charge, the proton creates a tiny magnetic moment when it spins around itself. This magnetic moment points in the same direction as the proton's spin axis. Under normal conditions, the protons' spin axes will point in random directions, which mean that the same will apply for the magnetic moments. MR imaging does not measure the magnetic moments of single nuclei but the sum of all magnetic moments which is called the net magnetization. Thus, if we tried to measure the magnetic moments under these normal conditions, we would not be able to pick up any signal because the moments would cancel each other out. This is where the magnet enters the stage: If we put our sample (with the containing hydrogen atoms) into a magnetic field, the spins will start to revolve around an axis that is parallel to the magnetic field. This additional spin is called precession spin. You can think about a nucleus' behavior in the magnetic field and the two spins (the regular and the precession spin) as a spinning top (like the ones you may have used to play with as a kid). A spinning top does not only spin around its own axis, it also precesses around a second axis parallel to the earth's gravitational field. If you would watch

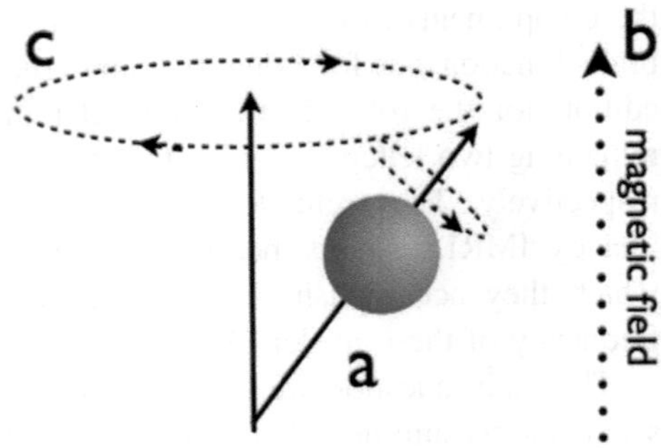

Fig. 20.1 Spin of a hydrogen nucleus around its own axis (**a**). When a magnetic field is applied (**b**), the nucleus falls into an additional precession spin in the transversal plane (**c**)

the spinning top from above you would see that the precession spin traces a circle perpendicular to the gravitational field. The precession spins of the nuclei behave in a similar way, only that they do not align with the earth's gravitational field but with the magnetic field applied by the MRI scanner. The axis around which the nuclei precess is called the longitudinal direction and the plane in which they precess is called the transversal plane (see Fig. 20.1).

The precession axes align with the magnetic field in two different ways: Either parallel or antiparallel to the magnetic field. The two states differ regarding their energy levels: The parallel state is a low-energy state and is therefore the preferred state of the nuclei. Nonetheless, at each point of time, many nuclei will also spin in the high-energy antiparallel state. Every now and then each nucleus will change its state and flip from the parallel to the antiparallel spin and vice versa.

The more nuclei spin in the parallel relative to the antiparallel state, the higher is the net magnetization in the sample. To get MRI to work we therefore need all (or most) of the nuclei in the parallel state. This can be accomplished by two means. The first approach would be to cool down the sample to the point where no or only little molecular motion occurs. This, however, would be way too cold for the living brain and is therefore not practical for our purpose. The other approach is the one used in MRI scanners: If we dramatically increase the field strength of our magnet, the vast majority of nuclei will align their precession spins with the magnetic field in parallel. The field strength of strong magnets is given in Tesla (T). MRI scanners approved for human research participants have field strengths between 1.5 and 9.4T (to give you an idea of how strong such magnetic fields are: the electromagnets used to lift cars in junk yards have field strengths of approximately 1 T). Fortunately, strong magnets do not harm biological tissue which make them safe to use in human research participants (as long as participants remove all ferromagnetic objects like glasses, belts, or certain jewelry).

20.2.2 Resonance

With the vast majority of spins in the parallel state, the net magnetization in the sample points into the same direction as the magnetic field. At this point, however, we have no chance to measure it. In order to do this, controlled changes in net magnetization need to be observed over time. This is where the resonance (the r in fMRI) comes into play: The idea behind the r is to attach energy to the nuclei which forces them leave the low-energy parallel state and flip toward the high-energy antiparallel state. This process is called excitation and is achieved by applying additional oscillating magnetic fields to the sample. It is important that the additional magnetic fields oscillate with the same frequency as the nuclei do. The spin frequency of a nucleus is called its lamor frequency. The lamor frequency depends on the amount of protons in the nucleus (which is the same in all hydrogen atoms) and the strength of the magnetic field. Because the magnet's field strength is known, the excitation signals can be adjusted to match the lamor frequency of hydrogen nuclei. As a result, energy is attached to the nuclei and they flip from their parallel spin toward the antiparallel spin. When the oscillating magnetic fields are switched off again, the nuclei will start to flip back into the parallel state while emitting the attached energy. This energy can be measured by reception coils in the MRI scanner. The emitted signal is affected by different tissue types and physiological processes. From the behavior of the nuclei returning into the parallel state, we can infer on properties of the brain tissue. Therefore, it allows for inferences in brain structure and function.

As outlined above, the nuclei precess around the longitudinal axis parallel to the magnetic field and precess in the transversal plane that is perpendicular to the magnetic field. The net magnetization (i.e., the sum of all magnetic moments) that is measured by MRI can be split up in longitudinal and in a transversal component. Without excitation of the spin system by oscillating magnetic fields, the transversal components of the net magnetization cancel each other out and only the longitudinal component parallel to the magnetic field prevails. The excitation pulses are usually designed to flip the net magnetization by 90° into the transversal plane. In consequence, the longitudinal component of the net magnetization is set to zero. As soon as the net magnetization is tipped into the transversal plane, the nuclei's precession spins will start their spins at the same starting point. In consequence, the transversal component of the net magnetization can be measured. After the excitation signals wear off, the nuclei will start to flip back into the parallel state. Two different components can be measured by the signal detection coils of the MRI scanner. First, the longitudinal component of the net magnetization will recover while the spins flip back. The longitudinal recovery is governed by a time constant that is labeled T_1. Second, the spins' coherence in the transversal plane will start to dephase until the transversal component of the net magnetization cannot be measured anymore. The transverse relaxation is governed by a time constant labeled T_2. Different tissue types (grey matter, white matter, cerebrospinal fluid, blood vessels, and bone) lead to different T_1 recovery and T_2 relaxation values. In order to construct images, spatial information must be provided along with the information on recovery or relaxation. You may recall that the lamor frequency of nuclei depends on the field strength of the magnet. Additional gradients that vary the field strength gradually across space can therefore be combined with excitation pulses at different frequencies to allow for a space dependent coding of the signal. This approach ensures that one two-dimensional slice of the brain is measured at a time. A three-dimensional image of the brain can be mathematically reconstructed from the spatial distribution of T_1 or T_2 values across different slices.

20.2.3 From Physics to Physiology

MRI protocols that are sensitive to T_1 or T_2 contrasts provide anatomical images of the brain. To measure brain function, however, a different signal is needed. Recall that after the application of excitation pulses, the spins start to precess at the same starting point in the transversal plane, thus giving rise to the transversal component of the net magnetization. The dephasing of the spins that leads to transversal relaxation depends on interactions between the spins of nuclei. This intrinsic factor is directly reflected in the loss of T_2 signal across time (T_2 decay). Additionally, the dephasing is also influenced by an extrinsic factor. Because the spin frequency (the lamor frequency) depends on the field strength, slight inhomogeneities in the external magnetic field do also contribute to dephasing. The combination of the intrinsic and extrinsic factor leads to a signal loss in transverse magnetization that is governed by a time constant labeled T_2^*. Local inhomogeneities in the external magnetic field can depend on physiological processes in the brain. Therefore, MRI protocols sensitive to T_2^* are the backbone of functional MRI.

How do physiological processes affect the local homogeneity of the magnetic field? To answer this question we need to discuss energy consumption of the brain. The cellular basis of psychological processes can be traced to the activity of nerve cells (neurons). Neurons communicate by short transient changes of their electric resting potential across the cell membrane. This process does not rely on external energy. What does require energy, however, are housekeeping tasks of neurons such as maintaining their resting potential and restoring the resting potential after an electric signal has traveled along the cell membrane. The energy currency of the brain is a tiny molecule called adenosine triphosphate (ATP). ATP is synthesized from glucose, a sugar absorbed from food sources. This synthesis is most efficient in the presence of oxygen. Both oxygen and glucose need to be delivered to the brain via the blood stream because the brain cannot store either of the molecules. Blood is pumped through the vascular system by the heart. On its way from the heart to the brain, blood is first circulated through the lungs where oxygen is bound to hemoglobin, the oxygen transport protein in red blood cells. Then, the blood with the oxygenated hemoglobin is pumped through arteries into all parts of the body including the brain. The brain is supplied by four major arteries. After entering the cranium, arteries branch out into smaller arteries that eventually become arterioles and then capillaries. The capillaries form a fine net of tiny blood vessels that enable the exchange of oxygen, glucose, and their metabolites between the bloodstream and nerve cells. At this point, the hemoglobin trades oxygen for waste carbon dioxide and becomes deoxygenated hemoglobin.

When you put your hand onto your neck you can feel the dilatation of the arteries in response to your heartbeat. Before entering the capillaries, the pulsatile blood supply needs to be slowed down by high-resistance blood vessels to ensure a steady blood flow. Otherwise, the fine capillaries would burst from peaks in blood pressure. Where supplying arteries branch out, muscular sphincters control the blood flow into arterioles and capillaries. When nerve cells in a circumscribed region increase their activity level and thereby their energy consumption, the sphincters expand the arterioles to increase blood flow into respective regions in order to meet the temporally enhanced requirements for glucose and oxygen. That is, locally confined neural activity leads to a locally confined increase in blood flow with blood that is rich in oxygenated hemoglobin. fMRI exploits the fact that hemoglobin has different magnetic properties that depend on the binding of oxygen. Oxygenated hemoglobin is diamagnetic while deoxygenated hemoglobin is paramagnetic. Generally, objects with paramagnetic properties cause spin dephasing when introduced into a magnetic field. An increase in blood flow leads to an increase in oxygenated hemoglobin relative to deoxygenated hemoglobin. In turn, it leads to less spin dephasing and in consequence to slower transversal relaxation and a stronger T_2^* signal. That is, MR protocols sensitive to T_2^* can use oxygen as an intrinsic contrast agent of the brain for the mapping of neural activity. In this case, we speak of blood oxygen level dependent fMRI, or in brief, of BOLD fMRI.

Because the recorded signal in fMRI relies on blood flow dynamics in response to changing neural events, the signal is called the hemodynamic response. The typical hemodynamic response as revealed by BOLD fMRI starts with a temporal offset of 1–2 s to the neural activity that triggered the response. The reason for this time lag reflects the time window until the feedback loop between active neurons and their supplying blood vessels has increased the local blood flow. After a steep rise, the hemodynamic response peaks about 4–5 s later and then falls steadily over another 5–6 s until it falls below baseline 12–13 s after the triggering neural activity,. The BOLD signal returns to baseline level approximately 20 s after the onset of the neural events. From this timing information, we can see that the hemodynamic response lags the neural events behind and is rather slow compared to psychological processes that often take only a couple of hundred milliseconds to finish. Nevertheless, neural correlates of even short-lived psychological processes can be traced by BOLD fMRI, given individual experimental trials are sufficiently spaced.

The sampling rate of the MRI scanner needs to be set in a way that sufficient information on the hemodynamic response will be acquired. The time between successive excitation pulses of the scanner is called repetition time (TR) and quantifies the acquisition speed of the scanner in a given experiment. Modern fMRI scanner that uses echo planar imaging (EPI) pulse sequences can scan the majority of the brain with a TR of 1–2 s while retaining a sufficient spatial resolution (usually 3 mm^3 voxels). It has been demonstrated that sampling rates below 0.5–1 Hz do not substantially improve the measurement. That is, a TR of 1–2 s provides an appropriate temporal resolution for fMRI even if the examined psychological processes follow a faster time scale.

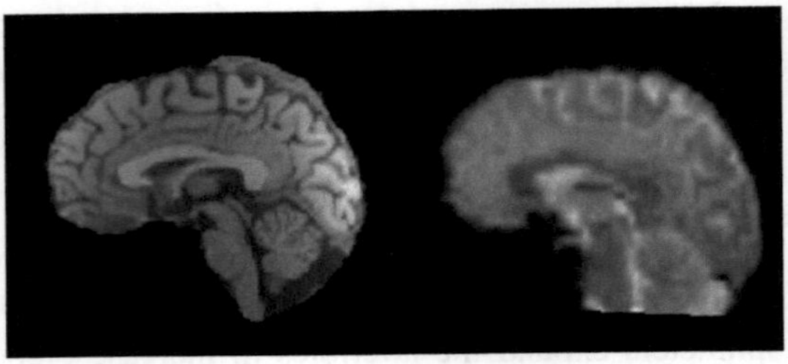

Fig. 20.2 A high-resolution T1-weighed anatomical MRI scan and a BOLD fMRI image (T_2^*-weighed) from the same participant

The EPI pulse sequences acquire data in two-dimensional slices from which a three-dimensional image of the brain can be reconstructed. The brain spans about 12.5 cm from the brainstem to the most dorsal part of the parietal lobe. With 3-mm-slices spaced by a 0.3 mm gap, it would require 38 slices to image the entire brain. Because the TR depends critically on the number and the spacing of these slices, the temporal resolution can be improved by omitting parts of the brain during image acquisition, in most cases the brain stem and the cerebellum. This is a feasible approach especially in neuroeconomical studies, because cortical or midbrain structures lie in main interest of most investigations. In the following, we will refer to functional images acquired in one TR as "volumes" and not "brains" to emphasize that the processing steps are applied to the functional data irrespective of the degree to which the entire brain is covered during imaging. Figure 20.2 shows a T_1-weighted anatomical and a T_2^*-weighted functional volume from the same participant.

20.2.4 Summary

In the first part of this chapter, we have established that fMRI measures magnetic properties of brain. Furthermore, we have discussed how vascular activity in the brain gives rise to the blood oxygen-dependent signal that can be measured by MRI scanners and allows for inferences on neural activity. We have concluded with remarks on temporal and spatial properties of the hemodynamic response. This part was supposed to give a brief overview on the physical and physiological basis of fMRI. For more in-depth information on the physical and physiological basis of fMRI, we refer to the excellent textbook by Huettel et al. (2009). In the following, we will deal with the statistical analysis of the acquired volumes in the context of statistical parametric mapping.

20.3 Analysis of fMRI Data

The analysis of fMRI data can be separated into three consecutive steps: (a) preprocessing of functional images (b) first-level analysis of fMRI time series, and (c) second-level (or higher order) analysis. Preprocessing describes necessary analysis steps that are carried out to ensure that all data are in the same five-dimensional coordinate space which is a prerequisite for statistical analysis (we will explain the five dimensions in the following). Then, the first-level analysis is carried out separately for each participant. It outputs statistical parameters that are eventually fed into the second-level analysis that aggregates the data across participants for statistical inference on activation patterns between groups or in the population the sample of study participants has been drawn from. All three analysis steps can be carried out in freely available analysis software tools. The two most popular tools in the neuroimaging community are Statistical Parametric Mapping (SPM) that is issued by the Wellcome Trust Centre for Neuroimaging (http://www.fil.ion.ucl.ac.uk/spm/) and FSL, issued by the Oxford Center for Functional Imaging of the Brain (http://fsl.fmrib.ox.ac.uk/fsl/fslwiki/). Both imaging software packages can be downloaded for free from the respective websites and come with detailed documentation and example data sets.

Before we start to discuss preprocessing and first- and second-level analyses in more depth, we start with some general remarks on experimental design, as this is the prerequisite to understand what is going on during the analysis steps.

20.3.1 Experimental Design

One inherent property of the BOLD signal is that it is no absolute signal: We always need to compare the signal to some sort of baseline or control condition. This control condition can be implicit, that is the hemodynamic activity evoked by a task is compared to hemodynamic activity while there is no task. This, however, might come with the downside that the task and the control condition differ in many different aspects. Let us assume we are interested in hemodynamic activity evoked by the feedback about the second mover's behavior in the trust game. In the experimental condition, the research participants in the role of the proposer face an information screen that states whether the responder is defected or not. To ensure that the participants actually process the information on the display, they are asked to execute some sort of manual response to the information. The evoked hemodynamic response could be compared to a condition were the participants did nothing. Such a condition, however, would not only differ in the decisive variable (cooperation versus defection) but also in physical appearance of the display, the lack of a motor response, and the absence of a monetary outcome. Thus, hemodynamic activity associated with these factors cannot be easily disentangled from the actual activity inherent to the trust game. Therefore, it is a better idea to contrast the task condition with an explicit control task that differs only in one aspect critical to the study. In our example, this could be a computer-raffled lottery where the participants either loose or win but that most critically lacks the social component of the trust game (for example, see Delgado et al. 2005). This experimental design relies on the pure insertion principle inherent to subtractive experimental methods: Different processes are assumed to be additive. By subtracting the hemodynamic response during one process (computer lottery) from the hemodynamic response during another process (outcome of the trust game), the mere difference in hemodynamic activity between both processes survives the subtraction (neural response to cooperation or defection). This approach can be very powerful and useful, it disregards, however, the possibility of non-additivity in the sense of interactions between task conditions. One way to study such interactions is the use of factorial designs in which all combinations of two or more independent variables with two or more levels are administered. If, for example, one would want to study the effect of absolute versus relative income, possible independent variables are absolute income (receiving a high or a low amount of money) and relative income (receiving more, less or just as much as somebody else). In consequence, six different conditions arise that allow to disentangle additive and interactive effects of absolute and relative income on reward related hemodynamic activity (see Fliessbach et al. 2007, for example).

A further comparison strategy is parametric designs. In parametric designs, the covariation of the BOLD response with a parametrically manipulated independent variable is examined. If, for example, we are interested in neural correlates of decision utility during gambles, we could vary the possible gains associated with the gambles parametrically and examine whether the BOLD signal in a given brain region responds contingently (see for example Tom et al. 2007).

The previous considerations all dealt with comparison strategies. A further thriving issue concerns the temporal sequence of stimulus presentation. There are two main approaches to stimulus timing in BOLD fMRI experimental design: blocked versus event-related presentation. In a blocked design the research participants are asked to alternate between blocks of many trials in the experimental and the control task. Ideally, the length of the blocks corresponds to the length of the hemodynamic response (about 10 s). With shorter time intervals, the BOLD response cannot return to its baseline and the differences between experimental and control conditions become blurry. With longer blocks, on the other hand, scanner drift can inflate the differences between experimental and control blocks by introducing noise to the data. In blocked designs, many experimental trials contribute linearly to the recorded hemodynamic response. Therefore, blocked designs come with a high power to detect differences in activation between conditions. The downside, however, is that temporal information on the hemodynamic response cannot be analyzed. Furthermore, many research questions cannot be operationalized in blocked designs. Many research designs do not allow for an a priori specification of task and control conditions. Let as assume we are interested in examining hemodynamic activity associated with continued gambling to recover previous losses, a phenomenon called "chasing losses", which is maladaptive decision behavior common in pathological gambling (Cambell-Meiklejohn et al. 2008). Participants are confronted with the outcomes of gambles and decide whether they want to continue or quit gambling after experiencing losses. In such an experiment, the participants' decision behavior determines if a given experimental trial is assigned to the task (chasing losses) or to the control condition (quitting gambling). If this research question would be addressed in a blocked design, the researchers would need to tell participants to chase losses in one block of trials and to quit gambling in others. This, however, would eliminate the critical behavior under the researchers' scrutiny: Participants would not make the decision to continue gambling by themselves and no neural activity associated with the decision making process could be recorded. Because decision-making is one of the main research interests in neuroeconomics, most studies in the field adopt event-related designs. In event-related designs, the evoked hemodynamic response to single experimental trials is examined. The advantage of event-related designs is that events can be assigned to experimental conditions post hoc. It is also possible to exclude certain trials from the analysis, for instance error trials or trials in that the participant fails to respond in a given time window. Furthermore, event-related designs allow for a precise temporal characterization of the hemodynamic response. Compared to blocked designs, however, they lack a high degree of detection power. A further potential drawback is that subsequent presentation of the same events can introduce an artificial blocked design, where the BOLD response saturates and becomes equivocal to different task conditions. To counteract this problem, the interstimulus interval (ISI) should be large and jittered which means that it is randomly varied in its duration across experimental trials (e.g. a randomly chosen ISI between 3500 and 6000 ms).

Now that we have established, how experiments can be designed to be suitable for fMRI we can go on with a discussion how to analyze the data.

20.3.2 Preprocessing

As we have discussed earlier, the best way to think about fMRI data is in voxels. Each functional volume lies within a three-dimensional grid comprising a large number of voxels with one activation value assigned to each of them. Each fMRI run comprises many of these volumes which are acquired consecutively over time. This results in an activation time series for each voxel (see Fig. 20.3) and, if we look at all voxels and all time-points at once, one-four dimensional data set (x by y

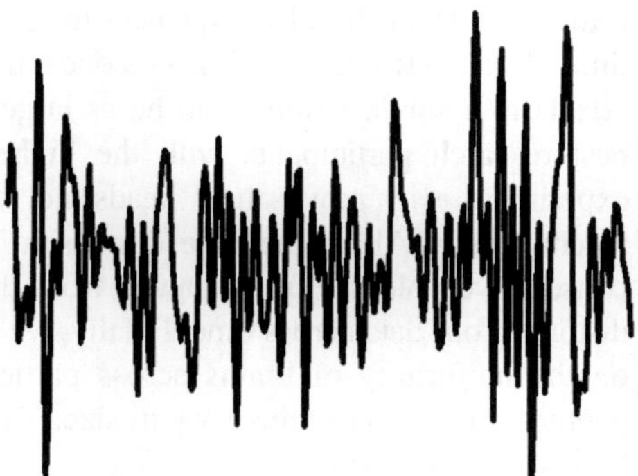

Fig. 20.3 The BOLD time series as measured from one gray matter voxel over a time period of approximately ten minutes

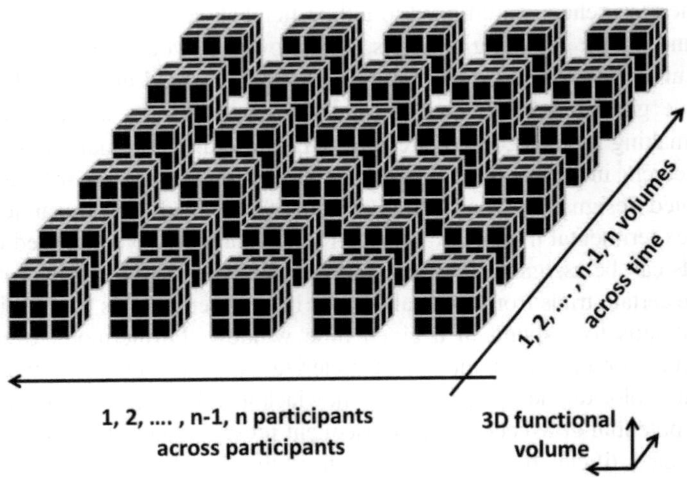

Fig. 20.4 The five-dimensional nature of BOLD fMRI experiments. The first three dimensions refer to the three-dimensional brain that is scanned across time (dimension 4). The fifth dimension refers to different participants who undergo the same imaging protocol

by z by time). We acquire such a four-dimensional data set for each participant in our experiment. That is, we can think of the data of an entire experiment as a five-dimensional matrix (a three dimensional brain scanned across time and across participants, see Fig. 20.4).

Now imagine a perfect world: We can expect various things from a perfect world: First, we would expect that all voxels in one volume were sampled at the same time (that is, at a given point in time, for a given participant, all data along the first three dimensions of our matrix are acquired simultaneously). Second, along the time dimension, we would expect that data measured in the same voxel merely reflects changes in activation (and not sampling error) at the same precise anatomical location (and not that of a neighboring location). And at last, we would expect the brains of all participants to match spatially, that is, we would expect that along the fifth dimension (the participants) a given voxel always corresponded to the same neural structure. Alas, the world is not perfect, especially not for neuroimagers: The EPI pulse sequences used for fMRI measure a volume one slice at a time. With a TR interval of two seconds, this implies that the temporal offset of two voxels in a single volume can be as large as two seconds. Furthermore, even the best research participants with the highest motivation to keep still during the experiment will move their heads no matter how tightly we constrain them mechanically. Along the time dimension, every millimeter of motion will move consecutive volumes further away from the first volume, slowly but dramatically distorting our data across time. Finally, we can intuitively agree that our expectation on the uniformity of brains across participants cannot hold: It is not only that peoples' heads and brains vary in size. There are also individual differences in the brains' gyri and sulci.

Given these constraints on our data, we need to come up with an idea to resolve these issue. Otherwise we would not be able to meaningfully analyze our data. Fortunately, there are powerful algorithms available to correct for temporal and spatial distortions within single volumes, across time and across participants. However, we should be aware that we substantially alter our data during preprocessing and that in consequence, statistical inferences on brain activation are not carried out based on data as they were initially measured. Therefore, all preprocessing should be applied carefully and with high caution. Preprocessing steps are nevertheless necessary means to correct for deviations from our perfect neuroimaging world: They ensure that subsequent statistical analysis will be meaningful.

Usually, preprocessing steps include (1) slice timing, (2) head motion correction, (3) coregistration/normalization, and (4) spatial smoothing. Temporal bandpass filtering and detrending can be applied as additional but not necessary steps.

Slice timing refers to the correction of acquisition delays between slices within the same volume. For the slice timing correction, the user specifies a reference slice for each volume. Then, an algorithm analyzes the time course in each voxel across volumes and interpolates how the data points in every other slice of a given volume would have looked like if they would have been acquired at the same time as the reference slice. The further away a given slice lies from the specified reference slice, the stronger the data are altered by the algorithm. Therefore, it is a wise idea to choose the slice in the middle of the volume as reference. In many cases, the slices of a single volume are not acquired in ascending or descending order but interleaved (that is, all even slice numbers are imaged before the odd ones). In this case, slice timing correction is always recommended. In cases where the scanner has acquired the slices consecutively, slice timing can be omitted if steps are taken to control for slice acquisition offsets later on during statistical analysis (and the TR is not too large).

Motion correction provides an algorithm to correct for spatial distortions across volumes within a single participant because of head motion. Fortunately, we can be sure that our participants' heads do not change their size or form during the relatively short fMRI run. That is, all gross deviations between heads across volumes are almost exclusively attributable to head motion. In the scanner, a head can move up and down, from left to right and back and forth. That is, it can translate along three dimensions. Furthermore, it and can rotate around three axes (nodding, shaking and tilting). We can therefore align the heads within all subsequent volumes with the head in the first volume with a six-parameter transformation. Because the head itself does not change during transformation, we call this a rigid-body transformation which has six degrees of freedom. Usually, the parameters for each volume are saved so that they can be used as covariates (nuisance regressors) during later statistical analysis. The motion parameters should also be examined for outliers. If single participants have moved too excessively during scanning it might be wise to exclude them from further analysis (usually, translations less than three millimeters and rotations less than three degrees are considered tolerable).

Coregistration and normalization are two consecutively applied preprocessing steps that ensure comparability of the data across participants. As mentioned earlier, peoples' brains to not only differ in size but also in gyrification and sulcification. Therefore, a rigid body transformation as discussed above or a linear transformation are both not suitable to match one brain with the others. The solution is an affine transformation with twelve degrees of freedom in total. That is, twelve parameters are needed to match an individual brain with a group. This process is called normalization. One way to do this would be to choose one representative participant from your sample (maybe the one who is closest to the mean of demographics like age or education and a member of the more frequent sex). The next step would be to apply the affine transformation. During this process all other brains are resized, squeezed and dragged until they match the reference brain most closely. After the transformation, each voxel in each volume will contain information on the activation level of exactly the same neural structure across participants. This approach of choosing a representative reference brain from the study sample, however, would come with two major downsides. On the one hand, the researcher would be in need of a high degree of anatomical skill to precisely identify what structures are activated during later analysis. On the other hand it would be quite tricky to compare results across different experiments and across publications. For that reason, neuroimagers have agreed on a standard reference brain in a standard coordinate space. This standard brain has been issued by the Montréal Neurological Institute (MNI). It unifies anatomical information from 152 white individuals and should therefore be a close reference template to most brains from healthy individuals of European descent. Other reference brains are available for individuals from other populations. Practically, we need to deal with one problem during normalization: As you can see from Fig. 20.2, the spatial resolution of functional EPI images is not as good as the spatial resolution of anatomical T_1-weighted scans. If the normalization algorithm was fed by anatomical information from the functional volumes alone, it would lack detailed information. Coregistration prior to normalization is a recommended resolution for this problem. Coregistration takes advantage of two facts: First, it is relatively easy to precisely align a high resolution anatomical T_1-weighted image with functional volumes from the same person because it requires only a rigid-body transformation (see above). Second, it is also relatively easy to normalize a T_1-weighted image of a study participant to a T_1-weighted reference brain in standard coordinate space because of the high anatomical detail of the individual T_1-image. For this purpose, a high resolution T_1-weighted anatomical image is usually acquired along with the functional volumes. Then, the functional volumes are aligned with the individual anatomical image. In a second step, the individual anatomical image is normalized to the reference brain and the resulting twelve transformation parameters are then applied to every single functional volume. In easier words, the individual anatomical image "piggybacks" the individual functional volumes to obtain the best normalization results possible.

Spatial smoothing is the last strongly recommended preprocessing step. The rationale behind smoothing is to use the information of neighboring voxels to smooth the signal from each voxel in a volume. As a result, the images become more blurry, but also increase in their signal-to-noise ratio. Smoothing leads to an increase in statistical power of subsequent statistical tests because the error terms of the test statistics are reduced and because single peak activation foci will more likely merge to robust activation clusters. Smoothing takes advantage of the fact that the time courses of adjacent voxels are highly intercorrelated because for most psychological processes, the spatial resolution of fMRI scanners exceeds the functional resolution required to image their neural correlates. Testing theory states that each measurement is additively composed of a true value and an error term. The error terms are thought to be independent from each other and have an expected value equal to zero. Thus, if we average the signal measured in neighboring voxels, we effectively decrease the error term while the true values are more or less left as they are. The averaging procedure during smoothing applies weighting to neighboring time series in a way that the time courses of more nearby voxels contribute more strongly. The weighting is accomplished by applying a three-dimensional Gaussian kernel that has its peak on the voxel to be smoothed. The size of the Gaussian kernel is given by its full width at half maximum (FWHM). The recommended size depends on the resolution of the functional data (and the neural structures and psychological processes that are imaged). In most cases, the kernel size varies between 6 and 12 mm FWHM.

This concludes the strongly recommended steps. Additional preprocessing can include detrending and temporal bandpass filtering. Detrending corrects for linear trends apparent across the entire time series because such linear trends are most likely attributable to MRI scanner drift. Bandpass filtering removes frequencies in a specified frequency band from the time series because certain frequencies do not reflect neural activity and are therefore most likely attributable to scanner artifacts or introduced by cardiorespiratory activity of the research participants. In most cases, a high pass filter of 0.008 Hz is applied, that is, all oscillations in the frequency bands below this threshold are removed from the data.

20.3.3 First-Level Analysis

A typical first-level analysis aims at the isolation of activation differences between our experimental conditions of interest on the level of single participants. In consequence, the first-level analysis is carried out for each participant separately. The most common approach in the context of statistical parametric mapping is to set up a statistical model that explains the acquired neurophysiological data best and then conduct inferences on activation differences based on the parameters from this model. In simple words, the activation data are correlated with the temporal sequence of experimental events voxel by voxel and inferences are subsequently conducted based on the correlation coefficients. Usually, a mass univariate approach is applied: Parameter estimation and statistical inference are conducted separately for each voxel and results are only combined in the very last step.

20.3.3.1 Model Specification and Parameter Estimation

The statistical model most widely applied in neuroimaging is called the general linear model (GLM). In a GLM analysis, a dependent variable (in our case the neurophysiological data) is predicted by a set of predictors that are linearly (additively) combined. In the most simple case, the model equation of a GLM is $y = mx + n$, where y refers to the dependent variable, x to one predictor, m to a weight attached to the predictor (the model parameter) and n to an intercept (a constant value added to the equation). You may notice that this equation resembles the linear equations you have solved in high school. Further predictors can be added to the model (like $y = mx_1 + px_2 + zx_n + n$) if we believe that this leads to a better prediction of y. In the context of fMRI, the dependent variable is the BOLD time series from one voxel. This time series has as many entries as there are volumes in our fMRI run (in the following we will refer to these time series as vectors). The model equation needs to be designed in a way that the predictors and their corresponding weights output values for y that come closest to the values in the time series vector. The experimental conditions in the experiment serve as predictors with one predictor for each condition. In the simplest case one experimental condition (for instance tapping with the right index finger) is compared with a control condition (doing nothing; most critically, no tapping with the right index finger). A basic GLM for this design would be $y = mx_{\text{tapping}} + px_{\text{nothing}} + n$. Let us assume that the participant in this example experiment alternated between the two tasks (tapping versus doing nothing) every 20 s for seven times in total. This experiment would last for 280 s. If we choose to image the brain at a TR of two seconds, we would acquire 140 functional volumes in the experiment, leaving us with a vector with 140 entries per voxel. These 140 observations per voxel serve as the criterion variable y. What do the predictors look like? As mentioned earlier, we plan to correlate the neurophysiological data with the temporal sequence of experimental events. Therefore, we need vectors for each predictor that have as many entries as the BOLD vectors holding the criterion variable. In the simple case of our example, we will code the onsets of the experimental condition with 1 and leave the value to 1 for the entire duration of the experimental block. That is, the vector of the first predictor will hold a 1 whenever the participants performed on the tapping task and a 0 whenever they did nothing. Thus, the first 10 entries of the vector will hold ones; the next 10 entries will hold zeros and so on. The onset vector for the second condition (resting) will hold ones whenever the participants are resting, and zeros whenever the participants perform on other tasks (in our case, tapping their right index fingers). Now that we have modeled all experimental conditions, further predictors could be added that are of no interest for the research question per se but might enhance the model fit by explaining systematic noise in the data. For example, we could use the six parameters from the motion correction during preprocessing as six additional predictors. Neuroimagers refer to these control predictors as covariates of no interest or as nuisance regressors. Note that we are still dealing with the data of one participant. Therefore, nuisance covariates such as age or gender that vary across participants cannot be entered to the model at this point.

After the specification of all predictors, we end up with an $m \times n$ matrix, where m refers to the number of functional volumes and n to the number of predictors. In this simple case, we have coded our experimental conditions in a binary mode (that is we have only used ones and zeros). If we believe that the BOLD signal increases stronger in different realizations of our predictor variable (for example, we measure brain reactivity to economic gambles and believe that BOLD activity increases with increasing gains), we could modulate the predictor's vector with a parameter. In the example with the gambles, one predictor would carry ones and zeros (whenever the gamble was shown) and a further predictor would carry the modulator (the potential gain of the gamble as an integer) whenever the former vector carried a 1.

There is reason to believe that the BOLD signal does not rise and fall in a way the binary predictors suggest. Figure 20.5 depicts a function that describes the hemodynamic response that is usually measured in response to stimulation. To achieve a better fit between the empirical data and the predictors from the GLM, the onset vectors are convolved with such a hemodynamic response function (HRF). In most cases, a canonical HRF is used that is distributed along with the analysis software packages. A further widely applied option is to additionally convolve the predictors with the HRF's first temporal derivative. This additional step is a good means to account for temporal variability of the onset of the hemodynamic response. In case that slice timing is omitted during preprocessing, this method is highly recommended.

After the GLM is specified and all regressors in the equation (experimental conditions, parametric modulators and nuisance covariates) are convolved with the canonical HRF (and its temporal derivative) the regression weights (called beta

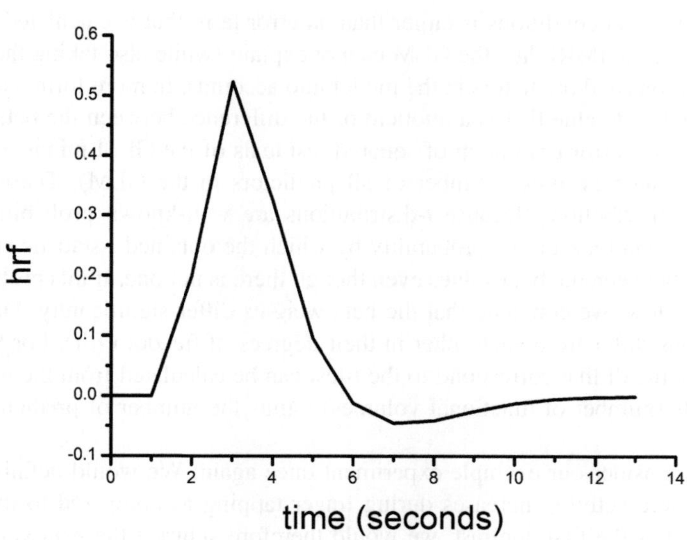

Fig. 20.5 The canonical hemodynamic response function as distributed alongside the SPM software package

weights) need to be estimated. This is done by a least square approach: The beta weights are set in a way that the sum of the squares of the deviations between the predicted values (the y values) and the empirical data (the BOLD time series) is minimized. At the end of this process, we end up with one beta weight for each predictor for each voxel. Call yourself in mind that we are still dealing with single subjects. The process of setting up the GLM and estimating the corresponding beta weights is repeated for each participant in the sample. In our simple example experiment, the temporal layout of experimental conditions is essentially the same for all participants; i.e., the same GLM with exactly the same predictors (except for nuisance covariates that are specific for individuals) can be used. In case of more complex experimental designs, (for instance an event-related design with events presented in random order), the GLM itself (that is the columns of the $m \times n$ matrix) would be the same, the onsets (and modulators), however, would be different for each participant.

20.3.3.2 Contrasts

Now that we have ended up with beta weights, we can conduct statistical inference on these weights. This is still carried out separately for each participant. The question addressed by first-level inferences is: In which voxels and to what extend do given experimental conditions lead to activation differences? In the most simple and most widely used case this question is answered by a t-test. A t-test is a statistical test that contrasts two measurements and assesses whether the difference is significantly larger as the general variability in the data set. In the case of a first-level inference, the t-test looks if the difference between the beta weights of two experimental conditions is larger than an error term that is calculated from the variability of the fMRI data the GLM cannot explain (while also taking the number of observations and predictors in the model into account). In more formal terms, the t-test outputs a t-value that is a quotient of the difference between the beta weights divided by this error term (sum of squared residuals of the GLM minus number of functional volumes minus number of all predictors in the GLM). These t-values follow a t-distribution. Because t-distributions are well-known probability distributions, we can look up the probability by which the obtained t-statistic suggests a difference between the beta values even though there is not one. If this probability is sufficiently low, we conclude that the beta weights differ significantly. Different t-distributions differ from each other in their degrees of freedom (df). For first-level inferences, the df that correspond to the t-test can be calculated from the number of data points (number of functional volumes) minus the number of predictors in the model.

Let us consider our example experiment once again: We would definitely want to look where activity increases during finger tapping as compared to the resting condition. For the first contrast, we would therefore subtract the beta value of the resting-regressor from the beta value of the tapping-regressor for each voxel and assign this difference to the nominator of the t-value. If we would also be interested in activity decreases during motor tapping, we would calculate a second contrast for which we subtract the beta value of the tapping-regressor from that of the resting-regressor. The error term is essentially the same for both contrasts, but differs slightly across voxels: For each data point in our dependent variable we would calculate the difference between the actual BOLD activity and the value predicted by our GLM (the residuals). These differences are squared to control for different signs and then summed up across data points. Finally, we would subtract the number of data points (240) and the number of regressors in the GLM (two conditions plus six motion regressors of no interest equals eight) from the sum of squared residuals. This whole term would then be assigned to the *t*-value's denominator. This step is repeated for all voxels in the brain and then for all participants in our sample. Thus, we would end up with one map per contrast and participant that maps the *t*-values on the brain. Finally, we would calculate the amount of degrees of freedom for our statistical test (240 data points minus eight regressors equals 232 degrees of freedom) and look up the corresponding probability values (*p*-values) to each *t*-value that tells us the probability by which the difference in the beta values can be attributed to chance. Because we want to be sure to not assume a significant difference between experimental conditions when there is none, we would want this probability to be low, for example below 0.001 %. With a *t*-distribution with 232 df, the *t*-value would need to be as high as 1.651 to be considered significant.

This, however, would us only leave with an assessment of statistical significance between experimental conditions within subjects. In order to assess significance across or between subjects, we would need to perform additional analyses.

20.3.4 Second-Level Analysis

The rationale behind the second-level analysis is to combine the results from the first-level analyses of all participants and assess statistical significance across participants. Various statistical inferences are possible: In the simplest case, one would want to examine whether neural activity differences between the conditions observed on the first-level are idiosyncratic to single participants or can be found in the majority of participants. This could be accomplished by a one-sample t-test. A one-sample t-test tests whether the mean of a dependent measure that is calculated across participants differs significantly from a given value. In the context of a second-level fMRI analysis, this t-test looks if the mean of activation in one voxel is significantly different from zero. Different statistical models can be set up, depending on the research question asked: If, for instance, a group of pathological gamblers is compared with a group of healthy control participants regarding their neural response to risky choice, a two-sample t-test that tests for differences in the mean of neural activity between the two groups is the appropriate statistical test. Multiple linear regressions and multifactorial models are also possible for more complex experimental paradigms. Please note that in the vast majority of cases, second-level analyses model participants as random effects. This statistical approach assumes that participants were sampled from an underlying population and ensures that any inference drawn from the data can be generalized to this population.

Up to this point, we have not specified on which values the second-level analysis is carried out. The first-level analysis outputs a *t*-value for each voxel. This *t*-value, however, is not the best means to quantify activation differences between experimental conditions. As outlined previously, the *t*-values depend not only on the difference in the beta values but also on the number of data points (i.e., functional volumes) and predictors in the model. It is a better idea to use a standardized measure to quantify activation differences. Such a standardized measure is the effect size that can be calculated for each statistical test on the first level. The effect size quantifies the magnitude of the activation differences in units of measurement and is fed into the statistical tests on the second level. For that purpose, the mean (or differences in means) and the variability of the effect sizes are calculated and compared in a term such as the *t*-term. The corresponding df for the *t*-distribution can be calculated from the sample size and the number of cells in the experimental design.

For the second-level analysis, we are more interested in assessing whether an observed activation difference in our sample reflects a true relationship in the population than in estimating the size of the effect. Therefore, the statistical parameters mapped for the visualization of results are usually the test statistics (such as *t*-values) rather than the effect sizes of the second-level tests. The test statistics are thresholded at a given probability (e.g., $p < 0.001$) which means that only parameters are considered that are high enough to let us assume that the observed activation difference cannot be attributed to chance. For visualization purposes, the statistical parameters that survived the statistical thresholding are color coded and projected onto a structural MR image in standard coordinate space. Besides from this height threshold, an extend threshold can be applied additionally. Because it is very likely that the statistical parameter of a single spatially isolated voxel exceeded the threshold by chance, it is recommended to specify a minimum cluster size in voxels (e.g., $k > 8$) and ignore clusters of adjacent voxels if the number of co-jointly activated voxels does not exceed this minimum cluster size.

At the end of all these processing steps (preprocessing, first- and second-level analysis including thresholding) stands one statistical parametric map that informs us about activation differences between experimental conditions. However, this approach comes with one major downside that will be discussed and resolved in the next paragraph.

20.3.4.1 Multiple Comparison Correction

In 2009, a wave of gloating newspaper articles was published in the popular media sarcastically criticizing functional MRI. The opinions were based on an a study that had been presented at the Annual Meeting of The Organization for Human Brain

Mapping (OHBM) earlier that year. What had happened? In the study that is available online as a conference poster (Bennett et al. 2009) the authors had put a dead Atlantic salmon (originally bought for food consumption) into the scanner and run structural and functional MRI on it. While the salmon was in the scanner, the authors presented visual stimuli and "asked" the salmon to perform a task. Even though the salmon did not respond to the task behaviorally (most certainly because it was dead), the authors still ended up with BOLD time series and temporal information on the sequence of task events that was eventually correlated with the neurophysiological data. The authors set up a GLM and processed the data up to the first level (because there was only one salmon involved) as discussed earlier, calculated a contrast between experimental conditions and thresholded the resulting t-map at $p < 0.001$ with an extend threshold of three adjacent voxels. Given that the salmon was not alive during scanning and the task more suitable for human research participants, results were strikingly surprising: Significant clusters lit up in both the salmon's brain and its spinal cord! Clearly, these results were easy bait for the popular media and fueled skepticism toward neuroimaging in general. How can we trust neuroimaging results in human research participant when even a dead fish shows brain activity which looks like it was evoked by an experimental task? To answer this question we need to take a closer look at a problem called the multiple comparison problem and how it can be resolved in the context of fMRI.

As outlined in the previous paragraph, the rationale behind statistical hypothesis testing is to calculate a probability by which the empirically observed difference (or relationship) in the data was obtained given that there is no such relationship in reality. If this probability is sufficiently low (e.g., below 0.001 %), we conclude that the observed effect must reflect a true relationship. However, we should be aware that in one out of 1000 cases, this conclusion would be wrong—simply because a very low probability only makes things become unlikely but does not rule them out entirely. A typical fMRI volume consists of ten-thousands of voxels. Because the analysis is run separately for each voxel, the number of statistical tests is as high as the number of voxels. Across all tests, the probability of erroneously assuming a true relationship because of a very low p-value increases dramatically. Such errors are called false positives (or alpha-errors) and describe a situation where we decide to accept a statistic as evidence for a difference between experimental conditions even though there is none in reality. The problem that arises from the massive amount of statistical tests run during fMRI analyses is called the multiple testing problem, or alpha error inflation. This is what happened in the dead salmon study: Because of the mere number of comparisons, some voxels lit up by chance because the high t-values suggested an activation difference even though the salmon was not paying attention to the task (which we can infer from the behavioral data and the fact that it was dead). Luckily for the neuroimaging community, there are various methods that can be used to eliminate (or at least minimize) the multiple comparison problem.

When we conduct statistical inferences on second-level data we want to make sure that we do not erroneously assume that a single voxel is active even though it is not. Furthermore, we want to make sure that across all statistical tests, the chance of obtaining a false positive result is very low as well. We can ensure this by correcting for the number of tests conducted. Because we control the alpha error probability for a family of statistical tests, this procedure has been labeled family wise error (FWE) rate correction. The standard method for this is the Bonferroni correction, which accepts a family of statistical tests (such as in the entire brain) as significant if the alpha error probability of each test in the family is below a specified significance level that is divided by the number of tests in the family. That is, if we consider a false positive rate of 0.05 % as acceptable and there are 10,000 voxels in our data set, we would consider a voxel active if the error probability of its t-value fell short of 0.05/10,000 = 5.0e − 6. Applying the Bonferroni procedure effectively minimizes the risk of false positive results but it comes with a major downside. It is very conservative: Whenever we conduct a statistical test, there is not only the risk of committing an alpha error (i.e., assuming a difference even though there is none in reality) but also the risk of committing a beta error (i.e., assuming no relationship although there is one in reality). Both types of error depend on each other: With an increasing statistical threshold, the probability of committing an alpha error decreases but beta errors on the other hand become more likely. Therefore, Bonferroni correction admittedly controls for alpha error commission but it also increases the chance of committing a beta error. In the context of fMRI there is reasonable doubt that the Bonferroni correction is the gold standard for multiple comparison correction. The Bonferroni correction is appropriate for families of independent statistical tests. In fMRI data, however, full independence between all tests cannot be assumed: As outlined earlier, data points from neighboring voxels are highly intercorrelated and these correlations are further amplified by spatial smoothing during preprocessing. Therefore, the number of test families that require correction is substantially lower than the amount of voxels in the brain, leaving Bonferroni correction as a too conservative approach. The more appropriate routine is the application of Gaussian random field theory (RFT) to control the family wise error rate. RFT is a research body in mathematics that deals with smooth statistical maps (such as our t-maps). RFT provides tools to estimate the overall smoothness of the fMRI data set which depends on the degree of spatial correlations between voxels in the raw data and the size of the Gaussian kernel applied during smoothing. If the overall smoothness is known, the number of resolution elements (resels) in the data set can be calculated. The number of resels equals the number of independent observation in the data set and gives the amount of test families for that we need to correct. From the number of resolution element, we can determine the expected Euler characteristic (EC). In the context of functional imaging, the EC gives the number of expected clusters in a smooth statistical map after thresholding. Because the expected EC depends on the statistical threshold, it is approximately equivalent to the probability of committing a family wise error. Thus, the statistical threshold for the second-level analysis that corrects for multiple comparisons by controlling the family wise error can be inferred from the EC. This method has been implemented in most analysis software packages. To come back to the dead salmon in the scanner: After controlling the FWE according to Gaussian RFT no active voxel could be observed in its central nervous system.

When we ask how we could trust neuroimaging results when even dead fish show neural activation contingent with experimental tasks, the answer clearly is: When we control for multiple comparisons!

20.4 Conclusion

The purpose of this chapter was to give an overview on how fMRI works. We have covered the physical and physiological basics of the BOLD signal and discussed processing steps and statistical analysis of BOLD time series in the context of statistical parametric mapping. It would have been beyond the scope of this chapter to discuss more advanced functional imaging methods than the mass univariate approach which is the most common way to analyze fMRI data. For more in-depth information on statistical parametric mapping, multivariate approaches to BOLD fMRI data and the analysis of functional connectivity between different brain regions during task performance, we would like to refer to the SPM textbook by Friston et al. (2007) and the textbook by Huettel et al. (2009).

References

Bennett CM, Miller MB, Wolford GL (2009) Neural correlates of interspecies perspective taking in the post-mortem Atlantic Salmon: an argument for multiple comparisons correction. Neuroimage, 47(Suppl 1):S125.

Campbell-Meiklejohn DK, Woolrich MW, Passingham RE, Rogers RD (2008) Knowing when to stop: the brain mechanisms of chasing losses. Biol Psychiatry 63(3):293–300

Delgado MR, Frank RH, Phelps EA (2005) Perceptions of moral character modulate the neural systems of reward during the trust game. Nat Neurosci 8(11):1611–1618

Fliessbach K, Weber B, Trautner P, Dohmen T, Sunde U, Elger CE, Falk A (2007) Social comparison affects reward-related brain activity in the human ventral striatum. Science 318 (5854):1305–1308

Friston KJ, Holmes AP, Worsley KJ, Poline JP, Frith CD, Frackowiak RSJ (1995) Statistical parametric maps in functional imaging: a general linear approach. Hum Brain Mapp 2(4): 189–210

Friston KJ, Ashburner JT, Kiebel SJ, Nichols TE, Penny WD (2007) Statistical parametric mapping: the analysis of functional brain images. Elsevier, Oxford

Huettel SA, Song AW, McCarthy G (2009) Functional magnetic resonance imaging. Sinauer Associates, Incorporated

Tom SM, Fox CR, Trepel C, Poldrack RA (2007) The neural basis of loss aversion in decision-making under risk. Science 315(5811):515–518

Chapter 21
Structural MRI: Morphometry

Christian Gaser

Abstract Human brains are characterised by considerable intersubject anatomical variability, which is of interest in both clinical practice and research. Computational morphometry of magnetic resonance images has emerged as the method of choice for studying macroscopic changes in brain structure. Magnetic resonance imaging not only allows the acquisition of images of the entire brain in vivo but also the tracking of changes over time using repeated measurements, while computational morphometry enables the automated analysis of subtle changes in brain structure. In this section, several voxel-based morphometric methods for the automated analysis of brain images are presented. In the first part, some basic principles and techniques are introduced, while deformation- and voxel-based morphometry are discussed in the second part.

21.1 Introduction

The Jena psychiatrist Hans Berger became famous for the discovery of electroencephalography. Less known, however, are Berger's imaginative studies of brain morphometry. He tried, for example, to estimate the cortical surface by gluing small metal plates onto a *post-mortem* brain. Since the area and weight of a single metal plate were known, the total weight of the plates was used in order to estimate the total area of the cortex. Nowadays, computer-based methods use the same idea with so-called triangulation. However, now the metal plates are replaced by small triangles forming a computerised mesh that renders the shape of the cortical surface and allows a reliable and accurate measurement. This new approach belongs to the recently developed methods for the automated analysis of brain structure that are referred to as 'computational morphometry' (Takao et al. 2010).

C. Gaser (✉)
Structural Brain Mapping Group, Departments of Neurology and Psychiatry,
Jena University Hospital, Am Klinikum 1, 07747 Jena, Germany
e-mail: christian.gaser@uni-jena.de

M. Reuter and C. Montag (eds.), *Neuroeconomics*, Studies in Neuroscience, Psychology and Behavioral Economics, DOI 10.1007/978-3-642-35923-1_21

Besides the use of computer algorithms, the availability of new imaging methods played a seminal role in morphometry. These new imaging methods not only allow the acquisition of images of the entire brain in vivo but also the tracking of changes over time using repeated measurements. Thus, they represented a real advance because previously *post-mortem* examinations were the only way to examine brain structures.

The first imaging method that allowed mapping of cerebral structures in vivo was pneumoencephalography. This procedure involved drainage of most of the cerebrospinal fluid (CSF) from around the brain and replacement with air. The ventricular system of the brain could then be identified on an X-ray of the skull. However, this method proved to be very invasive and painful. It took until the70s of the last century before computed tomography provided images of the brain in three dimensions. The real breakthrough in imaging techniques, however, came with magnetic resonance imaging (MRI), which allowed a much higher spatial resolution without ionising radiation. This method has become the standard tool of macroscopic anatomy, both in clinical practice and in research. Another advantage of this imaging method is that variable image contrasts can be achieved by using different parameters for longitudinal (T_1) and transverse (T_2) relaxation times and proton density. The signal intensities on T_1, T_2 and proton density relate here to specific tissue contrasts. The most commonly used imaging sequence for MR-morphometry is T_1-weighted imaging because of its high contrast for brain parenchyma (see Fig. 21.1). Other imaging sequences can be used to evaluate CSF spaces, oedema or subacute stroke (T_2 weighted), to enhance parenchymal abnormalities, such as low-grade glioma (fluid-attenuated inversion recovery [FLAIR]), or to visualise acute ischaemia (diffusion weighted).

In addition to the various methodological developments, morphometry has gained increasing importance in the field of neuroscience because completely new

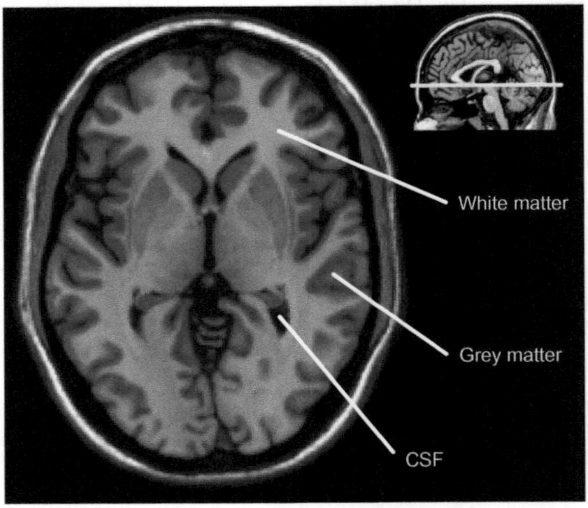

Fig. 21.1 T_1-weighted MRI scan. The small image (*top right*) shows the location of the axial slice (*main image*). This sequence reveals a high contrast for brain parenchyma and the different signal intensities relate to grey and white matter and cerebrospinal fluid (CSF) [modified from (Gaser 2005)]

applications have become possible. While in the early days the applications were limited to the quantification of global parameters such as brain weight or brain volume, nowadays a wide spectrum of applications is supported. This ranges from the investigation of local morphometric changes in certain diseases up to the detection of brain plasticity.

In this chapter, two different morphometric methods for the analysis of MR images of the brain are presented. In the first part, some basic principles and techniques are introduced, while two morphometric methods are discussed in the second part that both work on a voxel-wise level.

21.2 Basic Principles

21.2.1 Spatial Normalisation

Brains are characterised by considerable intersubject anatomical variability. In order to analyse brains across different subjects, an adjustment to a reference system using a stereotactic or spatial normalisation is required. This permits the analysis of brains in a standardised space or coordinate system. However, this procedure is also useful for brain morphometry and consequently a variety of methods based on this idea exists.

In order to spatially normalise brain images, it is first necessary to define a standardised coordinate system by using specific anatomical landmarks. The most widely used reference system is the Talairach atlas proposed by Talairach and Tournoux (1988). The basic idea is to define the anterior and posterior commissures and several points relative to them to align and scale a brain image. The anterior commissure is the origin of the coordinate system and all locations within the brain can now be defined with standardised coordinates in millimetres (Fig. 21.2). This allows the comparison of anatomical localisations between different brains and even different studies.

The adjustment due to spatial registration can be achieved in different ways (Fig. 21.3). The simplest procedure is to only correct the position of the images, for which displacements and rotations are applied. The image size (or brain size) remains unchanged, which is necessary, for example, for brain images of the same subject, in longitudinal (serial) measurements over time. Since image size is not changed, this special case is also referred to as 'rigid body transformation'. In contrast, images of different subjects need to be additionally corrected for image size by scaling or resizing the image. Furthermore, for a full affine transformation, an additional shearing of the image can be applied (Fig. 21.3). Since the adjustment is done for the entire image in the same way (or linearly), the term 'linear spatial normalisation' is used.

In contrast to linear normalisation, nonlinear normalisation also corrects for local differences between two brains. For this, images are locally warped (deformed)

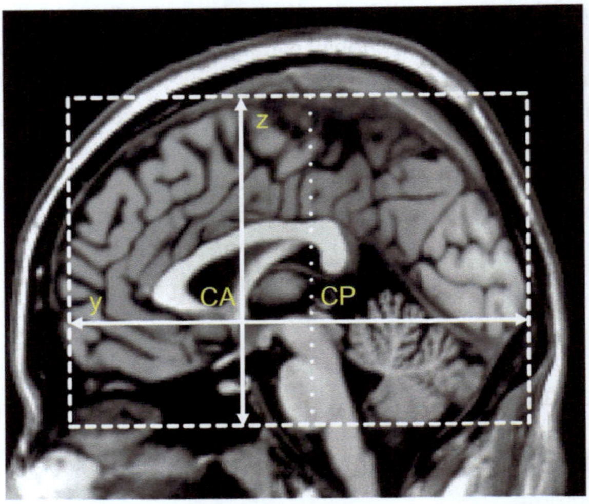

Fig. 21.2 Talairach coordinate system. The coordinate origin of the Talairach space is defined by the anterior commissure (CA). From here, all locations in the brain can be specified as coordinates in millimetres. The line through the anterior and posterior (CP) commissures is used for aligning the coordinate system. The image shows the extent in the y-direction (anterior-posterior) and z-direction (inferior-superior). The x-axis (not shown here) determines the left-right direction [modified from (Gaser 2005)]

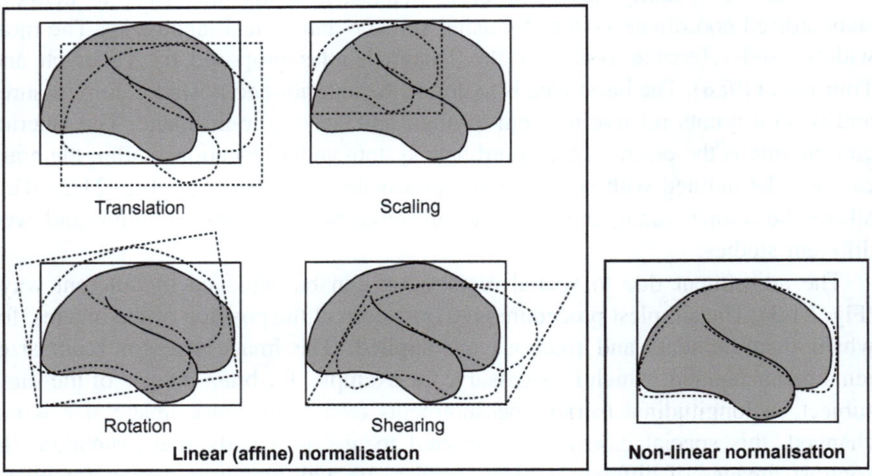

Fig. 21.3 Linear and nonlinear spatial registration. The *left side* of the figure shows the four possible linear transformations that are applied to the entire image. A special case is the so-called rigid body transformation. Here, the image is adjusted only by translations and rotations. An additional change in image size can be achieved by scaling and shearing the image. The aggregation of these linear transformations is known as 'affine normalisation'. In contrast to linear normalisation, nonlinear normalisation also corrects for local differences between two brains (*right side* of the figure). For this, images are locally warped (deformed) until the differences between them are minimised (modified from (Gaser 2005))

until the differences between them are minimised (Fig. 21.3). The cost of computing such local deformations is much higher and increases with the required spatial resolution of the deformations. The advantage of nonlinear normalisation, however, is the greater accurate adjustment of the brains to the reference brain.

Linear and nonlinear normalisation can be performed using different normalisation algorithms [a detailed overview is given in (Toga 1999)]. Landmark-based methods use manual label points (landmarks) in the brain. These corresponding points are defined in all brains and are then aligned. Contour-based methods use not just a few points or landmarks but the whole contour of a region, such as the outline of the corpus callosum in the sagittal plane or even the entire surface of the cortex as a three-dimensional contour (Thompson et al. 1997). Finally, intensity-based methods exist which use the local image intensity in order to achieve a spatial alignment between the images. Here, the squared sum of the signal intensity differences is used, for example, as an indicator of the similarity between two images. By minimising these intensity differences, an alignment of both images is achieved.

21.2.2 Segmentation

Segmentation algorithms are among the most commonly used methods in brain morphometry. The aim of these methods is to segment an image into separate anatomical tissue compartments, such as grey matter, white matter and CSF, after removing non-brain parts. With more sophisticated approaches, it is also possible to segment pathological changes, such as tumours, lesions or stroke-affected regions. However, in addition to the T_1 images, this usually requires MR sequences, such as T_2 weighting or FLAIR, where the pathological changes can be better differentiated.

A plethora of semi-automated and automated algorithms exists, such as intensity thresholding, region growing, classifiers, clustering, Markov random field models, artificial neural networks, deformable models or atlas-guided approaches (Pham et al. 2000). From all of these examples, one of the most commonly used methods will be presented here in detail: the Gaussian mixture model, which belongs to the group of classifiers (Ashburner and Friston 2005). First, an intensity histogram of the image is estimated that plots the frequencies of the image intensities on the y-axis (Fig. 21.4, bottom left). The simplified example in Fig. 21.4 shows only four different intensity distributions. Here, the smallest image intensities are assigned to the background (left part of the histogram), followed by CSF, grey and white matter with the highest image intensity in the right part of the histogram. Gaussian curves that can differ with regard to height and width are now fitted into this intensity distribution. The maximum of each of these four Gaussian curves represents the mean intensity value for the respective tissue compartment. For the example of grey matter, this means that at the peak maximum the probability that this image voxel belongs to grey matter is largest. The more the image intensity deviates from this

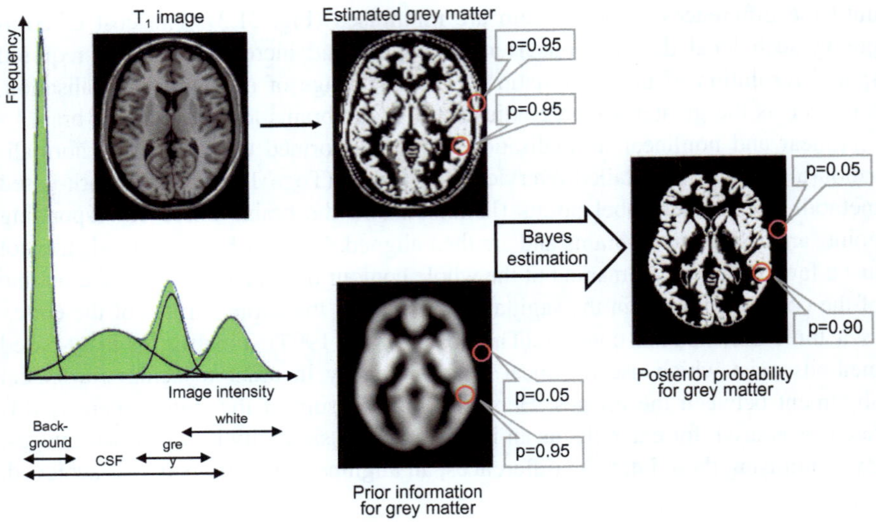

Fig. 21.4 Image segmentation using a priori information. First, the image intensities of the T_1 image (*upper left*) are used to plot their frequencies in a histogram. Several peaks—corresponding to different image intensities of the tissue compartments—can be differentiated. In the next step, Gaussian mixture curves for each tissue compartment are fitted into the histogram in order to estimate the probability that a voxel belongs to that tissue (*lower left*). A map for grey matter is shown (*upper right*) with the estimated probability for two selected locations (*red circles*). Based solely on a similar image intensity, the cerebral and extracranial circles exhibit a similar probability for belonging to grey matter. This can be adjusted by combining the image intensity-based information with prior information (*below*) using a Bayesian approach [modified from (Gaser 2005)]

value, the less likely it is grey matter and is rather CSF (at lower intensity) or white matter (with higher intensity).

The intensity distributions for each tissue compartment overlap because at a common voxel size of $1 \times 1 \times 1$ mm^3 any given voxel might contain more than one tissue. This is referred to as 'partial volume effect' and most often occurs at the border between brain parenchyma and CSF, at boundaries between grey and white matter, and in structures where white matter fibres cross the grey matter. These partial volume effects can be modelled explicitly in order to estimate a more accurate segmentation (Tohka et al. 2004).

To guide tissue segmentation, additional tissue probability maps can be used to consider prior anatomical knowledge about the spatial distribution of different tissues (Ashburner and Friston 2005). Image intensity and prior knowledge can then be combined via a Bayes estimator. In fact, prior anatomical knowledge is used to drive and restrict the tissue segmentation algorithm (Fig. 21.4, right). While this may be valuable as long as the prior probability maps match the subject's tissue distribution, it might lower segmentation accuracy in all populations that deviate from these maps (e.g. children, Alzheimer's disease patients) (Wilke et al. 2008).

21.3 Voxel-Based Methods

Voxel-based methods allow the analysis of each voxel in the MR data. This voxel-wise analysis is possible because all brains are adjusted by means of a spatial normalisation to a standard anatomical space. Thus, each voxel relates to the same corresponding anatomical structure across all brains, which can be assumed if a high-dimensional nonlinear spatial normalisation is applied.

While different voxel-wise measures can be used for that approach, the most common approach is to segment brains into different tissue compartments and analyse the local distribution for a specific tissue. This method is referred to as 'voxel-based morphometry.' Another approach is to analyse the deformations that are necessary in order to non-rigid deform a brain to adapt it to another brain. Because this approach is based on deformations, it is known as 'deformation-based morphometry.'

Prior to statistical analysis, the images have to be spatially smoothed (filtered) with a Gaussian kernel. The reason for this is threefold. First, parametric tests assume that the data follow a Gaussian distribution and after smoothing with a Gaussian kernel the data are more normally distributed according to the central limit theorem (Nichols and Hayasaka 2003). Second, smoothing accounts for small interindividual differences in local brain anatomy that remain after spatial normalisation. Finally, smoothing enables greater sensitivity for effects that approximately match the size of the smoothing kernel according to the matched filter theorem (Ashburner and Friston 2000).

In the next step, smoothed images can then be compared in each voxel (Fig. 21.5). For statistical analysis, usually a general linear model is used. This model—the equivalent of a multiple regression—incorporates a number of different statistical models ranging from simple correlation to repeated measures ANOVA in longitudinal designs. The result is a statistical parametric map, which allows a statistical statement about the initial hypothesis in each voxel. However, due to the mass-univariate approach, a correction for multiple comparisons has to be applied. The most frequently used correction is based on the Gaussian random field theory (Worsley et al. 1996) that enables a correction on the voxel or cluster level (although a correction on the more theoretical set level is also possible) (Friston et al. 1996). Another option for the consideration of the issue of multiple comparisons has become very popular in recent years. This method is based on the adaptive control of the false discovery rate (FDR) and was originally proposed for microarray data to identify genetic effects (Benjamini and Hochberg 1995). Finally, permutation tests do not assume normally distributed data and enable a correction for multiple comparisons particularly for small sample sizes. They use random shuffles of the data to attain a correct distribution of a test under a null hypothesis (Nichols and Holmes 2002). Again, a correction on the voxel or cluster level is possible. Another possibility is to use a correction based on threshold-free cluster enhancement (TFCE) that combines both levels by accumulating cluster-like local spatial support at a range of cluster-forming thresholds (Smith and Nichols 2009).

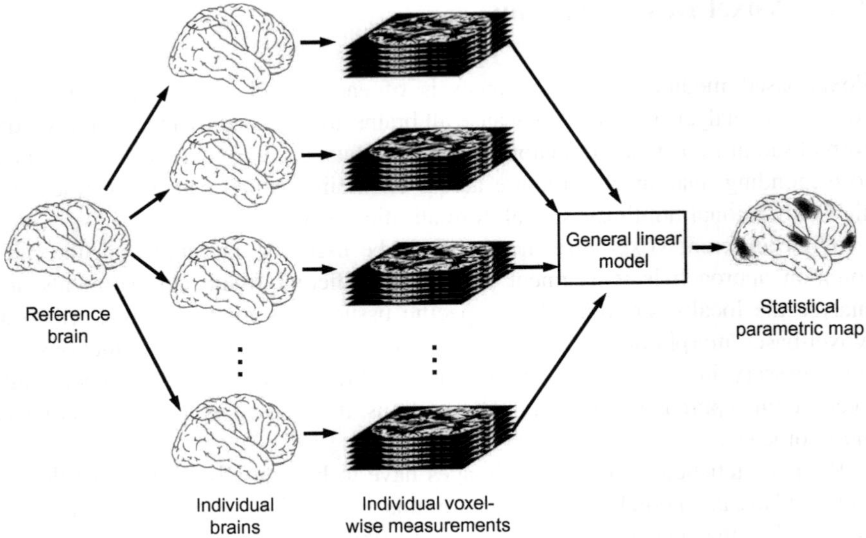

Fig. 21.5 Principle of a voxel-based analysis. For a voxel-wise analysis, it is first necessary to spatially register all brains to a reference brain. Now, in each voxel a morphometric parameter (e.g. grey matter volume) is estimated that can be statistically analysed using a general linear model. The result is a statistical parametric map which allows a statistical statement about the initial hypothesis in each voxel [modified from (Gaser 2005)]

Since the analysis is made on a voxel-wise level, this approach offers several advantages over conventional morphometry. One such advantage is the reduction of partial volume effects, since a structural change can be detected in each voxel of the brain and not only in the entire structure. Thus, structures that are only partially altered can be detected with higher sensitivity compared to region-based methods. Furthermore, an analysis can not only be carried out in predefined regions but also throughout the brain. Large sample numbers can be examined with high reliability due to the automated measurement. These advantages might explain the great popularity of these methods in recent times.

21.3.1 Deformation-Based Morphometry (DBM)

DBM is based on the application of nonlinear registration procedures to spatially normalise one brain to another one. The simplest case of spatial normalisation is to correct the orientation and size of the brains. In addition to these global changes, a nonlinear normalisation is necessary to minimise the remaining regional differences by means of local deformations. If this local adaptation is possible, the deformations now reveal information about the type and localisation of the structural differences between the brains and can undergo subsequent analysis (Fig. 21.6).

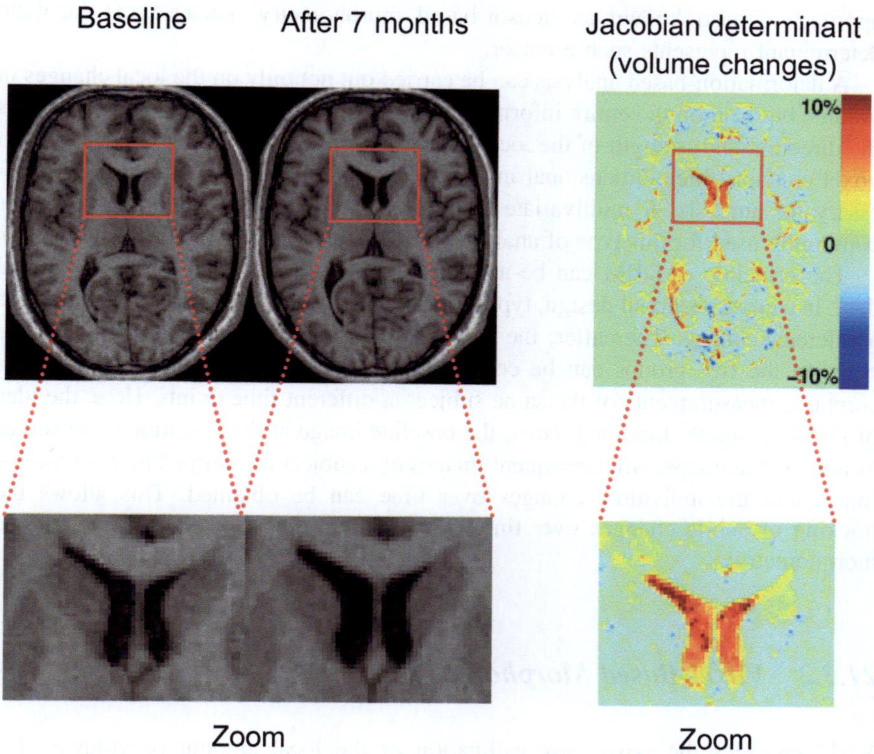

Fig. 21.6 Principle of DBM. *Left* This example shows two T_1 images of a male patient with schizophrenia at his first episode and a subsequent scan after 7 months. In the enlarged views shown underneath, the larger lateral ventricles at the second time point can be clearly seen. The principle of DBM is to warp the second scan to the baseline scan by introducing high-dimensional deformations. Once this is achieved, the differences between both images are encoded in the deformations applied for the warp. These deformations can then be used to calculate volume changes by using the Jacobian determinant (*right images*) [modified from (Mietchen and Gaser 2009)]

Figure 21.6 shows an example for a single patient with schizophrenia. A first baseline scan was acquired at the beginning of his first psychotic episode and a subsequent scan was acquired after 7 months where the enlarged ventricles are visible with the naked eye. The second image is warped to the baseline scan by introducing high-dimensional deformations. Differences between both images are minimised and are now coded in the deformations. Finally, a map of local volume changes can be quantified by a mathematical property of these deformations–the Jacobian determinant. This parameter is well known from continuum mechanics and is usually used for the analysis of volume changes in flowing liquids or gases. The Jacobian determinant allows a direct estimation of the percentage change in volume in each voxel and can be statistically analysed (Gaser et al. 2001). This approach is also known as 'tensor-based morphometry' because the Jacobian determinant represents such a tensor.

A deformation-based analysis can be carried out not only on the local changes in volume but also on the entire information of the deformations, which also includes the direction and strength of the local deformations (Gaser et al. 1999). Since each voxel contains three-dimensional information, a multivariate statistical test is necessary for analysis. A multivariate general linear model or Hotelling's T_2 test is commonly used for this type of analysis (Gaser et al. 1999; Thompson et al. 1997).

The principle of DBM can be applied to both cross-sectional and longitudinal data. In a cross-sectional design, typically brain images of two groups are warped to a reference image. Thereafter, the different deformations to the reference image between the two groups can be compared. On the other hand, longitudinal data comprise measurements of the same subject at different time points. Here, the idea of DBM is slightly modified. Now, the baseline image at the first time point serves as a reference image. All subsequent images of a subject are warped to this baseline image and the individual changes over time can be obtained. This allows the tracking of subtle changes over time, which cannot be detected by conventional morphometry.

21.3.2 Voxel-Based Morphometry (VBM)

VBM provides the voxel-wise estimation of the local amount or volume of a specific tissue compartment (Ashburner and Friston 2000). VBM is most often applied to investigate the local distribution of grey matter, but can also be used to examine white matter. However, the sensitivity for detecting effects in white matter is limited due to the low intensity contrast in large homogeneous white matter regions with only small changes in intensity. The concept of VBM incorporates different preprocessing steps: (1) spatial normalisation to a reference brain (template), (2) tissue classification (segmentation) into grey and white matter and CSF and (3) bias correction of intensity non-uniformities. Ashburner and Friston (2005) proposed an approach whereby all three steps are combined within the same generative model. This model is based on a mixture of Gaussians and additionally considers smooth intensity variations and nonlinear registration using tissue segmentations. This approach allows for more accurate and reliable results than simple serial applications of each single step.

Further improvement can be achieved if high-dimensional spatial registration techniques such as diffeomorphic registration approaches are used. Diffeomorphic registrations are based on a large-deformation framework and not only provide a number of elegant mathematical properties but generally allow for a better accuracy of the spatial registration (Ashburner 2007).

Local deformations are now used in order to reduce structural differences between original and template images. This facilitates a precise comparison within brain regions between different subjects. However, existing structural differences between the brains are now largely reduced and the sensitivity for detecting these effects in the statistical analysis is therefore minimised. Thus, the volume of a particular tissue within a voxel has to be preserved. This is attained by multiplying (or modulating) voxel values in the segmented images by the Jacobian determinants that are derived from the spatial registrations. This process is referred to as 'modulation.'

References

Ashburner J (2007) A fast diffeomorphic image registration algorithm. Neuroimage 38(1):95–113

Ashburner J, Friston KJ (2000) Voxel-based morphometry–the methods. Neuroimage 11(6 Pt 1):805–821

Ashburner J, Friston KJ (2005) Unified segmentation. Neuroimage 26(3):839–851

Benjamini Y, Hochberg Y (1995) Controlling the false discovery rate—a practical and powerful approach to multiple testing. J Roy Stat Soc B Met 57(1):289–300

Friston KJ, Holmes A, Poline JB, Price CJ, Frith CD (1996) Detecting activations in PET and fMRI: levels of inference and power. Neuroimage 4(3 Pt 1):223–235

Gaser C (2005) Morphometrie. In: Walter H (ed) Funktionelle Bildgebung in Psychiatrie und Psychotherapie. Schattauer Verlag, Stuttgart, pp 89–104

Gaser C, Volz HP, Kiebel S, Riehemann S, Sauer H (1999) Detecting structural changes in whole brain based on nonlinear deformations-application to schizophrenia research. Neuroimage 10 (2):107–113

Gaser C, Nenadic I, Buchsbaum BR, Hazlett EA, Buchsbaum MS (2001) Deformation-based morphometry and its relation to conventional volumetry of brain lateral ventricles in MRI. Neuroimage 13(6 Pt 1):1140–1145

Mietchen D, Gaser C (2009) Computational morphometry for detecting changes in brain structure due to development, aging, learning, disease and evolution. Front Neuroinform 3:25. doi:10.3389/neuro.11.025.2009

Nichols T, Hayasaka S (2003) Controlling the familywise error rate in functional neuroimaging: a comparative review. Stat Methods Med Res 12(5):419–446

Nichols TE, Holmes AP (2002) Nonparametric permutation tests for functional neuroimaging: a primer with examples. Hum Brain Mapp 15(1):1–25

Pham DL, Xu C, Prince JL (2000) Current methods in medical image segmentation. Annu Rev Biomed Eng 2:315–337

Smith SM, Nichols TE (2009) Threshold-free cluster enhancement: addressing problems of smoothing, threshold dependence and localisation in cluster inference. Neuroimage 44(1):83–98

Takao H, Abe O, Ohtomo K (2010) Computational analysis of cerebral cortex. Neuroradiology 52 (8):691–698

Talairach J, Tournoux P (1988) Co-planar stereotaxic atlas of the human brain: 3-dimensional proportional system—an approach to cerebral imaging. Thieme, New York

Thompson PM, MacDonald D, Mega MS, Holmes CJ, Evans AC, Toga AW (1997) Detection and mapping of abnormal brain structure with a probabilistic atlas of cortical surfaces. J Comput Assist Tomogr 21(4):567–581

Toga AW (ed) (1999) Brain warping. Academic Press, San Diego

Tohka J, Zijdenbos A, Evans A (2004) Fast and robust parameter estimation for statistical partial volume models in brain MRI. Neuroimage 23(1):84–97

Wilke M, Holland SK, Altaye M, Gaser C (2008) Template-O-matic: a toolbox for creating customized pediatric templates. Neuroimage 41(3):903–913

Worsley KJ, Marrett S, Neelin P, Vandal AC, Friston KJ, Evans AC (1996) A unified statistical approach for determining significant signals in images of cerebral activation. Hum Brain Mapp 4(1):58–73

Chapter 22
Diffusion Tensor Imaging (DTI) and Tractography

Theodor Rüber, Christian Erich Elger and Bernd Weber

Abstract Diffusion Tensor Imaging is a Magnet Resonance Imaging-technique that allows for the non-invasive in vivo assessment and delineation of white matter tracts. The Magnet Resonance signal is sensitized to the natural diffusion process of protons and a tensor model is fitted to the resulting data. By following the direction of the diffusion maximum across the brain, the process of tractography yields three-dimensional reconstructions of white matter tracts, which may be microstructurally analyzed by means of diffusivity parameters. Before the advent of Diffusion Tensor Imaging, studies of connectional anatomy in humans could only be conducted post-mortem. Diffusion Tensor Imaging has shifted neuroscientific attention from single brain regions to the networks connecting them. This, however, has not only extended our knowledge on brain networks but also on single loci being part of these networks. Although the biological substrates of Diffusion Tensor Imaging-derived parameters remain unclear to a certain extent, Diffusion Tensor Imaging and tractography have successfully informed many studies in the field of age- or training-related neuronal changes. In addition, it has been found to be applicable in clinical settings. However, Diffusion Tensor Imaging studies covering the fields of neuroeconomics and behavioral psychology are sparse. Two pioneer studies from these fields have successfully related personality characteristics of interest to Diffusion Tensor Imaging- and tractography-derived measures. This chapter tries to give an introduction to the mechanisms of Diffusion Tensor Imaging and will work to explain its utilization while highlighting limitations of Diffusion Tensor Imaging as well as examples of its successful applications.

22.1 Introduction

Diffusion Tensor Imaging (DTI) is a modern Magnet Resonance Imaging (MRI)-method which allows to examine the microstructural status of white matter in vivo and non-invasively. Tractography is a modeling technique which uses DTI data to

T. Rüber (✉) · C.E. Elger · B. Weber
Department of Epileptology, Bonn University Hospital, Bonn, Germany
e-mail: theodor.rueber@ukb.uni-bonn.de

M. Reuter and C. Montag (eds.), *Neuroeconomics*, Studies in Neuroscience, Psychology and Behavioral Economics, DOI 10.1007/978-3-642-35923-1_22

three-dimensionally reconstruct white matter pathways. The advent of DTI and tractography has revolutionized neuroscience in the last 25 years by opening an avenue for the study of white matter connections in living human subjects. Before that, white matter studies in humans could only be conducted *post-mortem*. Tracing neuronal connections in the brain dates back to the work of the nineteenth century neuroanatomists: Carl Wernicke was among the first to relate clinical pathologies to lesions of neuronal connections (Wernicke 1874). Joseph Jules Dejerine used myelin stains to study (so-called Wallerian) degeneration in patients which happened anterograde and retrograde to focal lesions (Dejerine and Dejerine-Klumpke 1895). Modern techniques of tract tracing exploit the active axonal transport mechanisms of neurons in animals (Dauguet et al. 2007; Johansen-Berg and Rushworth 2009; Mesulam 2005). Here, a tracer substance including small fluorescent molecules, neurotoxins, latex microspheres, or viruses is injected into a region of interest and taken up by the cell. This procedure is followed by a waiting period of several days in which the tracer is either transported from the cell body to the axon terminals (anterograde tracing) or from the axon terminals to the cell body (retrograde tracing). The animal is then sacrificed; the brain is fixed, sectioned and post-processed to allow for visualization of the tracer substance (Johansen-Berg and Rushworth 2009). The results may be used to localize neurons of origin, visualize pathways "hidden" in white matter, and to spot terminal fields. For obvious reasons, this procedure cannot be applied to humans. Thus, core questions of human neuroanatomy had to remain unanswered. In their editorial published 1993 in *Nature*, Francis Crick and Edward Jones concluded that "Human neuroanatomy is so backward because we cannot use the common tracer methods on humans […]" and underlined the need for new techniques to the study of the human brain (Crick and Jones 1993). DTI has been named as "most concrete response to their appeal" (Catani 2007). Functional MRI (fMRI) had shifted attention to cortical areas and their functions leading to what has been termed a "new sort of phrenology" (Mesulam 2005), and to the consideration of white matter only as passive infrastucture. DTI and tractography have pushed white matter back in the focus of neuroscientific attention after a long time of stagnation. This is voiced in the common alteration of the colloquialism "gray matters" to "white matters" in the neuroscience community.

A detailed introduction to the mechanisms and applications of DTI is beyond the scope of the current chapter. Rather, this chapter aims to provide a preliminary understanding of DTI, allowing the reader to value the opportunities but be aware of the limitations of this recent imaging modality. Emphasis of this chapter will be analysis approaches to DTI data and applications of DTI and tractography in order to give it a practical thrust. The principles of DTI will be briefly reviewed before introducing its most commonly used measures. Tractography will then be discussed and common approaches of DTI-tractography data analysis will be presented, including pitfalls and limitations of this method. Furthermore, common DTI applications for clinical purposes as well as research studies of anatomy and plasticity will be reviewed. Finally, two DTI-studies related to the field of neuroeconomics will be discussed and a conclusion from this chapter will be drawn.

22.2 Basic Principles

DTI is a specific modeling technique of diffusion-weighted datasets (Basser et al. 1994b; Pierpaoli and Basser 1996). Diffusion-Weighted Imaging (DWI) is a variant of conventional MRI (Le Bihan et al. 1986; Taylor and Bushell 1985). DWI/DTI may be performed with existing MRI technology without the additional use of new equipment or contrast agents. The advantages related to MRI, thus, also apply to DTI: It is non-invasive, may be performed in vivo, is harmless to the subject as long as security guidelines are being followed, is widely available and relatively inexpensive. In DWI, the MRI signal is sensitized to the tissue water diffusion rate. From the atomistic point of view, molecular diffusion is defined as a random walk of molecules that is driven by thermal energy. Diffusion of particles was first observed by the Scottish botanist Robert Brown in 1827 who interpreted the movement as "essence of life"; scientific explanation of molecular diffusion was given by Albert Einstein almost 80 years later (Einstein 1905). The process of diffusion may well be pictured by means of a simple experiment: Some drops of ink are carefully poured into a glass of water. At first, the ink will be clearly distinguishable in one area of the glass but will then diffuse all over and be uniformly dissolved. In the middle of the glass, molecules are not hindered by the container. Their diffusion process is nondirectional, whereas close to the container walls, diffusion is directional. Directionally unrestricted diffusion is described as isotropic, whereas directional diffusion is described as anisotropic. It is easy to see how diffusion will be anisotropic in coherent white matter bundles, such as in the internal capsule, where connections descend from the cortex to the spine, and how it will be isotropic in the fluid-filled ventricles. By means of a magnetic field gradient, the MR signal may be sensitized to diffusion. The most basic form of diffusion imaging, DWI, relies on a model of a three-dimensional isotropic Gaussian distribution and only yields a scalar measure of the diffusion rate, the so-called apparent-diffusion coefficient (ADC), indicated on a grayscale specifically for every voxel. A voxel (volumetric pixel) is one of many rectangular volume elements constituting a regularly segmented three-dimensional space. In brain imaging, one full measurement of the head consists of many thousands voxels and is termed *volume*. In a diffusion-weighted volume, fluid-filled ventricles appear dark, whereas regions of restricted diffusivity such as gray matter or white matter fiber bundles look bright. However, diffusion-weighted datasets may also be used to model the underlying diffusivity characteristics in each voxel. The most common three-dimensional model mapping of anisotropic diffusion as a function of spatial location is the diffusion tensor, introduced by Basser et al. (1994a, b). In this model, the tensor is fully described by a 3 × 3 matrix of variances and covariances.

$$D = \begin{pmatrix} D_{xx} & D_{xy} & D_{xz} \\ D_{yx} & D_{yy} & D_{yz} \\ D_{zx} & D_{zy} & D_{zz} \end{pmatrix}$$

Anisotropic diffusion has six degrees of freedom. This is why at least six diffusion measurements of the whole brain in several different directions (i.e., six volumes) are needed to derive the diffusion tensor matrix. However, it is intuitive that diffusion measurements in more than six directions will yield a more accurate diffusion tensor model and are most commonly applied (Hasan et al. 2001; Papadakis et al. 2000). In any case, one additional volume without diffusion-weighting, the *b*0, has to be acquired in every sequence which serves as a reference image. Once the tensor model is fitted, one can estimate the diffusivity in any arbitrary direction. Diagonalization of the tensor matrix results in a set of three orthogonal eigenvectors $\hat{e}_1$, $\hat{e}_2$, and $\hat{e}_3$ representing the major, medium and minor principle axes of diffusivity modeled. The corresponding eigenvalues λ_1, λ_2, and λ_3 quantify the apparent diffusivities along these axes. It is the most basic premise underlying DTI that molecular diffusion happens easier along the axis of coherent white matter bundles, but will be hindered by axon membranes and other white matter barriers perpendicular to it (Hagmann et al. 2006; Pierpaoli et al. 1996). The major principal eigenvector $\hat{e}_1$ indicating the direction of the greatest apparent diffusivity λ_1 is thus thought to be parallel to tract orientation (Moseley et al. 1990). Since the three vectors are orthogonal to each other, the medium and minor principle eigenvectors $\hat{e}_2$ and $\hat{e}_3$ are perpendicular to the tract orientation. The diffusion tensor is best visualized by an ellipsoid pointing in the direction of maximal diffusivity. In this illustration, the eigenvectors ($\hat{e}_1$, $\hat{e}_2$, $\hat{e}_3$) define the directions of the principle axes and the respective radii are defined by the eigenvalues (λ_1, λ_2, λ_3). The ellipsoid will be orbital if diffusion is isotropic and may be cigar-shaped if diffusion is anisotropic (Fig. 22.1).

In this context, it is important to note that the same combination of eigenvalues (i.e., an ellipsoid with the same shape) may be caused by different neuroanatomical underpinnings. Anisotropy, as modeled by the diffusion tensor, may be low because of "truly" isotropic diffusion (such as in the fluid-filled ventricles) or because of orthogonal but highly coherent fiber bundles crossing within the respective voxel. From this perspective, it becomes obvious how limits are set to the validity of tractography by the inadequacy of the tensor model (please see paragraph on BIOLOGICAL SUBTRATES OF DTI). Another problem concerns the display as well as the interpretation of a 3 × 3 matrix for every one of several thousand voxels within a brain volume. It is often tried to simplify things by converting the diffusion tensor matrix to a scalar measure. The most commonly used scalar DTI measures include fractional anisotropy (FA), mean diffusivity (MD), axial and radial diffusivity (AD, RD). Importantly, these measures may inform researchers about the underlying white matter status since the diffusion tensor from which they are derived does not only model the orientation of anisotropic diffusion (which these scalar measures do not contain) but also the magnitude and the degree of diffusion anisotropy. In neuroscience research, FA is the most commonly used scalar. It is rotationally invariant, ranges between 0 and 1 and describes the degree of the anisotropy of the diffusion process.

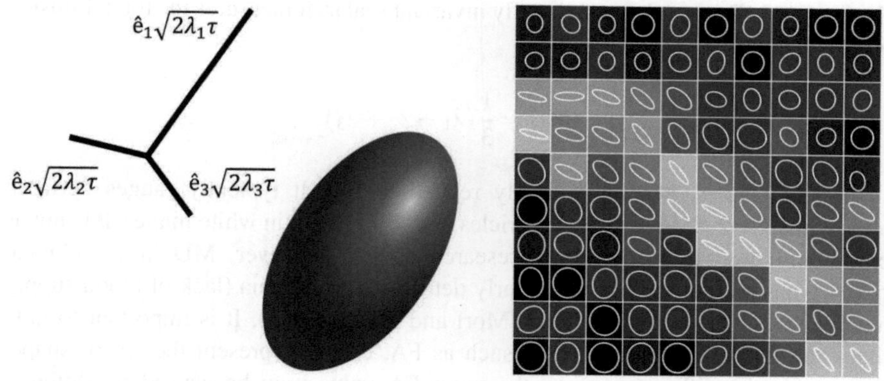

Fig. 22.1 Illustrations of the diffusion eigenvectors (*left*), the corresponding tensor ellipsoid (*middle*), and a schematic grid of voxels (*right*). The grid of voxels illustrates a fractional anisotropy map with additional tensor ellipsoids which are depicted as *ellipses*. The *grayscale* indicates fractional anisotropy ranging from close to zero (*black*) to almost one (*bright gray*). Diffusion in the *upper right corner* is isotropic and apparent diffusivity is rather low as it would be expected in *gray matter*. In the *lower left corner* diffusion also is isotropic; however, here apparent diffusivity is rather high as it would be expected in the ventricles. From the *upper left corner* to the *lower right corner*, diffusion is anisotropic and most major principal eigenvectors are oriented along this diagonal axis. This is how a diffusion tensor image of a fiber tract descending this way may look

$$\mathrm{FA} = \sqrt{\frac{1}{2}}\frac{\sqrt{(\lambda_1-\lambda_2)^2+(\lambda_2-\lambda_3)^2+(\lambda_3-\lambda_1)^2}}{\sqrt{\lambda_1^2+\lambda_2^2+\lambda_3^2}}$$

Values close to 0 indicate that diffusion is isotropic ($\lambda_1 \approx \lambda_2 \approx \lambda_3$), whereas values close to 1 mean that it is anisotropic $[(\lambda_1 \gg \lambda_2 \wedge \lambda_3) \vee (\lambda_1 \wedge \lambda_2 \gg \lambda_3)]$. However, these extreme values are not reached in physiological environments. In the brain, minimal FA values (approx. 0.1) are normally measured in the ventricles and maximal FA values (approx. 0.85) in the splenium of the corpus callosum (Pierpaoli and Basser 1996). FA is claimed to sensitively detect white matter alterations such as degeneration or plastic remodeling (Pierpaoli and Basser 1996; Pierpaoli et al. 1996). A common albeit not always correct notion is that aging and a loss of cognitive functions is associated with decreased FA values, whereas learning and rehabilitation is accompanied by increased FA values (Bengtsson et al. 2005; Engvig et al. 2012; Landi et al. 2011; Lebel et al. 2012; Lindenberg et al. 2010). This FA decrease/increase is usually explained by increased/decreased diffusivity perpendicular to the principal diffusion direction. It is thought to reflect decreased/increased fiber organization or myelination on a microstructural level. However, as outlined below one should be wary of inferences on neuroanatomical underpinnings of DTI measures (please see paragraph on BIOLOGICAL SUBTRATES OF DTI).

Beside FA, MD is another rotationally invariant scalar. It measures the total diffusion within a voxel.

$$\mathrm{MD} = \frac{1}{3}(\lambda_1 + \lambda_2 + \lambda_3)$$

In many cases, MD is inversely related to FA. It typically ranges between approximately 2×10^{-3} in the ventricles and 0.7×10^{-3} in white matter. It is not as commonly used in neuroscience research as FA, however, MD in the clinical practice has proven useful for the early detection of ischemia (lack of blood supply leading to stroke) in white matter (Mori and Barker 1999). It is important to note that even measures of anisotropy, such as FA, cannot represent the tensor shape. Just as for the diffusion tensor, the same FA value may be caused by different combinations of eigenvalues. For example, a disc-shaped and a cigar-shaped ellipsoid are properly described by the same FA value. See Fig. 22.2 for an illustration of the described DTI indices.

In the search of scalar measures being more tightly connected to the tensor shape and distribution, the use of AD and RD has been suggested (Song et al. 2002). AD is equivalent to the major principal eigenvalue and RD is equivalent to the average of the medium and minor principal eigenvalues.

$$\mathrm{AD} = \lambda_1$$

$$\mathrm{RD} = \frac{\lambda_2 + \lambda_3}{2}$$

AD and RD are often subjects to a second look at the data which is taken to shed light on the microstructural correlates of previously observed FA alterations. However, for reasons given below, their use has been heavily criticized (please see paragraph on BIOLOGICAL SUBTRATES OF DTI).

22.3 Tractography

Just as diffusion-weighted data becomes a lot more interesting when tensor models are applied, diffusion tensor data becomes a lot more interesting with the application of tractography (Basser et al. 2000; Jones et al. 1999; Mori et al. 1999; Mori and van Zijl 2002). Tractography is a powerful aid for the interpretation of DTI data. It has been developed "to clarify the orientational architecture of tissues by integrating pathways of maximum diffusion coherence" (Hagmann et al. 2006). Tractography is built on DTI's most basic premise that the major principal eigenvector $\hat{e}_1$ is oriented parallel to tract orientation. By following the direction of the diffusion maximum from voxel to voxel across the brain, the process of tractography yields beautiful three-dimensional models of white matter pathways

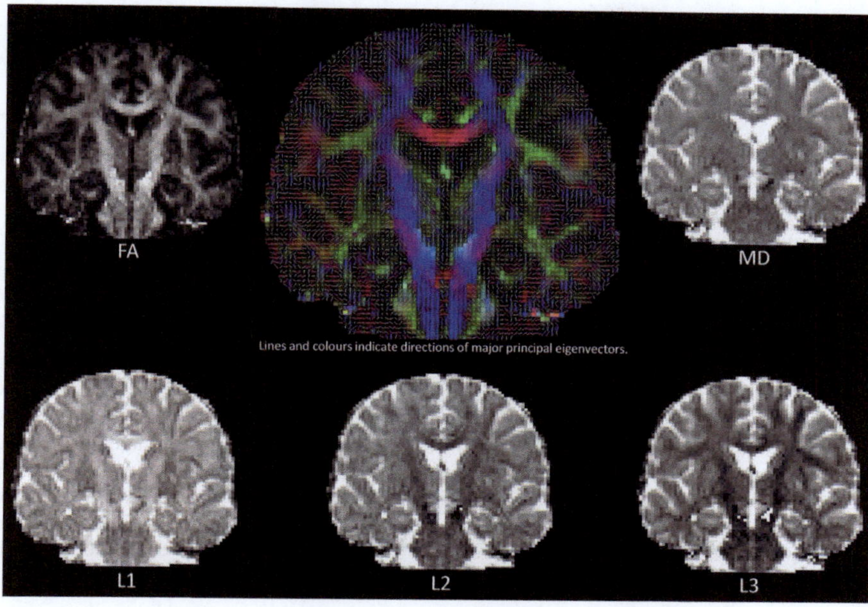

Fig. 22.2 Different displays of a DTI brain volume of a 25 year old healthy female. In the upper, middle brain the *lines* as well as the *colors* indicate the orientation of the major principal eigenvectors. The colors do so according to the most basic red-green-blue (RGB) color coded scheme (Pajevic and Pierpaoli 1999). Voxels in which the principal direction of diffusion is oriented along the *left-right*, *anterior–posterior*, or *superior–inferior* axes are visualized in *red*, *green*, or *blue*. Abbreviations: *FA* fractional anisotropy; *L1*: λ_1; *L2*: λ_2; *L3*: λ_3; *MD* mean diffusivity. Please note how coherent and unidirectional fiber bundles (such as the corticospinal tract) appear *bright* in *L1* but rather *dark* in *L2* and *L3* (i.e., it shows high axial but low radial diffusivity values)

(see Fig. 22.3) which show striking correspondence with classical neuroanatomical descriptions (Catani et al. 2002; Catani and Thiebaut de Schotten 2008).

Tractography is, however, not only a means of visualization, but may, even more importantly, be used as delineator for regions of interest (ROI) in the analysis of DTI measures (see paragraph on ANALYSIS OF DTI DATA). Furthermore, tractography is not only a method for qualitative analysis (that is, answering the question whether a certain white matter voxel belongs to a tract or not) but also a quantitative method: As set out below, tractography yields measures quantifying the degree of connectivity. The process of tractography may be structured along three keywords (Soares et al. 2013): seeding, propagation, and termination. Seeding describes the delineation of one or several voxels as a ROI from which the tracts will be drawn. ROIs may be delineated manually or derived from brain atlases and registered to the individual brain (exactly as it is done for quantitative analysis of diffusivity parameters; see paragraph on ANALYSIS OF DTI DATA). Once tractography is started, fiber pathways, also termed “streamlines,” are initiated in every voxel included in the seed ROI and gradually propagated through the tensor

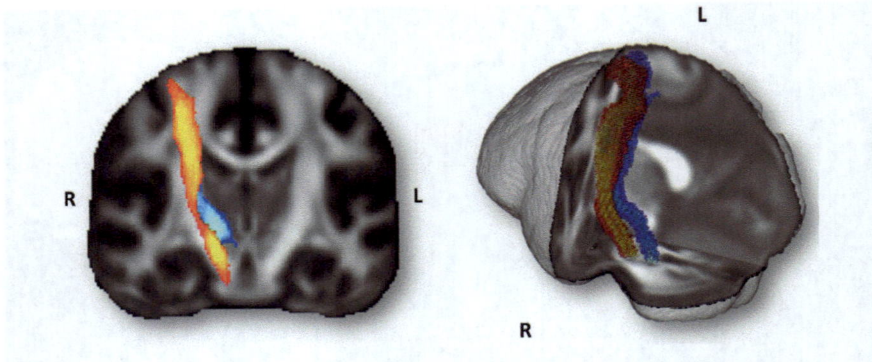

Fig. 22.3 Two-dimensional (*left*) and three-dimensional (*right*) rendering of two corticospinal fiber tracts reconstructed by means of probabilistic tractography. The tracts presented are canonical tracts: probabilistic tractography had been applied to several subjects and the resulting individual tracts were normalized, converted to binary images and then summed. *Color brightness* indicates the degree of voxel-by-voxel overlap of the individual normalized tracts. The canonical tracts are superimposed onto a template brain from the Montreal Neurological Institute. FSL's diffusion toolbox (FDT) was used for image preprocessing and tractography (Behrens et al. 2003b; Smith et al. 2004)

field. Common tractography algorithms fall in two main categories: deterministic and probabilistic tractography (Chung et al. 2011; Descoteaux et al. 2009; Mori and van Zijl 2002). In deterministic tractography, streamlines simply follow the direction of the major principal eigenvector $\hat{e}_1$ in three dimensions. Whenever the streamline reaches the edge of another voxel, it is reoriented according to $\hat{e}_1$ of this other voxel. However, signal noise, subject movement, and artifacts as well as the inherent inadequacy of the tensor model (see paragraphs on BASIC PRINCIPLES above and BIOLOGICAL SUBSTRATES OF DTI below) produce uncertainty, which is ignored by deterministic algorithms. Probabilistic tractography on the other hand incorporates the uncertainty of the estimation. Hence, it does not result in well-defined binary maps ("tract" or "no tract") but in a probability map representing the likelihood of any voxel to be connected with the seed region. These probability maps are most commonly generated by iterations of the streamline process. In other words, the local tensor is interpreted as a probability density distribution and many random samples are drawn from this distribution (Behrens et al. 2003b; Koch et al. 2002). The more streamlines emanating from the seed region cross a certain voxel on their way through the tensor field, the higher the so-called structural connectivity between this voxel and the seed region is. However, it should be kept in mind that these connectivity values are "best possible guesses" and that they should not be interpreted as anatomical fiber count, which is commonly done (Jones et al. 2013). One valuable feature of probabilistic tractography is that it may be applied with multi-fiber models (Behrens et al. 2007). These models have been developed to tackle the problem of fiber tracking in a multi-orientation field (e.g., regions where fibers cross). Anatomical regions represented by voxels with several "major" principal eigenvectors are completely meaningless as a subject to deterministic tractography. It is important to note that in the majority of cases more than one seed ROI is necessary to single out a specific tract of interest. For this purpose, additional ROIs have to be defined and related to the seed ROI with logical operators such as "AND," "OR," and "NOT." For example, in FSL's diffusion toolbox (Behrens et al. 2003b; Smith et al. 2004) one may define a "waypoint mask," "exclusion mask," or a "termination mask." "Waypoint masks" are related with a logical "AND" to the seed ROI. That is, all tracts have to pass this ROI on their way through the tensor field, while all others are being discarded. If several "waypoint masks" have been defined, the logical "AND" may be set to a logical (inclusive) "OR," meaning that all tracts have to pass at least one of the "waypoint masks" in order to be kept. "Waypoint masks" and "exclusion masks" are both connected to the seed region with a logical "NOT," however, there is one crucial difference: Whereas pathways that have come in touch with "exclusion masks" will be discarded as a whole, pathways that have contact with "termination masks" will only be stopped but not discarded. That is, the pathway from the seed ROI to the "termination mask" will be kept as result but cannot propagate any further. Obviously, pathways are not only stopped by "termination masks" but streamlining is also aborted based on criteria inherent to the respective tractography algorithm used. These criteria have been introduced to prevent streamlines from propagating through vectorial fields with great uncertainty of estimation. Common termination criteria include a minimum FA threshold of 0.2 and a turning angle threshold of roughly 60°.

22.4 Biological Substrates of DTI

It is unquestionable that DTI and tractography yield useful insights into neuroanatomy. However, big challenges lie in the verification of virtual fiber bundles as well as in the interpretation of diffusivity measures in terms of their underlying neuronal substrates. Approaches to face these challenges include DTI of objects of known structure, such as physical phantoms (Perrin et al. 2005) and validation against anatomical data acquired by the use of other methods (Johansen-Berg and Rushworth 2009; Lawes et al. 2008). It is easy to see how even the most complex physical phantom will fail as an adequate model of neuronal tissue. The direct validation of DTI indices and tractography performance against anatomical knowledge gained by the application of complementary methods is, thus, indispensable. One elegant study (Dyrby et al. 2007) has validated tractography in *post-mortem* brains of three minipigs by comparing the results of probabilistic tractography with results obtained from two established tracer methods. When inspecting virtual fiber bundles in the light of fiber trajectories identified by one of the tracers, overlaps of around 80 % were reported. Disjunct parts were due to false-negative as well as false-positive tractography results. Tractography had difficulties identifying the termination point of tracts originating from/terminating in subcortical gray matter, which lead to false-positive results. False-negative, "spurious" fibers were observed in anatomical regions of high fiber complexity (fibers crossing, kissing, spreading…). It should be noted that approximately one- to two-thirds of all voxels within a skull-stripped DTI brain volume has been estimated to cover these anatomical regions with high fiber complexity (Behrens et al. 2007; Jeurissen et al. 2012). This sets limits to tractography's role as in vivo dissector of white matter macrostructures. On a microstructural scale, the relation between diffusivity parameters such as FA, MD, AD, as well as RD and their biological substrates remains to be elucidated. However, it should be noted that the missing link between DTI indices and their microstructural underpinnings is not only due to a lack of research findings in this field; also constraints inherent to DTI and MRI such as its resolution (with each voxel containing several ten thousand neuronal connections) restrict the usage of DTI as physiological tool. It is important to keep in mind that there is no one-to-one relationship between any DTI index and a single white matter component (Johansen-Berg and Rushworth 2009). Distinct microstructural processes may lead to the same DTI observations. Alterations of DTI indices, thus, have to be interpreted with great caution. A common inference that is made in many DTI papers is that alterations in diffusivity mirror changes in the degree of myelination. However, as Jones and colleagues (Jones et al. 2013) correctly point out "the anisotropy in a region may also be lower because here is a larger axon diameter (Takahashi et al. 2002), a lower packing density (Takahashi et al. 2002)—both of which mean fewer barriers to diffusion in a given space—or it could be due to increased membrane permeability (reducing the effectiveness of a boundary)." Indeed, it soon became clear that myelination is no requirement for diffusion anisotropy in white matter (Beaulieu and Allen 1994). Rather one may think of several barriers hindering isotropic diffusion in white matter including microtubules, myelin, axon membranes and neurofilaments (Beaulieu 2002). On the contrary, damaged or absent myelin is reliably mirrored in reduced diffusion anisotropy (Blaschek et al. 2013; Klawiter et al. 2011; Mukherjee et al. 2001; Werring et al. 2000). Gulani and colleagues have compared excised spinal cords from myelin-deficient rats with spinal cords from age-matched controls (Gulani et al. 2001) and came to conclude that myelin may modulate FA by up to 20 %. One difficulty in determining the relative contribution of myelin to the measured anisotropy lies in the fact that any direct comparison between myelinated and non-myelinated nerve fibers is flawed because tissues differ in their whole structural composition (Assaf and Pasternak 2008). Axial and radial diffusivities may be helpful when trying to explain the emergence of observed FA alterations. Decreased axial diffusivity has been related to the growth of neurofibrils (Kinoshita et al. 1999), whereas several studies have linked a decrease in radial diffusivity to demyelination (Budde et al. 2007; Klawiter et al. 2011; Song et al. 2002, 2003). However, in their seminal paper, Wheeler-Kingshott and Cercignani argue that one should be wary of interpreting tensor eigenvalues as "axial" or "radial" and of relating these measures to microstructural properties, such as myelin (Wheeler-Kingshott and Cercignani 2009). The crucial point the authors make is that comparing eigenvalues across subjects without checking for the alignment of the corresponding eigenvector with the underlying tissue structures is pointless. In particular, this correspondence may be called into doubt when comparing tracts affected by some pathology with tracts from healthy controls and in voxels in which more than one principal direction is present. In the former case, it is well conceivable how a certain pathology may change the direction of the principal eigenvector from what would be expected in healthy tissue architecture; in the latter case, represented by an oblate diffusion ellipsoid, the "principal" eigenvalue may not differ much from the other eigenvalues (i.e., there are more than one principal eigenvectors), thus, making the analyses of "axial" and "radial" diffusivities completely meaningless. It may be concluded that DTI and tractography offer a unique way to gain insights into human neuroanatomy. However, over those "pretty pictures" (Johansen-Berg and Behrens 2006) and their tempting inferences on neuroanatomical underpinnings, one should not forget that DTI and tractography, like any other recently developed research method, urgently require validation and a precise definition of its scope.

22.5 Analysis of DTI Data

In the following, three common approaches to quantitative analysis of DTI data will be reviewed: ROI analysis (I), voxel-by-voxel analysis (II) and tract-based spatial statistics (TBSS; III). The first two analyses may be performed with most common analysis programs; TBSS is part of FSL (Smith et al. 2006). ROI analyses (I) can be performed in several ways but it is common to all of them that diffusivity parameters of every voxel within a two- or three-dimensional ROI are averaged and used for statistical analysis. ROIs may be defined manually on each individual brain (I.1), they may be taken from publicly available atlases (I.2) or they may be delineated by means of tractography (I.3). Manual definition of ROIs (I.1) is done directly on FA maps or on anatomical volumes and then transferred to diffusion space, depending on where the respective anatomical structure is best identifiable. When drawing ROIs on FA maps, it might be helpful to superimpose diffusion color maps indicating the main diffusion direction. ROI definition solely on FA maps, however, is prone to preferably include high FA voxels since they are best visible and, thus, are taken as landmarks for orientation (i.e., there is an influence of anisotropy on ROI borders). Squared or orbital ROIs are different in this regard and, furthermore, less time-consuming to define but these uniform ROIs also account less for the individual forms of anatomic structures. Overall, it may be said that its accountancy for individual anatomy is the most important advantage of manual ROI definition, whereas on the other side its operator-dependency as well as the fact that the manual definition work is time-killing and nerve-racking are among its disadvantages. The usage of atlas ROIs (I.2) is different in this regard. Here, anatomical template regions from publicly available atlases [Harvard-Oxford cortical and subcortical structural atlases (Caviness et al. 1996), John Hopkins University DTI-based white matter atlas (Hua et al. 2008), Freesurfer (Fischl et al. 2002)] are mapped on the individual brain. This approach is neither operator-dependent nor particularly time-consuming; however, it has other difficulties: The anatomical template region is generated from one or (in most cases) more brains and made available in a three-dimensional standardized, or "stereotaxic," coordinate space. It has to be transferred to individual, or "native," space in order to align with individual anatomy using a mapping function. Individual's brain anatomy, however, may not correspond to the anatomy of the template brain; moreover, the mapping function will never work perfectly leading to misalignment. [See paper by Evans et al. (2012) for a recent review of brain atlases and their problems]. When considering the pros and cons of manually defined ROIs versus atlas-derived ROIs, in many cases ROIs delineated by tractography (I.3) will yield the best of both worlds (Fig. 22.4): This procedure is not as operator-dependent as manual ROI definition (making it somewhat more "objective") and it does consider individual anatomy, since it is run for every single subject. Tracts resulting from tractography are thresholded (excluding low probability voxels) before diffusivity parameters of all voxels within the respective maps are averaged. However, it is obvious that not all anatomic regions which may be subject to DTI analysis can possibly be reconstructed using tractography. In general, it shall not be dismissed that ROI analyses have two systematic problems that may not be overcome: First, depending on the size of the ROI, variation of diffusivity parameters within it is lost since all voxel values are averaged, second, it is not feasible to investigate the whole brain expressing a need for strong a priori hypotheses.

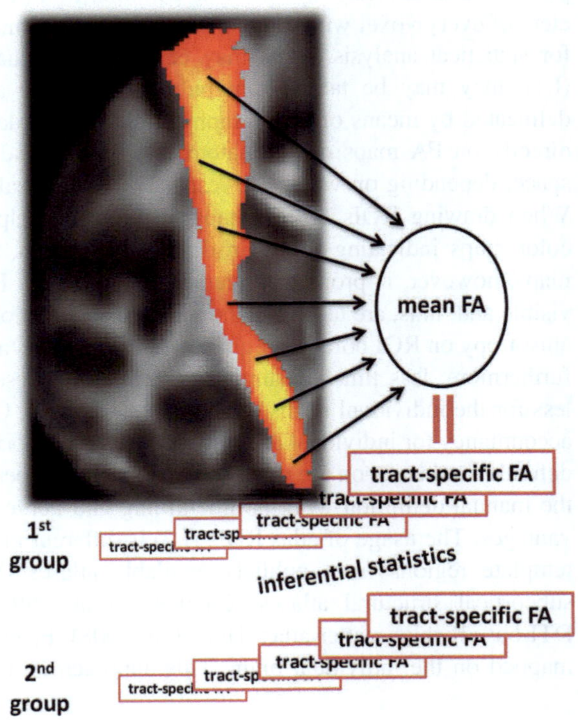

Fig. 22.4 Illustration of tractography-based Region of Interest analysis of fractional anisotropy (FA). Tracts resulting from tractography are thresholded (excluding low probability voxels) before FA values of all voxels within the respective tracts are averaged [yielding "tract specific FA"] and used for statistical analysis

Voxel-wise analysis (II) as well as TBSS (III, see below) resolves these issues. In these analyses, all DTI volumes are normalized. Normalization describes the process of registering an individual brain to a template brain using a mapping function. Every voxel is assigned a specific "address" (Snook et al. 2007). The crucial premise is that, after normalization a certain "address" corresponds to a certain anatomical structure. That is, all brains and their parts align well to each other. The standard method of brain normalization includes two steps (Ashburner and Friston 2000; Jenkinson and Smith 2001; Snook et al. 2007): The first step, a linear transformation, warps the brain in order to account for global brain differences (size and position); the second step applies a nonlinear transformation which estimates internal brain deformations. After normalization, analyses are run independently including every voxel with the same "address," under the assumption that all voxels with the same "address" represent the same anatomic location across subjects (Fig. 22.5). This enables comparison of diffusivity parameters between groups and correlation analysis with variables of interest. The advantages of voxel-wise analyses are obvious: They are fast, fully automated, operator-independent and investigate the whole brain without the need for a priori ROI definition. However, voxel-wise analyses are also connected with disadvantages. It is not surprising that the registration of healthy brains will never work perfectly, let alone the registration of lesioned brains (Jones and Cercignani 2010; Mukherjee et al. 2008). To alleviate some of the registration problems faced, one normally smoothes the normalized brains using a low pass filter. It shall not be dismissed, that smoothing FA maps is prone to cause problems, since most white matter tracts are thin, might thus be "washed out" by the smoothing and the selection of the right low pass filter is rather arbitrary but influences the results of voxel-wise analyses (Jones et al. 2005). Additionally, it is obvious that the need for brain normalization of low-resolution, high-contrast FA maps introduces problems: Smallest shifts between groups may incorrectly be interpreted as FA differences between groups. Furthermore, effects of partial volumes may challenge the analysis. Partial volume effects occur when a single voxel contains two or more tissue types but is presumed to only contain one that is white matter (Assaf and Pasternak 2008). This is especially likely to occur in white matter regions near the ventricles where high FA voxels are located just on the side of low FA voxels. Certain patient groups are particularly vulnerable to partial volume effects: Schizophrenic patients, for example, exhibit enlarged ventricles as compared to healthy controls (Honea et al. 2005). Another challenge of voxel-wise analysis concerns the sheer amount of statistical tests being performed. A skull-stripped FA map of somewhat good resolution contains approximately 250,000 voxels. A whole-brain voxel-wise analysis, thus, comes along with a gigantic problem of multiple comparisons (Loring et al. 2002). This problem may be tackled with statistical correction methods. In analysis of MRI data, most commonly correction for family-wise error (FWE) is applied (Friston et al. 1994; Worsley 2005). A complementary, though not alternative way, is to apply voxel-wise analysis only within certain ROIs. One might, for example, reconstruct white matter tracts by means of tractography in native space; normalize FA maps and use the resulting transformation matrices to also normalize white matter tracts.

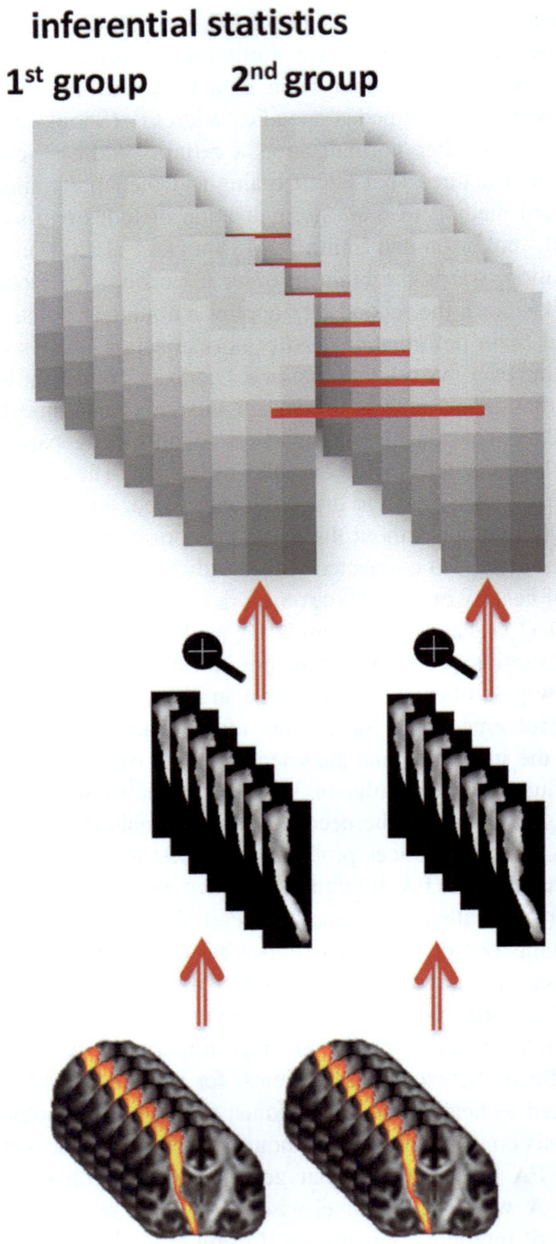

Fig. 22.5 Illustration of voxel-wise analysis of fractional anisotropy. After normalization of all volumes, statistical analyses are run independently for every voxel with the same "address" of a skill-stripped volume. The number of statistical comparisons may be reduced by only running voxel-wise analysis within a Region of Interest (which may be delineated by tractography as illustrated here)

These normalized tracts are then used to build a canonical tract, which may be thresholded based on the voxel-by-voxel overlap of the individual normalized tracts (Rüber et al. 2012). Another way to reduce the number of statistical comparisons, is to restrict voxel-wise analysis to white matter by using white matter masks. These masks may be derived from anatomical volumes with FreeSurfer (Dale et al. 1999; Fischl et al. 1999) or other automated tools for segmentation and then transferred to diffusion space.

TBSS (III) has been developed to face two of the biggest challenges of conventional voxel-wise DTI analysis (Smith et al. 2006): multiple comparison and alignment. Voxel-wise analysis in TBSS is only confined to the FA skeleton, which represents the centers of all white matter tracts common to the study group. This FA skeleton is used as an alignment-invariant feature to mitigate any residual misalignment after a common mapping function has been applied. The TBSS processing works as follows: First, individual FA maps are normalized applying a nonlinear transformation (similarly as described above). Based on all normalized brains, a mean FA map (in standard space) is generated and used to create the pseudo-anatomical white matter skeleton (it is *pseudo*-anatomical since white matter is segmented only based on FA and not on anatomical volumes such as T1 or T2). Next, the skeleton is projected on the individual FA map (in standard space), but is deformed based on a constrained local search for maximal FA values (Fig. 22.6). Lastly, voxel-wise testing may be performed within the skeleton as described above. Hence, TBSS is built on the strategy that the alignment of a skeleton containing maximal FA voxels to local FA maxima in the individual FA maps will mitigate the residual misalignment of previous nonlinear registrations. However, it has been criticized that this strategy primarily leads to correspondence of FA values but not to anatomical correspondence (Zalesky 2011). In other words, local FA maxima do not necessarily correspond to the same anatomical locations across subjects. In addition, if crude registration errors occur, they are difficult to

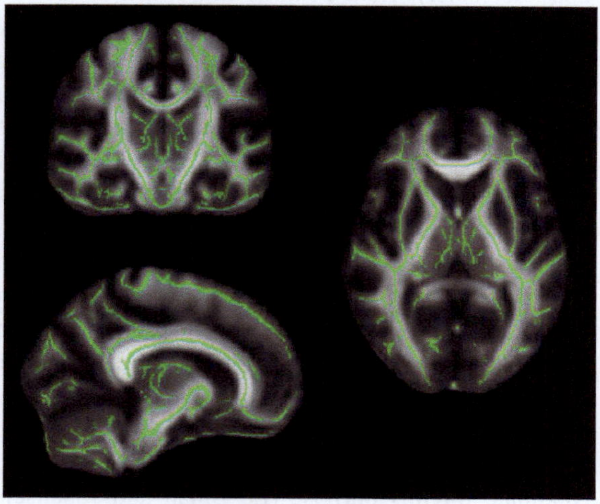

Fig. 22.6 Fractional anisotropy (FA) template brain and the respective FA skeleton as generated by TBSS. Please note how the FA skeleton aligns to the white matter but only covers the centers and, thus, reduces the number of multiple comparisons

identify in the skeleton (Jones and Cercignani 2010). Another disadvantage of TBSS concerns the fact that white matter regions exhibiting low FA values are not included in the analysis. Also, crossing fiber regions exhibit low FA values but are crucial from a neuroanatomical perspective. In newer versions of TBSS, one may choose to use models that incorporate fiber-specific measures (Jbabdi et al. 2010) which enhances analysis of crossing fiber regions. It may be concluded that TBSS is a powerful tool that successfully deals with some problems that come along with traditional analysis approaches to DTI data.

It is important to acknowledge that TBSS is not the only attempt to deal with the shortcomings of brain registration. However, one general problem of common registration algorithms is their dependency on "unidimensional" scalars, such as FA (Zalesky 2011). Conversely, DTI yields ellipsoids that are fully described by six independent parameters, and these parameters are lost when vectors are mapped to rotationally invariant scalars. Thus, anatomical information made available by DTI is not considered in common brain registration algorithms. Multichannel registration algorithms which depend on all six parameters of the tensor have been proposed (Guimond et al. 2001), but are not widespread.

22.6 Connectional Anatomy

DTI is a boon to modern neuroscience not only for allowing the examination of white matter as such, but also for tractography and its potential to study connectional anatomy. The human brain consists of 100 billion neurons of which 20 billion have been estimated to be located in the cerebral cortex (Azevedo et al. 2009; Noctor et al. 2007). On average, every cortical neuron builds 7000 connections with other neurons and 150,000–180,000 km of myelinated fibers are believed to be part of the brain (Drachman 2005). Connectional anatomy, however, does not solely yield insights into white matter circuits. Keeping in mind that "nothing defines the function of a neuron more faithfully than the nature of its inputs and outputs" (Mesulam 2005), it becomes clear that structure and function go together in neuroscience (Blits 1999) and how DTI may inform the study of neuronal mechanisms underlying cognitive functioning (ffytche and Catani 2005). Without any doubt, in the last decade has neuroscientific attention shifted from single loci to the networks connecting them. This has not only extended our knowledge on brain networks but also on single loci being part of these networks. For example, network analysis has opened an avenue for using tractography to provide information of gray matter parcellation (Behrens and Johansen-Berg 2005). Traditionally, cortical or subcortical regions are defined based either on cytoarchitectonic boundaries or on gross-anatomical landmarks. However, cytoarchitectonic boundaries do not necessarily correspond to gross-anatomical landmarks (Geyer et al. 2000). Thus, another basis for the definition of great matter boundaries is favored and connectivity-based parcellation has been named as a convincing candidate for this purpose (Le Bihan and Johansen-Berg 2012). The underlying

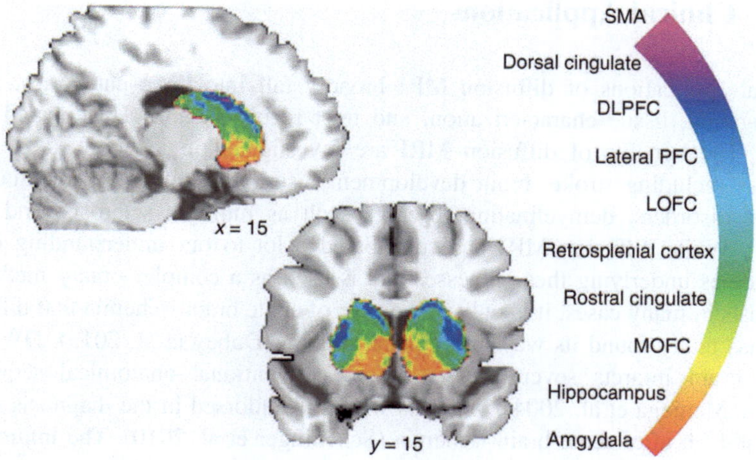

Fig. 22.7 Connectivity-based parcellation of the striatum according to its connections to several cortical and subcortical regions [Fig. 22.1 from Cohen et al. (2009); permission for reproduction by Nature Publishing Group]

notion of connectivity-based parcellation is that cortical and subcortical subregions may be distinguished according to their so-called connectivity fingerprint (Passingham et al. 2002). Connectivity fingerprints delineate clusters of voxels with similar connection patterns (Fig. 22.7). Several studies have compellingly demonstrated the potential of this novel approach by yielding connectivity-based parcellations of the thalamus (Behrens et al. 2003a), amygdalae (Bach et al. 2011), basal ganglia (Menke et al. 2010), cingulate cortex (Beckmann et al. 2009), of Broca's Area (Anwander et al. 2007), of the lateral premotor cortex (Tomassini et al. 2007), and supplementary motor area (Johansen-Berg et al. 2004). Results obtained by means of fMRI have yielded good correspondence with some of these parcellations, providing validation of this approach and underlining the functional relevance of connectivity-based boundaries (Johansen-Berg et al. 2005; Schubotz et al. 2010; Tomassini et al. 2007).

Tractography pushed DTI to another new field of study called *connectomics* (Hagmann et al. 2008; Sporns et al. 2005), the comprehensive studies of neural connections in the brain. In 2009, NIH has announced a Request for Applications for the so-called Human Connectome Project. One year later, grants totaling $40 million were awarded to two collaborating consortia. The overarching purpose of the project is to acquire and share data about human brain connectivity. It is hoped that "a deeper understanding of human brain connectivity and its variability will provide valuable insights into what makes us uniquely human and what accounts for the great diversity of behavioral capacities and repertoires in healthy adults" (Van Essen et al. 2012).

22.7 Clinical Applications

Clinical applications of diffusion MRI broadly fall into three categories: lesion interrogation, tissue characterization, and tract reconstruction (Huston and Field 2013). The benefits of diffusion MRI are investigated in the context of many diseases including stroke, brain development disorders, epilepsy, dementia, psychiatric disorders, demyelinating diseases such as multiple sclerosis, and brain tumors. While diffusion MRI has contributed a lot to our understanding of the pathologies underlying these illnesses and is used as a complementary method of diagnosis in many cases, it is only in the case of acute brain ischemia that diffusion MRI has really found its way into clinical routine (Dubey et al. 2013). DWI may detect brain infarcts several hours before conventional anatomical sequences (Munoz Maniega et al. 2004), which is why it is endorsed in the diagnosis within the first 12 h after acute brain ischemia (Schellinger et al. 2010). The infarct core volume, as estimated by DWI, has also been suggested to be an important predictor of clinical outcome and treatment efficacy (Gonzalez 2012; Schellinger et al. 2010). Even more than the volume, its degree of overlap with the corticospinal tract or the arcuate fasciculus seems to be a predictor of motor and language impairment, respectively (Marchina et al. 2011; Zhu et al. 2010). Indeed, diffusivity parameters of the corticospinal tract could reliably be related to the degree of motor impairment (Lindenberg et al. 2010; Schaechter et al. 2009). As mentioned above, diffusion MRI may detect pathologic tissue alterations with a higher sensitivity as conventional anatomical sequences in some cases. However, the low pathologic specificity of its measures has set boundaries to its clinical usefulness in tissue characterization (Huston and Field 2013). Its application in pre-surgical diagnostics seems to be more promising: In neurosurgery, preoperative reconstructions of white matter tracts may lead to a better understanding of their individual anatomy and may, thus, facilitate their preservation during resection. A recent study with patients undergoing anterior temporal lobe resection (Yogarajah et al. 2009) revealed that the extent of one part of the optic radiation as determined by preoperative tractography was related to the degree of complication by superior visual field deficits. Even more convincing, Wu et al. (2007) found that 118 patients with a highly aggressive tumor survived on average 7.2 months longer than 120 control patients with the same tumor if preoperative tract mapping had been performed.

22.8 Genetics/Environment/Age

White matter status may be thought to depend on genes, environment and age. It is an intuitive notion that white matter alterations happen within the framework of genes. This framework changes across the lifespan leaving more or less room for the white matter effects of environment in development and adulthood. DTI may detect even subtle white matter alterations; however, it remains challenging to specifically attribute these alterations to the influence of genetic factors, environment or age. Especially, it is difficult to differentiate between the influence of genes and environment (so-called *nature–nurture debate*) as will be shown below.

The influence of genes may well be studied through comparison between identical and nonidentical twins. In these studies, the excess in the pair correlation of identical twins over the pair correlation of nonidentical twins is attributed to genetic effects. Chiang and colleagues (Chiang et al. 2009) found that close to 80 % of FA variance may be explained by genetic factors, especially in parietal brain regions. Furthermore, the authors reported correlations between measures of intellectual performance and FA values. An overlapping set of genes seems to influence IQ scores as well as diffusivity parameters, since the authors were able to more accurately predict IQ by using FA values in the group of identical twins than they were in the group of nonidentical twins. In another DTI twin study (Jahanshad et al. 2010), 20–40 % of FA brain asymmetry was found to be genetically determined. Using a group of 705 twins and their siblings, Chiang et al. (2011) investigated factors moderating the heritability of white matter features as measured by FA. They found that higher heritability was associated with younger age, male sex, higher intellectual performance, and higher socioeconomic status. Some of the causal mechanisms behind their finding could only be subject to speculation. The observation of white matter heritability decreasing along age may be interpreted as increasing environmental influences on white matter microstructure.

Neuroplasticity is a term first introduced by the polish neuroscientist Jerzy Konorski in 1948 to describe the brain's ability to functionally and structurally adapt to changing environmental demands. Long enough, the brain was believed to be static after development but this conception has been substantially revised in the last decades (Jancke 2009): "[…] there is a continuous interaction between experiential process and brain structure. Over time, experiences are sedimented in the form of organic habits, dispositions and interactive schemes that eventually constitute the individual's personality." (Fuchs 2009) Additionally, white matter is no longer seen as passive infrastructure but as a plastic component of the brain. Using DTI, "neuroarcheology" (Le Bihan and Johansen-Berg 2012) may be performed to study the lasting effects of past experiences and to keep track of their neuronal correlates. Impressive associations between diffusivity characteristics of white matter and experience have been reported in highly trained experts in sensorimotor skills (Dayan and Cohen 2011). Bengtsson and colleagues reported higher FA values in professional pianists as compared to non-musicians in the posterior limb of the internal capsule (Bengtsson et al. 2005). Even more, the authors found several white matter regions exhibiting a positive correlation between the amount of hours practiced and FA values. DTI-studies have also reported evidence for white matter plasticity in response to the acquisition of other sensorimotor skills, such as ballet dancing or playing golf. Furthermore, neuronal white matter correlates of high cognitive abilities have been found by means of cross-sectional DTI-studies. Floel et al. (2009) found that individual subject's performance in an artificial learning task was related to FA values surrounding Broca's area (a cortical region in the left frontal lobe which traditionally has been implicated in language processing) and within fibers arising from this area. Tsang et al. (2009) showed that variations in performance in a mental arithmetic task reflect variability of FA values in a central chunk of the anterior superior longitudinal fascicle. This tract connects the inferior parietal lobe with precentral and inferior frontal regions—cortical regions that have been shown to be coactive by means of fMRI in subjects performing mental arithmetic tasks. As the studies presented above are cross-sectional, it remains unclear whether the white matter alterations reported are the cause or the consequence of the respective expertise (i.e., are people with altered diffusivity characteristics more likely to play golf or do the alterations occur as people pick up golf playing skills). The results of cross-sectional studies, thus, always have to be discussed in the light of the so-called *nature–nurture debate* (Stiles 2011) and no ultimate inferences on the direction of causality can be made. Longitudinal studies are clearer in this regard and have become the gold standard for neuroimaging studies examining neuroplasticity. Some longitudinal DTI-studies have been run and offer intriguing insights, since they enable the observation of the brain during the acquisition of new skills. One of the first longitudinal DTI-studies reported an increase of FA values in the subcortical white matter underlying the intraparietal sulcus after subjects learned to juggle within 6 weeks (Scholz et al. 2009). FA values in the cluster reported increased slightly after 4 weeks without juggling but still were higher than before subjects learned to juggle. Longitudinal studies have also found FA changes after intensive memory training (Engvig et al. 2012; Landi et al. 2011; Takeuchi et al. 2010). Keller and Just (2009) found lower FA values in the anterior left cerebral white matter in a group of eight to 12 year old children classified as poor readers compared to good readers at the same age. Poor readers then were randomly assigned to either an intervention group undergoing an intensive 100 h program of remedial reading instruction or to a control group. Children who had undergone the 100 h program showed significantly increased FA values in the anterior left cerebral white matter. Even more, this FA change correlated with improvement in phonological decoding. Children in the control group did not show significant FA changes between the first and the second scan. Schlaug et al. (2009) found that effective melodic intonation therapy of aphasic patients after stroke was related to an increase in structural connectivity in the contralesional arcuate fasciculus (a white matter pathway implicated in language processing). A recent study (Langer et al. 2012) investigated patients whose right arm injury required limb immobilization for at least 14 days. They applied DTI to examine structural integrity of the corticospinal tracts and, furthermore, measured cortical thickness of sensorimotor regions based on high-res anatomical volumes, they had acquired. The authors found reduced cortical thickness and decreased FA values in the left sensorimotor regions and left corticospinal tract, respectively. The opposite changes were observed on the right side, which may indicate a skill transfer from the left hemisphere (controlling the right injured hand) to the right hemisphere (controlling the left non-injured hand) being caused by a compensatory use of the non-deprived hand.

It is an interesting notion that most of the training-induced plastic white matter alterations described above are represented by comparably higher FA values and, hence, contrast an age-related decrease in FA values observed after the maturational peak. Many recent studies employ DTI to investigate the neurobiology of aging determining this maturational peak, when maturation turns into degeneration (Kochunov et al. 2012; Lebel et al. 2012; Sullivan et al. 2010; Westlye et al. 2010). A link between white matter changes in the aging brain, cognitive decline and degenerative diseases such as Alzheimer's disease have been established, commonly referred to as "disconnection hypothesis" (Bartzokis 2004; O'Sullivan et al. 2001). Generally, an increase in FA/a decrease in MD until the third and fourth decade are observed which is followed by a decrease in FA/an increase in MD. White matter volume seems to reach its peak ten to 20 years later in life than FA (Salat et al. 2009; Westlye et al. 2010). However, diffusivity parameters generally seem to be the better indicator of white matter maturation. It has been proposed that the inverted u-shaped (quadratic) course of FA along the lifespan is largely dependent on the degree of white matter myelination (Abe et al. 2002; Bartzokis et al. 2010; Gao et al. 2009). Keeping in mind that cognitive efficiency relies on the myelination of white matter fibers, this proposition provides a possible link between the changes of diffusivity parameters on one end and the alterations of cognitive functions in the aging human. Indeed, the zenith of FA values largely concurs with the age range in which cognitive performance has been found to reach its peak (Salthouse 2009). Indeed, alterations of cognitive processing abilities, verbal memory, and motor skills along the lifespan have been related to diffusivity parameters (Bartzokis et al. 2010; Hedden et al. 2005; Salthouse 2000). Recently, it was shown that age-related over-recruiting of frontal lobe structures when completing a task-switching paradigm were negatively correlated to FA of these structures (Zhu et al. 2013). Further studies are warranted in this field which derives its importance from the therapeutic potentials based on a better understanding of the aging brain.

22.9 DTI-Studies in Neuroeconomics

DTI-tractography has rarely been used in the field of neuroeconomics. In the following, two exemplary studies on associations between DTI measures and personality traits will be discussed: Olson et al. (2009) found several frontal and temporal white matter regions which showed correlations between FA values and results of a delay discounting task in a group of 79 adolescents. Using DTI-tractography, Cohen et al. (2009) showed that structural connectivity between the striatum and certain cortical as well as subcortical regions was related to self-reported individual differences in personality traits such as novelty seeking and reward dependence. Both studies are based on two premises: (1) That white matter structural architecture reflects personality characteristics (as determined by delay discounting tests and self-reports, respectively) and (2) that this may be measured by means of DTI. Whereas the first premise appeals to an intuitive notion (behavioral differences have to have a structural equivalent *somewhere* in the brain), it is more surprising that they can actually be measured (second premise) by comparably coarse-grained methods such as DTI. In the study by Olson and colleagues,

79 adolescents between the age of 9 and 23 underwent DTI and, among other tests, a delay discounting test. Delay discounting describes individual's choice to prefer an immediate reward over a larger delayed reward (Ainslie 1975). Mature delay discounting behavior emerges in adolescent's development and has been shown to correlate with age and verbal IQ (Olson et al. 2007). Equally, myelination of white matter is thought to occur as part of neuronal maturation (Arain et al. 2013) in adolescence and may be measured by diffusivity measures (Song et al. 2002). The aim of the current study was to relate delay discounting behavior to diffusivity measures. Previous fMRI studies had implicated the role of frontal, parietal and temporal regions in delay discounting (McClure et al. 2004; Wittmann et al. 2007) which is why the authors expected to find structure-behavior correlations in these regions. However, they performed a whole-cerebrum voxel-wise regression analysis. Results were not corrected for multiple comparisons, but the threshold of significance was set to $p < 0.001$ and an extent threshold of 25 voxels was applied. The application of neuroimaging data analysis not corrected for multiple comparisons has been heavily criticized (Bennett et al. 2009). It shall, however, not be dismissed that the authors applied a hypothesis-free whole-brain approach albeit having well-founded a priori hypotheses on where structural–behavioral correlations may be found which partly legitimates less conservative significance testing. Indeed, the authors could report clusters indicating a positive FA—delay discounting correlation in several white matter regions primarily in the frontal and temporal lobes: The higher the FA values, the more the subject preferred higher delayed rewards over lower immediate rewards. In addition, regions with associations between MD and delay discounting were found. These correlations were negative: The lower the MD values, the more the subject preferred higher delayed rewards over lower immediate rewards. Based on these analyses, it cannot be ruled out that, delay discounting as well as diffusivity measures both vary under the influence of a third (confounding) variable but are not actually related with each other. The authors mention age and verbal IQ as possible confounding variables and enter both as covariates into the regression equation. Several clusters remained significant after controlling for age and/or verbal IQ, thus, giving evidence that the observed relation between diffusivity parameters and delay discounting is not completely due to the general influence of neurodevelopmental maturation indicated by age or part of a more general cognitive process measured by a verbal IQ test. Consequently, observed structural–behavioral associations may be seen as partly age-dependent and partly age-independent. To rule out that these associations do reflect general processes of decision-making, the cluster-wise averages of FA/MD were correlated with the outcome of a probability discounting task which did, except for one cluster, not yield significant results. Observed white matter diffusivity characteristics may, thus, be viewed as specific microstructural correlate of delay discounting. Unfortunately, the authors missed to perform additional regression analyses for directional diffusivities. This would have been of special interest under the premise of radial diffusivity reflecting the degree of myelination and with regard to the fact that the frontal lobe becomes increasingly myelinated during adolescence. Lastly, the authors related the anatomical location of clusters to

common white matter tracts implicating their role in delay discounting. The tracts were not individually modeled using tractography but probability masks from a common tractography white matter atlas (Wakana et al. 2004) were used. It may be summarized, that the study by Olson and colleagues compellingly illustrates some of the structural white matter correlates of delay discounting and their development in adolescence.

The second exemplary study discussed also suggests white matter circuitry to underlie certain personality traits. However, this study by Cohen and colleagues uses a different methodological approach: Variations in diffusivity measures are not found to be relevant here, rather structural connectivity as determined by tractography was found to correlate with novelty seeking and reward dependence in 20 healthy adults. This study as well as the study by Olson and colleagues relied on previous studies linking the personality traits of interest to neuronal circuitry. Cohen and colleagues mention evidence from studies linking novelty seeking and reward processing to a hippocampus-ventral striatal-midbrain loop and to prefrontal-striatal connections, respectively (Hollerman et al. 2000; Lisman and Grace 2005). They, thus, applied tractography to parcellate the striatum according to its connections to several cortical and subcortical regions (see this chapter's paragraph on connectional anatomy and Fig. 22.7). These regions were defined using a common atlas (Tzourio-Mazoyer et al. 2002). Voxels in the striatum were then labeled according to the region they were most strongly connected to. The connectional strength between these voxels and their respective regions were found to correlate with novelty seeking in the case of connections to the amygdala as well as the hippocampus and with reward dependence in the case of a network to cortical regions including medical and lateral orbitofrontal cortices, dorsolateral prefrontal cortex and supplemental motor area. Connections which correlated with novelty seeking did not correlate with reward dependence and vice versa. As intriguing as these findings are, it remains unclear which microstructural characteristics cause the altered connection strength. Even functional inferences of these microstructural characteristics in terms of neuronal efficiency have to be made with great caution. The authors conclude that "although it is tempting to speculate that individuals with higher tract strength values can transmit information more efficiently between regions, further work is needed to confirm this interpretation." The explanatory gap between white matter features as measured by DTI and character traits as defined by psychology remains unsurpassed also by the most advanced DTI-studies.

22.10 Conclusion

The advent of DTI and tractography has paved the way to one of the most flourishing fields of today's neuroscience: The non-invasive in vivo assessment and delineation of white matter tracts. As this chapter has tried to show, DTI has made immense contributions to distinct areas of neuroscientific research just as it has changed our perspective on the brain as a whole. However, 25 years of experience with DTI has forced the community to increasingly recognize the limits of this imaging modality. In particular, the interpretation of DTI findings in terms of their neuronal underpinnings has been proven difficult. In their very readable review, Jones and colleagues point out that "the only thing that we can say with any certainty in diffusion MRI is that we measure a signal change when a motion-sensitizing gradient is applied along a given axis. Inferring anything else is dependent on the quality of the model and the quality of the data" (Jones et al. 2013). As outlined above, the tensor model is thought to be inadequate in some regards. More promising models of diffusion data are on hand; however, they are not in common use since most of them are connected to special hardware requirements as well as long acquisition times and a lack of experience in ongoing research. Hence, some of the main reasons for the continuously growing success of DTI may well be called pragmatic. Nonetheless, MRI hardware development will continue to push the field of diffusion imaging forward which most likely will compensate for some of the drawbacks of newer diffusion models. Approaches other than the tensor model will, therefore, play a more prominent role in future times: Diffusion Spectrum Imaging (DSI), High Angular Resolution Diffusion Imaging (HARDI), and Q-Ball Imaging (QBI) have all been named as promising candidates (Tournier et al. 2011).

This chapter has provided an introduction to the mechanisms of DTI and worked to explain its utilization while highlighting its limitations and examples of successful applications. Notably, DTI has not yet found an established place in neuroeconomics. However, there is no reason why this should not be the case in times to come. Even higher cognitive functions have already been shown to be at least partly constituted by white matter, which makes DTI a promising tool in behavioral psychology. It is our hope that this chapter will encourage neuroeconomists to use DTI in their research.

References

Abe O, Aoki S, Hayashi N, Yamada H, Kunimatsu A, Mori H, Ohtomo K (2002) Normal aging in the central nervous system: quantitative MR diffusion-tensor analysis. Neurobiol Aging 23 (3):433–441

Ainslie G (1975) Specious reward: a behavioral theory of impulsiveness and impulse control. Psychol Bull 82(4):463–496

Anwander A, Tittgemeyer M, von Cramon DY, Friederici AD, Knosche TR (2007) Connectivity-based parcellation of Broca's area. Cereb Cortex 17(4):816–825. doi:10.1093/cercor/bhk034

Arain M, Haque M, Johal L, Mathur P, Nel W, Rais A, Sharma S (2013) Maturation of the adolescent brain. Neuropsychiatr Dis Treat 9:449–461. doi:10.2147/NDT.S39776

Ashburner J, Friston KJ (2000) Voxel-based morphometry—the methods. Neuroimage 11(6 Pt 1):805–821. doi:10.1006/nimg.2000.0582

Assaf Y, Pasternak O (2008) Diffusion tensor imaging (DTI)-based white matter mapping in brain research: a review. J Mol Neurosci 34(1):51–61. doi:10.1007/s12031-007-0029-0

Azevedo FA, Carvalho LR, Grinberg LT, Farfel JM, Ferretti RE, Leite RE, Herculano-Houzel S (2009) Equal numbers of neuronal and nonneuronal cells make the human brain an isometrically scaled-up primate brain. J Comp Neurol 513(5):532–541. doi:10.1002/cne.21974

Bach DR, Behrens TE, Garrido L, Weiskopf N, Dolan RJ (2011) Deep and superficial amygdala nuclei projections revealed in vivo by probabilistic tractography. J Neurosci 31(2):618–623. doi:10.1523/JNEUROSCI.2744-10.2011

Bartzokis G (2004) Age-related myelin breakdown: a developmental model of cognitive decline and Alzheimer's disease. Neurobiol Aging 25(1):5–18; author reply 49–62

Bartzokis G, Lu PH, Tingus K, Mendez MF, Richard A, Peters DG, Mintz J (2010) Lifespan trajectory of myelin integrity and maximum motor speed. Neurobiol Aging 31(9):1554–1562. doi:10.1016/j.neurobiolaging.2008.08.015

Basser PJ, Mattiello J, LeBihan D (1994a) Estimation of the effective self-diffusion tensor from the NMR spin echo. J Magn Reson B 103(3):247–254

Basser PJ, Mattiello J, LeBihan D (1994b) MR diffusion tensor spectroscopy and imaging. Biophys J 66(1):259–267

Basser PJ, Pajevic S, Pierpaoli C, Duda J, Aldroubi A (2000) In vivo fiber tractography using DT-MRI data. Magn Reson Med 44(4):625–632

Beaulieu C (2002) The basis of anisotropic water diffusion in the nervous system—a technical review. NMR Biomed 15(7–8):435–455. doi:10.1002/nbm.782

Beaulieu C, Allen PS (1994) Determinants of anisotropic water diffusion in nerves. Magn Reson Med 31(4):394–400

Beckmann M, Johansen-Berg H, Rushworth MF (2009) Connectivity-based parcellation of human cingulate cortex and its relation to functional specialization. J Neurosci 29(4):1175–1190. doi:10.1523/JNEUROSCI.3328-08.2009

Behrens TE, Johansen-Berg H (2005) Relating connectional architecture to grey matter function using diffusion imaging. Philos Trans R Soc Lond B Biol Sci 360(1457):903–911. doi:10.1098/rstb.2005.1640

Behrens TE, Johansen-Berg H, Woolrich MW, Smith SM, Wheeler-Kingshott CA, Boulby PA, Matthews PM (2003a) Non-invasive mapping of connections between human thalamus and cortex using diffusion imaging. Nat Neurosci 6(7):750–757. doi:10.1038/nn1075

Behrens TE, Woolrich MW, Jenkinson M, Johansen-Berg H, Nunes RG, Clare S, Smith SM (2003b) Characterization and propagation of uncertainty in diffusion-weighted MR imaging. Magn Reson Med 50(5):1077–1088. doi:10.1002/mrm.10609

Behrens TE, Berg HJ, Jbabdi S, Rushworth MF, Woolrich MW (2007) Probabilistic diffusion tractography with multiple fibre orientations: what can we gain? Neuroimage 34(1):144–155. doi:10.1016/j.neuroimage.2006.09.018, S1053-8119(06)00936-0 [pii]

Bengtsson SL, Nagy Z, Skare S, Forsman L, Forssberg H, Ullen F (2005) Extensive piano practicing has regionally specific effects on white matter development. Nat Neurosci 8 (9):1148–1150. doi:10.1038/nn1516, nn1516 [pii]

Bennett CM, Wolford GL, Miller MB (2009) The principled control of false positives in neuroimaging. Soc Cogn Affect Neurosci 4(4):417–422. doi:10.1093/scan/nsp053

Blaschek A, Keeser D, Muller S, Koerte IK, Sebastian Schroder A, Muller-Felber W, Ertl-Wagner B (2013) Early white matter changes in childhood multiple sclerosis: a diffusion tensor imaging study. AJNR Am J Neuroradiol 34(10):2015–2020. doi:10.3174/ajnr.A3581

Blits KC (1999) Aristotle: form, function, and comparative anatomy. Anat Rec 257(2):58–63

Budde MD, Kim JH, Liang HF, Schmidt RE, Russell JH, Cross AH, Song SK (2007) Toward accurate diagnosis of white matter pathology using diffusion tensor imaging. Magn Reson Med 57(4):688–695. doi:10.1002/mrm.21200

Catani M (2007) From hodology to function. Brain 130(Pt 3):602–605. doi:10.1093/brain/awm008

Catani M, Thiebaut de Schotten M (2008) A diffusion tensor imaging tractography atlas for virtual in vivo dissections. Cortex 44(8):1105–1132. doi:10.1016/j.cortex.2008.05.004, S0010-9452(08)00123-8 [pii]

Catani M, Howard RJ, Pajevic S, Jones DK (2002) Virtual in vivo interactive dissection of white matter fasciculi in the human brain. Neuroimage 17(1):77–94

Caviness VS Jr, Meyer J, Makris N, Kennedy DN (1996) MRI-based topographic parcellation of human neocortex: an anatomically specified method with estimate of reliability. J Cogn Neurosci 8(6):566–587. doi:10.1162/jocn.1996.8.6.566

Chiang MC, Barysheva M, Shattuck DW, Lee AD, Madsen SK, Avedissian C, Thompson PM (2009) Genetics of brain fiber architecture and intellectual performance. J Neurosci 29 (7):2212–2224. doi:10.1523/JNEUROSCI.4184-08.2009

Chiang MC, McMahon KL, de Zubicaray GI, Martin NG, Hickie I, Toga AW, Thompson PM (2011) Genetics of white matter development: a DTI study of 705 twins and their siblings aged 12 to 29. Neuroimage 54(3):2308–2317. doi:10.1016/j.neuroimage.2010.10.015

Chung HW, Chou MC, Chen CY (2011) Principles and limitations of computational algorithms in clinical diffusion tensor MR tractography. AJNR Am J Neuroradiol 32(1):3–13. doi:10.3174/ajnr.A2041

Cohen MX, Schoene-Bake JC, Elger CE, Weber B (2009) Connectivity-based segregation of the human striatum predicts personality characteristics. Nat Neurosci 12(1):32–34. doi:10.1038/nn.2228, nn.2228 [pii]

Crick F, Jones E (1993) Backwardness of human neuroanatomy. Nature 361(6408):109–110. doi:10.1038/361109a0

Dale AM, Fischl B, Sereno MI (1999) Cortical surface-based analysis. I. Segmentation and surface reconstruction. Neuroimage 9(2):179–194. doi:10.1006/nimg.1998.0395, S1053-8119(98)90395-0 [pii]

Dauguet J, Peled S, Berezovskii V, Delzescaux T, Warfield SK, Born R, Westin CF (2007) Comparison of fiber tracts derived from in-vivo DTI tractography with 3D histological neural tract tracer reconstruction on a macaque brain. Neuroimage 37(2):530–538. doi:10.1016/j.neuroimage.2007.04.067

Dayan E, Cohen LG (2011) Neuroplasticity subserving motor skill learning. Neuron 72(3):443–454. doi:10.1016/j.neuron.2011.10.008, S0896-6273(11)00918-4 [pii]

Dejerine J, Dejerine-Klumpke AM (1895) Anatomie Des Centres Nerveux. Rueff et Cie, Paris

Descoteaux M, Deriche R, Knosche TR, Anwander A (2009) Deterministic and probabilistic tractography based on complex fibre orientation distributions. IEEE Trans Med Imaging 28 (2):269–286. doi:10.1109/TMI.2008.2004424

Drachman DA (2005) Do we have brain to spare? Neurology 64(12):2004–2005. doi:10.1212/01.WNL.0000166914.38327.BB

Dubey P, Pandey S, Moonis G (2013) Acute stroke imaging: recent updates. Stroke Res Treat 2013:767212. doi:10.1155/2013/767212

Dyrby TB, Sogaard LV, Parker GJ, Alexander DC, Lind NM, Baare WF, Jelsing J (2007) Validation of in vitro probabilistic tractography. Neuroimage 37(4):1267–1277. doi:10.1016/j.neuroimage.2007.06.022

Einstein A (1905) Über die von der molecularkinetischen Theorie der Wärme geforderte Bewegung von in ruhenden Flüssigkeiten suspendierten Teilchen. Ann Phys 17:549–569

Engvig A, Fjell AM, Westlye LT, Moberget T, Sundseth O, Larsen VA, Walhovd KB (2012) Memory training impacts short-term changes in aging white matter: a longitudinal diffusion tensor imaging study. Hum Brain Mapp 33(10):2390–2406. doi:10.1002/hbm.21370

Evans AC, Janke AL, Collins DL, Baillet S (2012) Brain templates and atlases. Neuroimage 62 (2):911–922. doi:10.1016/j.neuroimage.2012.01.024

ffytche DH, Catani M (2005) Beyond localization: from hodology to function. Philos Trans R Soc Lond B Biol Sci 360(1456):767–779. doi:10.1098/rstb.2005.1621

Fischl B, Sereno MI, Dale AM (1999) Cortical surface-based analysis. II: inflation, flattening, and a surface-based coordinate system. Neuroimage 9(2):195–207. doi:10.1006/nimg.1998.0396, S1053-8119(98)90396-2 [pii]

Fischl B, Salat DH, Busa E, Albert M, Dieterich M, Haselgrove C, Dale AM (2002) Whole brain segmentation: automated labeling of neuroanatomical structures in the human brain. Neuron 33 (3):341–355

Floel A, de Vries MH, Scholz J, Breitenstein C, Johansen-Berg H (2009) White matter integrity in the vicinity of Broca's area predicts grammar learning success. Neuroimage 47(4):1974–1981. doi:10.1016/j.neuroimage.2009.05.046

Friston KJ, Worsley KJ, Frackowiak RSJ, Mazziotta JC, Evans AC (1994) Assessing the significance of focal activations using their spatial extent. Hum Brain Mapp (1):214–220

Fuchs S (2009) Some aspects of speech and the brain. Peter Lang, Frankfurt am Main; New York

Gao W, Lin W, Chen Y, Gerig G, Smith JK, Jewells V, Gilmore JH (2009) Temporal and spatial development of axonal maturation and myelination of white matter in the developing brain. AJNR Am J Neuroradiol 30(2):290–296. doi:10.3174/ajnr.A1363

Geyer S, Matelli M, Luppino G, Zilles K (2000) Functional neuroanatomy of the primate isocortical motor system. Anat Embryol (Berl) 202(6):443–474

Gonzalez RG (2012) Clinical MRI of acute ischemic stroke. J Magn Reson Imaging 36(2): 259–271. doi:10.1002/jmri.23595

Guimond A, Roche A, Ayache N, Meunier J (2001) Three-dimensional multimodal brain warping using the demons algorithm and adaptive intensity corrections. IEEE Trans Med Imaging 20 (1):58–69. doi:10.1109/42.906425

Gulani V, Webb AG, Duncan ID, Lauterbur PC (2001) Apparent diffusion tensor measurements in myelin-deficient rat spinal cords. Magn Reson Med 45(2):191–195

Hagmann P, Jonasson L, Maeder P, Thiran JP, Wedeen VJ, Meuli R (2006) Understanding diffusion MR imaging techniques: from scalar diffusion-weighted imaging to diffusion tensor imaging and beyond. Radiographics 26(Suppl 1):S205–223. doi:10.1148/rg.26si065510

Hagmann P, Cammoun L, Gigandet X, Meuli R, Honey CJ, Wedeen VJ, Sporns O (2008) Mapping the structural core of human cerebral cortex. PLoS Biol 6(7):e159. doi:10.1371/journal.pbio.0060159

Hasan KM, Parker DL, Alexander AL (2001) Comparison of gradient encoding schemes for diffusion-tensor MRI. J Magn Reson Imaging 13(5):769–780

Hedden T, Lautenschlager G, Park DC (2005) Contributions of processing ability and knowledge to verbal memory tasks across the adult life-span. Q J Exp Psychol A 58(1):169–190. doi:10.1080/02724980443000179

Hollerman JR, Tremblay L, Schultz W (2000) Involvement of basal ganglia and orbitofrontal cortex in goal-directed behavior. Prog Brain Res 126:193–215. doi:10.1016/S0079-6123(00)26015-9

Honea R, Crow TJ, Passingham D, Mackay CE (2005) Regional deficits in brain volume in schizophrenia: a meta-analysis of voxel-based morphometry studies. Am J Psychiatry 162 (12):2233–2245. doi:10.1176/appi.ajp.162.12.2233

Hua K, Zhang J, Wakana S, Jiang H, Li X, Reich DS, Mori S (2008) Tract probability maps in stereotaxic spaces: analyses of white matter anatomy and tract-specific quantification. Neuroimage 39(1):336–347. doi:10.1016/j.neuroimage.2007.07.053

Huston JM, Field AS (2013) Clinical applications of diffusion tensor imaging. Magn Reson Imaging Clin N Am 21(2):279–298. doi:10.1016/j.mric.2012.12.003

Jahanshad N, Lee AD, Barysheva M, McMahon KL, de Zubicaray GI, Martin NG, Thompson PM (2010) Genetic influences on brain asymmetry: a DTI study of 374 twins and siblings. Neuroimage 52(2):455–469. doi:10.1016/j.neuroimage.2010.04.236

Jancke L (2009) The plastic human brain. Restor Neurol Neurosci 27(5):521–538. doi:10.3233/RNN-2009-0519, E1068WQ4LH1354P0 [pii]

Jbabdi S, Behrens TE, Smith SM (2010) Crossing fibres in tract-based spatial statistics. Neuroimage 49(1):249–256. doi:10.1016/j.neuroimage.2009.08.039

Jenkinson M, Smith S (2001) A global optimisation method for robust affine registration of brain images. Med Image Anal 5(2):143–156

Jeurissen B, Leemans A, Tournier JD, Jones DK, Sijbers J (2012) Investigating the prevalence of complex fiber configurations in white matter tissue with diffusion magnetic resonance imaging. Hum Brain Mapp. doi:10.1002/hbm.22099

Johansen-Berg H, Behrens TE (2006) Just pretty pictures? What diffusion tractography can add in clinical neuroscience. Curr Opin Neurol 19(4):379–385

Johansen-Berg H, Rushworth MF (2009) Using diffusion imaging to study human connectional anatomy. Annu Rev Neurosci 32:75–94. doi:10.1146/annurev.neuro.051508.135735

Johansen-Berg H, Behrens TE, Robson MD, Drobnjak I, Rushworth MF, Brady JM, Matthews PM (2004) Changes in connectivity profiles define functionally distinct regions in human medial frontal cortex. Proc Natl Acad Sci USA 101(36):13335–13340. doi:10.1073/pnas.0403743101

Johansen-Berg H, Behrens TE, Sillery E, Ciccarelli O, Thompson AJ, Smith SM, Matthews PM (2005) Functional-anatomical validation and individual variation of diffusion tractography-based segmentation of the human thalamus. Cereb Cortex 15(1):31–39. doi:10.1093/cercor/bhh105

Jones DK, Cercignani M (2010) Twenty-five pitfalls in the analysis of diffusion MRI data. NMR Biomed 23(7):803–820. doi:10.1002/nbm.1543

Jones DK, Simmons A, Williams SC, Horsfield MA (1999) Non-invasive assessment of axonal fiber connectivity in the human brain via diffusion tensor MRI. Magn Reson Med 42(1):37–41

Jones DK, Symms MR, Cercignani M, Howard RJ (2005) The effect of filter size on VBM analyses of DT-MRI data. Neuroimage 26(2):546–554. doi:10.1016/j.neuroimage.2005.02.013

Jones DK, Knosche TR, Turner R (2013) White matter integrity, fiber count, and other fallacies: the do's and don'ts of diffusion MRI. Neuroimage 73:239–254. doi:10.1016/j.neuroimage.2012.06.081

Keller TA, Just MA (2009) Altering cortical connectivity: remediation-induced changes in the white matter of poor readers. Neuron 64(5):624–631. doi:10.1016/j.neuron.2009.10.018

Kinoshita Y, Ohnishi A, Kohshi K, Yokota A (1999) Apparent diffusion coefficient on rat brain and nerves intoxicated with methylmercury. Environ Res 80(4):348–354. doi:10.1006/enrs.1998.3935

Klawiter EC, Schmidt RE, Trinkaus K, Liang HF, Budde MD, Naismith RT, Benzinger TL (2011) Radial diffusivity predicts demyelination in ex vivo multiple sclerosis spinal cords. Neuroimage 55(4):1454–1460. doi:10.1016/j.neuroimage.2011.01.007

Koch MA, Norris DG, Hund-Georgiadis M (2002) An investigation of functional and anatomical connectivity using magnetic resonance imaging. Neuroimage 16(1):241–250. doi:10.1006/nimg.2001.1052

Kochunov P, Williamson DE, Lancaster J, Fox P, Cornell J, Blangero J, Glahn DC (2012) Fractional anisotropy of water diffusion in cerebral white matter across the lifespan. Neurobiol Aging 33(1):9–20. doi:10.1016/j.neurobiolaging.2010.01.014

Landi SM, Baguear F, Della-Maggiore V (2011) One week of motor adaptation induces structural changes in primary motor cortex that predict long-term memory one year later. J Neurosci 31 (33):11808–11813. doi:10.1523/JNEUROSCI.2253-11.2011

Langer N, Hanggi J, Muller NA, Simmen HP, Jancke L (2012) Effects of limb immobilization on brain plasticity. Neurology 78(3):182–188. doi:10.1212/WNL.0b013e31823fcd9c

Lawes INC, Barrick TR, Murugam V, Spierings, Natalia, Evans DR, Song M, Clark CA (2008) Atlas-based segmentation of white matter tracts of the human brain using diffusion tensor tractography and comparison with classical dissection. Neuroimage 39(1):62–79

Le Bihan D, Johansen-Berg H (2012) Diffusion MRI at 25: exploring brain tissue structure and function. Neuroimage 61(2):324–341. doi:10.1016/j.neuroimage.2011.11.006

Le Bihan D, Breton E, Lallemand D, Grenier P, Cabanis E, Laval-Jeantet M (1986) MR imaging of intravoxel incoherent motions: application to diffusion and perfusion in neurologic disorders. Radiology 161(2):401–407. doi:10.1148/radiology.161.2.3763909

Lebel C, Gee M, Camicioli R, Wieler M, Martin W, Beaulieu C (2012) Diffusion tensor imaging of white matter tract evolution over the lifespan. Neuroimage 60(1):340–352. doi:10.1016/j.neuroimage.2011.11.094

Lindenberg R, Renga V, Zhu LL, Betzler F, Alsop D, Schlaug G (2010) Structural integrity of corticospinal motor fibers predicts motor impairment in chronic stroke. Neurology 74(4):280–287. doi:10.1212/WNL.0b013e3181ccc6d9, 74/4/280 [pii]

Lisman JE, Grace AA (2005) The hippocampal-VTA loop: controlling the entry of information into long-term memory. Neuron 46(5):703–713. doi:10.1016/j.neuron.2005.05.002

Loring DW, Meador KJ, Allison JD, Pillai JJ, Lavin T, Lee GP, Dave V (2002) Now you see it, now you don't: statistical and methodological considerations in fMRI. Epilepsy Behav 3 (6):539–547

Marchina S, Zhu LL, Norton A, Zipse L, Wan CY, Schlaug G (2011) Impairment of speech production predicted by lesion load of the left arcuate fasciculus. Stroke 42(8):2251–2256. doi:10.1161/STROKEAHA.110.606103

McClure SM, Laibson DI, Loewenstein G, Cohen JD (2004) Separate neural systems value immediate and delayed monetary rewards. Science 306(5695):503–507. doi:10.1126/science.1100907

Menke RA, Jbabdi S, Miller KL, Matthews PM, Zarei M (2010) Connectivity-based segmentation of the substantia nigra in human and its implications in Parkinson's disease. Neuroimage 52 (4):1175–1180. doi:10.1016/j.neuroimage.2010.05.086

Mesulam M (2005) Imaging connectivity in the human cerebral cortex: the next frontier? Ann Neurol 57(1):5–7. doi:10.1002/ana.20368

Mori S, Barker PB (1999) Diffusion magnetic resonance imaging: its principle and applications. Anat Rec 257(3):102–109

Mori S, van Zijl PC (2002) Fiber tracking: principles and strategies—a technical review. NMR Biomed 15(7–8):468–480. doi:10.1002/nbm.781

Mori S, Crain BJ, Chacko VP, van Zijl PC (1999) Three-dimensional tracking of axonal projections in the brain by magnetic resonance imaging. Ann Neurol 45(2):265–269

Moseley ME, Cohen Y, Kucharczyk J, Mintorovitch J, Asgari HS, Wendland MF, Norman D (1990) Diffusion-weighted MR imaging of anisotropic water diffusion in cat central nervous system. Radiology 176(2):439–445

Mukherjee P, Miller JH, Shimony JS, Conturo TE, Lee BC, Almli CR, McKinstry RC (2001) Normal brain maturation during childhood: developmental trends characterized with diffusion-tensor MR imaging. Radiology 221(2):349–358. doi:10.1148/radiol.2212001702

Mukherjee P, Chung SW, Berman JI, Hess CP, Henry RG (2008) Diffusion tensor MR imaging and fiber tractography: technical considerations. AJNR Am J Neuroradiol 29(5):843–852. doi:10.3174/ajnr.A1052

Munoz Maniega S, Bastin ME, Armitage PA, Farrall AJ, Carpenter TK, Hand PJ, Wardlaw JM (2004) Temporal evolution of water diffusion parameters is different in grey and white matter in human ischaemic stroke. J Neurol Neurosurg Psychiatry 75(12):1714–1718. doi:10.1136/jnnp.2003.033852

Noctor SC, Martinez-Cerdeno V, Kriegstein AR (2007) Contribution of intermediate progenitor cells to cortical histogenesis. Arch Neurol 64(5):639–642. doi:10.1001/archneur.64.5.639

Olson EA, Hooper CJ, Collins P, Luciana M (2007) Adolescents' performance on delay and probability discounting tasks: contributions of age, intelligence, executive functioning, and self-reported externalizing behavior. Pers Individ Dif 43(7):1886–1897. doi:10.1016/j.paid.2007.06.016

Olson EA, Collins PF, Hooper CJ, Muetzel R, Lim KO, Luciana M (2009) White matter integrity predicts delay discounting behavior in 9- to 23-year-olds: a diffusion tensor imaging study. J Cogn Neurosci 21(7):1406–1421. doi:10.1162/jocn.2009.21107

O'Sullivan M, Jones DK, Summers PE, Morris RG, Williams SC, Markus HS (2001) Evidence for cortical "disconnection" as a mechanism of age-related cognitive decline. Neurology 57 (4):632–638

Pajevic S, Pierpaoli C (1999) Color schemes to represent the orientation of anisotropic tissues from diffusion tensor data: application to white matter fiber tract mapping in the human brain. Magn Reson Med 42(3):526–540

Papadakis NG, Murrills CD, Hall LD, Huang CL, Adrian Carpenter T (2000) Minimal gradient encoding for robust estimation of diffusion anisotropy. Magn Reson Imaging 18(6):671–679

Passingham RE, Stephan KE, Kotter R (2002) The anatomical basis of functional localization in the cortex. Nat Rev Neurosci 3(8):606–616. doi:10.1038/nrn893

Perrin M, Poupon C, Rieul B, Leroux P, Constantinesco A, Mangin JF, Lebihan D (2005) Validation of q-ball imaging with a diffusion fibre-crossing phantom on a clinical scanner. Philos Trans R Soc Lond B Biol Sci 360(1457):881–891. doi:10.1098/rstb.2005.1650

Pierpaoli C, Basser PJ (1996) Toward a quantitative assessment of diffusion anisotropy. Magn Reson Med 36(6):893–906

Pierpaoli C, Jezzard P, Basser PJ, Barnett A, Di Chiro G (1996) Diffusion tensor MR imaging of the human brain. Radiology 201(3):637–648. doi:10.1148/radiology.201.3.8939209

Rüber T, Schlaug G, Lindenberg R (2012) Compensatory role of the cortico-rubro-spinal tract in motor recovery after stroke. Neurology 79(6):515–522. doi:10.1212/WNL.0b013e31826356e8

Salat DH, Greve DN, Pacheco JL, Quinn BT, Helmer KG, Buckner RL, Fischl B (2009) Regional white matter volume differences in nondemented aging and Alzheimer's disease. Neuroimage 44(4):1247–1258. doi:10.1016/j.neuroimage.2008.10.030

Salthouse TA (2000) Aging and measures of processing speed. Biol Psychol 54(1–3):35–54

Salthouse TA (2009) When does age-related cognitive decline begin? Neurobiol Aging 30(4):507–514. doi:10.1016/j.neurobiolaging.2008.09.023

Schaechter JD, Fricker ZP, Perdue KL, Helmer KG, Vangel MG, Greve DN, Makris N (2009) Microstructural status of ipsilesional and contralesional corticospinal tract correlates with motor skill in chronic stroke patients. Hum Brain Mapp 30(11):3461–3474. doi:10.1002/hbm.20770

Schellinger PD, Bryan RN, Caplan LR, Detre JA, Edelman RR, Jaigobin C, Technology Assessment Subcommittee of the American Academy of, Neurology (2010) Evidence-based guideline: the role of diffusion and perfusion MRI for the diagnosis of acute ischemic stroke: report of the Therapeutics and Technology Assessment Subcommittee of the American Academy of Neurology. Neurology 75(2):177-185. doi:10.1212/WNL.0b013e3181e7c9dd

Schlaug G, Marchina S, Norton A (2009) Evidence for plasticity in white-matter tracts of patients with chronic Broca's aphasia undergoing intense intonation-based speech therapy. Ann N Y Acad Sci 1169:385–394. 10.1111/j.1749-6632.2009.04587.x, doi:NYAS04587 [pii]

Scholz J, Klein MC, Behrens TE, Johansen-Berg H (2009) Training induces changes in white-matter architecture. Nat Neurosci 12(11):1370–1371. doi:10.1038/nn.2412, nn.2412 [pii]

Schubotz RI, Anwander A, Knosche TR, von Cramon DY, Tittgemeyer M (2010) Anatomical and functional parcellation of the human lateral premotor cortex. Neuroimage 50(2):396–408. doi:10.1016/j.neuroimage.2009.12.069

Smith SM, Jenkinson M, Woolrich MW, Beckmann CF, Behrens TE, Johansen-Berg H, Matthews PM (2004) Advances in functional and structural MR image analysis and implementation as FSL. Neuroimage 23(Suppl 1):S208–219. doi:10.1016/j.neuroimage.2004.07.051, S1053-8119(04)00393-3 [pii]

Smith SM, Jenkinson M, Johansen-Berg H, Rueckert D, Nichols TE, Mackay CE, Behrens TE (2006) Tract-based spatial statistics: voxelwise analysis of multi-subject diffusion data. Neuroimage 31(4):1487–1505. doi:10.1016/j.neuroimage.2006.02.024

Snook L, Plewes C, Beaulieu C (2007) Voxel based versus region of interest analysis in diffusion tensor imaging of neurodevelopment. Neuroimage 34(1):243–252. doi:10.1016/j.neuroimage.2006.07.021

Soares JM, Marques P, Alves V, Sousa N (2013) A hitchhiker's guide to diffusion tensor imaging. Front Neurosci 7:1. doi:10.3389/fnins.2013.00031

Song SK, Sun SW, Ramsbottom MJ, Chang C, Russell J, Cross AH (2002) Dysmyelination revealed through MRI as increased radial (but unchanged axial) diffusion of water. Neuroimage 17(3):1429–1436. doi:S105381190291267X [pii]

Song SK, Sun SW, Ju WK, Lin SJ, Cross AH, Neufeld AH (2003) Diffusion tensor imaging detects and differentiates axon and myelin degeneration in mouse optic nerve after retinal ischemia. Neuroimage 20(3):1714–1722. doi:S1053811903004403 [pii]

Sporns O, Tononi G, Kotter R (2005) The human connectome: a structural description of the human brain. PLoS Comput Biol 1(4):e42. doi:10.1371/journal.pcbi.0010042

Stiles J (2011) Brain development and the nature versus nurture debate. Prog Brain Res 189:3–22. doi:10.1016/B978-0-444-53884-0.00015-4, B978-0-444-53884-0.00015-4 [pii]
Sullivan EV, Rohlfing T, Pfefferbaum A (2010) Quantitative fiber tracking of lateral and interhemispheric white matter systems in normal aging: relations to timed performance. Neurobiol Aging 31(3):464–481. doi:10.1016/j.neurobiolaging.2008.04.007
Takahashi M, Hackney DB, Zhang G, Wehrli SL, Wright AC, O'Brien WT, Selzer ME (2002) Magnetic resonance microimaging of intraaxonal water diffusion in live excised lamprey spinal cord. Proc Natl Acad Sci USA 99(25):16192–16196. doi:10.1073/pnas.252249999
Takeuchi H, Sekiguchi A, Taki Y, Yokoyama S, Yomogida Y, Komuro N, Kawashima R (2010) Training of working memory impacts structural connectivity. J Neurosci 30(9):3297–3303. doi:10.1523/JNEUROSCI.4611-09.2010
Taylor DG, Bushell MC (1985) The spatial mapping of translational diffusion coefficients by the NMR imaging technique. Phys Med Biol 30(4):345–349
Tomassini V, Jbabdi S, Klein JC, Behrens TE, Pozzilli C, Matthews PM, Johansen-Berg H (2007) Diffusion-weighted imaging tractography-based parcellation of the human lateral premotor cortex identifies dorsal and ventral subregions with anatomical and functional specializations. J Neurosci 27(38):10259–10269. doi:10.1523/JNEUROSCI.2144-07.2007
Tournier JD, Mori S, Leemans A (2011) Diffusion tensor imaging and beyond. Magn Reson Med 65(6):1532–1556. doi:10.1002/mrm.22924
Tsang JM, Dougherty RF, Deutsch GK, Wandell BA, Ben-Shachar M (2009) Frontoparietal white matter diffusion properties predict mental arithmetic skills in children. Proc Natl Acad Sci USA 106(52):22546–22551. doi:10.1073/pnas.0906094106
Tzourio-Mazoyer N, Landeau B, Papathanassiou D, Crivello F, Etard O, Delcroix N, Joliot M (2002) Automated anatomical labeling of activations in SPM using a macroscopic anatomical parcellation of the MNI MRI single-subject brain. Neuroimage 15(1):273–289. doi:10.1006/nimg.2001.0978
Van Essen DC, Ugurbil K, Auerbach E, Barch D, Behrens TE, Bucholz R, Consortium W, U-Minn HCP (2012) The Human Connectome Project: a data acquisition perspective. Neuroimage 62(4):2222–2231. doi:10.1016/j.neuroimage.2012.02.018
Wakana S, Jiang H, Nagae-Poetscher LM, van Zijl PCM, Mori S (2004) Fiber tract-based atlas of human white matter anatomy. Radiology 230(1):77–87. doi:10.1148/radiol.2301021640
Wernicke C (1874) Der Aphasische Symptomencomplex. Cohn and Weigert, Breslau
Werring DJ, Toosy AT, Clark CA, Parker GJ, Barker GJ, Miller DH, Thompson AJ (2000) Diffusion tensor imaging can detect and quantify corticospinal tract degeneration after stroke. J Neurol Neurosurg Psychiatry 69(2):269–272
Westlye LT, Walhovd KB, Dale AM, Bjornerud A, Due-Tonnessen P, Engvig A, Fjell AM (2010) Life-span changes of the human brain white matter: diffusion tensor imaging (DTI) and volumetry. Cereb Cortex 20(9):2055–2068. doi:10.1093/cercor/bhp280
Wheeler-Kingshott CA, Cercignani M (2009) About "axial" and "radial" diffusivities. Magn Reson Med 61(5):1255–1260. doi:10.1002/mrm.21965
Wittmann M, Leland DS, Paulus MP (2007) Time and decision making: differential contribution of the posterior insular cortex and the striatum during a delay discounting task. Exp Brain Res 179(4):643–653. doi:10.1007/s00221-006-0822-y
Worsley KJ (2005) An improved theoretical P value for SPMs based on discrete local maxima. Neuroimage 28(4):1056–1062. doi:10.1016/j.neuroimage.2005.06.053
Wu JS, Zhou LF, Tang WJ, Mao Y, Hu J, Song YY, Du GH (2007) Clinical evaluation and follow-up outcome of diffusion tensor imaging-based functional neuronavigation: a prospective, controlled study in patients with gliomas involving pyramidal tracts. Neurosurgery 61 (5):935–948; discussion 948–939. doi:10.1227/01.neu.0000303189.80049.ab
Yogarajah M, Focke NK, Bonelli S, Cercignani M, Acheson J, Parker GJ, Duncan JS (2009) Defining Meyer's loop-temporal lobe resections, visual field deficits and diffusion tensor tractography. Brain 132(Pt 6):1656–1668. doi:10.1093/brain/awp114

Zalesky A (2011) Moderating registration misalignment in voxelwise comparisons of DTI data: a performance evaluation of skeleton projection. Magn Reson Imaging 29(1):111–125. doi:10.1016/j.mri.2010.06.027

Zhu LL, Lindenberg R, Alexander MP, Schlaug G (2010) Lesion load of the corticospinal tract predicts motor impairment in chronic stroke. Stroke 41(5):910–915. doi:10.1161/STROKEAHA.109.577023, STROKEAHA.109.577023 [pii]

Zhu Z, Johnson NF, Kim C, Gold BT (2013) Reduced frontal cortex efficiency is associated with lower white matter integrity in aging. Cereb Cortex. doi:10.1093/cercor/bht212

Chapter 23
Molecular Genetics

Martin Reuter, Andrea Felten and Christian Montag

Abstract This chapter presents a concise introduction to molecular genetics. While readers familiar with the field of genetics may find this article trivial, those readers less experienced in the area may consider this a kind of 'refresher', still others may be new to the area, having not heard about this topic since their biology classes at school. Our aim is to provide all of our readers with new knowledge—or at least a different perspective—on molecular genetics, with a particular emphasis on the neuroeconomics framework. Being mindful of the varying levels of familiarity with genetics our readership may have, we want to provide a 'crash course' that starts with the very basics in genetics and concludes with the most recent developments and perspectives in the field. Our own experience from interdisciplinary cooperative projects is that our colleagues from other disciplines, such as economics, often ask for a simple introduction to behavioral genetics. Unfortunately, we were unaware of any reference to date that fits this description. It is for this reason we decided to write this chapter and to ask experts in other neuroscientific fields to do the same in this book for methods like EEG, MRI, etc.

M. Reuter (✉) · A. Felten
Department of Psychology, University of Bonn, Bonn, Germany
e-mail: martin.reuter@uni-bonn-diff.de

A. Felten
e-mail: andrea.felten@uni-bonn-diff.de

M. Reuter · A. Felten
Center for Economics and Neuroscience (CENs), University of Bonn, Bonn, Germany

C. Montag
Department of Molecular Psychology, Institute of Psychology and Education, Ulm University, Ulm, Germany
e-mail: christian.montag@uni-ulm.de

C. Montag
Key Laboratory for NeuroInformation, Center for Information in Medicine, School of Life Science and Technology, University of Electronic Science and Technology of China, Chengdu, People's Republic of China

M. Reuter and C. Montag (eds.), *Neuroeconomics*, Studies in Neuroscience, Psychology and Behavioral Economics, DOI 10.1007/978-3-642-35923-1_23

23.1 Introduction

Molecular genetics is a very broad field encompassing a group of different methods that are applied by different scientific disciplines with diverse research foci. Whereas some researchers are interested in the structure of the genes and the genetic code, others want to unravel how the transcription ("reading") of the genes is influenced or how protein biosynthesis is accomplished. Another focus of research, known as *behavioral genetics,* investigates how genes influence our behavior. The later aspect is relevant for neuroeconomists who are, by definition, interested in the biological basis of human decision making—hence also in behavior.

A large proportion of the population holds a rather naïve view of genes as something obscure, unchangeably fixed before birth, that determines our lives in a fatalistic way. Most people are not aware that genes simply represent a blueprint for constructing all kinds of proteins that are necessary for our organism to function properly. This implies that at every moment of our life genes in different tissues of our body are active in order to produce "gene products" (e.g. proteins, enzymes, hormones, neurotransmitters, etc.). This production process—known as protein biosynthesis—does not work blindly, but is regulated by the demands of our body. In turn, what our body needs largely depends on environmental stimuli. For example, eating a huge portion of spaghetti at lunch demands that carbohydrates must be metabolized. The hormone responsible for this is insulin. Therefore, the external stimulus of having spaghetti for lunch initiates the production of insulin. In our genome, on the short arm of chromosome 11, we have a gene that provides the relevant information on how to produce insulin. This gene is necessary so that our body 'knows' how to build insulin. Our genome is the entirety of our genetic information. This information is mainly present in our genes. The genes are located on the chromosomes.[1] A gene is a distinct part of a chromosome that provides the information for the synthesis of a specific gene product (e.g. insulin). So if we think back to our spaghetti example, the need for insulin initiates certain molecules, referred to as transcription factors that bind to the promoter region of the insulin gene. The promoter is a part of a gene that functions like a switch. If this switch is turned on by the transcription factors, it initiates the reading (transcription) of the gene. Our chromosomes are located in the nucleus of each cell of our body, however, the "factories" that build up the gene products, called ribosomes, are

[1]The human genome consists of 23 pairs of chromosomes, which are divided into gonosomes and autosomes. Gonosomes (the X and the Y chromosomes) refer to the chromosomes defining human sex. The XX genotype marks the female and the XY genotype the male gender. Autosomes are all other chromosomes which are not gonosomes. There is also additional genetic information on mitochondrial DNA.

located in the cytoplasm of the cells. Therefore, the blueprints showing how to produce the gene product (i.e. which amino acids have to be chained up) must be transferred from the nucleus into the cytoplasm.[2] The transfer of information is done by the messenger ribonucleic acid (mRNA), a single stranded copy of the deoxyribonucleic acid (DNA). The DNA is a double-stranded molecule that provides the genetic information within genes. This information is written in a rather simple language, the genetic code. The genetic code consists of only four letters, A (adenine), T (thymine), C (cytosine), G (guanine) that are in fact nucleotides, molecules composed of a nucleobase, a five-carbon sugar and at least one phosphate group that form the DNA. The gene transcription is the process of reading and translating those parts of a gene (exons and introns, the latter spliced out in mature mRNA) into mRNA that are essential for the synthesis of the gene product. Here it is important to know that the DNA consists of two antiparallel strands forming the DNA double helix, stabilized by hydrogen bonds and base stacking: The nucleotide A is always paired with a T and C is always paired with a G (see Fig. 23.1). The transcription of DNA into mRNA means that if there is an A on the DNA the mRNA will contain uracil (U), a G is translated into a C, a C into a G, and a T is translated into A. The structure of uracil is very similar to that of thymine. In the open reading frame of exonic regions, the sequence of three nucleotides codes for a distinct amino acid. Such a triplet of nucleotides/bases is called a codon. With the exception of three stop codons, each codon codes for a distinct amino acid, however, an amino acid can sometimes be coded by different codons.

The short mRNA that conveys the information for the synthesis of one protein at its maximum leaves the nucleus of the cell and migrates into the cytoplasm. At the ribosomes, the information of the mRNA must be translated into a chain of amino acids forming a protein. Here another form of RNA, the transfer RNA (tRNA) comes into play. The tRNA has two functions, first it carries an anticodon that is the complementary nucleotide sequence to a given codon on the mRNA, and second the tRNA picks up the appropriate amino acid in the cytoplasm that is coded by the mRNA. Step by step the codons are translated into amino acid chains until the gene product, i.e., the protein, is built up. In sum, a protein consists of a chain of amino acids and the order of this amino acid chain, is provided by the blueprint that is our genes.

---

[2]Cytoplasm refers to both cytosol and the organelles. Cytosol is the fluid substance in the cell separated from the outer cell wall by the cell membrane. Organelles refer to structures such as ribosomes within the cell.

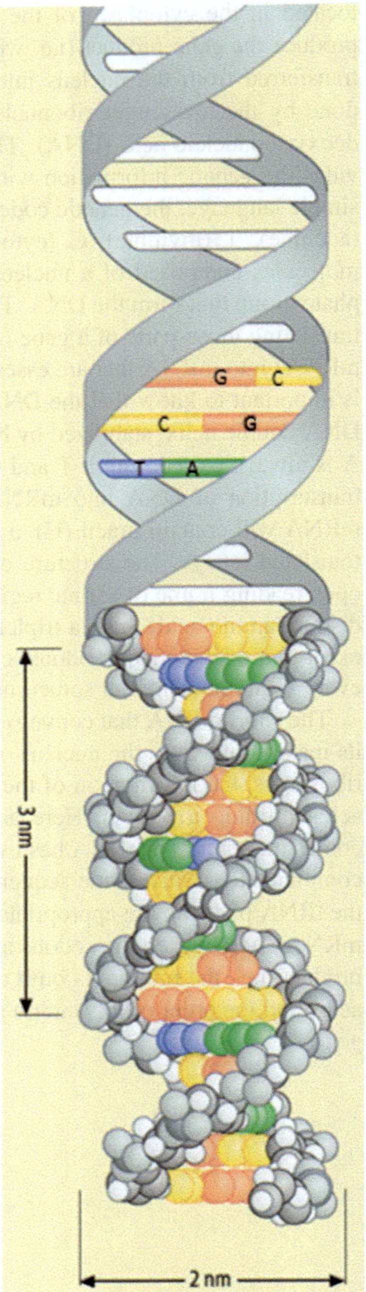

Fig. 23.1 The structure of the DNA double helix as first identified by James Watson and Francis Crick in 1953. The DNA double strand that comes in the form of a wound helix gets its stability by knitting the two single strands together. The nucleotide A is always paired with a T and C is always paired with a G

23.2 Genetic Variability—Polymorphisms

This rough overview of the protein biosynthesis is crucial to our understanding of how genetic variations can cause individual differences in behavior. It must first be mentioned that there is a 99.9 % overlap in the genome of two unrelated humans. This means that most of the sequences of base pairs constituting our genome (a total of 3.2 billion bases/nucleotides/letters) are invariant across individuals! In other words, tiny variations among the remaining 0.1 % (corresponding to 3,000,000 bases) may have huge effects. This great similarity in the genetic code across individuals is the prerequisite for belonging to the same species, i.e., we have comparable metabolic, motor, cognitive, etc., processes. However, the small differences in the DNA sequence across individuals cause the thrilling differences in personality, abilities, and behavior. If two people, let us call them Peter and Bob, met to compare their DNA sequences, they would only detect differences on average in every thousand letters. For example, at position 1003 of a certain gene, Peter has an A and Bob has a T. Adenine (A) and Thymine (T) are called alleles at position 1003. Loci on the DNA where there is no variation in the population are called monomorph and have only one possible allele. However, a gene locus where there is variability is called polymorphism if it occurs at least in 1 % of the population. If the prevalence of such a base exchange is lower than 1 %, it is called "mutation." In other words, polymorphisms are mutations that are rather widespread in a population. A mutation occurs by chance through a deficient self-reproduction of genetic material. Only when such a "mistake" of nature occurs in the gametes they are transferred to the next generation, i.e., they are heritable. The "Out of Africa Hypothesis" postulates that every living human being is descended from a small group we immodestly call *Homo sapiens* ("wise human") in Africa, who then dispersed across the globe and successfully conquered the world (while other forms of the genus *homo* such as *Homo erectus* or the newly discovered *Homo denisova* or *Homo naledi* did not survive to the present day). The migration of *Homo sapiens* started about 70,000 years ago. Since this time, the *Homo sapiens* population has dramatically increased and accordingly, the more humans that exist, the higher is the probability that a "mistake" in DNA reduplication or DNA repair takes place. If an individual inherits a mutation that is not adaptive for survival or reproduction, the probability that this individual will pass this mutation on to the next generation is low. This is the process known as natural selection. Therefore, polymorphisms, which are widely distributed mutations, cannot be completely disastrous. The opposite might even be more accurate—these mutations will bring advantages in certain environmental niches.

There exist different kinds of polymorphisms. By far, the most widespread form is the single nucleotide polymorphism (SNP) that is defined by an alteration in a single nucleotide/base (as in the example of Peter and Bob). The deletion of a base can be fatal if it occurs in a coding region (located in an exon) that is translated into an amino acid sequence. By the deletion of a single base, the reading frame is destroyed, i.e., all ensuing codons are translated incorrectly or the translation

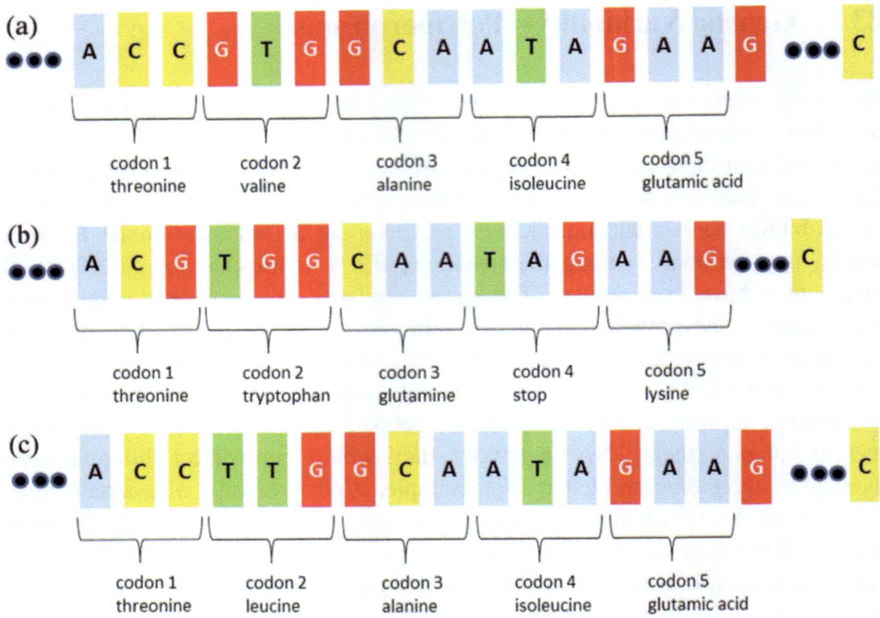

Fig. 23.2 Consequences of a single nucleotide polymorphism (SNP). Depicted is a small sequence from the coding region of a gene. **a** DNA strand with exclusively wild-type alleles; **b** deletion SNP in the third base of codon 1. Although the first amino acid of the polypeptide remains unchanged, because both codons—ACC and ACG—code for the amino acid threonine, the rest of the codons lead to other amino acids. Most severe, the transcription ends with codon 4 because TAG is a stop codon. Lysine is no longer considered. The protein cannot be synthesized **c** a SNP where there is a transition from G to T in the first base of the second codon. Leucine instead of valine is inserted into the polypeptide. Typically, such a SNP does not prevent the production of the protein, but alters its functionality

process ends too early (see Fig. 23.2b in comparison to Fig. 23.2a: a cytosine at position 3 has been deleted). Therefore, deletion polymorphisms in exons that change the reading frame are very rare, because the consequences are so disastrous that the organism can likely not survive without the production of this protein. Base pair transitions as depicted in the comparison of Fig. 23.2a, c (there is a G to T transition at codon No. 2, resulting in production of an alternative amino acid) are rather frequent and mostly do not prevent the production of the protein. However, this base pair transition may have consequences on the structure and functionality of the protein (e.g. reducing its thermostability[3]). SNPs do not only occur in coding regions but can also be observed all over the genome. We have just observed that polymorphisms in the coding region of a gene are very likely to have an effect on the gene product if it is related to an amino acid exchange. An exception is the presence of a silent mutation, as depicted in the comparison of Fig. 23.2b, c. Some

[3]Thermostability describes how stable a certain structure is in the face of rising temperature.

amino acids can be coded by different codons. The codons ACC and ACG both code for threonine (1st codon). Silent mutations are less likely to result in alternative gene products with altered functionality, although this may happen.

Another hot candidate for changes in functionality of a gene is the promoter region. The promoter initiates the transcription of a gene. Polymorphisms in the promoter region of a gene may have an influence on the efficiency of transcription factors to bind to the DNA, disturbing transcription complex assembly, and recruitment of RNA Polymerase II. Therefore, polymorphisms in the promoter region can affect the mRNA expression. Lower levels of mRNA expression are often related to a lower amount of the gene product (e.g. number of expressed neurotransmitter receptors in the brain).

Another form of polymorphism is the variable number of tandem repeats polymorphism (VNTR). Here, a sequence of bases is repeated several times on a gene. Individuals may differ in the number of these repeats. The dopamine transporter polymorphism DAT1-VNTR belongs to this family of polymorphisms. The most common alleles observable are a 9 and a 10 times repeat of a sequence of 40 bases in the 3′ un-translated region (3′UTR) of the dopamine transporter gene (SLC6A3), however, the number of repeats possible can range from 3–11 (Vandenbergh et al. 1992). Although results are controversial, the 10-repeat allele is most commonly related to an increased level of DAT1 expression (Heinz et al. 2000). Within the category of VNTRs which belongs to the simple sequence repeats (SSRs) polymorphism, also known as microsatellite polymorphism or short tandem repeats (STRs). Here, very few bases, normally 2–6 bases, are repeated many times. The SSRs are often analyzed for kinship or population diversity analyses. A prominent example for the role of SSRs for social behavior is the microsatellite polymorphism in the promoter region of the arginine vasopressin 1A receptor (AVPR1A) gene, that influences partnership satisfaction in humans and explains bonding differences between the monogamous prairie vole and the promiscuous meadow vole (Walum et al. 2008; Lim et al. 2004).

Copy-number variations (CNVs) constitute alterations in the genome characterized by a variation in the number of copies of one or more sections of the DNA. CNVs correspond to relatively large regions of the genome that have been deleted (fewer than the normal number of copies) or duplicated (more than the normal number of copies). Reasons for CNVs are incorrect DNA duplications or insertional transpositions (gain in copy number) or deletions (losses of copies), as well as more complex rearrangements of the DNA. CNVs account for about 12 % of human genomic DNA and each variation may range from about one kilobase (1 kb = 1000 bases/nucleotides) to several megabases (1 Mb = 1,000,000 bp) in size. The functionality of a gene can be altered by CNVs through disruption of a regulatory element (e.g. an enhancer or suppressor site) or by causing functional loss of a gene by a deletion. As with traditional polymorphisms, some CNVs do not have phenotypic effects, e.g., the CNV does not result in an observable change at the level of behavior. Meanwhile, some prominent CNVs have been successfully related to psychiatric diseases like schizophrenia, attention-deficit hyperactivity disorder

(ADHD), autism spectrum disorders (ASDs) or Williams–Beuren syndrome (WBS) (for an overview see Cook and Scherer 2008).

23.3 Extraction of DNA

Before we can analyze polymorphisms located on the genome, the DNA has to be extracted. Eukaryotic organisms (humans, animals, plants, fungi, and protists) store most of their DNA inside the nucleus of the cells (genomic DNA) and some of their DNA in organelles (e.g. mitochondria in humans and chloroplasts in plants). Therefore, the cell wall and the cell nucleus must be separated prior to analysis. This is done using an enzyme, proteinase K, in the presence of a chaotropic salt, which immediately inactivates all nucleases (nucleases can degrade the DNA). After cell lysis, the sample is pipetted into a reaction tube containing a filter composed of glass fibers. The DNA binds selectively to the glass fibers in the tube. Bound DNA is purified in a series of rapid "wash-and-spin" steps to remove contaminating cellular components (e.g. organelles). Finally, low salt elution releases the purified DNA from the glass fibers. Today DNA extraction kits containing the filter tubes and all necessary reagents are commercially available. In addition, only pipettes and a centrifuge are necessary to conduct the DNA extraction. In modern labs, DNA extraction is typically performed by robots. These automated DNA extraction machines rely on alternate techniques that do not use filter tubes but instead rely on magnetic beads or vacuum units.

23.4 Genotyping

Genotyping is the process of determining the given alleles at a certain polymorphic locus in our genome. Due to the fact that we have a diploid set of chromosomes, each individual has two alleles at a distinct gene locus. For example, with respect to a SNP characterized by an A → G transition, an individual can have two identical alleles across the two chromosomes (AA or GG) or different alleles (AG) on each chromosome. AA, AG, and GG represent the three possible genotypes at this polymorphic site and the two alleles are named A and G.

The revolutionary invention of the polymerase chain reaction (PCR) by Kary Mullis in 1983 provided the starting point for molecular genetics research by making genotyping possible. The main problem in genotyping a polymorphic region on the genome is how tiny the information is that one is interested in: In the case of a SNP one out of three billion bases is exchanged. Furthermore, we have to keep in mind that the DNA is located in the small nucleus of a cell. In other words, the information we are interested in is unbelievably small. In order to solve this problem we have to amplify the "signal" to enable us to decide whether a certain polymorphic region contains, for example, an A or a G at a particular locus. PCR is conducted by a PCR machine that is simply a copy machine for nucleic acids (DNA or RNA). In brief, the principle of PCR—a series of repetitive chemical reactions running in a reaction tube—is as follows:

At first, you must mark the region on the DNA where the polymorphism is located. This is done by designing two oligonucleotides (a chain of about 17–25 bases), called primers, that are complementary to the original DNA strand. The forward primer binds to the $3' \rightarrow 5'$ strand and the reverse primer to the $5' \rightarrow 3'$ strand. The two primers mark a region of about 200 base pairs, along which the SNP is located (i.e. the target region). The experimenter has to verify that the sequence of the primers—although very small—does not occur anywhere else in the genome, i.e., the primer sequence is specific to the region of interest. The aim of the PCR is to make more than a billion copies of the target region. So if you want to build up new copies you need to insert nucleotides (A, C, T, and G) into your reaction tube. As with other PCR reagents, these synthetic nucleotides are also commercially available from biotech companies. Reduplication of DNA is a process that is constantly ongoing in the living organism. PCR imitates this process. The crucial ingredient of a PCR reaction mix is a polymerase that does the amplification process. Most common is the *Taq* polymerase.[4] Finally, you must put a few ng/µl of DNA (the template) into the tube. The PCR contains the repetition of at least 30 PCR cycles. Each PCR cycle consists of three phases; denaturation, annealing, and elongation, each requiring a different temperature. During denaturation the temperature is raised to 95 °C to divide the two complementary DNA strands. Once the DNA strands are separated, the primers can anneal to the single DNA strands at a temperature of about 60 °C. Note, the exact annealing temperature can vary depending on the PCR protocol used and the sequence of the bases and primers in the target region. Then the temperature is raised up to 72 °C allowing the *Taq* polymerase to work.[5] The *Taq* now elongates its own sequence by incorporating the synthetic nucleotides previously added to the PCR reaction mix. The PCR starts with double-stranded template DNA. After the first PCR cycle we have two copies, after the second cycle four copies, after the third cycle eight copies, and so forth (see Fig. 23.3). The number of PCR copies per PCR run is calculated as 2^n, with n indicating the number of PCR cycles. Therefore, a PCR run with 30 cycles would result in over 1 billion copies of the target region. The aim of a PCR is to amplify the signal of the polymorphic regions by producing billions of copies. However, a conventional PCR does not provide the genotype information, but only creates the prerequisites for genotyping. Genotyping is achieved by putting

[4]This thermostable enzyme called *Thermus aquaticus* (*Taq*) has not been "invented" by a scientist, but has been extracted naturally from a hot thermal spring called *Morning Glory Pool* in the US national park Yellowstone.

[5]The great variance of temperatures in a PCR explain why a thermostable enzyme is needed and scientists are lucky to have found such an enzyme in nature.

Fig. 23.3 Schematic overview on polymerase chain reaction (PCR). **a** The starting point is a double-stranded DNA template. The lens glass highlights the region on the genome where a single nucleotide polymorphism (SNP; *red letter*) is located. It is apparent that the two DNA strands are complementary to each other (a G is paired with a C and an A is paired with a T). **b** The DENATURATION phase is portrayed. Heating the reaction tube up to 95 °C separates the two complementary DNA strands. **c** The ANNEALING phase is depicted. Lowering the temperature to about 60 °C allows the primers to bind to the single-stranded DNA. The bases of the primers are complementary to the DNA strands. **d** The ELONGATION phase is portrayed. At 72 °C the *Taq* polymerase can elongate the primers by incorporating synthetic nucleotides available in the reaction tube. **e** The status after the first PCR cycle is depicted. A double-stranded DNA template has been copied and results in a double-stranded DNA

the DNA amplicon (resulting from the described PCR process) on a gel-electrophoresis unit. If the polymorphism constitutes a VNTR, this indicates that the alleles differ with respect to the amplified fragment length. For example, the 9 repeat allele of the dopamine transporter polymorphism DAT1 is 40 bases shorter than the 10 repeat alleles. Using an electric field, shorter PCR amplicons move faster through the matrix of the electrophoresis gel (often made out of agarose) than longer amplicons. By labeling the PCR product with a dye (typically ethidium bromide), the different bands on the gel can be visualized later by means of a UV light. Samples of subjects who are heterozygous exhibit two bands, one upper band indicates homozygosity for the long allele and one lower band indicates homozygosity for the short allele (see Fig. 23.4). Of note, the direction of flow of the DNA fragments in Fig. 23.4 goes from the pockets to the bottom of the picture. In order to use gel-electrophoresis for genotyping SNPs (the PCR amplicons do not differ in size if there is a base exchange at a certain gene locus) the PCR amplicon has to be treated with an enzyme digest. Here, a specific restriction enzyme is able to cut the DNA strand into two shorter pieces in the presence of a certain allele (e.g. the G allele), but does not affect the DNA in the presence of the alternate allele (e.g. A). By means of the enzyme digest we yield two different strand lengths dependent on the absence or presence of the respective allele and this can be visualized/genotyped by means of gel-electrophoresis (this works similar to the mechanism described in Fig. 23.4). Today, more elaborate techniques like Real-Time (RT) PCR are available that allow amplification and genotyping in one PCR run. The advantage of RT-PCR is that the time-consuming gel-electrophoresis is no longer necessary. In brief, during the RT-PCR amplification additional oligonucleotides labeled with fluorescent dyes (the so-called 'probes') are annealed to the target region. The probes are either complementary to the wild type (the typical allele) or to the mutant (non-standard) allele and by means of a melting curve analysis following the last PCR cycle, the genotypes can be determined with perfect reliability (for a detailed description of Real-Time PCR see Reuter et al. 2005). A disadvantage of RT-PCR is that mainly SNPs can be genotyped with this method.

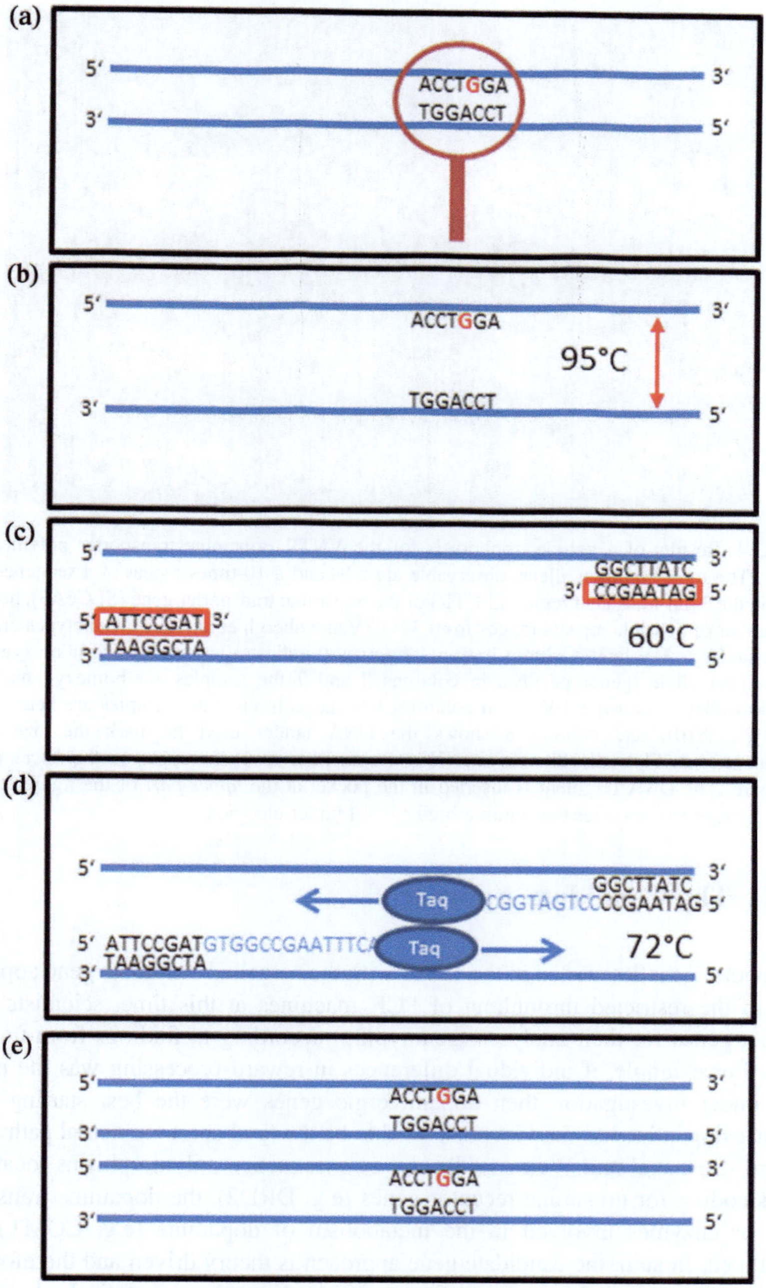
(a)
5′ 3′
ACCTGGA
TGGACCT
3′ 5′
(b)
5′ 3′
ACCTGGA
95°C
TGGACCT
3′ 5′
(c)
5′ 3′
GGCTTATC
3′ CCGAATAG 5′
5′ ATTCCGAT 3′
60°C
3′ TAAGGCTA 5′
(d)
5′ 3′
GGCTTATC
Taq
CGGTAGTCCCCGAATAG 5′
5′ ATTCCGATGTGGCCGAATTTCA
Taq
72°C
3′ TAAGGCTA 5′
(e)
5′ ACCTGGA 3′
3′ TGGACCT 5′
5′ ACCTGGA 3′
3′ TGGACCT 5′

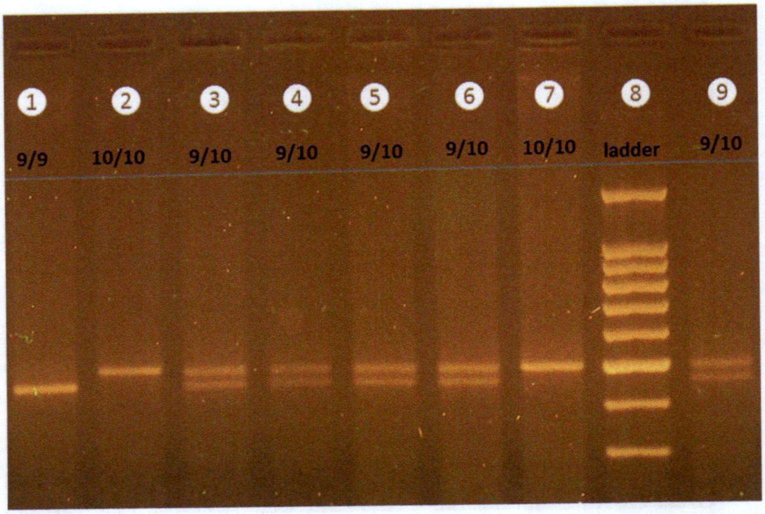

Fig. 23.4 Results of a gel-electrophoresis for the VNTR dopamine transporter polymorphism DAT1. The most common alleles observable are a 9- and a 10-times repeat of a sequence of 40 bases in the 3′ un-translated region (3′UTR) of the dopamine transporter gene (SLC6A3), however, the number of possible repeats ranges from 3–11 (Vandenbergh et al. 1992). Genotypes are made visible under an UV light. Column 1 (*from left to right*) indicates a sample being homozygous for the 9-repeat allele (genotype 9/9), in columns 2 and 7 the samples are homozygous for the 10-repeat allele (genotype 10/10), in columns 3–6 and column 9 the samples are heterozygous (genotype 9/10), and column 8 shows the DNA ladder used to mark the size of the PCR-amplicons. Flow direction of the DNA fragment is from the *upper* to the *lower part* of the figure. The DNA fragment is inserted in the pocket at the *upper part* of the figure and then floats through the gel after the running buffer is set under electricity

23.5 The Candidate Approach

Behavioral genetics started in the 1990s with the so-called candidate gene approach. Due to the restricted throughput of PCR machines at this time, scientists chose distinct genes for their analyses (genotyping) according to findings from the literature. For example, if individual differences in reward processing was the phenotype under investigation then dopaminergic genes were the best starting point, because dopamine has (and is) postulated to be the final common neural pathway of reward (Spanagel and Weiss 1999). As a consequence polymorphisms located on genes coding for dopamine receptor genes (e.g. DRD2), the dopamine transporter gene or enzymes involved in the metabolism of dopamine (e.g. COMT) were genotyped. In sum, the candidate gene approach is theory driven and therefore it is a deductive method (Montag and Reuter 2014). Nowadays, technical advances in PCR machines and PCR product analyzers make it possible to analyze many essential SNPs across whole genes in reasonable time.

23.6 Sequencing

Sequencing is a technique that allows reading of the original genetic code, i.e., through sequencing the precise order of bases within a DNA or RNA molecule can be determined. In contrast, genotyping only allows the determination of genotypes at a particular polymorphic region. In 1977, Frederick Sanger and colleagues developed a sequencing technique, the Sanger sequencing method, popularly used for over 25 years and which is still in use today. This method is based on the selective incorporation of chain-terminating dideoxynucleotides (ddGTP, ddATP, ddTTP, and ddCTP) by DNA polymerase during in vitro PCR. This means that after a dideoxynucleotide has been inserted in the nucleotide chain during the elongation process (see PCR), no further nucleotide can be incorporated and amplification is terminated. To get information on the order of the nucleotide sequence of a distinct DNA strand, the experimenter has to run four separate PCR reactions. Each reaction contains all four deoxynucleotides (dATP, dGTP, dCTP and dTTP—i.e. one dNTP for each base A, G, C and T), but only one of the four dideoxynucleotides. The results of the DNA amplification is made visible by putting the PCR products of the four PCR runs on separate lanes on an electrophoresis gel (see Fig. 23.5, left).

Sequencing was revolutionized by the invention of the dye-terminator sequencing method, which is currently used by high-throughput DNA sequence analyzers. By labeling the ddNTPs with fluorescent dyes, like in RT-PCR analysis, sequencing depends on just one rather than four PCR reactions. The fluorescent dyes emit light at different wavelengths that are measured by a sequencer machine (see Fig. 23.5, right).

23.7 Genome Wide Scan Technique

Besides following the already described candidate gene approach, genome wide scans (GWS) are often conducted to grasp as much variation as possible across the complete DNA to spot genetic variants associated with a phenotype of interest. For example, if researchers are interested in identifying which genetic variants are responsible for anxiety, they could draw participants from a large genetic databank based on their anxiety scores gathered via self-report (e.g. high and low anxiety scores). In a second step, researchers test the frequency of millions of allelic variants in the anxious versus the non-anxious group: significant higher/lower statistical prevalence rates of a particular SNP in the comparison of low and high anxious humans could be seen as an association with anxious behavior.

GWSs conducted to cover genetic variations across the genome which differs in terms of fine granularity. The greater coverage of the DNA required, the more expensive the procedures are. If the researchers aim for a complete coverage of genetic variation of a single person, the complete genome must be sequenced. This

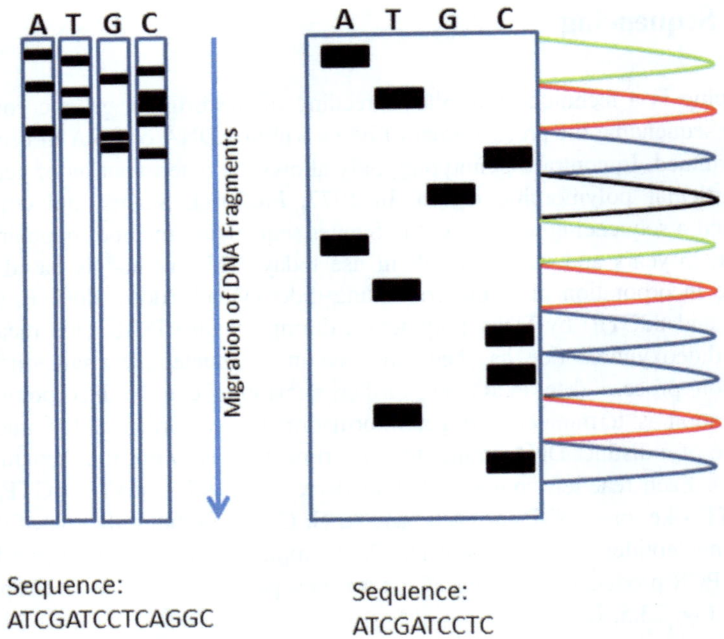

Fig. 23.5 Schematic overview of sequencing results. *Left* Sanger sequencing. For each of the four nucleotides (A, T, G, C) a separate analysis has to be run. The product of each reaction is spotted in a separate lane on an electrophoresis gel. The nucleic acid molecules are separated by applying an electric field to move the negatively charged molecules through a matrix of agarose or other substances like polyacrylamide. Shorter molecules move faster and migrate further than longer ones, because shorter molecules migrate more easily through the pores of the gel. *Right* Dye-terminator sequencing: The ddNTPs are labeled with different fluorescent dyes, which emit light at different wavelengths and are measured by an optical unit of the sequencing machine

means that basically every base of a person's DNA is read out (see Sequencing). Clearly this is a very costly endeavor, which is usually prohibitively expensive. Therefore, researchers need to choose between a range of chips (e.g. available from companies such as Illumina) giving information on up to five million SNPs of a single participant. Among these SNPs are up to 4.2 million TagSNPs. The first TagSNPs were derived by sequencing the DNA of $N = 270$ participants from four populations in the International HapMap project. In contrast to the major success of having sequenced the complete human genome of two persons in the year 2000, the HapMap project provided for the first time a sufficient number of participants to reliably detect common genetic variants on the human DNA. TagSNPs take advantage of the linkage disequilibrium between SNPs lying closely to each other on the genome. Specifically, SNPs located in adjacent areas are likely to be inherited together. A perfect linkage equilibrium is characterized by $r^2 = 1$. Knowing the genotypes of a series of TagSNPs covering a larger genomic area gives researchers information on which genetic variants to expect without having genotyped every single one of the known SNPs in this area. TagSNPs can be inferred by means of haplotype analyses. These are mathematical algorithms that calculate the linkage between SNPs in a certain gene region on one chromosome.

The newest large international genomic project is called 1kGP and investigates the human genome of a much larger group consisting of $N = 2500$ participants from 25 populations around the globe. By sequencing this large number of participants it will be possible to get more detailed insights into the variability of human DNA, detecting alleles with a minor allele frequency of >1 % in the population. A large number of the aforementioned 4.2 million TagSNPs were detected by this new project. It has been suggested that it is possible to cover up to 35 million SNPs with 5 million TagSNPs. Among the 5 million SNPs covered, about 2 million SNPs can be found in genic regions. The rest of the covered SNPs can be found in intergenic regions—the vast areas lying between genes. These areas have been known as 'junk DNA' for a long time, because researchers thought that intergenic regions are evolutionary relics on our DNA and no longer of importance. This view has changed in recent years, as these intergenic regions appear responsible for much of the regulatory processes of gene activity.

23.8 Matrix Assisted Laser Desorption Ionization—Time of Flight Mass Spectrometry (MALDI-TOF-MS)

MALDI-TOF-MS is a technique that can be used for high-throughput genotyping. However, the throughput, i.e., the number of genotypes assessed per run (about 15,000), is much lower than in GWS studies (GWS; see above). Therefore, the choice of which method to use for genotyping (PCR, Sequencing, MALDI-TOF-MS or GWS) depends on the size and the aim of the planned project. Clearly, the available funding for a project also plays an important rule, because GWS—although providing fine-grained information on each participant—are still more costly than analyzing a number of SNPs of a person by means of MALDI-TOF-MS. Results of GWS studies often come up with only a very small number of polymorphisms that hold the massive correction for multiple testing. Often nothing is known about the gene on which these SNPs are located. The straightforward strategy is now to screen these genes by covering at least all TagSNPs. In addition, a further approach would be the investigation of all relevant genes of a certain neural pathway by means of Tag-SNP analysis. For example, if a researcher is interested in reward processing, he or she could investigate all pathway-based genes of relevance for dopaminergic signaling. Here, MALDI-TOF-MS is the method of choice.

Before conducting a MALDI-TOF-MS analysis there are several preprocessing steps. The scientist has to set up a list of SNPs he/she is interested in. By software supported in silico[6] analysis the primers for the PCR analysis and so-called

[6]In silico means that the designs are conducted on a computer.

extension primers for an ensuing extension reaction have to be designed and tested for compatibility. The best possible result is to "plex" up to 40 different SNPs in one reaction. When using a 384-well PCR plate it is possible to get 40 by 384 (=15,360) genotype results. Before the actual analysis, the preprocessed samples (preprocessing encompasses three different biochemical reactions including PCR) must be spotted by a nanodispenser on a microchip, which is then inserted into the mass spectrometer. Using a laser each well of the microchip is then fired and the molecules fly high through the wind tunnel of the mass spectrometer, which is under a vacuum. The time it takes for the molecules to fly to the top of the wind tunnel (time-of-flight) gives information on the size of the molecules, e.g., smaller molecules fly high faster. Using earlier plex designed in silico, it has been ensured that each of the genotypes of the approximately 40 SNPs has a distinct molecular size. The mass spectrometer is capable of detecting DNA fragments within a range of 4500–9000 Da, and can distinguish between analytes separated by 16 Da or more (analytes are typically 18–25 nucleotides).

23.9 Epigenetics

Genetic research had—and often still has—a bad image among the public. Although people are aware that genetic research can prospectively promote the development of new medical therapies, the fear of the misuse of genetic information remains prevalent in the minds of many. For example, the perception that important phenotypes such as Intelligence, which is proven to be causally related to professional success, are strongly and genetically determined (heritability estimates for intelligence range between 0.50 and 0.70; Bouchard et al. 1990), makes many people uncomfortable, because it means that our own influence on our fate is limited. A strong heritability for psychiatric diseases has also been proven (McGuffin et al. 2001). However, the fact that many disorders, such as depression, are influenced by many genes makes identification of the polymorphisms involved very complicated. The same is true for intelligence and other complex phenotypes, such as those studied in Neuroeconomics. In contrast to monogenetically caused disorders, such as the neurological condition Huntington's disease, a complex concert of hundreds of genes influences the aforementioned psychological phenotypes. It can be assumed that each of the single genetic markers merely represents a genetic risk factor, which only comes into play when additional risk factors—not just genetic ones—are also present. These risk factors can be environmentally conditioned. For example, stress, financial worries, relationship problems, the illness of one's child, etc., can trigger the onset of a psychiatric disease if a genetic risk factor is given.

On the other hand, we know that environmental factors, e.g., treatments, trainings, or therapies can ameliorate or cure a disease. It could be shown that therapy outcome can be dependent on the genetic makeup of an individual. For example, Knuts et al. (2014) reported that a sample of agoraphobia patients those carrying the

5-HTTLPR low expression genotypes had a more favorable response to exposure therapy two weeks following treatment, compared to patients with other genotypes. Using a slightly different approach, Caspi et al. (2003) demonstrated that the risk for depression is dependent on 5-HTTLPR genotype but only in the presence of critical life events. The risk for depression increases only exponentially in carriers of the 5-HTTLPR ss-genotype. If no critical life events are present, carriers of the ss-, the sl-, and the ll-genotype do not differ with respect to their risk for depression. The study by Caspi et al. is one of the first and most prominent studies demonstrating a gene by environment interaction. At this point in time the molecular mechanisms of how environmental factors can influence genes were totally unknown. Interestingly, the term "epigenetics" had already been coined in 1942 by the British developmental biologist Conrad Hal Waddington. At this time even the structure of DNA was still unknown. Epigenetics denotes processes that act outside the regulation of gene expression, but that affect it (Epi stems from an old Greek word meaning "over" or "outside of"). Epigenetics is a field of biology that investigates what factors determine the activity of a gene and thus the development of the cell temporarily and whether some of these changes are inherited by the next generation. In molecular genetics, epigenetic analyzes focus mostly, in addition to histone modifications, on methylation analysis. DNA methylation is a modification of DNA that occurs through the addition of a methyl group to the 5-carbon position of a cytosine ring (see Fig. 23.6a). Notably, DNA methylation only occurs if the cytosine base (C) is followed by a guanine base (G). Adjacent cytosine and guanine bases are knitted together by a phosphodiester bond (p). This is the reason why the sequence of a cytosine-phosphoguanine is designated as CpG-site (see Fig. 23.6b). CpG islands are regions in the human genome where the prevalence of CpG-sites is very high (i.e. >50 %) and the observed-to-expected CpG ratio is greater than 60 %. CpG islands represent about 1 % of the genome (see Fig. 23.6c). A strongly methylated gene has a low or no expression rate. Hypermethylated genes are densely "packed," which prevents the transcription factors from binding. This is the reason why genes that are not expressed in a certain cell type are hypermethylated. For example, the insulin gene is strongly methylated in a neuronal cell in which insulin is not produced. Therefore, cell differentiation is closely related to methylation.

Two methods are widely used to assess the percentage of CpG methylation in a given DNA region: Bisulfite-Sequencing and MALDI-TOF-MS. Both techniques demand bisulfite treated DNA. Treatment of DNA with bisulfite converts cytosine residues to uracil (U), but leaves 5-methylcytosine residues unaffected. In other words, if a CpG site is methylated the 5-methylcytosine is unchanged, whereas in the case of no methylation, a uracil replaces the cytosine base. After the DNA sequence is bisulfite treated, the Bisulfite Sequencing is run as in normal sequencing. However, MALDI-TOF-MS needs a special chemistry to finally allow quantification of the percentage of methylation for discrete CpG sites. However, the principle of MALDI-TOF-MS remains unchanged.

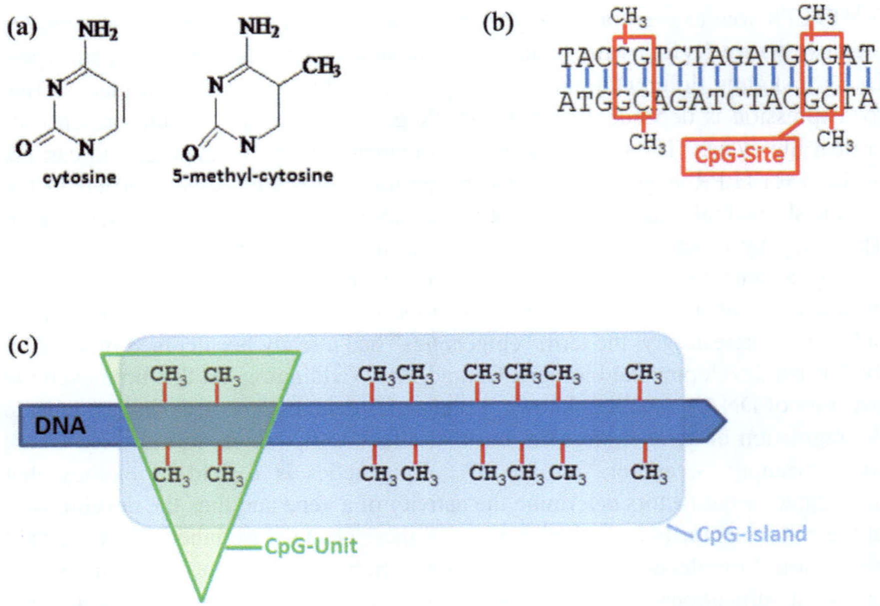

Fig. 23.6 Methylation in epigenetics: **a** DNA methylation is a modification to DNA that occurs through the addition of a methyl group to the 5 carbon position of a cytosine ring. DNA methylation only occurs if the cytosine base (C) is followed by a guanine base (G). **b** Adjacent cytosine and guanine bases are knitted together by a phosphodiester bond (p). This is why the sequence of a cytosine-phosphoguanine is designated as CpG-site. **c** CpG islands are regions in the human genome where the prevalence of CpG-sites is very high (i.e. >50 %) and the observed-to-expected CpG ratio is greater than 60 %. CpG islands represent about 1 % of the genome

References

Bouchard TJ Jr, Lykken DT, McGue M, Segal NL, Tellegen A (1990) Sources of human psychological differences: the Minnesota study of twins reared apart. Science 250(4978):223–228

Caspi A, Sugden K, Moffitt TE, Taylor A, Craig IW, Harrington H, McClay J, Mill J, Martin J, Braithwaite A, Poulton R (2003) Influence of life stress on depression: moderation by a polymorphism in the 5-HTT gene. Science 301(5631):386–389

Cook EH Jr, Scherer SW (2008) Copy-number variations associated with neuropsychiatric conditions. Nature 455(7215):919–923

Heinz A, Goldman D, Jones DW, Palmour R, Hommer D, Gorey JG, Lee KS, Linnoila M, Weinberger DR (2000) Genotype influences in vivo DAT availability in human striatum. Neuropsychopharmacology 22:133–139

Knuts I, Esquivel G, Kenis G, Overbeek T, Leibold N, Goossens L, Schruers K (2014) Therapygenetics: 5-HTTLPR genotype predicts the response to exposure therapy for agoraphobia. Eur Neuropsychopharmacol 24(8):1222–1228

Lim MM, Wang Z, Olazábal DE, Ren X, Terwilliger EF, Young LJ (2004) Enhanced partner preference in a promiscuous species by manipulating the expression of a single gene. Nature 429(6993):754–757

McGuffin P, Riley B, Plomin R (2001) Genomics and behavior. Toward behavioral genomics. Science 291(5507):1232–1249

Montag C, Reuter M (2014) Disentangling the molecular genetic basis of personality: from monoamines to neuropeptides. Neurosci Biobehav Rev 43:228–239

Reuter M, Küpper Y, Schmitz A, Breuer JP, Wend U, Hennig J (2005) Detection of new single nucleotide polymorphisms by means of Real Time PCR. J Genet 84:341–345

Spanagel R, Weiss F (1999) The dopamine hypothesis of reward: past and current status. Trends Neurosci 22(11):521–527

Vandenbergh DJ, Persico AM, Hawkins AL, Griffin CA, Li X, Jabs EW, Uhl GR (1992) Human dopamine transporter gene (DAT1) gene maps to chromosome 5p15.3 and displays a VNTR. Genomics 14:1104–1106

Walum H, Westberg L, Henningsson S, Neiderhiser JM, Reiss D, Igl W, Ganiban JM, Spotts EL, Pedersen NL, Eriksson E, Lichtenstein P (2008) Genetic variation in the vasopressin receptor 1a gene (AVPR1A) associates with pair-bonding behavior in humans. Proc Natl Acad Sci USA 105:14153–14156

Chapter 24
Hormones

Robert Miller and Clemens Kirschbaum

Abstract With the advent of readily available, convenient, and cost-efficient biochemical tools, the analysis of hormones has become a valuable approach to gain deeper insight into physiological processes, which influence human behavior. Neuroeconomic research has benefited from this development as indicated by a continuously growing number of publications during the past 20 years, relating variation in various endocrine systems to human decision-making. However, an informed reception and interpretation of such research relies heavily on knowledge about endocrine physiology and methods. Therefore, the present chapter aims to serve as a primer for endocrine methods by providing an overview about the most relevant aspects of design and data acquisition, as well as biochemical, and statistical methodology for hormone analyses. Proceeding from the brief delineation of endocrine systems' properties, various approaches to the measurement of hormone concentrations and to the inference on endocrine process components are presented in order to enable the reader to take part in this promising field of research.

24.1 Introduction

With the advent of readily available, convenient, and cost-efficient biochemical tools, the analysis of hormones has become a valuable approach to gain deeper insight into physiological processes, which influence human behavior. Neuroeconomic research, in particular, has benefited from this development as indicated by a continuously growing number of publications during the past 20 years, which is exemplarily illustrated in Fig. 24.1. This growing research interest is discussed by Amos Nadler and Paul Zack in Chap. 3 of the present book,

R. Miller (✉) · C. Kirschbaum
Department of Psychology, Technische Universität Dresden, Dresden, Germany
e-mail: robert.miller@tu-dresden.de

C. Kirschbaum
e-mail: clemens.kirschbaum@tu-dresden.de

M. Reuter and C. Montag (eds.), *Neuroeconomics*, Studies in Neuroscience, Psychology and Behavioral Economics, DOI 10.1007/978-3-642-35923-1_24

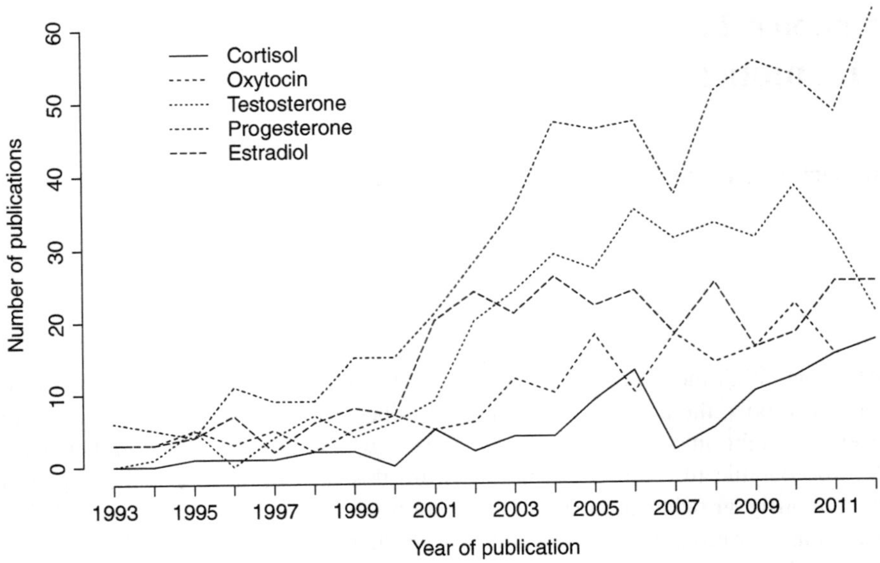

Fig. 24.1 Number of hormone-related publications on decision making per year as listed by Scopus (Elsevier, 2013)

who provide a comprehensive, content-related review on the most relevant findings relating variation in various endocrine systems to human decision-making.

However, an informed reception and interpretation of the respective research results, and even more importantly, active participation in this promising field of research, relies heavily on knowledge about endocrine physiology and methods. Therefore, the present chapter aims to serve as a primer for endocrine methods by providing an overview about the most relevant aspects of design and data acquisition, as well as biochemical, and statistical methodology for hormone analyses. As such, we do not intend to discuss specific endocrine systems and functions, but occasionally illustrate their features by selected examples.

24.2 Characteristics of Endocrine Systems

Endocrine systems are mostly defined by complex interaction patterns of different molecules that serve, in contrast to neuronal signal transduction, to unfold some *lagged* physiological impact over time. Some of these molecules, which are involved in endocrine signal transduction, are called hormones. Hormones differ in their physical and chemical properties due to their specific molecular structure (see Table 24.1). They are secreted by various glands and distributed by the blood in order to serve as chemical messengers altering cell function at remote target sites, mostly by activation of their associated intracellular or membrane-bound receptor

Table 24.1 List of selected hormones (adopted from Knox et al. 2011)

Chemical formula	Hormone	Also known as	Acronym
Peptide/protein hormones (large molecules, water-soluble)			
$C_{257}H_{387}N_{65}O_{76}S_6$	Insulin	–	–
$C_{43}H_{67}N_{15}O_{12}S_2$	Vasopressin	Arginine vasopressin, antidiuretic hormone	AVP/ADH
$C_{207}H_{308}N_{56}O_{58}S$	Corticotropin	Adrenocorticotropic hormone	ACTH
$C_{990}H_{1532}N_{262}O_{300}S_7$	Somatropin	Somatotropin, growth hormone	GH
$C_{208}H_{344}N_{60}O_{63}S_2$	Corticoliberin	Corticotropin-releasing hormone/factor	CRH/CRF
$C_{55}H_{75}N_{17}O_{13}$	Gonadoliberin	Luliberin, gonadorelin, gonadotropin-releasing hormone, luteinizing-hormone-releasing hormone	GnRH/LHRH
$C_{1014}H_{1609}N_{287}O_{294}S_{27}$	Lutropin	Luteinizing hormone	LH
α: $C_{437}H_{681}N_{121}O_{135}S_{13}$ β: $C_{975}H_{1513}N_{267}O_{304}S_{26}$	Follitropin	Follicle-stimulating hormone	FSH
$C_{21}H_{40}Cl_2O_2S_2$	Prolactin	Lactropin, lactotrophic hormone	PRL/LTH
$C_{43}H_{66}N_{12}O_{12}S_2$	Oxytocin	–	OXT
α: $C_{975}H_{1513}N_{267}O_{304}S_{26}$	Thyrotropin	Thyroid-stimulating hormone	TSH
Amine hormones (small molecules, rather water-soluble)			
$C_9H_{13}NO_3$	Adrenaline	Epinephrine	–
$C_8H_{11}NO_3$	Noradrenaline	Norepinephrine	–
$C_{15}H_{11}I_4NO_4$	Thyroxine	–	T4
$C_{15}H_{12}I_3NO_4$	Triiodothyronine	–	T3
Steroid hormones (small molecules, lipid-soluble)			
$C_{21}H_{28}O_5$	Aldosterone	–	–
$C_{21}H_{30}O_5$	Cortisol	Hydrocortisone, compound F	F
$C_{19}H_{28}O_2$	Dehydroepiandrosterone	Androstenolone, prasterone	DHEA
$C_{18}H_{24}O_2$	Estradiol	Oestradiol	E2
$C_{18}H_{22}O_2$	Estrone	Oestrone	E1
$C_{18}H_{24}O_3$	Estriol	Oestriol	E3
$C_{21}H_{30}O_2$	Progesterone	–	P4
$C_{19}H_{28}O_2$	Testosterone	–	–

molecules. After secretion and distribution, hormones are bound by specific plasma proteins (i.e., binding globulins) and/or continuously degraded to their respective metabolites by enzymes. Finally, they are cleared from the blood and excreted by the kidneys (renal clearance) and/or the liver (biliary clearance).

All endocrine systems involve the secretion and elimination of several hormones that serve "communication" between their respective physiological components. Such components interact by feed-forward and feedback loops, enabling self-regulation toward a systemic equilibrium. For instance, the hypothalamic-pituitary-adrenal (HPA) axis, is formed by the three endocrine organs: (1) the paraventricular nucleus of the hypothalamus, (2) the pituitary gland, and (3) the adrenal cortex, which interact by CRH, ACTH, and cortisol. Upon deflection from the equilibrium state (e.g., by stress experience or external *zeitgebers*), the hypothalamus releases CRH into the pituitary, which in turn releases ACTH that signals the adrenal cortex to synthesize cortisol. Thus, a feed-forward loop is established. Cortisol, however, inhibits further release of CRH and ACTH by a feedback loop. In consequence, complex and oscillating signal cascades are established, which are present not only in the HPA axis, but in any endocrine system. For a more detailed in-depth discussion of specific endocrine systems the reader shall be referred to Barrett (2005).

An explicit consideration of these properties is crucial, if one intends to investigate the influence of hormones on human decision-making processes. The major issues that arise from these properties and that ought to be taken into account for designing such a study shall be summarized as follows:

(a) *Operationalization of endocrine constructs*: Any study investigating hormone effects, needs to appropriately quantify the construct of interest. Endocrine activity, however, is a continuous process, rather than a fixed state, which involves many physiological subprocesses that unfold their impact across time (e.g., the speed and the magnitude of hormone secretion, and the speed of hormone metabolization and excretion). Thus, only extensive sampling and hormone measurement can provide a basis for inference on such subprocesses (although it also bears the danger of having to deal with huge amounts of data, which can be analyzed and interpreted in various ways). If one, for instance, assumed that an experimental intervention altered exclusively the amount of a secreted hormone, the sampling of two specimens might be considered sufficient in order to assess this hypothesis; a first specimen at the concentration nadir and a second at the concentration peak (as illustrated in Fig. 24.2). If we, however, concurrently wanted to investigate competing hypotheses (e.g., that the same intervention did not alter the amount of a secreted hormone, but rather the secretory speed), the sampling of two specimens would be insufficient. This is due to the presumed convolution of multiple endocrine subprocesses; the change of hormone concentration in between two sampling occasions could be interpreted as a difference in the amount of secreted hormone although it was actually generated by another mechanism (Fig. 24.2, blue line). To avoid such interpretative biases, an appropriate study design is required that adequately considers available knowledge about the hormone kinetics and dynamics, and rules out alternative mechanistic effect explanations.

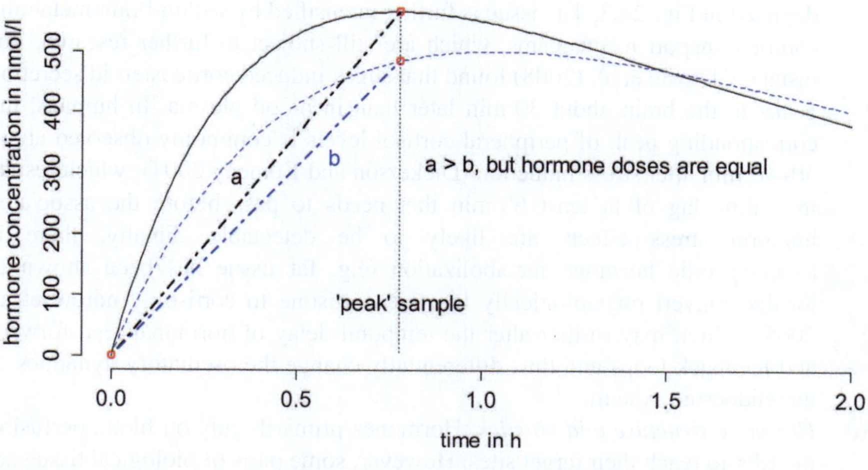

Fig. 24.2 Example of how an inadequate operationalization of endocrine constructs may promote misleading data interpretation. The *dot-dashed lines* represent the change of hormone concentrations (*a* and *b*) in between two specimens (*red dots*) that are sampled in each of two experimental conditions (*black and blue lines*). As *a* differs from *b*, one might infer that the amount of secreted hormone (doses) differs between both conditions. However, a varying speed of hormone secretion caused the observable difference by selectively shifting the concentration peak in the blue condition, whereas the hormone doses actually remained unchanged

(b) *Pacemakers and zeitgebers*: Most endocrine systems exhibit pronounced ultradian, circadian, and infradian (e.g., terrestrial or lunar revolution) rhythms, which presumably serve to maintain physiological adaptability of the investigated organism to environmental changes. In consequence, all measured hormone concentrations fluctuate and therefore encompass only a low amount of trait variance. This is an inherent problem of hormone measurement that can only be partly alleviated by rigid monitoring and controlling for possible confounding state variables (e.g. sampling time, season, menstrual cycle and the intake of oral contraceptives in case of female participant sex, time of awakening, currently and previously encountered phases of stress, or recent intake of meals or legal drugs; Kirschbaum et al. 1997), are well-known sources for deflected hormone signals. Another countermeasure to account for such unwanted variance components is to perform repeated hormone sampling in similar settings, as has been for instance proposed for the reliable assessment of trait-specific variability in the acute HPA axis activity (Hellhammer et al. 2007; Kirschbaum et al. 1990).

(c) *Time lags*: Delays between hormone secretion and corresponding psychophysiological responses restrict the time window, in which humoral effects on information processing in the central nervous system could be observed. Such delays are in particular problematic, when the physiological effect mediators are located more "downstream" from the measured/manipulated hormone. For the HPA axis, such an endocrine signal cascade is exemplarily depicted in Fig. 24.3. The issue is further intensified by within-brain metabolic and/or transport mechanisms, which are still subject to further research. For instance, Droste et al. (2008) found that stress-induced corticosteroid secretion peaks in the brain about 30 min later than in blood plasma. In humans, the corresponding peak of peripheral cortisol levels is commonly observed about 30–40 min after stress induction (Dickerson and Kemeny 2004), which results in a time lag of at least 60 min that needs to pass before the associated humoral stress effects are likely to be detectable. Finally, there is tissue-specific hormone metabolization (e.g. fat tissue has been shown to locally convert physiologically inactive cortisone to cortisol; Andrew et al. 2005), which may further alter the temporal delay of hormonal feed-forward and feedback loops and thus differentially change the oscillatory dynamics of the endocrine system.

(d) *Hormone structure and kinetics*: Hormones primarily rely on blood perfusion in order to reach their target sites. However, some parts of biological tissue are more richly perfused than others (e.g. fat vs. muscle tissue) which may result in a varying concentration of hormones at different target sites (depending on the hormone's molecular structure and corresponding properties). Small, lipophilic hormones can easily penetrate tissue and therefore bypass their exclusive distribution by the blood stream to a certain degree. By contrast, comparably large, hydrophilic peptide hormones (e.g. Oxytocin) cannot benefit from this property, which has serious implications for specimen selection and the availability of the hormone to target sites such as brain tissue, which is

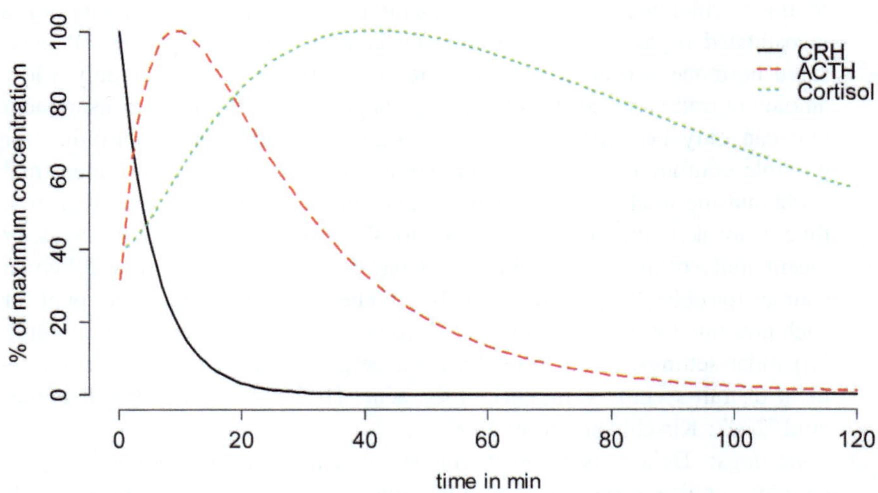

Fig. 24.3 Transit-compartment model (Bonate 2011) illustrating the kinetics of HPA hormones after stimulation by bolus CRH administration. The model was informed by data from Schürmeyer et al. (1984). Assuming a direct correspondence of CRH/ACTH elimination to ACTH/cortisol secretion, ACTH is predicted to peak 9 min after challenge whereas the cortisol blood concentration peaks 41 min after challenge

probably most relevant to the field of Neuroeconomics. In this regard, the blood brain barrier that is formed by endothelial cells covering the interior surface of blood vessels, should be taken into particular account. This barrier prevents large molecules, and peptide hormones in particular, to enter the blood vessels of the brain from the peripheral blood stream and vice versa. Consequently, central peptide hormone levels differ from levels in peripheral specimens (e.g. saliva) and ought to be measured in specimens such as the cerebrospinal fluid. Conversely, central peptide hormone levels cannot be altered by peripheral pharmacological challenge. Instead, the blood brain barrier needs to be bypassed as for instance by intranasal substance administration (e.g. Guastella et al. 2013).

24.3 Measurement of Hormone Concentrations

24.3.1 Selection of Specimen and Analytes

In analytical chemistry, the term specimen is used to refer to any sample of a certain compound, which is to be analyzed. Any specimen contains a certain amount of hormones (i.e., the analyte), which is to be determined by some measuring system. The remains of the specimen form the so-called matrix, which can in turn influence the measurement of the analyte. Proceeding from the cursorily outlined properties of endocrine systems (see Sect. 24.2), there are various options of which and how many specimens within a certain time period are to be assayed for what analyte by what measurement system. Regarding specimen and analyte selection, the inclined researcher has to consider three design parameters, which determine the adequacy of endocrine activity assessment.

First, specimen selection is supposed to rest upon theoretical considerations. The most proximate choices for specimens comprise biochemical compounds containing an analyte that is most likely indicative for activity of the investigated endocrine system. According to the physiological knowledge about endocrine systems in general, blood and excrements (urine and feces) are the most reasonable options. While blood is presumably the most suitable type of specimen for monitoring currently bioavailable hormone concentrations, it is barely suited for monitoring long-term hormone exposure. Conversely, excrements can serve quite well for such a purpose, but any measured analyte concentration within these specimens is dependent on the volume of food and fluid being ingested by the participant. Consequently, these confounders have to be controlled. Depending on the availability of knowledge about specific metabolic processes, however, other specimen might as well be taken into consideration. The most prominent example for such a specimen forms saliva for the monitoring of current hormone concentrations. Saliva is particularly useful to determine steroid hormones, but larger hormones have also been found to be present in saliva although these rely on active transport mechanisms (Gröschl 2008; Vining et al. 1983). Furthermore, many salivary hormones do not seem to be (mostly steroids, but no peptides; Hofman 2001) influenced by the flow rate of saliva. The popularity of saliva sampling is mostly due to the ease of its collection (by commercially available sampling tools; e.g. Gröschl et al. 2008) and its noninvasiveness, which results in a substantially facilitated implementation process as compared to blood sampling. Hair has also been shown to contain considerable amounts of steroid analytes (e.g. Gao et al. 2013). With a barely interindividually varying growth rate of 1 cm per month has been proposed to reflect cumulative hormone expose across time periods (Stalder and Kirschbaum 2012), that are substantially longer than those covered by feces or urine. However, the mechanism of analyte incorporation into hair needs still to be unveiled.

The second design parameter concerns to the number of specimen, which should be sampled from each subject. This is not only of practical importance, as hormone determination for many specimen and analytes can be quite costly (see Fig. 24.4), but also closely linked to the choice of specimen, as the number of specimen per subject is (given a reliable measurement) only relevant with regard to the time period required to infer on the construct of interest (as has been mentioned in the previous section). Example given, 3 cm of hair are supposed to contain an analyte concentration, which reflects cumulative hormone expose during the last 3 months. If this cumulative hormone exposure is the construct of interest, only one specimen is required. If one, by contrast, intends to infer on the change of cumulative hormone exposure due to some experimental intervention at least two specimens need to be taken at an appropriate time interval. Analogous deliberation is particularly important, if subprocesses of endocrine activity are to be investigated (e.g. the speed of hormone secretion and concurrent elimination). Such a research question needs at least three specimens covering reasonably distant sampling points and containing analytes, which reflect the organism's current hormonal status (i.e. blood or saliva). With complex endocrine systems acquisition of only two specimens bears the risk, that the construct of interest is not reasonably assessed. This is due to

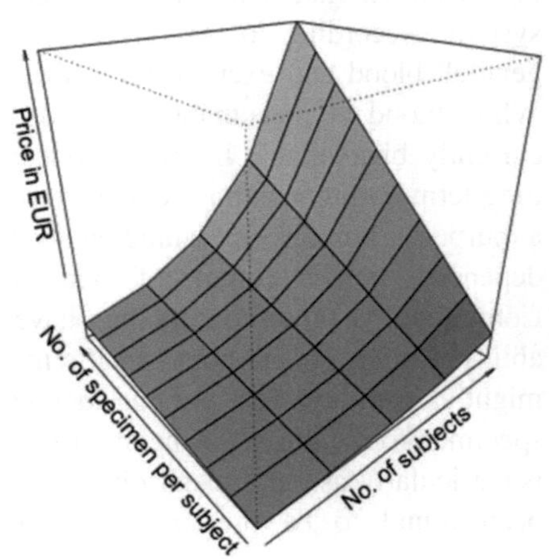

Fig. 24.4 Relation between study cost, number of sampled specimens per subject, and number of tested subjects. While the costs increase linearly with the number of sampled specimens, more subjects increase the study cost in a quadratic fashion

the fact that endocrine activity can hardly be, at least from a physiological point of view, characterized by only two parameters (i.e., the mean and variance of hormone level change in between two sampling occasions). Similarly, research questions that involve the assessment of hormonal rhythms, should similarly consider at least three specimens during the phase of rising hormone levels, around the expected peak, and during the phase of decreasing hormone levels.

The third design parameter concerns the choice of analyte in the specimens, which is to be measured. Apart from the directly measuring the hormones being involved signal transduction of the investigated endocrine system, endocrine activity is not only indicated by hormones that directly unfold their associated physiological effects but also by their metabolites, which are concurrently present within the same specimen. Such considerations may become important, if the kinetics of the respective target hormone require special precautions, which complicate the process of specimen collection. Adrenaline and noradrenaline form prominent examples for such a case, as their half-lives amount to less than five minutes, which requires that specimens are to be preprocessed and frozen immediately after collection. Their metabolites metanephrine and normetanephrine, however, persist much longer and may also serve to indicate activity of the sympathetic nervous system. In a similar fashion, it is reasonable to choose those analytes that are directly involved in endocrine signal transduction and easily obtainable, but yield a longer half-life than the other hormones being involved in the respective endocrine system (e.g., cortisol is eliminated much slower than CRH or ACTH). Another occasion, where metabolite assessment may become handy, concerns study designs, in which endocrine activity is induced pharmacologically and the specimens are to be obtained from the same route that is used for substance administration (e.g. oral administration of cortisol and saliva collection; see Perogamvros et al. 2010).

24.3.2 Laboratory Methods

The determination of hormones from blood, saliva, urine, or any other specimen requires highly sensitive biochemical methods, since the concentrations of these analytes are typically in the nanomolar to picomolar (10^{-9}–10^{-12} M) range. Thus, even minute quantities of endocrine substances must be captured with high precision and reproducibility. Following seminal work by the nobel-laureate Rosalyn Yalow in the 1970s, a rapid acceptance and increasing use of immunological approaches, which show just these characteristics, has been seen (Wild 2013). Today, almost any endocrine parameter can be assessed by so-called immunoassays and literally hundreds of commercial test kits are available on the market. Immunoassays are today probably the most frequently employed method of measuring hormone levels in a given matrix.

The measurement principle of any immunoassay is based on the use of proteins generated by the B lymphocytes of the immune system, i.e., the antibodies. In our body, these proteins bind foreign materials ("antigens") that may jeopardize the integrity and health of an organism. Millions of different antibodies can be generated by the immune systems, allowing us to become "immune" to virtually every pathogen or foreign material on earth. With certain biochemical tricks, antibodies against any hormone can be generated by injecting the hormone into a different species (like sheep or swine). After repeated injection, or "immunization," high concentrations of antibodies produced against the particular human hormone can be found in this host. Attaching a biochemical label to such an antibody now allows for its use to quantify the amount of the hormone in plasma, serum, urine, feces, saliva, cell extracts, or supernatants. The nature of the biochemical label gives the name to the specific immunoassay: Attaching a radioactive compound to the antibody will result in a *radioimmunoassay* (RIA), enzymes as labels will produce *enzyme immunoassays* (EIA or ELISA), depending on the specific composition of the test, while light-emitting labels will result in fluorescence or chemiluminescence immunoassays (FIA or CLIA), respectively.

Immunoassays typically require sample volumes between 10 and 200 microliters per determination, which are used in singlet or duplicate (with the mean of the two replicates serving as the measured concentration). The reproducibility of an immunoassay result is reflected in the intra- and interassay coefficients of variation. Defined as the standard deviation of repeated assays performed from the same specimen divided by the average concentration measured, these coefficients should not exceed 12–15 % for reliable and reproducible results. Most modern assay systems are able to deliver much better performance, the better kits give interassay CVs below 5 %.

While immunoassays are the first choice for reliable and rather inexpensive analysis, they sometimes lack the required specificity. Since a given antibody will not bind exclusively to the molecule it has been raised against, a certain crossreactivity with matrix components of similar molecular structure is usually seen. Crossreactivity may be sufficiently low so that the 'true' analyte concentration is not significantly overestimated. For example, most immunoassay for cortisol will also detect its metabolite, cortisone, with a crossreactivity between less than 1 % to more than 10 % (the latter posing a significant source of measurement error; see Miller et al. 2013). An even more difficult situation arises with a much higher crossreactivity to certain glucocorticoids used for the treatment of certain medical conditions (e.g., inflammatory disorders). One of such glucocorticoids, prednisolone, is detected by anticortisol antibodies with 50–80 % crossreactivity! If a reliable differentiation between those steroids is required, other analytical methods need to be employed.

Among the most specific biochemical methods available to date for analysis of molecules in a complex matrix, the measurement of hormones using liquid chromatography coupled with tandem mass spectrometry (LCMS/MS) is becoming the gold standard for such analytic purpose. With this method, the analytes are not detected after binding, but rather identified on the basis of its molecular mass. It is therefore superior to immunoassays due to much greater specificity. Despite this clear advantage, this methodology is unsuited for most smaller research labs, since LCMS/MS equipment is still prohibitively expensive, requires highly trained personnel, and allows for rather low throughput.

24.4 Data Analysis

Given that alteration of an endocrine system can be achieved by simple experimental manipulations (e.g. by intranasal injection of Oxytocin; Guastella et al. 2013, or by brief inhalation of carbon dioxide; Wetherell et al. 2006), and that one intends to use some measured hormone concentration *C* only as means to check for intervention efficacy (i.e., to use *C* as a dependent variable), analyses of endocrine data are comparably straightforward. Here, the general linear model (GLM, see Everitt 2009) is particularly straightforward to implement, and can be easily interpreted as long as its prerequisites are considered (Miller and Plessow 2013). If one, however, intends to use endocrine activity as an explanatory variable, data analyses becomes much more complex, as an adequate quantification of the construct of interest requires specification an appropriate statistical (measurement) model. In biobehavioral studies, the application of two major options for quantification of endocrine activity is observable: (a) data-driven quantification approaches, which can in most cases be related to the calculation of change scores, and (b) theory based approaches (including area-under-the-concentration-curve calculations), which have been originally developed for pharmacological studies but are also about to become more prominent in the field of Neuroeconomics.

24.4.1 Data-Driven Quantification of Endocrine Activity

Considering that hormone levels fluctuate continuously across time due to the inherent properties of the respective endocrine system, a large variety of models have been proposed to infer on the status of the investigated system from a limited number of specimens, which have been sampled at discrete points of time. Such data are called panel data. Statistical models for the analysis of panel data can be roughly differentiated into models in which the change of hormone concentration per unit of time is considered explicitly (being referred to as continuous-time models) and models in which time is considered implicitly (being referred to as discrete-time models).

Continuous-time models commonly incorporate estimation of a time-related parameter. In the present context, the simplest form of such a model is a linear regression (see formula I below) that relates *C* at two different points in time t_0 and t_1 to the time on its original scale.

$$C(t_0, t_1) = a_a \times \text{time} + c_a$$

Such a simple model, however, assumes that the change that occurs in between t_0 and t_1 is the same for each time unit, which is quite unrealistic considering the inherent nonlinear kinetics of endocrine markers. In case a third data point $C(t_2)$ is added to the present example, which deviates from the projected linear trajectory of C across time, one could sufficiently account for it by adding an additional quadratic term (see formula II below). This model also implies that any continuous change that occurs in between t_0 and t_2, can be reasonably approximated by the given equation.

$$C(t_0, t_1, t_2) = a_b \times \text{time} + b_b \times \text{time}^2 + c_b$$

By contrast, discrete-time models do not assume a continuous trajectory of C across time. Proceeding from the examples mentioned above, the simplest model for analyzing panel data of such kinds is a repeated-measures analysis of variance (ANOVA). As such ANOVA also belongs to the GLM-class, its model equation is quite similar to the listed regression models, but uses dummy-coded sampling times as predictors. Consequently, the equation for the $C(t_0, t_1)$ example is the same as in formula I, but different for the $C(t_0, t_1, t_2)$ example represented by formula II.

$$C(t_0, t_1, t_2) = a_c \times \text{time}(t_0, t_1) + b_c \times \text{time}(t_0, t_2) + c_c$$

These outlined examples illustrate, that the distinction between continuous- and discrete-time models is of a rather conceptual kind, because all examples belong to the same model class. Differences between them mostly relate to the coding of time and thus to the interpretation of the model coefficients. From a content perspective, however, continuous-time modeling yields a major advantage as compared to discrete-time modeling of hormone time series: Given an adequate theoretical foundation, they enable a precise quantification of endocrine activity at any point in time, which is formed by occasional (or forced) secretory events and concurrent hormone elimination. By contrast, endocrine parameters that have been obtained by discrete-time models can only be used to infer on endocrine activity at the time of sampling, which restricts their comparability with the results reported by other studies (i.e., its interpretability) for the benefit of a more convenient usability.

Any statistical model for hormone time series relies on appropriate theoretical assumptions on the construct being investigated. In the most basic and popular, but rather data-driven case, change of C can be formalized as $\Delta C(t_i) = C(t_i) - C(t_{i-1})$ occurring in between two points in time t_i and t_{i-1}, at which specimens have been sampled. Although ΔC can serve as a convenient and easily interpretable indicator of endocrine activity, it relies on discrete points in time (i.e., any informative change of in-between and outside of bounds of specimen sampling is disregarded). Although this problem can be alleviated by sampling of more than two specimens, the analytical complexity would concurrently increase because the number of possible ΔCs grows in a quadratic fashion the more specimens are sampled.

Continuous-time modeling of ΔC can provide a solution to this problem, although no mechanistic and physiological knowledge about endocrine functioning used. A key aspect of the continuous-time modeling of endocrine activity rests on the idea that any $\Delta C(t_i)$ within a certain, fixed time window ($\Delta t = t_i - t_{i-1}$) also contains information about ΔC within any other Δt. By rescaling $\Delta C(t_i)$ according to length of its corresponding time window Δt_i, a difference quotient $\Delta C_i/\Delta t$ can be obtained that denotes any change of hormone levels within one unit of its time scale. This quotient can be interpreted similar to the regression parameter a_a in formula I (see above). Furthermore, any ΔC_i can be modeled by an autoregression that predicts $C(t_i)$ by $C(t_{i-1})$ [or its rearranged form $C(t_i - \Delta t)$; (see Bollen and Curran 2004)]. The integration of both trains of thought implies the difference equation given below, where A as a matrix containing all autoregression parameters relating the difference quotient $\Delta C_i/\Delta t$ to any sampling point:

$$\frac{\Delta C(t_i)}{\Delta t} = A \times C(t_i - \Delta t)$$

If we now let Δt converge toward zero, the autoregression matrix A will define the slope of C at any point in time, and thus become a matrix containing the local change (or drift) parameters of $C(t_i)$. This is formalized by the deterministic differential equation given below. Solving this formula yields a nonlinear regression function where t_0 represents the beginning of the time series (i.e., the initial sampling point):

$$\frac{\mathrm{d}C(t)}{\mathrm{d}t} = A \times C(t)$$

$$C(t) = C(t_0)\mathrm{e}^{A(t-t_0)}$$

Implementation of such a model and its extension by randomly occurring deflections of C that can be related it to the change of concurrently assessed psychological constructs, has been extensively elaborated by Voelkle et al. (2012) and shall therefore not be discussed any further.

24.4.2 *Theory-based Quantification of Endocrine Activity*

Although the previously presented model may render useful to merely describe the change of C across time, it fails to adequately account for endocrine activity in response to a certain experimental intervention due to lack of the incorporation of knowledge that may cause corresponding change of C (by its underlying subprocesses). As has been mentioned above, the adequate formalization of such mechanisms relies on substantive theory. For the modeling of endocrine time-series in particular, the pharmacokinetic *compartment theory* might serve as a reasonable starting point. This theory has originally been developed for the analysis of substance absorption, distribution, metabolism, and excretion within and between different physiological compartments (Bonate 2011; Gabrielsson and Weiner 2007). Each of such compartments (e.g., the adrenal glands and the blood) contains a specific amount of the investigated substance at each point in time, which is determined by the exchange of substance between those compartments (being indicated by various kinetic parameters k_i). Proceeding from this reasoning, one can directly specify differential equations for each compartment that can be integrated into a single physiologically plausible model of the whole system: If we assume, for instance, that a certain amount of substance within the adrenal glands (C_1: compartment 1) entered the blood (C_2: compartment 2) unidirectionally, we would specify two differential equations; with $C_i(t)$ representing the substance concentration at any time in the respective compartment i, k_a representing the substance's transfer (or absorption) rate from C_1 to C_2, and k_{el} representing the substance's elimination rate from C_2.

$$\frac{dC_1(t)}{dt} = -k_a C_1(t)$$
$$\frac{dC_2(t)}{dt} = k_a C_1(t) - k_{el} C_2(t)$$

Notably, specimen sampling from C_1 is optional, as k_a and $C_1(t)$ can be estimated as well from C_2 specimens being sampled across time. Using Laplace transforms (Mayersohn and Gibaldi 1970), both differential equations can be solved and integrated to a nonlinear regression function. Similar to the solution of the data-driven continuous-change differential equation in the previous section, such a function relies on information about the initial state of the system. This is indicated by $C_1(t_0)$; the need to estimate an initially available concentration of target substance in C_1.

$$C_2(t) = \frac{C_1(t_0)\ k_a}{(k_a - k_{el})} \times \left(e^{-k_{el}t} - e^{-k_a t}\right)$$

Adopting the compartment model outlined above for hormone analysis by assuming that C_2 corresponded to the blood compartment, from which hormones were measured [i.e., $C(t) = C_2(t)$], C_1 could be regarded as a latent compartment containing the fraction of hormone to be secreted (before feedback inhibits any further feed-forward hormone release). However, its implementation is (yet) implausible, because the blood compartment $C(t_0)$ would contain no hormone at all at the beginning of specimen sampling. Such an assumption is inconsistent with the empirically observable patterns of continuously oscillating endocrine activity (e.g., Brown et al. 2001; Klerman et al. 2003) that cause hormone concentrations to deviate substantially from zero at any point in time. Within a conventional panel design comprising blood sampling in temporal proximity to a single secretory event, we could use a generalization of the derived formula for estimating $C(t)$ after

n previous substance administrations (Gabrielsson and Weiner 2007), that is, $C(t_0)$ is assumed to converge toward some steady state gives a fixed phase τ (that defines the elapsed time period before $C_1(t_0)$ is reinstated). The corresponding equation is provided below:

$$C_2(t) = C(t) = \frac{C_1(t_0)k_a}{(k_a - k_{\text{el}})} \times \left(\left[\frac{1 - \mathrm{e}^{-nk_{\text{el}}\tau}}{1 - \mathrm{e}^{-k_{\text{el}}\tau}} \right] \mathrm{e}^{-k_{\text{el}}t} - \left[\frac{1 - \mathrm{e}^{-nk_a\tau}}{1 - \mathrm{e}^{-k_a\tau}} \right] \mathrm{e}^{-k_a t} \right)$$

As n increases toward infinity and given the substance has been completely transferred to C_2 as t approaches the phase duration τ, the minimal $C(t_0)$ can be calculated by the following formula, with t_0 denoting a point in time that is found between the initial sample and the first observable rise of hormone levels.

$$C(t_0) = \frac{C_1(t_0)k_a}{(k_a - k_{\text{el}})} \times \left(\frac{\mathrm{e}^{-k_{\text{el}}\tau}}{1 - \mathrm{e}^{-k_{\text{el}}\tau}} \right)$$

Fitting of such a model can nowadays be conveniently be implemented via commercially available nonlinear regression tools (e.g. R, NONMEM, or SAS) or utilization of generic Markov-Chain Monte-Carlo methods (e.g. WinBUGS; see Lunn et al. 2002, or STAN). We illustrate such a model in Fig. 24.5, that has been fitted to the mean cortisol concentrations being obtained before, during, and after stress-induction by von Dawans et al. (2011).

Proceeding from such a model, one can also estimate the overall extent of endocrine activity by calculating the amount of hormone that is additionally available as compared to a scenario, where endocrine stimulation was unsuccessful. Such a measure can be quantified as the difference between the area under the concentration curve (AUC_C) given a pulse was present (within the time period from $t_0 = 0$ to phase duration τ, see solid line in Fig. 24.4) and AUC_C given no pulse was present (i.e., the time period from $t_0 = \tau$ to twice the phase duration 2τ, see dashed line in Fig. 24.4):

$$\begin{aligned} \text{AUC}_C = {} & \frac{C_1(t_0)k_a}{(k_a - k_{\text{el}})} \\ & \times \left(\int_0^{\tau} \left[\frac{1 - \mathrm{e}^{-nk_{\text{el}}\tau}}{1 - \mathrm{e}^{-k_{\text{el}}\tau}} \right] \mathrm{e}^{-k_{\text{el}}t} - \left[\frac{1 - \mathrm{e}^{-nk_a\tau}}{1 - \mathrm{e}^{-k_a\tau}} \right] \mathrm{e}^{-k_a t} \mathrm{d}t \right. \\ & \left. - \int_{\tau}^{2\tau} \left[\frac{1 - \mathrm{e}^{-nk_{\text{el}}\tau}}{1 - \mathrm{e}^{-k_{\text{el}}\tau}} \right] \mathrm{e}^{-k_{\text{el}}t} - \left[\frac{1 - \mathrm{e}^{-nk_a\tau}}{1 - \mathrm{e}^{-k_a * \tau}} \right] \mathrm{e}^{-k_a t} \mathrm{d}t \right) \end{aligned}$$

Conditional on the validity of the proposed model for mapping the kinetics of C, this AUC_C to actually quantifies endocrine activity on a theoretical basis. In contrast

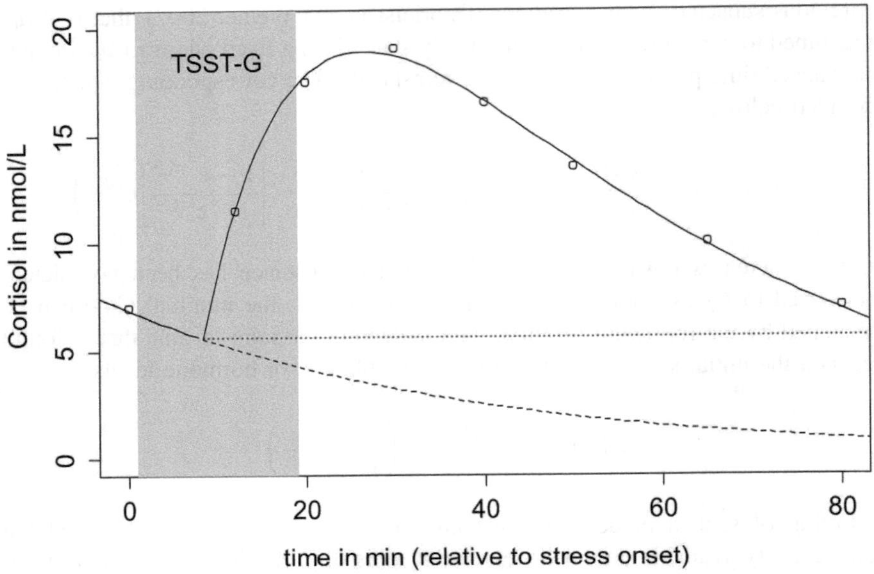

Fig. 24.5 Empirical salivary cortisol concentrations and the predicted salivary cortisol trajectory before, during, and after HPA stimulation by a stress-protocol (TSST-G; see von Dawans et al. 2011). Model fitting yielded the following parameters: the amount of cortisol being secreted relative to the volume of distribution $C_1(t_0) = 26.06$ nmol/L, the rate of cortisol secretion $k_a = 0.083$, and the rate of cortisol elimination $k_e = 0.026$. The *dotted line* denotes the steady state $C(t_0)$, that is, the concentration that needs to be reached before the cortisol in $C_1(t_0)$ is secreted again (after τ minutes). The *dashed line* shows the predicted trajectory of cortisol concentrations in case no endocrine response would have been elicited due to the stress-protocol

to the ΔC operationalization of endocrine activity, which crucially depends on the time specimens are sampled, the AUC is easily interpretable and represents a more holistic operationalization of individual endocrine activity. Due to the complexity of its calculation from such parametric models, however, nonparametric AUC variants have been proposed (e.g., AUC_i and AUC_g being calculated from trapezoidal C decomposition; Fekedulegn et al. 2007; Pruessner et al. 2003), which are based on linear interpolation between all discrete points in time. Consequently, these variants are no real, but rather discrete-time indicators of endocrine activity, as the sampling time of specimens is crucial to recover the data generating mechanism. Despite of this shortcoming, they can be regarded as convenient data-driven approximations to the information being provided by continuous-time AUCs (given a sufficient number of available specimens). Finally, their utility with regard to measuring endocrine activity has been supported by (Fekedulegn et al. 2007), who showed that a combination of AUC_i and AUC_g may serve well to account for the major fraction of time-series variance in hormone concentrations.

References

Andrew R, Westerbacka J, Wahren J et al (2005) The contribution of visceral adipose tissue to splanchnic cortisol production in healthy humans. Diabetes 54:1364–1370

Barrett EJ (2005) The endocrine system. In: Boron WF, Boulpaep EL (eds) Medical physiology: a cellular and molecular approach. Elsevier Saunders, Philadelphia

Bollen KA, Curran PJ (2004) Autoregressive latent trajectory (ALT) models a synthesis of two traditions. Sociol Method Res 32:336–383

Bonate PL (2011) Pharmacokinetic-pharmacodynamic modeling and simulation, 2nd edn. Springer, New York City

Brown EN, Meehan PM, Dempster AP (2001) A stochastic differential equation model of diurnal cortisol patterns. Am J Physiol Endocrinol Metab 280(3):E450–E461

Dickerson SS, Kemeny ME (2004) Acute stressors and cortisol responses: a theoretical integration and synthesis of laboratory research. Psychol Bull 130:355–391

Droste SK, de Groote L, Atkinson HC et al (2008) Corticosterone levels in the brain show a distinct ultradian rhythm but a delayed response to forced swim stress. Endocrinology 149:3244–3253

Everitt BS (2009) Multivariable modeling and multivariate analysis for the behavioral sciences. CRC Press, Boca Raton, FL

Fekedulegn DB, Andrew ME, Burchfiel CM et al (2007) Area under the curve and other summary indicators of repeated waking cortisol measurements. Psychosom Med 69:651–659

Gabrielsson J, Weiner D (2007) Pharmacokinetic/pharmacodynamic data analysis, 4th edn. Swedish Pharmaceutical Press, Stockholm

Gao W, Stalder T, Foley P et al (2013) Quantitative analysis of steroid hormones in human hair using a column-switching LC-APCI-MS/MS assay. J Chromatogr B Anal Technol Biomed Life Sci 928:1–8

Gröschl M (2008) Current status of salivary hormone analysis. Clin Chem 54:1759–1769

Gröschl M, Köhler H, Topf HG et al (2008) Evaluation of saliva collection devices for the analysis of steroids, peptides and therapeutic drugs. J Pharmaceut Biomed 47:478–486

Guastella AJ, Hickie IB, McGuinness MM et al (2013) Recommendations for the standardisation of oxytocin nasal administration and guidelines for its reporting in human research. Psychoneuroendocrinology 38:612–625

Hellhammer J, Fries E, Schweisthal OW et al (2007) Several daily measurements are necessary to reliably assess the cortisol rise after awakening: state-and trait components. Psychoneuroendocrinology 32:80–86

Hofman LF (2001) Human saliva as a diagnostic specimen. The Journal of Nutrition 131:1621S–1625S

Kirschbaum C, Steyer R, Eid M et al (1990) Cortisol and behavior: 2. application of a latent state-trait model to salivary cortisol. Psychoneuroendocrinology 15:297–307

Kirschbaum C, Gonzalez Bono E, Rohleder N et al (1997) Effects of fasting and glucose load on free cortisol responses to stress and nicotine. J Clin Endocrinol Metab 82:1101–1105

Klerman EB, Adler GK, Jin M et al (2003) A statistical model of diurnal variation in human growth hormone. Am J Physiol Endocrinol Metab 285:E1118–E1126

Knox C, Law V, Jewison T et al (2011) DrugBank 3.0: a comprehensive resource for 'omics' research on drugs. Nucleic Acids Res 39:D1035–D1041

Lunn DJ, Best N, Thomas A et al (2002) Bayesian analysis of population PK/PD models: general concepts and software. J Pharmacokinet Pharmacodyn 29:271–307

Mayersohn M, Gibaldi M (1970) Mathematical methods in pharmacokinetics. I. Use of Laplace transform in solving differential rate equations. Am J Pharm Educ 34:608–614

Miller R, Plessow F (2013) Transformation techniques for cross-sectional and longitudinal endocrine data: application to salivary cortisol concentrations. Psychoneuroendocrinology 38:941–946

Miller R, Plessow F, Rauh M et al (2013) Comparison of salivary cortisol as measured by different immunoassays and tandem mass spectrometry. Psychoneuroendocrinology 38:50–57

Perogamvros I, Keevil BG, Ray DW et al (2010) Salivary cortisone is a potential biomarker for serum free cortisol. J Clin Endocrinol Metab 95:4951–4958

Pruessner JC, Kirschbaum C, Meinlschmid G et al (2003) Two formulas for computation of the area under the curve represent measures of total hormone concentration versus time-dependent change. Psychoneuroendocrinology 28:916–931

Schürmeyer TH, Avgerinos PC, Gold PW et al (1984) Human corticotropin-releasing factor in man: pharmacokinetic properties and dose-response of plasma adrenocorticotropin and cortisol secretion. J Clin Endocrinol Metab 59:1103–1108

Stalder T, Kirschbaum C (2012) Analysis of cortisol in hair—state of the art and future directions. Brain Behav Immun 26:1019–1029

Vining RF, McGinley RA, Symons RG (1983) Hormones in saliva: mode of entry and consequent implications for clinical interpretation. Clin Chem 29:1752–1756

Voelkle MC, Oud JHL, Davidov E et al (2012) An SEM approach to continuous time modeling of panel data: Relating authoritarianism and anomia. Psychol Methods 17:176–192

von Dawans B, Kirschbaum C, Heinrichs M (2011) The trier social stress test for groups (TSST-G): a new research tool for controlled simultaneous social stress exposure in a group format. Psychoneuroendocrinology 36:514–522

Wetherell MA, Crown AL, Lightman SL (2006) The four-dimensional stress test: psychological, sympathetic-adrenal-medullary, parasympathetic and hypothalamic-pituitary-adrenal responses following inhalation of 35 % CO_2. Psychoneuroendocrinology 31:736–747

Wild D (2013) The immunoassay handbook: theory and applications of ligand binding, ELISA and related techniques, 4th edn. Elsevier, Oxford

Chapter 25
Eye Movements

Ulrich Ettinger and Christoph Klein

Abstract The study of eye movements may provide a "window to the soul", that is, a unique opportunity to investigate the mind and brain of humans and animals. In this chapter we argue that the study of eye movements may also be of use in neuroeconomics research. We first describe the types of eye movements that are of relevance in this context and outline their neural correlates. We then outline the key oculographic methods that are likely to be applied in neuroeconomics experiments and point out the advantages of oculomotor research over manual motor experiments. We address the long researched issue of the association between eye movements and visual attention. That research supports the benefit of studying eye movements as a concurrent level of analysis in addition to manual responses in order to better understand the temporal and spatial features of attention. We then review key literature on the pupil, which shows a close relationship between pupil dynamics and various cognitive and affective states. Finally, we summarise some important findings from oculomotor research in the field of neuroeconomics. It is concluded that the study of eye movements represents a convenient, objective and reliable method that may yield important additional data to better understand economic decision-making processes as well as their neurophysiological correlates.

U. Ettinger (✉)
Department of Psychology, University of Bonn, Kaiser-Karl-Ring 9, 53111 Bonn, Germany
e-mail: ulrich.ettinger@uni-bonn.de

C. Klein
Department of Child and Adolescent Psychiatry and Psychotherapy, University of Freiburg, Hauptstr. 8, 79104 Freiburg, Germany
e-mail: christoph.klein.kjp@uniklinik-freiburg.de; c.klein@bangor.ac.uk

C. Klein
Department of Child and Adolescent Psychiatry, Psychosomatics and Psychotherapy, University of Cologne, Cologne, Germany

C. Klein
School of Psychology, Bangor University, Penrallt Road, LL57 2AS Bangor, UK

M. Reuter and C. Montag (eds.), *Neuroeconomics*, Studies in Neuroscience, Psychology and Behavioral Economics, DOI 10.1007/978-3-642-35923-1_25

25.1 Introduction

It has long been suggested that the eye is the "window to the soul". Wade and Tatler (2005, p. 1) in their comprehensive history of oculomotor research quote De Laurens in a writing from the sixteenth century as stating "*Orpheus* called the eyes, the looking glasses of nature: *Hesichius*, the doores for the Sunne to enter in by: *Alexander* the Peripatecian, the windowes of the mind, because that by the eyes we doe cleerely see what is in the same, we pearce into the deepe thoughts thereof, and enter into the privities of his secret chamber".

The notion that the eye may offer privileged access to the study of the mind, i.e. to human and animal cognition, motivation, emotion, ability and personality as well as psychopathology, has been reiterated in modern times (Stern and Dunham 1990). The eye offers a wealth of information about ongoing cognitive and emotional processes, first and foremost of course about where the person is directing their overt visual attention (see below). Additionally, the measurement of eye blinks as well as the characterisation of the more subtle features of the eye such as the size of the pupil can be used profitably to make inferences about state variations in the viewer's arousal levels as well as demands on information processing (see below).

Together, the eye and its internal as well as external movements presents an easily accessible model system that can be studied across species—a microcosm within which to explore the interactive processes of perception, cognition and motor control as well as their neural substrates, their modulation by emotion and their impairments in disease (Leigh and Zee 2006; Liversedge et al. 2011).

Historically, the scientific study of eye movements can be broadly structured into four eras (Duchowski 2002; Rayner 1998). First, in the nineteenth century and in the early twentieth century, basic facts about the oculomotor system were discovered. These discoveries include fundamental eye movement characteristics such as saccadic latencies (reaction times) as well as the phenomenon of saccadic suppression, i.e. the reduction of visual perception during saccades which facilitates a stable representation of the visual world despite changes to the retinal image during fast ocular movements. The second phase, from the 1930– 1950s, a time when psychology was dominated by the behaviourist paradigm, saw a stronger focus of applications of eye movement research. A third major impetus into oculomotor research could be observed in the 1970s to the late 1990s, due to the increasing availability of ever more accurate oculographic methods. Fourth, we are today in the fortunate position to be able to draw upon a range of easily available and highly sophisticated oculographic hardware and software systems. These allow us not only to record eye movements with never seen temporal and spatial resolution but also to perform interactive experiments, for example, by modifying the visual input on the basis of eye movements or eye position (Duchowski 2002). An additional development with major impact on applied as well as basic research is the measurement of eye movements outside the laboratory in everyday life situations (Holmqvist et al. 2011). Finally, significant recent advances in the development of oculographic methods include the availability of eye-trackers that are compatible with magnetic resonance imaging (MRI) scanners, thus allowing precise recording of eye movements simultaneously to the acquisition of brain images.

In this chapter, it will be proposed that the study of eye movements may inform neuroeconomics research. To do so, different types of eye movements will first be described and fundamental methodological issues in their study will be detailed. The key oculographic methods are then reviewed before it will be argued, thus the study of eye movements offers significant advantages in neurocognition research in comparison to standard manual motor experiments. Next, the processes of visual perception and attention and their interactive relationships with movements of the eyes will be outlined. Finally, literature on eye movements and neuroeconomics will be discussed.

25.2 Types of Eye Movements

The human eye movement repertoire can be classified into different types of eye movements along various lines. A widely recognised functional classification by Leigh and Zee (2006) distinguishes between saccades, fixation, smooth pursuit, vestibular eye movements, optokinetic eye movements, the quick phase of nystagmus and vergence eye movements.

Another commonly made distinction is that between gaze stabilising and gaze shifting eye movements (Gilchrist 2011). Gaze stabilizing eye movements are those that maximise stability of foveal processing during target movement, head movement or both, such as vestibular, optokinetic and smooth pursuit eye movements. Fixation, although not characterised by overtly observable eye movements, is also considered a mechanism of oculomotor gaze stabilisation. The primary gaze shifting movements on the other hand are saccades, i.e. rapid eye movements that serve to bring the image of an object of interest onto the fovea. Of most relevance to oculomotor paradigms in experimental psychology and cognitive neuroscience are saccades, smooth pursuit eye movements and fixation.

Saccades (from the French word "la saccade", meaning "jerk") are fast, jerk-like and ballistic eye movements with durations of 20–50 ms and a maximal velocity of up to 600°/s. In everyday situations humans efficiently explore their visual environment by making approximately 3–4 saccades per second (or about 12,000 per hour) with maximal amplitudes of approximately 15°, beyond which gaze shifts are more likely to be implemented by head movements (Gilchrist 2011). Saccades can also be elicited very easily in experimental settings by instructing the observer to shift their gaze from one (e.g. central) target to a sudden-onset target at another location (e.g. in the periphery). This basic "prosaccade" task, which is considered to involve largely automatic processing, can be experimentally modified in order to tap specific cognitive processes (Hutton 2008). For example, spatial working memory can be assessed by asking a saccade to the location of a briefly flashed peripheral target to be executed after a short delay interval (Crawford et al. 1989; Ettinger et al. 2005; Sawaguchi and Goldman-Rakic 1991). Temporal and/or spatial prediction abilities can be challenged by requiring the execution of saccades to a target that is displayed in different locations in a temporally and/or spatially predictable manner (Allman et al. 2012; Joiner and Shelhamer 2006). Another variant is the so-called antisaccade task, in which a saccade to the opposite location of a peripheral target is requested, thus requiring cognitive control functions (Hutton and Ettinger 2006; Munoz and Everling 2004). Importantly, non-human primates can also be trained to make saccades in different experimental paradigms, allowing the development of cross-species model systems of cognitive processes (see below).

Prosaccades are controlled by a neural network involving posterior parietal cortex, frontal and supplementary eye fields, superior colliculus, striatum, thalamus and cerebellum, whereas volitional saccades recruit increased activation levels in these areas and involve additional activations, such as in the dorsolateral and ventrolateral prefrontal cortices (for reviews, see Johnston and Everling 2008; McDowell et al. 2008; Muri and Nyffeler 2008). See Fig. 25.1 for an illustration of some of the areas mediating prosaccades.

Smooth pursuit serves to keep the image of a slowly moving object on the fovea by matching smooth eye velocity to target velocity as closely as possible (Barnes 2008, 2011). Pursuit typically involves a combination of smooth eye movements and saccades. The latter may be compensatory, reducing position error when eye velocity is higher or lower than target velocity or intrusive, occurring in segments of otherwise good pursuit, for example by anticipating future target movement. The neural control of smooth pursuit eye movements (Fig. 25.2) involves motion sensitive neurons in the visual cortex (V5 in humans, MT/MST in monkeys), the posterior parietal cortex, the frontal and supplementary eye fields, the cerebellum and pontine, vestibular and oculomotor nuclei (for reviews, see Ilg and Thier 2008; Lencer and Trillenberg 2008; Sharpe 2008).

Fixation occurs in periods when the eyes are relatively still and when most visual information is gained from the environment, primarily by foveal but also parafoveal and peripheral visual processing. Movements of the eyes do in fact occur during

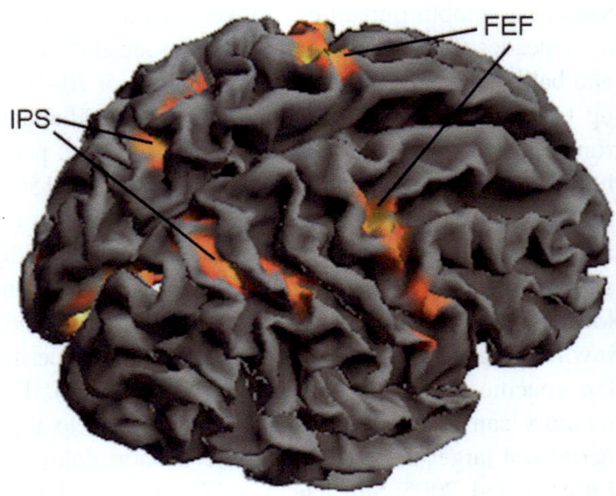

Fig. 25.1 Neural correlates of saccadic eye movements. The figure shows clusters of significant activation ($p < 0.05$ corrected cluster level) in a prosaccade task in healthy humans measured with fMRI at 3T field strength. Unpublished data (Herweg, Weber and Ettinger). *FEF* frontal eye fields; *IPS* intraparietal sulcus

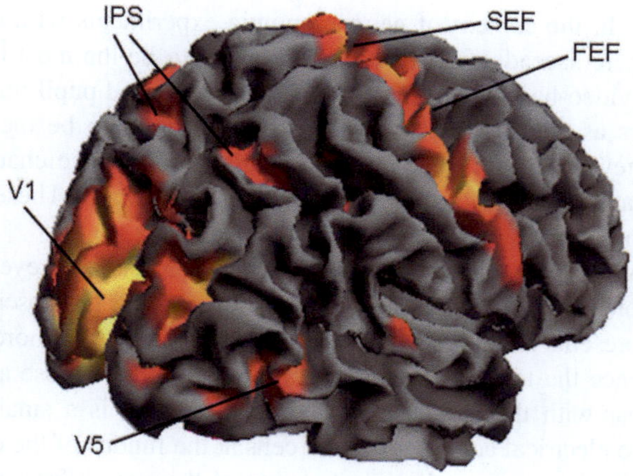

Fig. 25.2 Neural correlates of smooth pursuit eye movements. The figure shows clusters of significant activation ($p < 0.05$ corrected voxel level) in a smooth pursuit task in healthy humans measured with fMRI at 3T field strength. Unpublished data (Meyhöfer, Steffens, Weber and Ettinger). *FEF* frontal eye fields; *SEF* supplementary eye fields; *IPS* intraparietal sulcus; *V1* primary visual cortex; *V5* motion sensitive area in visual cortex

fixation, particularly tremor, drift and microsaccades (Martinez-Conde et al. 2013). These are small movements, however, and are not usually noticed by the viewers themselves or by untrained observers.

The reason for the interest of experimental psychology in saccades, pursuit and fixations lies primarily in the close relationship of these types of eye movements with overt visual attention. The link between these will be elaborated upon later in this chapter, after oculographic recording methods and their advantages have been delineated.

25.3 Oculographic Methods

At present,[1] researchers interested in the study of eye movements can choose between a range of different recording methods. Good current overviews of these oculographic methods have recently been published (Eggert 2007; Haslwanter and Clarke 2010; Holmqvist et al. 2011; Shelhamer and Roberts 2010); older but still insightful reviews can be found in Yarbus (1967) and Young and Sheena (1975). The methods most commonly used in psychology and neuroscience include video-based combined corneal reflection and pupil tracking, infrared limbus oculography, electro-oculography (EOG), dual Purkinje tracking and the scleral search

[1]The history of methods in oculomotor research is summarised in detail by Wade and Tatler (2005).

coil technique. In the context of neuroeconomic experiments with eye movement recordings, where saccades and fixations are likely to be the most relevant oculomotor events, video-based combined corneal reflection and pupil trackers, infrared limbus trackers as well as the older EOG are suggested to be the most suitable methods, as they have good temporal and spatial performance characteristics, are non-invasive and relatively quick to set up and calibrate. These methods are described in the following.

EOG involves the placing of surface electrodes lateral to each eye as well as to a reference region such as the forehead. The method exploits the observation that the human eye represents an electrical dipole, with the retina being more negative than the cornea (hence the term "corneo-retinal potential"), and whose axis is approximately collinear with the optical axis of the eye. There is a small difference in potential due to electrical activity of retina cells at the fundus of the eye. Given that the electrical dipole moves with every rotation of the eye, differences in electrical potential at the skin surface can be used to detect changes in eye position.

Advantages of EOG include the accurate recording of horizontal eye movements over a relatively large amplitude range (up to ±40°; Leigh and Zee 2006) whilst being non-invasive and causing only minimal discomfort to participants. Disadvantages primarily include the presence of (electrical, electro-cortical and electro-myographic) noise as well as eyelid artefacts. Additionally, vertical eye movements are measured much less reliably than horizontal eye movements due to the concurrent movement of the eye lid.

Infrared limbus oculography (e.g. Reulen et al. 1988) uses emitter/detector arrays placed directly in front of the eye, usually by being fixed to a headset. The emitters shine infrared light onto the eye, from where it is reflected back onto the detectors. The intensity of the reflected light differs between iris and sclera and, accordingly, the distribution of light across the array of sensors of reflected light changes when the eye moves. This signal can be used to measure eye movements with high temporal and spatial performance characteristics. Disadvantages of the method include artefacts due to movements of the head or the headset, discomfort from wearing the headset and blocking of the visual field due to the emitter/detector arrays.

Video-based combined corneal reflection and pupil trackers are currently amongst the most widely used oculographic methods in psychology. Unlike other methods, these eye-trackers work on a combination of two different signals, viz. the image of the pupil and the reflection of infrared light from the cornea (the first Purkinje image; see Fig. 25.3). These signals together yield a precise and accurate measure of the point of regard with excellent temporal and spatial resolution. Such eye-trackers often work remotely, that is the infrared light source and camera are placed away from the participant, e.g. below the computer screen that displays the experimental stimuli (see Fig. 25.4 for an example), thus posing zero discomfort to participants. A further advantage of these eye-trackers is the possibility of not only obtaining eye position data but simultaneously also a measure of pupil size with every sample (see below).

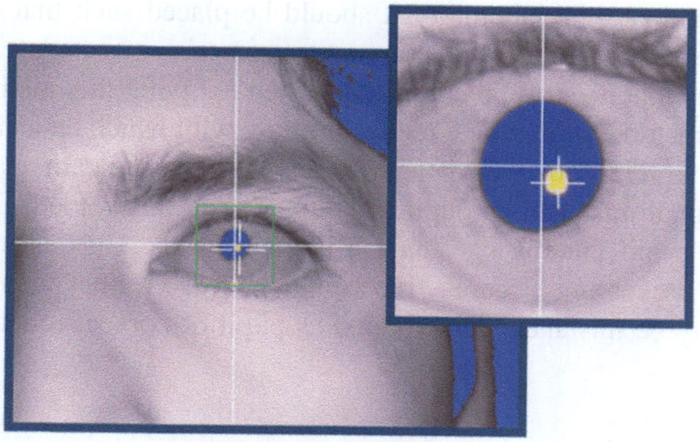

Fig. 25.3 Pupil and corneal reflection images. The pictures show images of the pupil (in *blue*) and the first corneal reflection (in *yellow*) obtained by a video-based combined pupil and corneal reflection tracker (EyeLink 1000; SR Research Ltd., Ottawa, Ontario, Canada). Picture used with permission from SR Research Ltd.

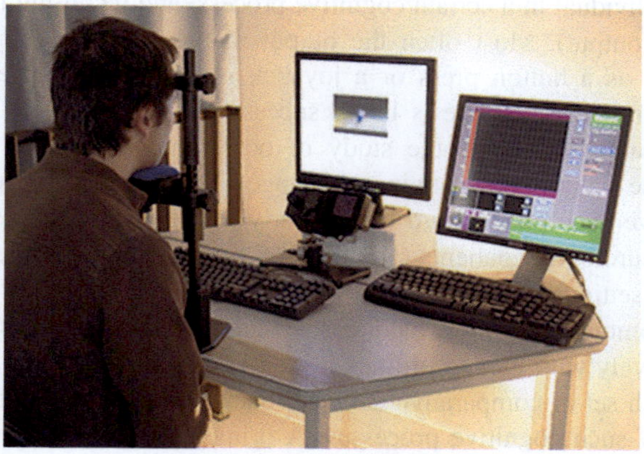

Fig. 25.4 Setup of a remote eye-tracker. The picture shows as an example of a laboratory setup the EyeLink 1000 eye-tracker (SR Research Ltd.). The infrared light source and camera are placed remotely, beneath the monitor that displays the stimulus for the participant. A chinrest is used to minimise head movements. Picture used with permission from SR Research Ltd.

Video-based combined corneal reflection and pupil trackers tend to be user friendly and quick to set up. However, this scientific method still requires precise operating and attention to detail in order to yield valid data. In this context, a number of laboratory conditions have to be considered (Holmqvist et al. 2011). In brief, an ideal eye-tracking laboratory is a quiet room with dim but strictly controllable light conditions. Complete darkness should be avoided as it may lead to enlarged and more variable pupil size, which can pose problems for pupil tracking. Additionally, the infrared light source may become visible in complete darkness and pose a source of distraction for participants. When a light source is used to moderately illuminate the laboratory it should be placed such that no direct light falls onto the eye, the camera or the computer screen.

Other factors that need to be considered when applying this method are certain participant characteristics that may reduce the ease with which data can be collected or the quality of the collected data. Makeup such as mascara may lead to the detection of signal mistaken as pupil and should thus be avoided. Droopy eyelids may cover the pupil thereby causing partial signal loss. Glasses may usually be worn without preventing accurate recordings; however, they can also produce signals that may be mistakenly interpreted as pupil or corneal reflection.

25.4 Methodological Advantages of Oculomotor Research

Psychologists and cognitive neuroscientists wishing to study certain cognitive processes often face the problem of operationalising their measurement. A typical experimental setup involves presenting a stimulus (input), which then arguably engages the individual in a certain cognitive process, and recording the response to that stimulus (output). Most often the output that is studied is a manual motor response, such as a button press or a joystick movement, or a verbal response. Whilst this standard procedure is being successfully employed in the pursuit of numerous research questions, the study of ocular movements as an alternative motor response may offer certain advantages over manual responses. These advantages have been put forward most clearly in the context of research on psychiatric and neurological patients (Klein and Ettinger 2008), but they also hold in other research settings.

First, oculomotor responses and associated cognitive processes have been studied intensively with unit recordings in non-human primates and other animals, providing a rich set of comparative data and background knowledge on the neural mechanisms of such cognitive processes. A prime example of this is the investigation of spatial working memory and its correlation in the dorsolateral prefrontal cortex (DLFPC) using the oculomotor delayed response task in non-human primates (Sawaguchi and Goldman-Rakic 1991) as well as humans (Sweeney et al. 1996). The ability to perform the same tasks in both experimental animals and humans provides a comprehensive and systematic description of the relevant cognitive and neural processes.[2]

Second, eye movements can be observed directly and without much technical artefacts. Specifically, movements of the eyes, which involve only six muscles, are more easily recorded and interpreted than movements of the limbs, which involve dozens of muscles. The high precision of measurement stems from the fact that the eyeball can rotate only within the socket. The resultant oculographic recordings allow for the derivation of a manifold of meaningful, specific and reliably

[2]See also the work of Glimcher et al. (2005) on saccadic decision making discussed below.

measurable parameters (Holmqvist et al. 2011). This holds both for eye movements recorded in the laboratory and in the "field", that is in ecologically valid situations such as during domestic tasks, whilst moving through the environment, whilst driving a car or operating a plane or when engaged in sports (for reviews, see Land and Tatler 2009; Duchowski 2002).

Third, oculomotor tasks are hands- and language-free, they typically utilise simple stimuli and instructions and they are shorter than complex neuropsychological assessments of cognitive function. These tasks are therefore culture fair and they have proven highly acceptable to a diverse range of individuals from the general population, including children and older participants as well as cognitively impaired patients.

Fifth, nowhere in the human brain can the complex interaction of perception and action—elaborated in the "Gestaltkreis" theory by von Weizsäcker (1996)—be studied more directly than in the oculomotor system. And due to the dynamic nature of this interplay an important feature of the interaction between "mind" and "world" becomes visible and analysable that remains completely concealed during the simple measurement of manual reaction times: the process nature of these interactions. In order to understand the structure of such processes, however, it is required to understand the interplay between vision, attention and eye movements (see below).

Together these methodological advantages of oculomotor research reiterate its importance in the operationalisation of the measurement of specific cognitive processes in basic and applied research situations where manual motor responses would otherwise be used to indicate the output of the organism.

25.5 Eye Movements, Attention and Vision

Before we investigate the interaction of eye movements, attention and vision in more detail, we need to ask why we move our eyes at all. Or, as Delgado-García (2000) put it: "Why move the eyes if we can move the head"? The answer to this question lies in the observation that eye movements serve at least two important purposes: (1) By shifting the point of gaze, they allow swift, high-acuity, foveal processing of visual stimuli without the need to perform conspicuous—and comparatively clumsy—head movements. (2) They stabilise visual perception by compensating for head or body movements.

Both arguments point to the important role of eye movements in supporting visual perception, a point expanded by Findlay and Gilchrist (2003) in relation to the importance of eye movements in visual attention. The primary role of the eye amongst all human sensory organs, in turn, is buttressed by the dual observations that the eyes take in more information (bits) than any other sensory organ and that approximately 50 % of human neocortex responds to changes in the visual environment. Thus, eye movements serve to maximise the quality and structure of the

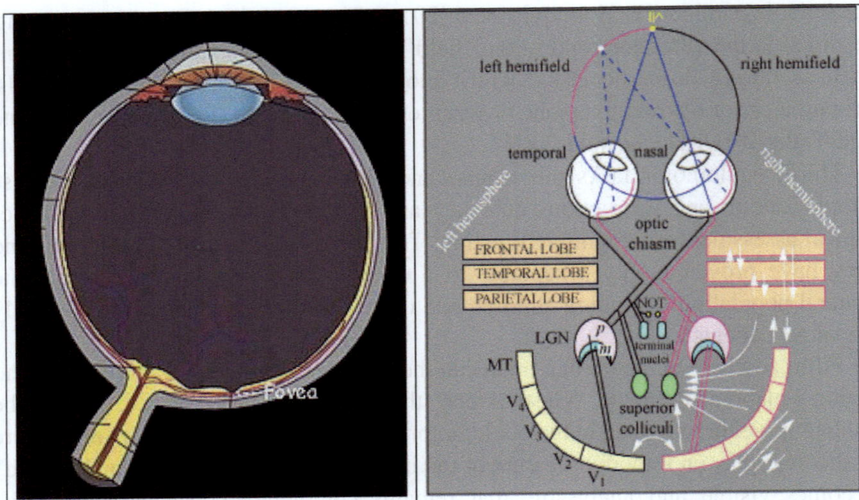

(a) Fovea centralis: **Source:** http://en.wikipedia.org/wiki/Fovea_centralis	(b) Diagrammatic representation of the visual system in primates, showing the visual information processing pathways. (Image by Prof. Peter Schiller. Used with permission.) **Source:** MIT Open Courseware

Fig. 25.5 The visual system

information intake of the most important of human sensory modalities. But why should this be the case? The answer is revealed by the mere anatomy of the eyes.

Figure 25.5 provides an illustration of the visual system of primates. Through the pupil, light reaches the photoreceptors (i.e. rods (dim light) and cones (bright light, colours)) on the retina. Transduction in the photoreceptors translates light information from a spotlight into an electric signal. The highest density of cones enabling the most precise viewing can be found in the central fovea, a roughly 1.5 mm small area located about 5° temporally to the visual axis. From the photoreceptors the neural impulses are sent via retinal ganglion cells to the lateral geniculate nucleus of the thalamus and from there to the primary visual cortex V1 and further to the visual association cortices V2–V5, from where visual information spreads out to be processed by various cortical and sub-cortical regions. With regard to the intricate link between eye movements and visual attention, the mere fact that the central fovea comprises less than 1 % of the retinal size but projects to over 50 % of the visual cortex to process the light that comes from only the central 2° of the visual field makes clear how strongly the selective nature of visual attention is anatomically "embodied".

Whilst these (functional) anatomic considerations underline the importance of the visual system as a system of information selection, many of the close and complex relationships between eye movements and visual attention have over the decades been unveiled by experimental research in human and non-human primates (for a recent review primarily of the experimental literature, see Kowler 2011). Here, two fundamental questions can be asked. First, how do "top-down" and "bottom-up" factors interact; and second, how important are low-level features such as contrast, colour or motion compared to other higher level stimulus structures such as objects or gist (Einhäuser et al. 2008)?

One of the answers to the second question has been the concept of "saliency maps" suggested by Koch and Ullman (1985) as a spatial distribution of local contrasts of features such as colour, luminance or movement. The higher the physical salience of a location in the visual field, the greater is the likelihood that it is attended and fixated. The corresponding rather low-level perceptual analyses are thought to take place at early stages of information processing in the visual system. The concept of saliency maps is certainly intriguing as a theoretical approach that enables precise predictions of fixation locations. On the one side, such models have indeed been shown to outperform random models of fixation location (Foulsham and Underwoord 2008). On the other side, however, multiple other factors in addition to such "outstanding" features are involved in determining our gaze position that emphasise in different ways top-down aspects of gaze control, including strategic decisions (Najemnik and Geisler 2005), specific interests (Birmingham et al. 2009; Einhäuser et al. 2008) or specific tasks (Buswell 1935; Yarbus 1967).

Amongst the different top-down factors, *strategic decisions* may aim for optimising task performance. This may include gazing at locations that maximise the probability of finding a searched target (Najemnik and Geisler 2005) or aligning gaze near the centre position in a scene to identify large portions of the scene with this single optimal viewing position (Tatler 2007). *Specific interests* such as interest in relevant social information, in addition, may override perceptual saliency and direct an observer's gaze to heads and eyes in complex social scenes (Birmingham et al. 2009). Similarly, *objects* that can be identified as "interesting" because they can be recalled retrospectively by the participants, explain the allocation of attention and fixations better than early perceptual saliency which, however, may impact on an object's saliency and thus contribute to an object's "interestingness" indirectly (Einhäuser et al. 2008). In line with the implied interaction of top-down and bottom-up factors, Cerf et al. (2008), too, found that the combined influences of face preference (top-down) and low-level salience (bottom-up) improve the prediction of gaze positions. The aforementioned examples thus indicate that visual attention is in fact controlled both by exogenous bottom-up processes of perceptual saliency and endogenous top-down processes (Chica et al. 2013).

One of the crucial issues in understanding the dynamics of eye movements refers to the spatial sequencing of saccades during the exploration of visual scenes. According to Kowler (2011), two concepts have been important here. The first, the "winner-takes-all" principle, defines how saccadic goals are chosen; the second principle, called "inhibition of return", specifies how the currently fixated area is left again to fixate another goal. According to the first principle, the region with the momentarily highest "strength" (e.g., due to its saliency) attracts the attention, but quickly loses its strengths for other regions to be fixated, according to the second principle. During such sequences of saccades, (covert) visual attention moves to the location that is about to be fixated next before the eyes move to that location (Henderson et al. 1989). The strength of the coupling of movements of attention and of the eyes has also been shown by experiments demonstrating that it is not possible to orient attention to one location whilst moving the eyes to another (Hoffman and Subramanian 1995).

This sequence of alternations between attending a certain location, followed by its inhibition and the focussing of the next salient location, generates a scan path of fixations and saccades across a stimulus of interest (Liechty et al. 2003). Scan paths have been described since Noton and Stark's (1971) seminal observation that when participants look at visual stimuli (here: different patterns) their gazes follow repeatedly a fixed path that is characteristic of an individual subject. These researchers also observed that different individuals develop different scan paths for the same pattern, and each individual develop different scan paths for different patterns, but that each scan path remained consistent across different presentations of the same pattern. The "replication" of scan paths developed during viewing conditions in the absence of the stimulus under conditions of visual imagery is certainly amongst the clearest demonstrations of top-down factors in gaze control.

According to some authors, there are two different states of covert visual attention during the exploration of complex stimuli, viz. local and global (Liechty et al. 2003). During states of local visual attention, short saccades are employed to extract information from specific and adjacent locations of a stimulus; during states of global visual attention, long saccades are used to extract and integrate information from various locations of a complex stimulus (Liechty et al. 2003). Local and global visual attention is thought to be controlled by the inferior temporal and posterior parietal cortex, respectively; their interaction is thought to be controlled by the prefrontal cortex (Liechty et al. 2003).

To conclude, the anatomy of the human eye with the central fovea as a small area of outstanding visual acuity "dictates" the dynamics of visual attention and eye movements as a temporal-spatial process that is co-determined both by perceptual bottom-up factors and cognitive-motivational top-down influences. Needless to say that no technique is better suited to investigating these dynamics of visual information processing than oculographic methods.

25.6 Pupillometry

Information processing through eye movements is inevitably associated with complex cognitive and emotional processes and their central and autonomous correlates. Modern eye-trackers can measure one particular facet of this complex response in addition to the proper eye movements, namely the pupil size and its changes and fluctuations. This classical psychophysiological parameter has been shown to be sensitive to a range of cognitive and affective states that may be relevant at the workplace or in confrontation with stimuli from an economics context, such as signals of financial rewards.

Pupillary responses manifest either in increase or decrease of the pupil diameter. The *increase* in pupil diameter is called dilation or "mydriasis" and is caused by the iris dilator. The iris dilator is a muscle bundle composed of radially arranged muscle fibres that is controlled by sympathetic innervation from the superior cervical ganglion, which is itself under tonic excitatory control descending from the hypothalamus (Kandel et al. 2000). The *decrease* in pupil diameter is called constriction or "miosis" and is effected by the iris sphincter muscle. This muscle is composed of circularly arranged muscle fibres and is innervated by the parasympathetic nervous system exerting its effects through the mesencephalic Nucleus Edinger-Westphal that innervates the ciliary ganglion cells through the oculomotor brain nerve III (Kandel et al. 2000).

More recently, pupil size has been shown to co-vary intra-individually with the firing rate of the locus coeruleus (LC), one of the brain's two origins of noradrenergic neurons. The LC, which is located in the pons, has been associated with general arousal through its widespread afferent projections but is not directly or anatomically linked with the pupil (Aston-Jones and Cohen 2005). According to Aston-Jones and Cohen (2005), the LC provides a tonic mode of firing that can vary between low activity in inattentive, non-alert states and high activity in distractible states that promote task shifting. This structure, in addition, provides a phasic mode of firing when the subject is engaged in a task. The tonic and phasic modes interact in that prominent phasic LC firing which takes place during phases of moderate tonic firing, that is, when the subject is neither drowsy nor distracted and shifting to another task. According to Aston-Jones and Cohen (2005), these patterns of LC firing are controlled by the anterior cingulate (ACC) and orbitofrontal cortices (OFC), both of which are thought to monitor task-related utility.

In one experiment, worthwhile to be described in some detail, these researchers presented human participants with a discrimination task, in which task difficulty and rewards for correct responses continually increased. Initially, the reward value increased faster than task difficulty and thus error rates; after several trials, however, the increase in task difficulty led to increased number of errors and reduced reward rate. In line with the tonic-phasic distinction of LC firing, the authors observed a steady rising of baseline (tonic) pupil diameter with increasing task difficulty, and a decrease in phasic discrimination-related pupil dilation. When discrimination success declined and participants abandoned the current task series to start a new one, baseline pupil diameter was greatest. This pattern of findings, which confirms Wilder's "Law of Initial Values" (Wilder 1958), exemplifies nicely some of the features of the pupillary dynamics: it is sensitive to both cognitive and affective/motivational processes (which may be as fundamental as those proposed by Aston-Jones and Cohen 2005); it encompasses both tonic and phasic responses; it is controlled both by more proximal (N. Edinger-Westphal) and more distal (hypothalamus) brain structures and may be correlated with the activity in anatomically not directly related brain regions (LC, ACC, OFC). Therefore, the dynamics of the human pupil can be expected to be sensitive to a wide range of tasks, situations and states and hence they may help in interpreting the findings of eye movement studies that employ the experimental variation of conditions, situations or states.

In line with this overall summary, dilation of the pupil, in comparison to appropriate control conditions, has been associated with high task load/difficulty (Beatty 1982a; Hess and Polt 1964; Kahneman and Beatty 1966; Moresi et al. 2008; Steinhauer et al. 2004), high stimulus presentation rate (Poock and Noel 1975), low stimulus probability (Reinhard and Lachnit 2002), processing of ambiguity (Ben-Nun 1986; Schluroff et al. 1986), exploratory versus exploitative gambling choices (Jepma and Nieuwenhius 2011), emotional excitement or emotional valence (Goldwater 1972; Partala and Surakka 2003), (un)pleasantness of verbal passages (White and Maltzman 1978), pleasantness of pictures (Metalis and Hess 2013), positive or negative emotional baby voices (Partala and Surakka 2003), interest value (Hess and Polt 1960) or aesthetic liking (Johnson et al. 2010). Conversely, decreases in task-evoked phasic responses or smaller tonic pupil diameters have been reported with increasing time-on-task (Beatty 1982b) or in insomniacs (Lichtstein et al. 1992). Sleepiness, however, is not only associated with miotic responses but also with faster oscillations of the pupil diameter (McLaren et al. 1992).

These examples amply illustrate that pupil recordings may provide valuable additional insights into the processing of difficulty or effort-related aspects of stimuli in a wide range of settings of relevance to decision making and experimental economics.

25.7 Oculomotor Research in Neuroeconomics

Neuroeconomics is an interdisciplinary scientific endeavour that aims to understand human decision making, particularly in the context of economic choices. The goal of neuroeconomics is thus to use neuroscientific data in order to better understand the deliberation process in decision-making situations with the ultimate goal of improving economic models.

Accordingly, the methods of neuroeconomics are drawn from its constituent scientific fields, namely neurobiology, behavioural and experimental economics as well as cognitive, social and evolutionary psychology (see Part 8 of this book). They include behavioural experiments in which participants can obtain financial rewards, "games" where participants' rewards depend on the behaviour of other (simulated or real) participants, recordings of heart rate and skin conductance, electroencephalography (EEG), magnetoencephalography (MEG), positron emission tomography (PET), functional magnetic resonance imaging (fMRI), animal studies involving single-neuron recordings and, of course, oculography, viz. the recording of eye movements.

Here we argue that the recording of eye movements (with or without concurrent recordings of neurophysiological activity) may be one particularly useful method to inform our understanding of the neural and cognitive processes taking place during economic decision making. This argument is based on the methodological advantages of oculography detailed above as well as our understanding of the association between attentional processes and eye movements and their neural mechanisms. Indeed, oculographic recordings in the study of decision making were first made in the 1970s (Russo and Rosen 1975) and have since been used profitably to elucidate the underlying processes. In the following we will not provide a comprehensive overview of the published work in neuroeconomics that has used oculographic methods; instead, we will give some examples selected to illustrate the richness of oculographic approaches in neuroeconomics and refer the reader at this point to further relevant articles (see e.g. Causse et al. 2011; Costa-Gomes et al. 2001; Glöckner and Herbold 2011; Krajbich et al. 2010; Lohse and Johnson 1996; Middlebrooks and Sommer 2011; Reutskaja et al. 2011; Ross et al. 2011; Stritzke et al. 2009).

In one recent example, Arieli et al. (2011) recorded eye movements in a decision task where participants had to choose amongst two lottery scenarios with differing prizes and probabilities. They aimed to compare two possible procedures that may be implemented in the decision-making process, viz. the holistic and component procedures. In the holistic procedure, participants treat the two alternatives holistically, for example by evaluating the certainty equivalent of each alternative and selecting the one with the higher certainty equivalence or by computing the expectation of each of the two alternatives and selecting the one with the higher expectation. In the component procedure, it is thought that participants compare the prizes and the probabilities as separate components. Using eye movement recordings, Arieli et al. (2011) found that the more difficult of these problems were associated with saccades between the same components of the two alternatives, suggesting a component procedure, whereas in easier problems saccades between elements within the same alternative were more frequent, suggesting a holistic processing approach. That study thus nicely demonstrates how the investigation of eye movements can provide valuable information on the likely cognitive processes taking place during decision making.

Similarly, Camerer and colleagues (e.g. Bhatt and Camerer 2005) have used oculographic methods to distinguish between different neuroeconomic theories which make similar predictions at the behavioural level. As these theories could not be distinguished purely in terms of the participants' behavioural decisions, the recording of eye movements provided an additional level of analysis that helped to rule out alternative explanations.

In addition to serving as a tool to enrich behavioural experiments by providing an additional level of explanation, eye movements have also been studied in the search for the neural substrate of economic decision making in non-human primates. An impressive example of this approach is the work by Glimcher and colleagues (see e.g. Glimcher et al. 2005; Glimcher 2009). These researchers have developed a neurobiological model of decision making that is based on utility theory from economics and makes prominent use of eye movement behaviour and neurophysiology in monkeys.

In this model, the final stages of saccadic decision making involve the selection of that action from a number of alternatives that is judged to be most desirable in a "winner-takes-all" process, resulting in a saccade in a certain direction. Work by Schall and colleagues (Hanes and Schall 1996; Schall and Thompson 1999) has shown that frontal eye field (FEF) neurons display topographic firing over the stimulus position in a "winner-takes-all" fashion that encodes the direction of the saccade; when the level of activity at that location exceeds a certain threshold, a saccade is executed. Of importance to an understanding of saccadic decision making are, however, also the processing steps that lead up to this decision in the FEFs.

For example, in studies using moving stimuli, Newsome and colleagues (Parker and Newsome 1998) observed that the topographically arranged neurons in MT are motion sensitive. It was also shown that the outputs of these neurons are passed on to neurons in the posterior parietal cortex, where they are mathematically integrated, thereby generating a topographic map of motion direction from which they extract a decision variable which is then passed on to the FEFs.

This process is not just a model of primitive, low-level decision making, but likely a model of much more complicated decisions with relevance to neuroeconomics was shown in a study by Platt and Glimcher (1999). In that study, monkeys were trained to make saccades to the right or left, and both the likelihood and the magnitude of reward for these saccades were manipulated experimentally. It was found that the firing of neurons in the lateral intraparietal area (LIP) accurately represented the expected value of a saccade, thus yielding a directly measureable neural substrate of economic choice. In other words, area LIP represents a topographic map in which the relative expected value or utility of each possible saccade is coded.[3]

Glimcher and colleagues also draw upon work by Schultz et al. (1997) on dopamine release in the striatum in order to explain from where LIP neurons receive information about the physiological expected utilities. Schultz et al. (1997) observed that dopamine neurons code what is called reward prediction error in learning theory (Sutton and Barto 1998), that is, these neurons receive inputs on expected and on actual reward, calculate the difference (the so-called reward prediction error) and transmit the results to cortical and basal ganglia neurons using dopamine as neurotransmitter. As such, the outputs of the computations of these neurons may explain how physiological expected utilities are generated in parietal neurons.

On the basis of the observation of neuronal encoding in LIP of relative expected utility under such fairly straightforward experimental conditions, Glimcher and

[3]Of note, a recent study by Leathers and Olson (2012) showed that LIP neurons encode not just the reward value of saccades but also an incurred penalty, suggesting that these neurons represent not the action value but the motivational salience of the stimulus (see, however, Newsome et al. 2013).

colleagues went on to investigate how their firing patterns would respond when an action, i.e. the choice of saccade direction, was less clearly related to an expected value than in the above studies.

In a series of experiments, Dorris and Glimcher (2004) engaged both humans and monkeys in a version of the inspection game, a strategic game from game theory where one participant, the inspector, verifies the another participant, the inspectee, adheres to certain rules. The monkeys played against a computer in a saccadic choice version of the game. At the behavioural level, the experiments showed that the monkeys' dynamic behaviour closely resembled that of humans, supporting the cross-species validity of the oculomotor model of this game. At the neurophysiological level, the research showed that the brain topographically encodes, in directly measurable neural activity patterns, the relative desirabilities of all possible actions (which are of course directly and unambiguously measurable via oculographic methods). As Glimcher et al. (2005) argue, it may be these topographic maps in LIP which represent the neural substrate of economic decision making.

The implications of the development of oculomotor model systems of economic choice are wide ranging. Given the availability of cross-species models and the above described advantages of studying the eye movement system of further models could be developed allowing the study of a wide range of influences on economic decision making, from state factors such as pharmacological challenges to trait factors such as genetics (cf. the chapters by Reuter and Montag in this book).

Glimcher and colleagues also draw conclusions from these studies concerning the historical tension between prescriptive and descriptive approaches to the study of economic decision making (see Chap. 1, by Reuter and Montag, in this book). Prescriptive approaches emphasise the optimal and efficient decision making of *homo economicus*, whereas descriptive approaches, based on data collection, often obtain evidence of irrational economic decisions leading to poor utility maximising. Neuroeconomics as a discipline may contribute to the debate between these approaches by studying the mechanisms of decision making in the brain using experimental model systems, some of which may involve saccadic eye movements as the means to indicate the outcome of the decision-making process. According to Glimcher et al. (2005), the neurobiological work on primate saccadic decision making provides evidence against the dual-systems view held by some economists, that decisions arise out of a conflict between two interacting systems, one being clearly utility based (and evolutionarily recent) and the other being irrational (and evolutionarily old). Instead, the work by Glimcher and others has identified direct evidence of neural activity representing—in many situations—classical expected utility.

25.8 Conclusions

The present chapter has provided an introduction to the study of eye movements as a method in neuroeconomics research. We have described different recording techniques, highlighted the history of oculomotor research, and listed advantages of oculography over other behavioural measurements of cognition. An important aspect of oculomotor research concerns the complex interplay between eye movements and visual attention that underlies the cycle of saccades and fixation. Furthermore, we touched upon the sensitive dynamics of pupil responses, the measurement of which can be obtained "for free" when using modern eye-trackers, and gave examples of how neuroeconomic research may benefit from eye movement recordings. Overall, we hope to have raised interest in this method in researchers in neuroeconomics.

What can be concluded from the work described in this chapter? We will phrase our answers to this question in the form of theses and proposals for future eye movement research in the field of neuroeconomics.

1. The study of eye movements provides a rich body of behavioural and neuroscientific evidence that can be incorporated into neuroeconomics research. This applies especially to the work on attention and decision making as well as their neural correlates.
2. Neuroeconomics research may benefit from the systematic integration of eye movement recordings in more or less all research settings, be it the behavioural lab, the MRI scanner or the field. As described, such recordings may provide an additional level of analysis with which it may be possible to separate competing accounts of economic decision making.
3. The availability of cross-species models of economic decision making provides a highly innovative model system which builds upon animal neurophysiology, behavioural oculomotor research and human neuroeconomics. Such work provides an important foundation for extending research into utility theory and neuroeconomic decision making into animal behaviour and neurophysiology, for example involving direct pharmacological interventions to modulate the reward system, by drawing upon the method of recording eye movements.

References

Allman AA, Ettinger U, Joober R, O'Driscoll GA (2012) Effects of methylphenidate on basic and higher-order oculomotor functions. J Psychopharmacol 26(11):1471–1479

Arieli A, Ben-Ami Y, Rubinstein A (2011) Tracking decision makers under uncertainty. Am Econ J Microecon 3:68–76

Aston-Jones G, Cohen JD (2005) An integrative theory of locus coeruleus-norepinephrine function: adaptive gain and optimal performance. Annu Rev Neurosci 28:403–450

Barnes GR (2008) Cognitive processes involved in smooth pursuit eye movements. Brain Cogn 68:309–326

Barnes GR (2011) Ocular pursuit movements (Chapter 7). In: Liversedge SP, Gilchrist I, Everling S (eds) The Oxford handbook of eye movements. Oxford University Press, Oxford

Beatty J (1982a) Task-evoked pupillary responses, processing load, and the structure of processing resources. Psychol Bull 91:276–292

Beatty J (1982b) Phasic not tonic pupillary responses vary with auditory vigilance performance. Psychophysiology 19(2):167–172

Ben-Nun Y (1986) The use of pupillometry in the study of on-line verbal processing, evidence for depth of processing. Brain Lang 28:1–11

Bhatt M, Camerer CF (2005) Self-referential thinking and equilibrium as states of mind in games: fMRI evidence. Games Econ Behav 52:424–459

Birmingham E, Bischof WF, Kingstone A (2009) Saliency does not account for fixations to eyes within social scenes. Vision Res 49:2992–3000

Buswell GT (1935) How people look at pictures: a study of the psychology and perception in art. University Chicago Press, Oxford

Causse M, Baracat B, Pastor J, Dehais F (2011) Reward and uncertainty favor risky decision-making in pilots: evidence from cardiovascular and oculometric measurements. Appl Psychophysiol Biofeedback 36:231–242

Cerf M, Harel J, Einhäuser W, Koch C (2008) Predicting human gaze using low-level saliency combined with face detection. Adv Neural Inf Process 20:241–248

Chica AB, Bartolomeo P, Lupianez J (2013) Two cognitive and neural systems for endogenous and exogenous spatial attention. Behav Brain Res 237:107–123

Costa-Gomes M, Crawford VP, Broseta B (2001) Econometrica 69(5):1193–1235

Crawford TJ, Henderson L, Kennard C (1989) Abnormalities of nonvisually-guided eye movements in Parkinson's disease. Brain 112(Pt 6):1573–1586

Delgado-García JM (2000) Why move the eyes if we can move the head? Brain Res Bull 52 (6):475–482

Dorris MC, Glimcher PW (2004) Activity in posterior parietal cortex is correlated with the relative subjective desirability of action. Neuron 44(2):365–378

Duchowski AT (2002) A breadth-first survey of eye-tracking applications. Behav Res Methods Instrum Comput 34(4):455–470

Eggert T (2007) Eye movement recordings: methods. In: Straube A, Büttner U (eds) Neuro-Ophthalmology, vol 40. Karger Publishers, pp 15–34

Einhäuser W, Spain M, Perona P (2008) Objects predict fixations better than early saliency. J Vis 8:1–26

Ettinger U, Hejda S, Flak V, Corr PJ (2005) Prepulse inhibition of the acoustic startle reflex and oculomotor control. Psychophysiology 42:473–482

Findlay JM, Gilchrist ID (2003) Active vision: the psychology of looking and seeing. Oxford University Press, Oxford

Foulsham T, Underwoord G (2008) What can saliency models predict about eye movements? Spatial and sequential aspects of fixations during encoding and recognition. J Vis 8:1–17

Gilchrist I (2011) Saccades (Chapter 5). In: Liversedge SP, Gilchrist I, Everling S (eds) The Oxford handbook of eye movements. Oxford University Press, Oxford

Glimcher PW (2009) Choice: towards a standard back-pocket model. In: Glimcher PW, Camerer C, Poldrack RA, Fehr E (eds) Neuroeconomics: decision making and the brain. Academic Press, New York, pp 501–519

Glimcher PW, Dorris MC, Bayer HM (2005) Physiological utility theory and the neuroeconomics of choice. Games Econ Behav 52(2):213–256

Glöckner A, Herbold AK (2011) An eye-tracking study on information processing in risky decisions: evidence for compensatory strategies based on automatic processes. J Behav Decision Mak 24:71–98

Goldwater BC (1972) Psychological significance of pupillary movements. Psychol Bull 77:340–355

Hanes DP, Schall JD (1996) Neural control of voluntary movement initiation. Science 274 (5286):427–430

Haslwanter T, Clarke AH (2010) Eye movement measurement, electro-oculography and video-oculography. Handb Clin Neurophysiol 9:61–79

Henderson JM, Pollatsek A, Rayner K (1989) Covert visual attention and extrafoveal information use during object identification. Percept Psychophys 45:196–208

Hess EH, Polt JM (1960) Pupil size as related to interest value of visual stimuli. Science 132:349–350

Hess EH, Polt JM (1964) Pupil size in relation to mental activity during simple problem-solving. Science 143:1190–1192

Hoffman JE, Subramanian B (1995) The role of visual attention in saccadic eye movements. Percept Psychophys 57:787–795

Holmqvist H, Nyström M, Andersson R, Dewhurst R, Jarodzka H, van de Weijer J (2011) Eye tracking: a comprehensive guide to methods and measures. Oxford University Press, Oxford

Hutton SB (2008) Cognitive control of saccadic eye movements. Brain Cogn 68(3):327–340

Hutton SB, Ettinger U (2006) The antisaccade task as a research tool in psychopathology, a critical review. Psychophysiology 43:302–313

Ilg UJ, Thier P (2008) The neural basis of smooth pursuit eye movements in the rhesus monkey brain. Brain Cogn 68:229–240

Jepma M, Nieuwenhius S (2011) Pupil diameter predicts changes in the exploration-exploitation trade-off, evidence for the adaptive gain theory. J Cogn Neurosci 23:1587–1596

Johnson MG, Muday JA, Schirillo JA (2010) When viewing variations in paintings by Mondrian, aesthetic preferences correlate with pupil size. Psychol Aesthet Creativity Arts 4:161–167

Johnston K, Everling S (2008) Neurophysiology and neuroanatomy of reflexive and voluntary saccades in non-human primates. Brain Cogn 68:271–283

Joiner WM, Shelhamer M (2006) An internal clock generates repetitive predictive saccades. Exp Brain Res 175:305–320

Kahneman D, Beatty J (1966) Pupil diameter and load on memory. Science 154:218–219

Kandel ER, Schwartz JH, Jessell TM (2000) Principles of neural science, 3rd edn. Prentice Hall, London

Klein C, Ettinger U (2008) A hundred years of eye movement research in psychiatry. Brain Cogn 68:215–218

Koch C, Ullman S (1985) Shifts in selective visual attention, towards the underlying neural circuitry. Hum Neurobiol 4:219–227

Kowler E (2011) Eye movements: the past 25 years. Vis Res 51:1457–1483

Krajbich I, Armel C, Rangel A (2010) Visual fixations and the computation and comparison of value in simple choice. Nat Neurosci 13(10):1292–1298

Land MF, Tatler BW (2009) Looking and acting: vision and eye movements in natural behaviour. Oxford University Press, Oxford

Leathers ML, Olson CR (2012) In monkeys making value-based decisions, LIP neurons encode cue salience and not action value. Science 338(6103):132–135

Leigh RJ, Zee DS (2006) The neurology of eye movements, 4th edn. Oxford University Press, Oxford

Lencer R, Trillenberg P (2008) Neurophysiology and neuroanatomy of smooth pursuit in humans. Brain Cogn 68:219–228

Lichtstein KL, Johnson RS, Gupta S, O'Laughlin DL, Dykstra TA (1992) Are insomniacs sleepy during the day? A pupillometric assessment. Behav Res Ther 30(3):283–292

Liechty J, Pieters R, Wedel M (2003) Global and local covert visual attention: evidence from a Bayesian Hidden Markov Model. Psychometrika 68:519–541

Liversedge SP, Gilchrist I, Everling S (2011) The Oxford handbook of eye movements. Oxford University Press, Oxford

Lohse GL, Johnson EJ (1996) A comparison of two process tracing methods for choice tasks. Organ Behav Hum Decis Process 68(1):28–43

Martinez-Conde S, Otero-Millan J, Macknik SL (2013) The impact of microsaccades on vision: towards a unified theory of saccadic function. Nat Rev Neurosci 14(2):83–96

McDowell JE, Dyckman KA, Austin BP, Clementz BA (2008) Neurophysiology and neuroanatomy of reflexive and volitional saccades, evidence from studies of humans. Brain Cogn 68:255–270

McLaren JW, Erie JC, Brubaker RF (1992) Computerized analysis of pupillograms in studies of alertness. Invest Ophthalmol Vis Sci 33(3):671–676

Metalis SA, Hess EH (2013) Pupillary response/semantic differential scale relationships. J Res Pers 16:201–216

Middlebrooks PG, Sommer MA (2011) Metacognition in monkeys during an oculomotor task. J Exp Psychol Learn Mem Cogn 37(2):325–337
Moresi S, Adam JJ, Rijken J, Van Gerven PWM, Kuipers H, Jolles J (2008) Pupil dilation in response preparation. Int J Psychophysiol 67:124–130
Munoz DP, Everling S (2004) Look away, the anti-saccade task and the voluntary control of eye movement. Nat Rev Neurosci 5:218–228
Muri RM, Nyffeler T (2008) Neurophysiology and neuroanatomy of reflexive and volitional saccades as revealed by lesion studies with neurological patients and transcranial magnetic stimulation (TMS). Brain Cogn 68:284–292
Najemnik J, Geisler WS (2005) Optimal eye movement strategies in visual search. Nature 434:387–391
Newsome WT, Glimcher PW, Gottlieb J, Lee D, Platt ML (2013) Comment on "In monkeys making value-based decisions, LIP neurons encode cue salience and not action value". Science 340(6131):430
Noton D, Stark L (1971) Scanpaths in saccadic eye movements while viewing and recognizing patterns. Vis Res 11:929–942
Parker AJ, Newsome WT (1998) Sense and the single neuron: probing the physiology of perception. Annu Rev Neurosci 21:227–277
Partala T, Surakka V (2003) Pupil size variation as an indication of affective processing. Int J Hum Comput Stud 59:185–198
Platt ML, Glimcher PW (1999) Neural correlates of decision variables in parietal cortex. Nature 400(6741):233–238
Poock GK, Noel CE (1975) Effects of layout of workplace on pupillary diameter. Percept Motor Skill
Rayner K (1998) Eye movements in reading and information processing: 20 years of research. Psychol Bull 124:372–422
Reinhard G, Lachnit H (2002) The effect of stimulus probability on pupillary response as an indicator of cognitive processing in human learning and categorization. Biol Psychol 60:199–215
Reulen JP, Marcus JT, Koops D, de Vries FR, Tiesinga G, Boshuizen K, Bos JE (1988) Precise recording of eye movement: the IRIS technique. Part 1. Med Biol Eng Comput 26(1):20–26
Reutskaja E, Nagel R, Camerer CF, Rangel A (2011) Search dynamics in consumer choice under time pressure: an eye-tracking study. Am Econ Rev 101:900–926
Ross M, Lanyon LJ, Viswanathan J, Manoach DS, Barton JJ (2011) Human prosaccades and antisaccades under risk: effects of penalties and rewards on visual selection and the value of actions. Neuroscience 196:168–177
Russo EJ, Rosen LD (1975) An eye fixation analysis of multialternative choice. Memory Cogn 3:267–276
Sawaguchi T, Goldman-Rakic PS (1991) D1 dopamine receptors in prefrontal cortex: involvement in working memory. Science 251(4996):947–950
Schall JD, Thompson KG (1999) Neural selection and control of visually guided eye movements. Annu Rev Neurosci 22:241–259
Schluroff M, Zimmermann T, Freeman RB, Hofmeister K, Lorscheid T, Weber A (1986) Pupillary responses to syntactic ambiguity of sentences. Brain Lang 27:322–344
Schultz W, Dayan P, Montague PR (1997) A neural substrate of prediction and reward. Science 275:1593–1599
Sharpe JA (2008) Neurophysiology and neuroanatomy of smooth pursuit, lesion studies. Brain Cogn 68:241–254
Shelhamer M, Roberts DC (2010) Magnetic scleral search coil. Handb Clin Neurophysiol 9:80–87
Steinhauer SR, Siegle GJ, Condray R, Pless M (2004) Sympathetic and parasympathetic innervation of pupillary dilation during sustained processing. Int J Psychophysiol 52:77–86
Stern JA, Dunham DN (1990) The ocular system. In: Cacioppo JT, Tassinary LG (eds) Principles of psychophysiology: physical, social, and inferential elements. Cambridge University Press, Cambridge, pp 513–553

Stritzke M, Trommershäuser J, Gegenfurtner KR (2009) Effects of salience and reward information during saccadic decisions under risk. J Opt Soc Am A: Opt Image Sci Vis 26 (11):B1–13

Sutton R, Barto A (1998) Reinforcement learning. MIT Press, Cambridge

Sweeney JA, Mintun MA, Kwee S, Wiseman MB, Brown DL, Rosenberg DR, Carl JR (1996) Positron emission tomography study of voluntary saccadic eye movements and spatial working memory. J Neurophysiol 75:454–468

Tatler BW (2007) The central fixation bias in scene viewing, selecting an optimal viewing position independently of motor biases and image feature distributions. J Vis 7:1–17

von Weizsäcker V (1996) Gestaltkreis - Theorie der Einheit von Wahrnehmen und Bewegen (Bearb. v. Peter Achilles). Suhrkamp Verlag, Berlin

Wade NJ, Tatler BW (2005) The moving tablet of the eye: the origins of modern eye movement research. Oxford University Press, Oxford

White GL, Maltzman I (1978) Pupillary activity while listening to verbal passages. J Res Pers 12:361–369

Wilder J (1958) Modern psychophysiology and the law of initial value. Am J Psychother 12:199–221

Yarbus AL (1967) Eye movements and vision. Plenum Press, New York

Young LR, Sheena D (1975) Eye-movement measurement techniques. Am Psychol 30:315–330

Appendix A
Neuroanatomy

See Figs. A.1, A.2, A.3, A.4, A.5, A.6, A.7, A.8 and A.9.

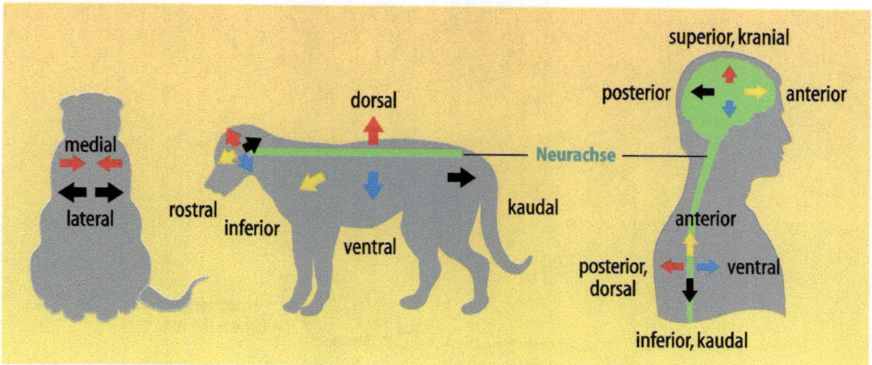

Fig. A.1 Explanation of the orientation in the human brain

M. Reuter and C. Montag (eds.), *Neuroeconomics*, Studies in Neuroscience, Psychology and Behavioral Economics, DOI 10.1007/978-3-642-35923-1

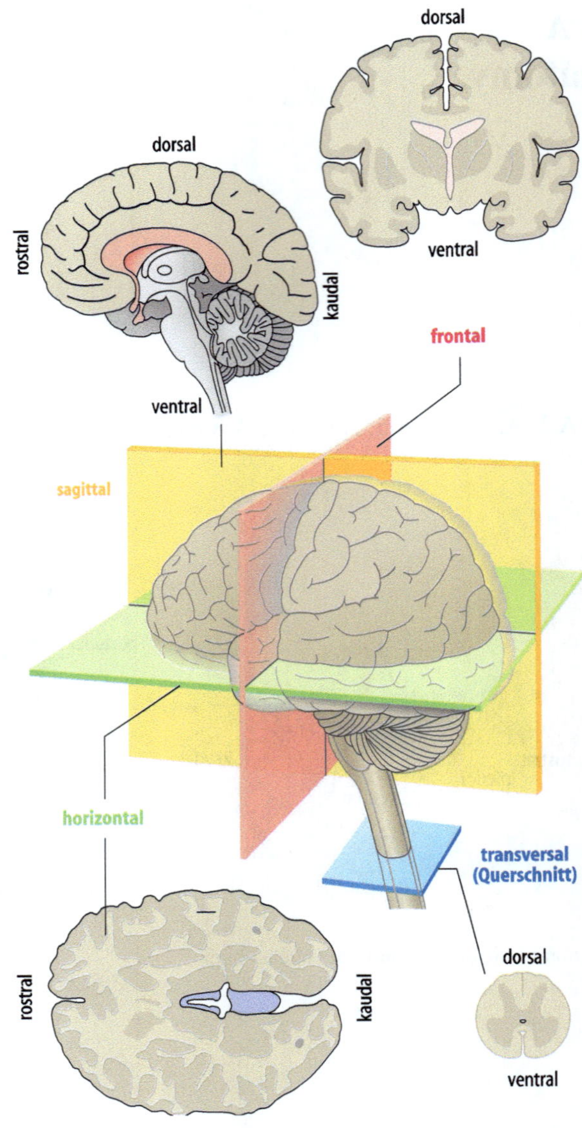

Fig. A.2 Different perspectives on the human brain

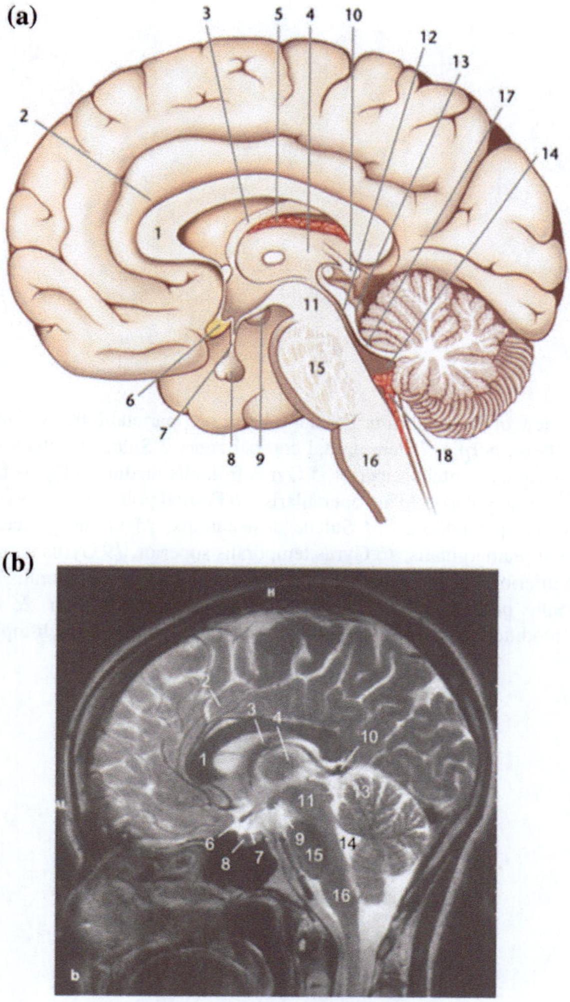

Fig. A.3 Medial sagittal section through the brain. **a** Midsagittal section. **b** MRI image (midsagittal section). *1* Corpus callosum (callosal commissure), *2* Gyrus cinguli (cingulate cortex), *3* Fornix, *4* Thalamus, *5* Chiasma opticum, *6* Infundibulum hypophysis, *7* Hypophysis, *8* Corpus mamillare, *9* Epiphysis, *10* Mesencephalon (midbrain), *11* Aqueductus mesencephali (only in **a**), *12* Lamina tecti, *13* Ventriculus quartus, *14* Pons, *15* Medulla oblongata, *16* Velum medullare superius (MRI image from the University clinic in Freiburg, by Dr. J. Klisch, Department of Neuroradiology)

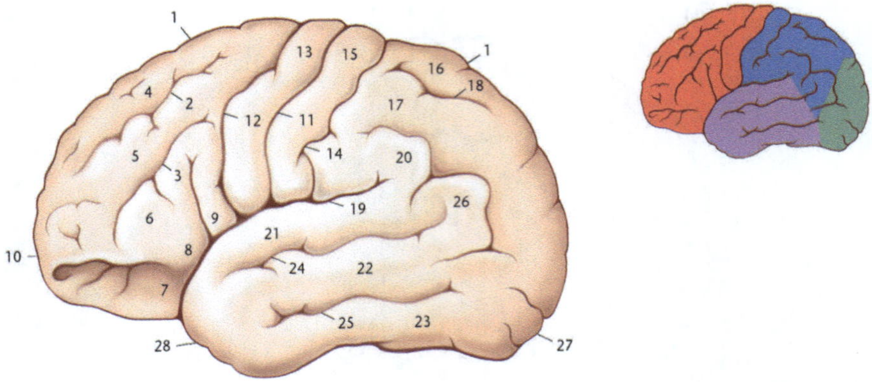

Fig. A.4 Lateral view of the cerebrum. Frontal lobe in *red*, parietal lobe in *blue*, occipital lobe in *green*, temporal lobe in *purple*. *1* Parasagittal cortical zone, *2* Sulcus frontalis superior, *3* Sulcus frontalis inferior, *4* Gyrus frontalis superior, *5* Gyrus frontalis medius, *6* Gyrus frontalis inferior, *7* Pars orbitalis, *8* Pars triangularis, *9* Pars opercularis, *10* Frontal pole, *11* Sulcus centralis, *12* Sulcus precentralis, *13* Gyrus precentralis, *14* Sulcus postcentralis, *15* Gyrus postcentralis, *16* Sulcus lateralis, *17* Gyrus supramarginalis, *18* Gyrus temporalis superior, *19* Gyrus temporalis medius, *20* Gyrus temporalis inferior, *21* Sulcus temporalis superior, *22* Sulcus temporalis inferior, *23* Gyrus angularis, *24* Lobulus parientalis superior, *25* Lobulus parientalis inferior, *26* Occipital pole, *27* Temporal pole (modified according to Spitzer, in Duus: Neurologisch-topische Diagnostik, Thieme 1990)

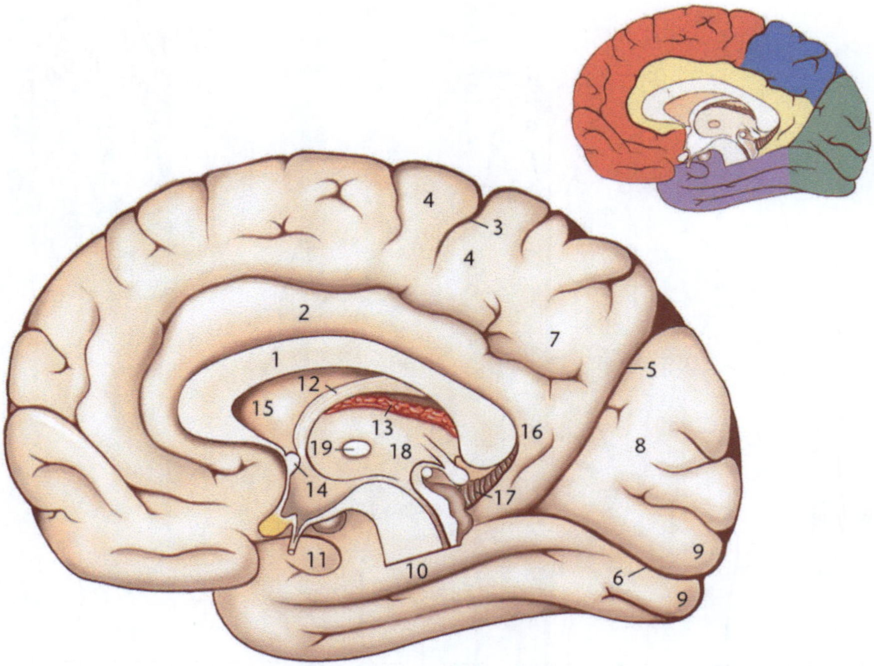

Fig. A.5 Medial view of the cerebrum. Frontal lobe in *red*, parietal lobe in *blue*, occipital lobe in *green*, temporal lobe in *purple*, Cyrus cinguli in *yellow*. *1* Corpus callosum, *2* Gyrus cinguli, *3* Sulcus centralis, *4* Lobulus paracentralis, *5* Sulcus pararietooccipitalis, *6* Sulcus calcarinus, *7* Precuneus, *8* Cuneus, *9* Visual cortex, *10* Gyrus parahippocampalis, *11* Uncus, *12* Fornix, *13* Tela choroidea, *14* Commissura anterior, *15* Septum pellucidum, *16* Isthmus gyri cinguli, *17* Gyrus dentatus, *18* Thalamus, *19* Adhesio interthalamica (modified according to Spitzer, in Duus: Neurologisch-topische Diagnostik, Thieme 1990)

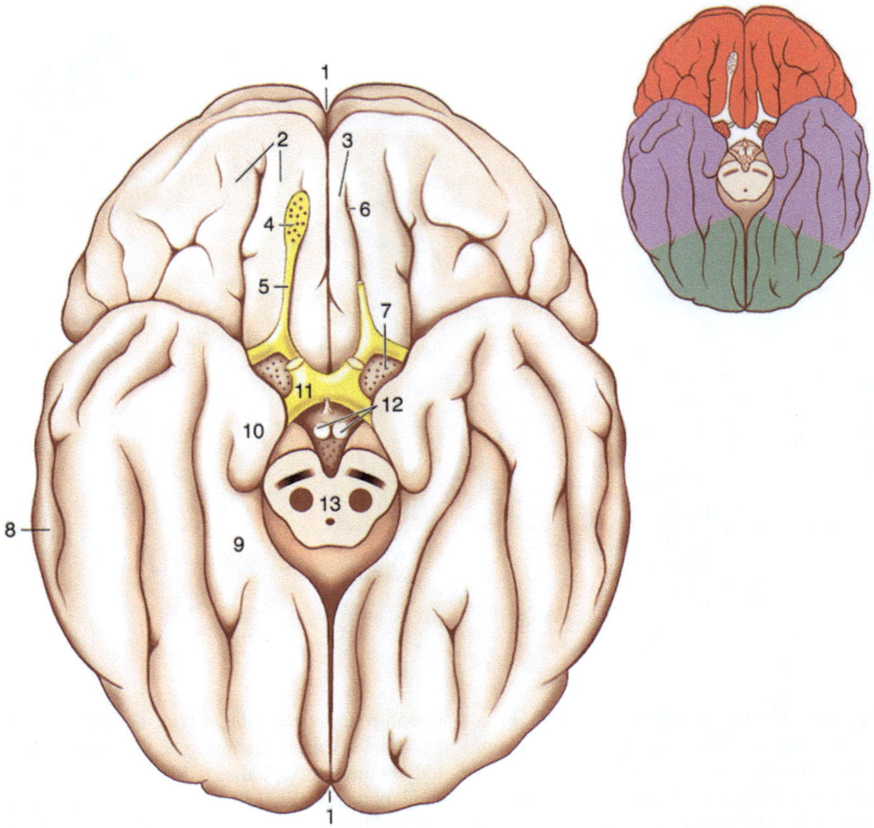

Fig. A.6 Basal view of the cerebrum. Frontal lobe in *red*, occipital lobe in *green*, temporal lobe in *purple*. *1* Fissura longitudinalis cerebri, *2* Gyri orbitales, *3* Gyrus rectus, *4* Bulbus olfactorius, *5* Tractus olfactorius, *6* Sulcus olfactorius, *7* Substantia perforata anterior, *8* Gyrus temporalis inferior, *9* Gyrus parahippocampalis, *10* Uncus, *11* Chiasma opticum, *12* Corpora mamillaria, *13* Midbrain (modified according to Spitzer, in Duus: Neurologisch-topische Diagnostik, Thieme 1990)

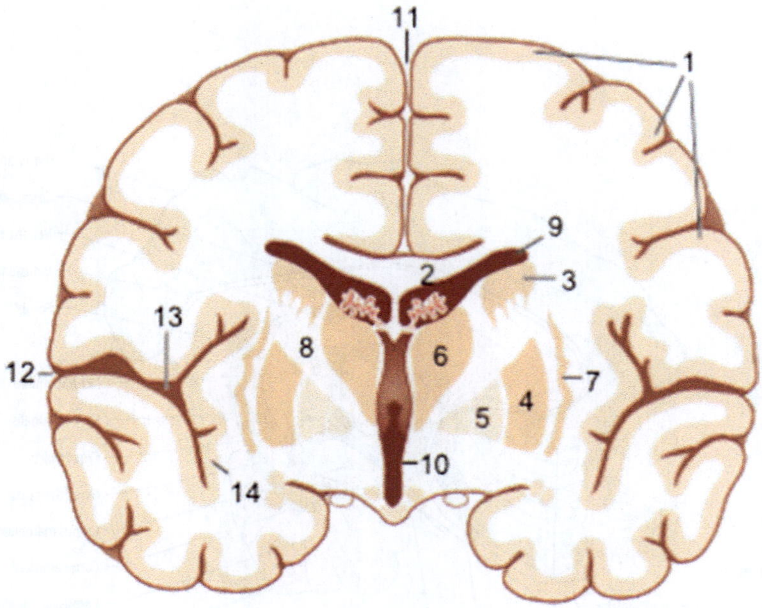

Fig. A.7 The most important inner structures of the cerebrum (frontal section). *1* Cerebral cortex, *2* Corpus callosum, *3* Ncl. caudatus, *4* Putamen, *5* Globus pallidus, *6* Thalamus, *7* Claustrum of insula, *8* Capsula interna, *9* Lateral ventricles, *10* Third ventricle, *11* Fissura longitudinalis cerebri, *12* Sulcus lateralis, *13* Fossa lateralis, *14* Insular cortex

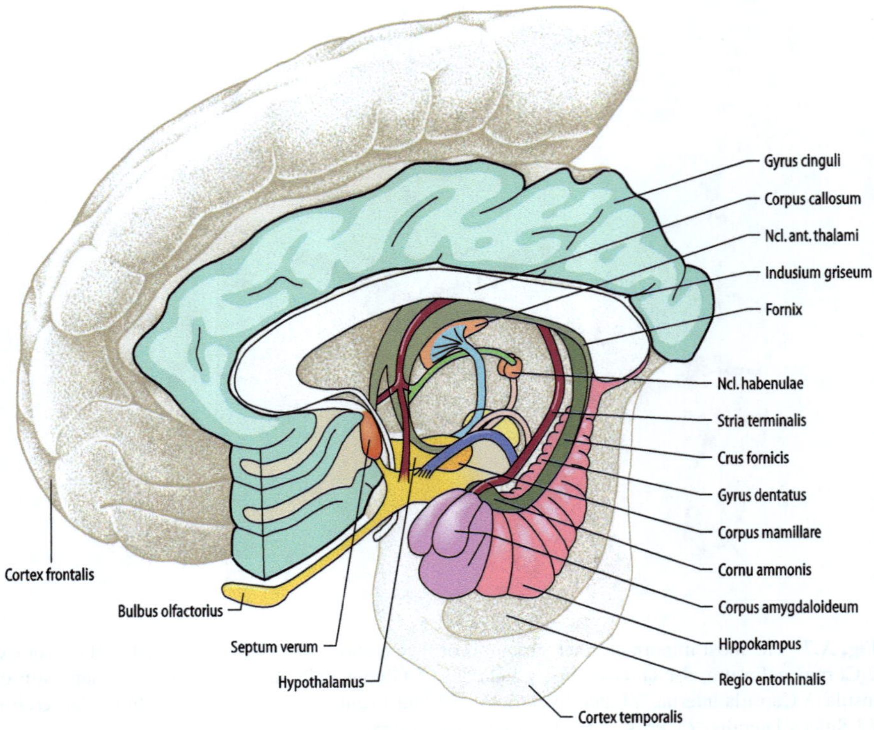

Fig. A.8 View on the limbic system

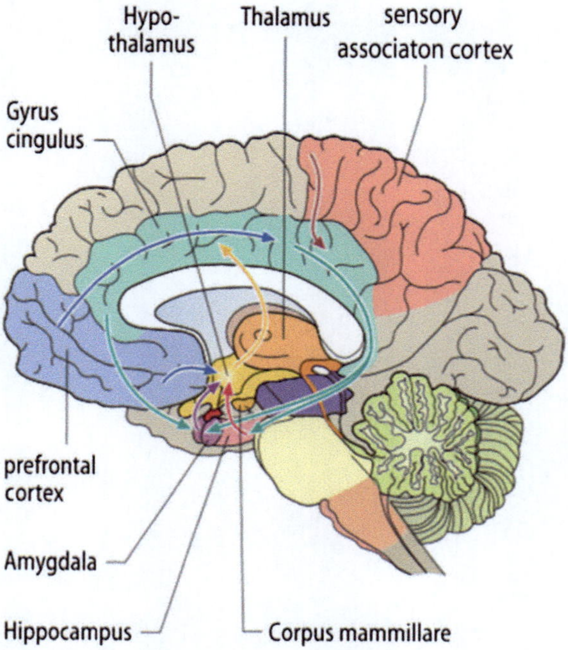

Fig. A.9 Projections of the limbic system